Methods in Molecular Biology

Series Editor
John M. Walker
School of Life Sciences
University of Hertfordshire
Hatfield, Hertfordshire, AL10 9AB, UK

For further volumes:
http://www.springer.com/series/7651

Human Pluripotent Stem Cells

Methods and Protocols

Edited by

Philip H. Schwartz

Neuroscience Laboratories, Centers for Neuroscience and Translational Research, Children's Hospital of Orange County Research Institute, Orange, CA, USA

Robin L. Wesselschmidt

Center for Applied Technology Development, Beckman Research Institute, City of Hope, Duarte, CA, USA

Humana Press

Editors
Philip H. Schwartz
Neuroscience Laboratories
Centers for Neuroscience
and Translational Research
Children's Hospital of Orange
County Research Institute
Orange, CA
USA
pschwartz@choc.org

Robin L. Wesselschmidt
Center for Applied Technology Development
Beckman Research Institute
City of Hope
Duarte, CA
USA
rwessel@coh.org

ISSN 1064-3745 e-ISSN 1940-6029
ISBN 978-1-4939-5802-3 ISBN 978-1-61779-201-4 (eBook)
DOI 10.1007/978-1-61779-201-4
Springer New York Dordrecht Heidelberg London

Softcover re-print of the Hardcover 1st edition 2011

Printed on acid-free paper

Humana Press is part of Springer Science+Business Media (www.springer.com)

Preface

The 1998 report describing the derivation of human embryonic stem cells (hESCs) set in motion an unprecedented wave of research and public discourse on the basic biology and the potential therapeutic application of these cells as well as the ethics surrounding their derivation and use. It seemed that everyone was interested in these cells derived from human blastocyst-stage embryos. The public debate and scientific interest continues today, 13 years later. Although there is no denying that the derivation of hESCs was a major breakthrough in our efforts to understand early human development and to treat human diseases, the science behind it was built on decades of solid research that began with the study of human and mouse teratomas and teratocarcinomas (see the reference list for a necessarily abridged compendium of seminal papers in this stem cell field).

Teratocarcinomas are the malignant form of teratomas, tumors that comprise a complex, disorganized mixture of cells and tissues representing cellular derivatives of all three of the embryonic germ layers. Importantly, the study of teratocarcinomas significantly increased our knowledge of embryonic development through pioneering work that led to recognition of the relationship between the differentiated cells and tissues that developed from the tumor and normal embryogenesis. Eventually, methodologies that allowed for the long-term in vitro culture of the stem cells of the teratocarcinomas were devised and, thus, the so-called embryonic (or embryonal) carcinoma cell (ECC) lines were established. Many ECC and other germ cell tumor lines were shown to be "pluripotent," which by definition means that they, like the tumors from which they were derived, were able to give rise to cellular derivatives of all three germ layers. Most of this groundwork in terms of basic culture protocols, basic characterization protocols, and basic in vitro and in vivo differentiation protocols developed using ECCs in the 1960s and 1970s was then used in early work to derive and characterize both mouse and human ESCs. In fact, the prototypical list of cell surface markers, TRA-1-60, TRA-1-81, SSEA-3, and SSEA-4, which are now commonly used to define hESCs as pluripotent, are the very same set originally used to define pluripotent human ECCs. The isolation of ECC lines from teratocarcinomas in the 1970s, therefore, provided the platform for the study of human pluripotent stem cells (PSCs), of which the hESC is but only one type.

While work was progressing on understanding the cause and nature of human teratocarcinomas, Stevens and Little at the Jackson Laboratory were developing a mouse model (the 129 strain) in which one could study the development and progression of teratomas in the mouse testis because this model had an unusually high occurrence of testicular teratomas. The 129 mouse strain ultimately became the strain of choice for those doing the methods development work that led to the derivation of ESCs and, in 1981, two reports describing the derivation of "embryo-derived pluripotent stem cells," which came to be known as mouse ESCs (mESCs), were published. Most of the mESC research during the 1980s focused on characterizing these diploid cells derived from the inner cell mass of the blastocyst. These highly unique cells were capable of prolonged in vitro culture, induced differentiation down all three germ lineages in vitro and in vivo, and, most importantly, when injected into mouse blastocysts, contributing to the germline of chimeric mice. The subsequent development of technologies to target specific genes in mESCs gave rise to an entirely new way to study the function of mammalian genes through the generation of "knockout" mice.

The use of normal pluripotent stem cells (i.e., mESCs) that could be precisely manipulated in vitro and then placed back into a mouse embryo to give rise to a live animal carrying the precise mutation, which could be passed on to its offspring, changed the face of biomedical research – it provided a superior model in which to study the function of genes in the context of mammalian physiology. These technologies had such a major impact on biomedical research that the 2007 Noble Prize in Medicine and Physiology was awarded to Martin Evans, Mario Capecchi, and Oliver Smithies for their work developing the "principles for introducing specific gene modifications in mice by the use of embryonic stem cells."

In our view, three key technological advances led to the widespread use of mESCs for the study of mammalian gene function: (1) the discovery that leukemia inhibitory factor inhibited spontaneous differentiation, improving the ease with which these cells could be cultured; (2) the use of isogenic DNA targeting constructs, improving gene targeting efficiency and lowering the time and cost associated with producing mouse knockouts; and (3) the derivation, systematic banking, and distribution of new mESC lines, improving the availability, access, and quality of the cells available to researchers.

When we review the evolution of hESC research, we can draw some comparisons to the progression of mESC research, where the first decade for both was spent optimizing culture conditions and methods for derivation, characterization, and maintenance. As we move into the second decade of hESC research, we are almost daily being presented with new technologies that may move the field toward its promise as stated in Thomson's 1998 paper "that these cell lines should be useful in studying human developmental biology, drug discovery, and transplantation medicine." We suggest that three enabling technologies are at hand for human ESCs: (1) directed reprograming of somatic cells, which eliminates many of the ethical issues associated with the derivation and use of hESCs, increases genetic diversity of the available human PSC lines, and gives rise to better in vitro human disease models; (2) the discovery that a Rho-associated protein kinase (ROCK) inhibitor allows for efficient single-cell passaging and cryopreservation, increasing the efficiency and reliability of hPSC culture; and (3) defined, animal component-free media, which lay the groundwork for simplified scale-up for therapeutic applications, differentiation protocols, and toxicology screens (All of these technologies are well-described in this book).

When we pause to consider combining technologies, such as the production of induced pluripotent stem cells with next generation sequencing technologies, we can glimpse the future that may yield highly effective, personalized, medical treatments. Whether or not any or all of our hopes for the future of this research are fully realized, we know we are part of a dynamic scientific community that has at its core the desire to further human knowledge and improve the human condition.

We are, therefore, pleased to provide *Human Pluripotent Stem Cells: Methods and Protocols* to our scientific community. This book is a compilation of 33 detailed protocols in six categories of PSC research that cover laboratory essentials and the derivation of new PSC lines, including induced PSC lines, as well as their growth, maintenance, characterization, genetic manipulation, and differentiation. This book, of course, would not have been possible without the generous contributions of our authors, all from leading research laboratories, who, working with their postdocs and students, have offered to share their validated and detailed protocols that we hope you, our colleagues, will find helpful in your own stem cell research programs.

Orange, CA, USA *Philip H. Schwartz*
Duarte, CA, USA *Robin L. Wesselschmidt*

References

Amit, M., Carpenter, M.K., Inokuma, M.S., Chiu, C.P., Harris, C.P., Waknitz, M.A., Itskovitz-Eldor, J., Thomson, J.A. (2000) Clonally derived human embryonic stem cell lines maintain pluripotency and proliferative potential for prolonged periods of culture. Dev. Biol. 227:271–278.

Andrews, P.W., Goodfellow, P.N., Shevinsky, L.H., Bronson, D.L., Knowles, B.B. (1982) Cell-surface antigens of a clonal human embryonal carcinoma cell line: Morphological and antigenic differentiation in culture. Int. J. Cancer 29:523–531.

Andrews, P.W., Banting, G., Damjanov, I., Arnaud, D., Avner, P. (1984) Three monoclonal antibodies defining distinct differentiation antigens associated with different high molecular weight polypeptides on the surface of human embryonal carcinoma cells. Hybridoma 3:347–361.

Aoi, T., Yae, K., Nakagawa, M., Ichisaka, T., Okita, K., Takahashi, K., Chiba, T., Yamanaka, S. (2008) Generation of pluripotent stem cells from adult mouse liver and stomach cells. Science 321:699–702.

Bongso, A., Fong, C.Y., Ng, S.C., Ratnam, S. (1994) Isolation and culture of inner cell mass cells from human blastocysts. Hum. Reprod. 9:2110–2117.

Bradley, A., Evans, M., Kaufman, M.H., Robertson, E. (1984) Formation of germ-line chimaeras from embryo-derived teratocarcinoma cell lines. Nature 309:255–256.

Brinster, R.L. (1974) The effect of cells transferred into the mouse blastocyst on subsequent development. J. Exp. Med. 140:1049–1056.

Burdon, T., Stracey, C., Chambers, I., Nichols, J., Smith, A. (1999) Suppression of SHP-2 and ERK signalling promotes self-renewal of mouse embryonic stem cells. Dev. Biol. 210:30–43.

Byrne, J.A., Pedersen, D.A., Clepper, L.L., Nelson, M., Sanger, W.G., Gokhale, S., Wolf, D.P., Mitalipov, S.M. (2007) Producing primate embryonic stem cells by somatic cell nuclear transfer. Nature 450:497–502.

Cartwright, P., McLean, C., Sheppard, A., Rivett, D., Jones, K., Dalton, S. (2005) LIF/STAT3 controls ES cell self-renewal and pluripotency by a Myc-dependent mechanism. Development 132:885–896.

Evans, M.J., Kaufman, M.H. (1981) Establishment in culture of pluripotential cells from mouse embryos. Nature 292:154–156.

Gachelin, G., Kemler, R., Kelly, F., Jacob, F. (1977) PCC4, a new cell surface antigen common to multipotential embryonal carcinoma cells, spermatozoa, and mouse early embryos. Dev. Biol. 57:199–209.

Hogan, B., Fellous, M., Avner, P., Jacob, F. (1977) Isolation of a human teratoma cell line which expresses F9 antigen. Nature 270:515–518.

Humphrey, R.K., Beattie, G.M., Lopez, A.D., Bucay, N., King, C.C., Firpo, M.T., Rose-John, S., Hayek, A. (2004) Maintenance of pluripotency in human embryonic stem cells is STAT3 independent. Stem Cells 22:522–530.

Kahan, B.W., Ephrussi, B. (1970) Developmental potentialities of clonal in vitro cultures of mouse testicular teratoma. J. Natl. Cancer Inst. 44:1015–1036.

Kannagi, R., Cochran, N.A., Ishigami, F., Hakomori, S., Andrews, P.W., Knowles, B.B., Solter, D. (1983) Stage-specific embryonic antigens (SSEA-3 and -4) are epitopes of a unique globo-series ganglioside isolated from human teratocarcinoma cells. EMBO J. 2:2355–2361.

Kleinsmith, L.J., Pierce, G.B. (1964) Multipotentiality of single embryonal carcinoma cells. Cancer Res. 24:1544–1551.

Labosky, P.A., Barlow, D.P., Hogan, B.L. (1994) Mouse embryonic germ (EG) cell lines: Transmission through the germline and differences in the methylation imprint of insulin-like growth factor 2 receptor (Igf2r) gene compared with embryonic stem (ES) cell lines. Development 120:3197–3204.

Ledermann, B., Burki, K. (1991) Establishment of a germ-line competent C57BL/6 embryonic stem cell line. Exp. Cell Res. 197:254–258.

Lowry, W.E., Richter, L., Yachechko, R., Pyle, A.D., Tchieu, J., Sridharan, R., Clark, A.T., Plath, K. (2008) Generation of human induced pluripotent stem cells from dermal fibroblasts. Proc. Natl. Acad. Sci. 105:2883–2888.

Ludwig, T.E., Levenstein, M.E., Jones, J.M., Berggren, W.T., Mitchen, E.R., Frane, J.L., Crandall, L.J., Daigh, C.A., Conard, K.R., Piekarczyk, M.S., et al. (2006) Derivation of human embryonic stem cells in defined conditions. Nat. Biotechnol. 24:185–187.

Maherali, N., Sridharan, R., Xie, W., Utikal, J., Eminli, S., Arnold, K., Stadtfeld, M., Yachechko, R., Tchieu, J., Jaenisch, R., et al. (2007) Directly reprogrammed fibroblasts show global epigenetic remodeling and widespread tissue contribution. Cell Stem Cell 1:55–70.

Martin, G.R. (1980) Teratocarcinomas and mammalian embryogenesis. Science 209:768–776.

Martin, G.R. (1981) Isolation of a pluripotent cell line from early mouse embryos cultured in medium conditioned by teratocarcinoma stem cells. Proc. Natl. Acad. Sci. 78: 7634–7638.

Martin, G.R., Evans, M.J. (1975) Differentiation of clonal lines of teratocarcinoma cells: Formation of embryoid bodies in vitro. Proc. Natl. Acad. Sci. 72:1441–1445.

Matin, M.M., Walsh, J.R., Gokhale, P.J., Draper, J.S., Bahrami, A.R., Morton, I., Moore, H.D., Andrews, P.W. (2004) Specific knockdown of Oct4 and β2-microglobulin expression by RNA interference in human embryonic stem cells and embryonic carcinoma cells. Stem Cells 22:659–668.

Matsuda, T., Nakamura, T., Nakao, K., Arai, T., Katsuki, M., Heike, T., Yokota, T. (1999) STAT3 activation is sufficient to maintain an undifferentiated state of mouse embryonic stem cells. EMBO J. 18:4261–4269.

Matsui, Y., Zsebo, K., Hogan, B.L. (1992) Derivation of pluripotential embryonic stem cells from murine primordial germ cells in culture. Cell 70:841–847.

Meissner, A., Wernig, M., Jaenisch, R. (2007) Direct reprogramming of genetically unmodified fibroblasts into pluripotent stem cells. Nat. Biotechnol. 25:1177–1181.

Miller, R.A., Ruddle, F.H. (1976) Pluripotent teratocarcinoma-thymus somatic cell hybrids. Cell 9:45–55.

Miller, R.A., Ruddle, F.H. (1977a) Properties of teratocarcinoma-thymus somatic cell hybrids. Somatic Cell Genet. 3:247–261.

Miller, R.A., Ruddle, F.H. (1977b) Teratocarcinoma X friend erythroleukemia cell hybrids resemble their pluripotent embryonal carcinoma parent. Dev. Biol. 56:157–173.

Okita, K., Ichisaka, T., Yamanaka, S. (2007) Generation of germline-competent induced pluripotent stem cells. Nature 448:313–317.

Park, I.H., Zhao, R., West, J.A., Yabuuchi, A., Huo, H., Ince, T.A., Lerou, P.H., Lensch, M.W., Daley, G.Q. (2008) Reprogramming of human somatic cells to pluripotency with defined factors. Nature 451:141–146.

Shamblott, M.J., Axelman, J., Wang, S., Bugg, E.M., Littlefield, J.W., Donovan, P.J., Blumenthal, P.D., Huggins, G.R., Gearhart, J.D. (1998) Derivation of pluripotent stem cells from cultured human primordial germ cells. Proc. Natl. Acad. Sci. 95: 13726–13731.

Solter, D., Knowles, B.B. (1978) Monoclonal antibody defining a stage-specific mouse embryonic antigen (SSEA-1) Proc. Natl. Acad. Sci. 75:5565–5569.

Solter, D., Skreb, N., Damjanov, I. (1970) Extrauterine growth of mouse egg-cylinders results in malignant teratoma. Nature 227: 503–504.

Stevens, L.C. (1962) The biology of teratomas including evidence indicating their origin form primordial germ cells. Annee Biol. 1:585–610.

Stevens, L.C. (1968) The development of teratomas from intratesticular grafts of tubal mouse eggs. J. Embryol. Exp. Morphol. 20:329–341.

Stevens, L.C. (1970) The development of transplantable teratocarcinomas from intratesticular grafts of pre- and postimplantation mouse embryos. Dev. Biol. 21:364–382.

Stevens, L.C., Little, C.C. (1954) Spontaneous testicular teratomas in an inbred strain of mice. Proc. Natl. Acad. Sci. 40:1080–1087.

Stojkovic, M., Lako, M., Stojkovic, P., Stewart, R., Przyborski, S., Armstrong, L., Evans, J., Herbert, M., Hyslop, L., Ahmad, S., et al. (2004) Derivation of human embryonic stem cells from day-8 blastocysts recovered after three-step in vitro culture. Stem Cells 22:790–797.

Strelchenko, N., Verlinsky, O., Kukharenko, V., Verlinsky, Y. (2004) Morula-derived human embryonic stem cells. Reprod. Biomed. Online 9:623–629.

Takahashi, K., Yamanaka, S. (2006) Induction of pluripotent stem cells from mouse embryonic and adult fibroblast cultures by defined factors. Cell 126:663–676.

Takahashi, K., Tanabe, K., Ohnuki, M., Narita, M., Ichisaka, T., Tomoda, K., Yamanaka, S. (2007) Induction of pluripotent stem cells from adult human fibroblasts by defined factors. Cell 131:861–872.

Thomson, J.A., Kalishman, J., Golos, T.G., Durning, M., Harris, C.P., Becker, R.A., Hearn, J.P.

(1995) Isolation of a primate embryonic stem cell line. Proc. Natl. Acad. Sci. 92:7844–7848.

Thomson, J.A., Kalishman, J., Golos, T.G., Durning, M., Harris, C.P., Hearn, J.P. (1996) Pluripotent cell lines derived from common marmoset (Callithrix jacchus) blastocysts. Biol. Reprod. 55:254–259.

Thomson, J.A., Itskovitz-Eldor, J., Shapiro, S.S., Waknitz, M.A., Swiergiel, J.J., Marshall, V.S., Jones, J.M. (1998) Embryonic stem cell lines derived from human blastocysts. Science 282:1145–1147.

Turnpenny, L., Brickwood, S., Spalluto, C.M., Piper, K., Cameron, I.T., Wilson, D.I., Hanley, N.A. (2003) Derivation of human embryonic germ cells: an alternative source of pluripotent stem cells. Stem Cells 21:598–609.

Vallier, L., Alexander, M., Pedersen, R.A. (2005) Activin/Nodal and FGF pathways cooperate to maintain pluripotency of human embryonic stem cells. J. Cell Sci. 118:4495–4509.

Wang, L., Schulz, T.C., Sherrer, E.S., Dauphin, D.S., Shin, S., Nelson, A.M., Ware, C.B., Zhan, M., Song, C.Z., Chen, X., et al. (2007) Self-renewal of human embryonic stem cells requires insulin-like growth factor-1 receptor and ERBB2 receptor signaling. Blood 110:4111–4119.

Watanabe, K., Ueno, M., Kamiya, D., Nishiyama, A., Matsumura, M., Wataya, T., Takahashi, J.B., Nishikawa, S., Nishikawa, S., Muguruma, K., et al. (2007) A ROCK inhibitor permits survival of dissociated human embryonic stem cells. Nat. Biotechnol. 25:681–686.

Wernig, M., Meissner, A., Foreman, R., Brambrink, T., Ku, M., Hochedlinger, K., Bernstein, B.E., Jaenisch, R. (2007) In vitro reprogramming of fibroblasts into a pluripotent ES-cell-like state. Nature 448:318–324.

Williams, R.L., Hilton, D.J., Pease, S., Willson, T.A., Stewart, C.L., Gearing, D.P., Wagner, E.F., Metcalf, D., Nicola, N.A., Gough, N.M. (1988) Myeloid leukaemia inhibitory factor maintains the developmental potential of embryonic stem cells. Nature 336:684–687.

Wilmut, I., Schnieke, A.E., McWhir, J., Kind, A.J., Campbell, K.H. (1997) Viable offspring derived from fetal and adult mammalian cells. Nature 385:810–813.

Xu, C., Rosler, E., Jiang, J., Lebkowski, J.S., Gold, J.D., O'Sullivan, C., Delavan-Boorsma, K., Mok, M., Bronstein, A., Carpenter, M.K. (2005) Basic fibroblast growth factor supports undifferentiated human embryonic stem cell growth without conditioned medium. Stem Cells 23:315–323.

Xu, R.H., Peck, R.M., Li, D.S., Feng, X., Ludwig, T., Thomson, J.A. (2005) Basic FGF and suppression of BMP signaling sustain undifferentiated proliferation of human ES cells. Nat. Methods 2:185–190.

Yu, J., Vodyanik, M.A., Smuga-Otto, K., Antosiewicz-Bourget, J., Frane, J.L., Tian, S., Nie, J., Jonsdottir, G.A., Ruotti, V., Stewart, R., et al. (2007) Induced pluripotent stem cell lines derived from human somatic cells. Science 318:1917–1920.

Contents

Part VI Differentiation

Contributors

LARISSA AGAPOVA • *International Stem Cell Corporation, Oceanside, CA, USA*
JACKELYN A. ALVA • *Lab of Dr. April D. Pyle, Department of Microbiology, Immunology, and Molecular Genetics, University of California, Los Angeles, Los Angeles, CA, USA*
LARA J. AUSUBEL • *Center for Biomedicine and Genetics, Center for Applied Technology Development, Beckman Research Institute of City of Hope, Duarte, CA, USA*
RUCHI BAJPAI • *Wysocka lab, Department of Chemical and Systems Biology, School of Medicine, Stanford University, Palo Alto, CA, USA*
STEVEN E. BATES • *Department of Biology, Beckman Research Institute, City of Hope National Medical Center, Duarte, CA, USA*
ROSITA BERGSTRÖM • *Department of Clinical Science, Intervention and Technology, Karolinska Institutet, Stockholm, Sweden*
DAVID J. BRICK • *Neuroscience Laboratories, Centers for Neuroscience and Translational Research, Children's Hospital of Orange County Research Institute, Orange, CA, USA*
JEROLD CHUN • *The Dorris Neuroscience Center, Molecular Biology Department, The Scripps Research Institute, La Jolla, CA, USA*
GREGORY M. COOPER • *Hudson Alpha Institute for Biotechnology, Huntsville, AL, USA*
LARRY A. COUTURE • *Center for Biomedicine and Genetics, Center for Applied Technology Development, Beckman Research Institute of City of Hope, Duarte, CA, USA*
SALLY A. COWLEY • *Sir William Dunn School of Pathology, University of Oxford, Oxford, UK*
SUJOY K. DHARA • *Indian Veterinary Research Institute, Izatnagar, Bareilly, UP, India*
MAHESH C. DODLA • *Animal and Dairy Science Department, University of Georgia, Athens, GA, USA*
PETER J. DONOVAN • *Departments of Biological Chemistry and Developmental and Cell Biology, Sue and Bill Gross Stem Cell Research Center, School of Biological Sciences, University of California, Irvine, Irvine, CA, USA*
PAULINE A.H. EDSER • *Molecular Neurodegeneration and Gene Therapy Research Group, Department of Physiology, Anatomy and Genetics, University of Oxford, Oxford, UK*
PAUL J. FAIRCHILD • *Sir William Dunn School of Pathology, University of Oxford, Oxford, UK*
ANIS FEKI • *Department of Obstetrics and Gynaecology, Geneve University Hospital, Geneva, Switzerland*

HELEN FONG • *Department of Biological Chemistry, Sue and Bill Gross Stem Cell Research Center, School of Medicine, University of California, Irvine, Irvine, CA, USA*
VICTORIA FOX • *Harley Norris Research Tower, University of Southern California Stem Cell Core, Los Angeles, CA, USA*
ANDREW FRYGA • *Flow Core, Australian Stem Cell Centre and Monash University, Clayton, Australia*
RIIKKA HÄMÄLÄINEN • *Nagy laboratory, SLRI, Mount Sinai Hospital, Toronto, ON, Canada*
MAYA HATCH • *Keirstead Research Group, Anatomy and Neurobiology, School of Medicine, University of California, Irvine, Irvine, CA, USA*
LYN HEALY • *UK Stem Cell Bank, Cell Biology and Imaging, National Institute for Biological Standards and Control, South Mimms, Hertfordshire, UK*
ALBERT J.R. HECK • *Biomolecular Mass Spectrometry and Proteomics Group, Bijvoet Center for Biomolecular Research, Utrecht, The Netherlands; Utrecht Institute for Pharmaceutical Sciences, Utrecht University, Utrecht, The Netherlands*
MELINDA K. HEXUM • *Department of Medicine and Stem Cell Institute, University of Minnesota, Minneapolis, MN, USA*
MICHAEL J. HITCHLER • *Department of Biochemistry and Molecular Biology, University of Southern California, Keck School of Medicine, Los Angeles, CA, USA*
MIRABELLE S.H. HO • *Department of Anatomy and Developmental Biology, Monash University, Clayton, Australia*
K.A. HOHENSTEIN ELLIOTT • *Peter Donovan Lab, Human Genetics and Molecular Biology, McKusick-Nathans Institute of Genetic Medicine, Johns Hopkins University School of Medicine, Baltimore, MD, USA*
FRIDA L. HOLM • *Department of Clinical Science, Intervention and Technology, Karolinska Institutet, Stockholm, Sweden*
OUTI HOVATTA • *Department of Clinical Science, Intervention and Technology, Karolinska Institutet, Stockholm, Sweden*
WILLIAM S. JAMES • *Sir William Dunn School of Pathology, University of Oxford, Oxford, UK*
JEFFREY JANUS • *International Stem Cell Corporation, Oceanside, CA, USA*
DAN S. KAUFMAN • *Hematology, Oncology and Transplantation, Department of Medicine and Stem Cell Institute, University of Minnesota, Minneapolis, MN, USA*
HANS KEIRSTEAD • *Reeve Irvine Research Center, Gillespie Neuroscience Research Facility, School of Medicine, University of California, Irvine, Irvine, CA, USA*
MARK KIBSCHULL • *Laboratory of Dr. S.J. Lye, Samuel Lunenfeld Research Institute, Mount Sinai Hospital, Toronto, ON, Canada*
LEONID KUZMICHEV • *Research Center of Obstetrics, Gynecology, and Perinatology, Moscow, Russia*
MICHAEL A. LAFLAMME • *Pathology, Institute for Stem Cell and Regenerative Medicine, University of Washington, Seattle, WA, USA*
ANDREW L. LASLETT • *Department of Anatomy and Developmental Biology, Monash University, Clayton, Australia*

JANE LEBKOWKSI • *Geron Corporation, Menlo Park, CA, USA*
PAUL LEROU • *Department of Newborn Medicine, Brigham & Women's Hospital/ Harvard Medical School, Harvard Institutes of Medicine, Boston, MA, USA*
QIONG LIN • *Institute for Biomedical Engineering, Department of Cell Biology, RWTH Aachen University Medical School, Universitätklinikum, Aachen, Germany*
YAN LIU • *Institute of Stem Cells and Regenerative Medicine, Department of Anatomy, Histology and Embryology, Shanghai Medical College, Fudan University, Shanghai, China*
YING LIU • *Department of Reproductive Medicine, UCSD Medical Center, University of California, San Diego, San Diego, CA, USA*
LESLIE F. LOCK • *Departments of Biological Chemistry and Developmental and Cell Biology, Sue and Bill Gross Stem Cell Research Center, School of Biological Sciences, University of California, Irvine, Irvine, CA, USA*
PATRICIA M. LOPEZ • *Center for Biomedicine and Genetics, Center for Applied Technology Development, Beckman Research Institute of City of Hope, Duarte, CA, USA*
JEANNE LORING • *Chemical Physiology, The Scripps Research Institute, La Jolla, CA, USA*
MICHELE M.P. LUFINO • *Molecular Neurodegeneration and Gene Therapy Research Group, Department of Physiology, Anatomy and Genetics, University of Oxford, Oxford, UK*
LIXIANG MA • *Department of Anatomy, Histology and Embryology, Shanghai Medical College, Fudan University, Shanghai, China*
ANIRBAN MAJUMDER • *Regenerative Bioscience Center, University of Georgia, Athens, GA, USA*
HEATHER C. MEFFORD • *Department of Pediatrics, University of Washington, Seattle, WA, USA*
MARIA MILEIKOVSKY • *Nagy laboratory, SLRI, Mount Sinai Hospital, Toronto, ON, Canada*
JUSTINE D. MILLER • *Dr. Lorenz Studer Laboratory, Developmental Biology Department, Gerstner Sloan-Kettering Graduate School, Memorial Sloan-Kettering Cancer Center, New York, NY, USA*
FRANZ-JOSEF MÜLLER • *Stem Cell Research, Zentrum für Integrative Psychiatrie ZIP gGmbH, Kiel, Germany*
JAVIER MUÑOZ • *Biomolecular Mass Spectrometry and Proteomics Group, Bijvoet Center for Biomolecular Research, Utrecht, The Netherlands; Utrecht Institute for Pharmaceutical Sciences, Utrecht University, Utrecht, The Netherlands*
ANDRAS NAGY • *Mount Sinai Hospital, Samuel Lunenfeld Research Institute, Toronto, ON, Canada*
HUBERT E. NETHERCOTT • *Neuroscience Laboratories, Centers for Neuroscience and Translational Research, CHOC Research Institute, Children's Hospital of Orange County Research Institute, Orange, CA, USA*
KEVIN P. NISHIMOTO • *Cell Biology, Regenerative Medicine, Geron Corporation, Menlo Park, CA, USA*
GABRIEL NISTOR • *Keirstead Research Group, Anatomy and Neurobiology, School of Medicine, University of California, Irvine, Irvine, CA, USA*

SUZANNE E. PETERSON • *Jeanne Loring Lab, Chemical Physiology, The Scripps Research Institute, La Jolla, CA, USA*
ANNA R. POPPLESTONE • *Molecular Neurodegeneration and Gene Therapy Research Group, Department of Physiology, Anatomy and Genetics, University of Oxford, Oxford, UK*
APRIL D. PYLE • *Molecular Biology Institute, Department of Microbiology, Immunology, and Molecular Genetics, Eli and Edythe Broad Center of Regenerative Medicine and Stem Cell Research at UCLA, University of California, Los Angeles, Los Angeles, CA, USA*
MAHENDRA RAO • *Primary and Stem Cell Systems, Research and Development, Life Technologies, Frederick, MD, USA*
ANITA REDDY • *Immunology and Translational Research, Cell Biology, Regenerative Medicine, Geron Corporation, Menlo Park, CA, USA*
ELENA REVAZOVA • *International Stem Cell Corporation, Oceanside, CA, USA*
JUDD C. RICE • *Department of Biochemistry and Molecular Biology, University of Southern California, Keck School of Medicine, Los Angeles, CA, USA*
THORSTEN M. SCHLAEGER • *hESC Core Facility, Hematology/Oncology and Stem Cell Program, Harvard Medical School, Children's Hospital Boston, Boston, MA, USA*
BERNHARD SCHULDT • *Graduate School AICES, RWTH Aachen University, Aachen, Germany*
PHILIP H. SCHWARTZ • *Neuroscience Laboratories, Centers for Neuroscience and Translational Research, Children's Hospital of Orange County Research Institute, Orange, CA, USA*
ANDREY SEMECHKIN • *International Stem Cell Corporation, Oceanside, CA, USA*
JASON SHARP • *Keirstead Research Group, Anatomy and Neurobiology, School of Medicine, University of California, Irvine, Irvine, CA, USA*
SEAN P. SHERMAN • *Lab of Dr. April D. Pyle, Molecular Biology Institute, University of California, Los Angeles, Los Angeles, CA, USA*
KATHRYN M. SILK • *Stem Cell Sciences Laboratory, Sir William Dunn School of Pathology, University of Oxford, Oxford, UK*
GLYN N. STACEY • *UK Stem Cell Bank, Cell Biology and Imaging, National Institute for Biological Standards and Control, South Mimms, Hertfordshire, UK*
STEVEN L. STICE • *Animal and Dairy Science Department, Rhodes Center for Animal and Dairy Sciences, College of Agricultural and Environmental Sciences, University of Georgia, Athens, GA, USA*
ALEXANDER E. STOVER • *Neuroscience Laboratories, Centers for Neuroscience and Translational Research, CHOC Research Institute, Children's Hospital of Orange County Research Institute, Orange, CA, USA*
SUSANNE STRÖM • *Department of Clinical Science, Intervention and Technology, Karolinska Institutet, Karolinska University Hospital, Hudiinge K 57 Stockholm, Sweden*
KAUSHALI THAKORE-SHAH • *Lab of Dr. April D. Pyle, Molecular Biology Institute, University of California, Los Angeles, Los Angeles, CA, USA*

XINGHUI TIAN • *Department of Medicine and Stem Cell Institute, University of Minnesota, Minneapolis, MN, USA*
SU-YI TSENG • *Cell Biology, Regenerative Medicine, Geron Corporation, Menlo Park, CA, USA*
NIKOLAY TUROVETS • *International Stem Cell Corporation, Oceanside, CA, USA*
BENJAMIN VAN BIBER • *Laflamme Lab, Pathology, Institute for Stem Cell and Regenerative Medicine, University of Washington, Seattle, WA, USA*
RICHARD WADE-MARTINS • *Molecular Neurodegeneration and Gene Therapy Research Group, Department of Physiology, Anatomy and Genetics, University of Oxford, Oxford, UK*
ROBIN L. WESSELSCHMIDT • *Center for Department of Applied Technology Development, Beckman Research Institute, City of Hope, Duarte, CA, USA*
KNUT WOLTJEN • *Woltjen Laboratory, Department of Reprogramming Sciences, Center for iPS Cell Research and Application (CiRA), Kyoto University, Sakyo-ku, Kyoto, Japan*
LESLEY YOUNG • *UK Stem Cell Bank, Cell Biology and Imaging, National Institute for Biological Standards and Control, South Mimms, Hertfordshire, UK*
SU-CHUN ZHANG • *Waisman Center, School of Medicine and Public Health, University of Wisconsin, Madison, WI, USA*
WEI-ZHONG ZHU • *Laflamme Lab, Pathology, Institute for Stem Cell and Regenerative Medicine, University of Washington, Seattle, WA, USA*

Part I

Laboratory Essentials

Chapter 1

The Stem Cell Laboratory: Design, Equipment, and Oversight

Robin L. Wesselschmidt and Philip H. Schwartz

Abstract

This chapter describes some of the major issues to be considered when setting up a laboratory for the culture of human pluripotent stem cells (hPSCs). The process of establishing a hPSC laboratory can be divided into two equally important parts. One is completely administrative and includes developing protocols, seeking approval, and establishing reporting processes and documentation. The other part of establishing a hPSC laboratory involves the physical plant and includes design, equipment and personnel.

Proper planning of laboratory operations and proper design of the physical layout of the stem cell laboratory so that meets the scope of planned operations is a major undertaking, but the time spent upfront will pay long-term returns in operational efficiency and effectiveness. A well-planned, organized, and properly equipped laboratory supports research activities by increasing efficiency and reducing lost time and wasted resources.

Key words: pluripotent stem cell laboratory, establishing a cell laboratory, cell culture, human PSC culture laboratory

1. Introduction

Establishing a well-functioning laboratory for the culture of human pluripotent stem cells (hPSCs) provides the foundation for successful culture and experimentation. This chapter will describe the major considerations for establishing a successful PSC research-grade laboratory (see Chapter 11 for considerations of a clinical-grade laboratory) (see Fig. 1). While the culture of hPSCs is carried out in a laboratory that is not much different than one used to culture other types of human cells (1, 2), due to the special status of these cells, there is a higher degree of oversight, review, and reporting. We have found that a nearly equal amount of time and effort is required to establish initial protocols and seek approval for culturing and obtaining human pluripotent cell lines as is required to design and equip the laboratory.

Philip H. Schwartz and Robin L. Wesselschmidt (eds.), *Human Pluripotent Stem Cells: Methods and Protocols*, Methods in Molecular Biology, vol. 767, DOI 10.1007/978-1-61779-201-4_1,

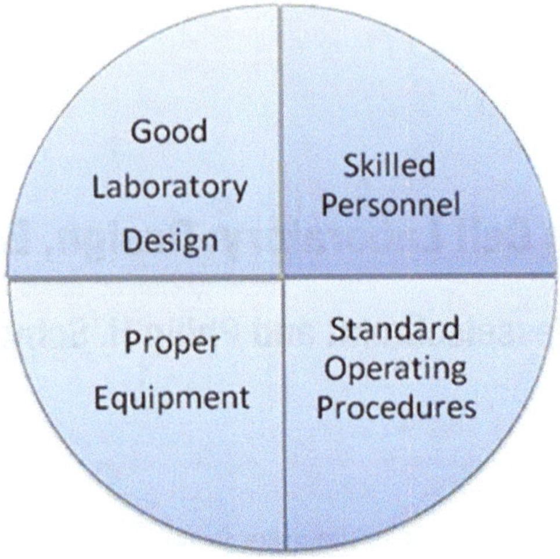

Fig. 1. The keys to operating a successful laboratory. This diagram indicates the interrelated nature of the four key elements in a stem cell laboratory. While we cannot underestimate the importance of skilled personnel and established standard operating procedures (SOPs), this chapter will focus on laboratory design and proper equipment (SOPs can be formulated using other chapters found in this book).

Whether retrofitting an existing laboratory or designing a new laboratory from shell space, the basic issues are the same (3–5). One must consider the equipment, number of people, and type of activities to be performed in order to achieve the best design (within budget constraints). Using a modular design, both in terms of laboratory benches and equipment, allows the laboratory to be expanded as needed by adding additional tissue culture modules and personnel to manage the work load (6). The key considerations when setting up the laboratory include (1) defining of the scope of the work which includes the numbers and types of cell lines to be cultured and (2) determining the number of people who will work in the laboratory and their specific tasks. One must consider, for example, how the lines will be maintained and characterized. If more than one PSC line will be cultured simultaneously, what safeguards will be put into place to prevent cross-contamination? How will new lines be introduced to the laboratory? Is there an area in which to quarantine these new lines? Accommodations must be made for the receipt of incoming materials and their testing and storage, as well as proper disposal of waste material. Culturing stem cells is very much like culturing other types of cells, with some additional special techniques. Specifically, the PSC research laboratory, at a minimum, requires space and equipment for: tissue culture, microscopy, and standard biochemistry and molecular biology.

2. Equipment

This section lists the minimal equipment required for basic culture and characterization of hPSCs. A given program may require specialized culture equipment, such as incubators that allow culture in low oxygen tension, or characterization and analysis equipment. This section lists the equipment required to establish a standard hPSC culture laboratory.

2.1. Tissue Culture Laboratory

1. Class II Biosafety Cabinet (BSC) (see Note 1).
2. CO_2 incubator (see Note 2).
3. Pipettors.
4. Vacuum flask/aspiration device.
5. Water bath (37°C).
6. Low-speed centrifuge (clinical grade, for spinning cells).

2.2. Microscopy

1. Phase–contrast microscope.
2. Dissecting microscope.

2.3. Storage

1. Cabinets and shelves for the storage of tissue culture supplies (see Note 3).
2. Refrigerator (4°C).
3. Freezer (−20°C, nondefrosting).
4. Low-temperature freezer (−70 to −85°C).
5. Cryogenic freezer (storage below −140°C, usually liquid nitrogen).

2.4. Molecular Biology Laboratory/Quality Control Laboratory

1. RT-PCR.
2. Flow cytometer (might be in a Core facility).
3. Fluorescence microscope (might be located in a Microscopy Core).
4. Confocal microscope (might be located in a Microscopy Core).

2.5. Quarantine Laboratory

1. Class II Biosafety cabinet.
2. CO_2 incubator.
3. Phase–contrast microscope.
4. Water bath (37°C).
5. Low-speed centrifuge (clinical grade, for spinning cells).
6. Pipettors.

7. Aspiration/vacuum flask.
8. Sink.

2.6. Additional Access to Common Equipment or Core Facilities

1. Microscopy.
2. Flow cytometry.
3. Microarray gene expression.
4. Genomics.
5. Proteomics.
6. Virus production.
7. Vivarium.

3. Methods

3.1. Laboratory Design and Layout

When building out new laboratory space, the design team usually includes an architect, a contractor, a builder, an electrician, a mechanic, a plumber, and a laboratory director or manager. The design team for retrofitting existing laboratory space is usually smaller and less formal, but is still comprised a number of people with specialized skills including the builder, electrician, mechanic, plumber, and the laboratory manager or director. Since the costs for building-out average laboratory space is 3–5 times that of office space, the budget is a key consideration and should be established as part of the preplanning.

The dynamic nature of biomedical research and the cost of laboratory construction have resulted in the prevalent use of modular design that allows reconfiguration of the laboratory as needed while keeping construction costs to a minimum.

3.1.1. Key Considerations for Planning and Design

1. Budget
   (a) How much money to spend and over what time?
   (b) Return on investment (what is expected to be gained through this investment).
2. Space
   (a) What type? Is it new construction, build-out of shell space, rehab of existing laboratory or office?
   (b) How much space is available?
   (c) Phased construction, how do the phases relate to each other? Should the plan include drawings for eventual build-out of all of the space?
3. Type and scope of work to be performed

   (a) Biosafety level of laboratory.
   (b) Major equipment.
   (c) Number of people.
4. Major functional areas
   (a) Tissue culture laboratory.
   (b) Quarantine laboratory (optional, but desired).
   (c) Molecular biology/quality control laboratory.
   (d) Microscopy laboratory.
5. Personnel
   (a) How many people?
   (b) In which functional area will they work?
   (c) How many offices?
   (d) How many desk spaces, shared or dedicated?
   (e) Break room.
6. Freezer rooms
   (a) How many and what types of freezers.
   (b) Back-up generator.
   (c) Alarm system.
   (d) A separate cryogenic freezer room with limited access.
7. Storage areas in and adjacent to the laboratory
   (a) Cabinets.
   (b) Shelves with 1-in. lip to prevent objects from falling.
   (c) Closets.
8. Information technology
   (a) IT closet.

3.1.2. Key Construction Considerations

Described and listed below are very high-level considerations that are meant to stimulate thinking by the laboratory manager and director as they begin working with their design or construction team to establish a new laboratory for the culture of human PSCs.

1. Mechanical/plumbing/engineering:
   (a) Back-up generators: dedicated or shared, alarms and controlled access.
   (b) CO_2 and LN_2 delivery: piped to the laboratory from "tank farm" or cylinders delivered to the laboratory and cryobank.
   (c) Vacuum systems.
2. HVAC

The numbers and types of BSCs will play a major role in the build-out of a tissue culture laboratory (7) and can add greatly to the cost of the build-out. Air handlers must be sized to accommodate the numbers and types of BSCs, as well as any other heat-producing equipment such as incubators and freezers. Efficient and comfortable conditioning of the air in the laboratory is one of the more challenging issues of the design and operation of the laboratory. It can be very uncomfortable working in laboratories, where the heat cannot be controlled due to the installation of HVAC system that is not of sufficient capacity to handle the heat produced by the BSCs and incubators. Air flow and its direction (negative, positive, or neutral) is a key consideration when designing the laboratory and is critical to the safe operation of the cell culture laboratory.

3. Electrical capacity and routing

   Accurately predicting the number of BSCs, freezers, and other major equipment and determining how much electricity they draw and heat they produce is critical to calculating the correct electrical capacity to maintain and safe working environment. The power requirements for all equipments as well as the location of power outlets, light switches, the determination of emergency power requirements, proximity of outlets to water faucets (GFC), and the coordination of the placement of outlets with the modular furniture designer is part of the detailed design. There need to be sufficient and dedicated circuits to handle all current equipments as well as future expansion. Incubators and freezers must be on back-up generator circuits and alarm systems.

4. Interior finishes:

   (a) Vinyl flooring.

   (b) Nonporous ceilings.

   (c) Washable, impermeable paint and coatings.

   (d) Impermeable bench-tops and furniture.

3.2. Tissue Culture Area

The tissue culture is comprised a minimal set of equipment, referred to here as the "tissue culture module" (see Table 1). It consists of one biosafety cabinet, one CO_2 incubator, one phase–contrast microscope, and a low-speed centrifuge, vacuum source, a water bath, two–2 L flasks, Pipet-Aid, micropipettors, and either a cabinet or cart next to the BSC for easy access to tissue culture supplies. If one keeps with the modular design concept, one can increase output by increasing the number of tissue culture modules in the laboratory design. In the hPSC laboratory, one tissue culture module can accommodate 1–2 technicians. Access to cryogenic storage, centrifuges, microscopes, refrigerators, and freezers is a must for a functional cell culture laboratory; however,

Table 1
Tissue culture module

Equipment	Considerations
Biosafety cabinet	What class and type? Most commonly: Class II Type A2
Incubator	Water or air jacketed One gas (CO_2 and air) Two gases (CO_2, N_2, and air) Three gasses (CO_2, N_2, and O_2)
Microscope	Phase–contrast Photo port
Water bath	37°C constant
Low-speed centrifuge	Refrigeration not required
Vacuum source	Portable Supplied by the building
2-L Erlenmeyer flasks, in series, with in-line filter	Collect aspirate and protect vacuum system
Pipet-Aid	Automatic pipettor, cordless, rechargeable
Micropipettors	2, 20, 200, and 1,000 μL

these frequently can be shared between multiple modules or laboratories depending on the goals of the overall laboratory program. If setting up a PSC laboratory contained within an existing cell culture laboratory, at least one tissue culture module is required. This will greatly improve productivity and reduce the chances of cross-contamination.

3.3. Molecular Biology Laboratory/Quality Control Laboratory

The molecular biology/quality control laboratory is comprised of the equipment and SOPs required to perform a predetermined list of characterization assays that allows one to systematically assess the quality of the cells in culture in the laboratory. Cultures are tested for the expression of specific markers, cytogenetic structure (karyotype or SKY), and the ability to differentiate, as described in many chapters throughout this book (and specifically in Chapter 2 describing the preparation of cell banks).

3.4. Quarantine Laboratory

The introduction of new cell lines into the laboratory is the major source by which cultures may become contaminated. Therefore, it is important that control systems are in place to minimize the potential for contamination. The quarantine control system can be a separate laboratory or cells can be quarantined through the use of a specific operating procedure where the incoming line is cultured at the end of the day by dedicated staff and grown in a dedicated incubator until it has been shown to be free of contaminates.

3.5. Storage

The proper import and storage of materials and reagents, including cell stocks, is key to the long-term success of the laboratory. When setting up a new laboratory, one has the opportunity to establish systems for logging incoming supplies and documenting their testing and use. Many of the reagents used in the PSC laboratory are derived from animal sources and therefore are subject to lot-to-lot variability that may necessitate in-house testing to determine which lot is suitable for use for specific applications in the laboratory. In addition, tracking materials for expiration dates and keeping the laboratory properly stocked is a critical function that can be facilitated through the use of databases and bar-coded inventory systems. The banking and retrieval of large numbers of small vials of frozen materials such as cell stocks and reagents can be a challenging and is greatly facilitated by developing an efficient easy-to-use tracking system.

3.6. Quality Control

The laboratory should be designed, equipped, staffed, and operated in a manner that allows for the production of reliable and reproducible experimental results. While the research laboratory must operate with enough flexibility to allow discovery to take place, establishing standard operating procedures and a quality control system can provide the foundation on which new discoveries can take place (see Fig. 2).

1. *Reliable techniques*: When establishing a PSC laboratory, identifying reliable techniques for the culture and characterization

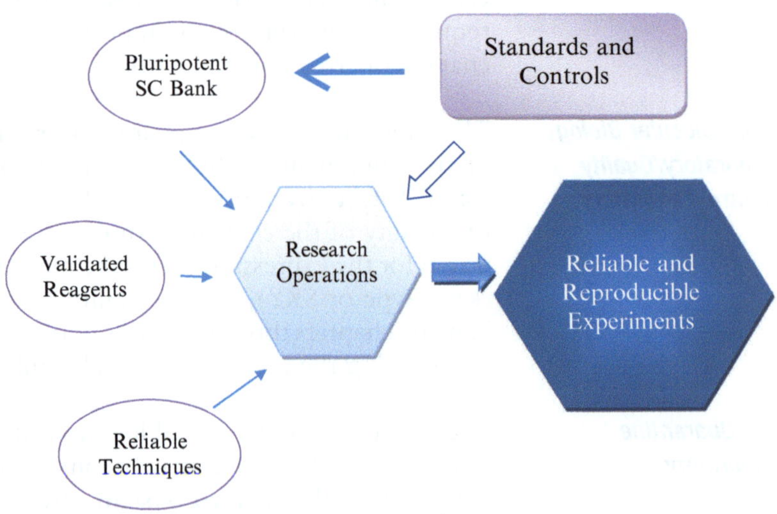

Fig. 2. Quality control in the stem cell laboratory. This figure shows the key quality control systems that allow one to achieve reliable and reproducible experimental results (adapted from (4)).

of cell lines is a critical function of the laboratory manager. Identifying techniques that are well-established and using cell lines that are well-characterized are the keys to establishing a solid foundation. New technologies and the rapid recent growth in the stem cell field require the laboratory manager to keep current with the literature, especially around the technologies that allow for the directed reprogramming of somatic cells to generate induced pluripotent stem cells (iPSCs). There is a growing belief that one can "simply" generate some iPSC lines and that will be all that is required to establish a PSC laboratory. This is far from true. We strongly encourage hands-on learning in an established laboratory, core, or training center in the art and science of human embryonic stem cell culture prior to embarking on the establishment of a hPSC laboratory. Also, we strongly recommend the maintenance of a well-characterized hESC line(s) in laboratory as the "gold standard" for all subsequent PSC work.

2. *Validated reagents*: It is critical that the reagents used to culture and characterize hPSCs are validated and shown to be reliable. As described in Subheading 3.5, the testing of reagent lots, especially animal-derived products such as fetal bovine serum (FBS), Knockout serum replacer (KSR), and mouse embryonic fibroblasts (MEFs), is critical to successful and efficient operations in the PSC laboratory.
3. *Quality PSCs*: It is critical that the PSCs used in the laboratory are from well-characterized stocks and have the morphologic and genetic characteristics of high-quality PSC lines as described in several chapters of this book. We emphasize the use of well-characterized hESC lines as the "gold standard" in laboratories establishing iPSC cultures. Having these bona fide hPSCs in the laboratory will allow both the testing of reagents and techniques and the verification of iPSC properties.
4. *Checklist for confirming quality*: Establishing a system that allows for the periodic testing of cultures against known standards will facilitate long-term success and provides confidence and reliability in the experimentation. As described in other chapters in this book, hPSCs need to be free of contamination, have a normal karyotype, express defined markers, and be capable of differentiating into cells derived from all three germ layers.

3.7. Oversight Approvals

When establishing a new laboratory for the culture of hPSCs, gaining the proper approval for the culture of the cells can be a lengthy and involved process. Where one is located, the country, state, city and institution, all play a role in the type of approval(s) required prior to initiating culture. In most cases, applications will have to be filed with review boards and specialized committees, material transfer agreements will have to be obtained and executed, and personnel will need to be trained, all before the first hPSC can be cultured in the new laboratory. For this reason, we suggest that one begin the approval process before or during the laboratory design process.

3.7.1. Typical Institutional Review and Oversight for US Academic Institutions

1. Stem Cell Research Oversight Committee (SCRO).
2. Institutional Review Board (IRB).
3. Biosafety Committee Review (IBC).
4. Institutional Animal Committee (IACUC).

4. Notes

1. The type of biosafety cabinet (BSC) that one installs depends on the type of work being conducted in the laboratory (7). The Class II BSC is typically utilized in the hPSC laboratory. Class II BSCs are partial barrier systems that rely on the directional movement of air to provide containment. They provide protection to both the worker and the material that is being manipulated in the cabinet when properly maintained. They provide the microbe-free work environment that is necessary for cell culture. Class II BSCs are available in four different types: Types A1 and A2 re-circulate the HEPA-filtered air back into the room. Types B1 and B2 are hard-ducted into a (preferably dedicated) exhaust system that carries the HEPA-filtered air to the outside. Class II A2 is the most common type of BSC installed in cell culture laboratories. Since the safe operation of the BSC relies on the HEPA filter, they are usually tested and certified on an annual basis or before any work commences following the relocation or movement of the BSC (see ref. 3 for an excellent review of the safe use and operation of BSCs). Generally, Class II BSCs can be used to work with *nonvolatile* and *nonhazardous* chemicals or gases; however, only type A2-exhausted or Types B1 and B2 can be used for preparing small amounts of volatile chemicals and minute quantities of hazardous chemicals, since they exhaust to the outside.
2. CO_2 incubators come with a variety of options. Incubators can come as two-gas (CO_2:air) or three-gas models (CO_2:O_2:N_2), the two-gas being the most common and least expensive. Chamber temperature is maintained by a water jacket or an air jacket. It is important that the shelving and the hardware securing it be easy to remove and clean. Although copper shelving and interior walls can inhibit the

growth of organisms, it is expensive and, unless all hardware components are of copper construction, thorough cleaning is still routinely needed. Routine cleaning with a disinfectant and ethanol rinse will help to reduce sources of contamination. The chamber, however, should be allowed to equilibrate overnight after a thorough cleaning prior to returning the cultures as the volatile components of the cleaning agents may kill the cells in culture. Use a portable RTD thermometer for accurate temperature determination and a Fyrite unit for CO_2 calibration.

3. Storage cabinets for tissue culture supplies are usually overlooked when designing the laboratory; however, sufficient storage allows for smooth and safe laboratory operations. Sufficient space should be made available that accommodates large bulky boxes so that supplies can be properly stored off of the floor and off of the equipment, allowing for easy cleaning and accessibility.

Acknowledgments

We would like to express our gratitude to all the people with whom we have built our laboratories and to all those in whose great laboratories we have had the privilege and pleasure of working.

References

1. Working with Human, NHP, and Other Mammalian Cells and Tissues (2009). *Appendix H* in Biosafety in Microbiological and Biomedical Laboratories. Fifth Edition, Department of Health and Human Services, Centers for Disease Control and Prevention, National Institutes of Health. HHS Publication No. (CDC) 21–1112.
2. Freshney I R (2005). Laboratory Design and Layout, in *Culture of Animal Cells: a manual of basic technique*, fifth edition, R. Ian Freshney. New Jersey, John Wiley & Sons, pp. 43–51.
3. Wesselschmidt RL and Thompson B (2006). Bioprocessing Tutorial: Embryonic stem cell research operations. *Genetic Engineering News*. **26**: 34–35.
4. Lyons I, Tan D, Schwartz PH, Rao M (2007). Setting up a facility for human embryonic stem cell research. *Human Stem Cell Manual: A Laboratory Guide*. First edition. Edited by Loring JF, Wesselschmidt RL, and Schwartz PH. New York, Elsevier, pp. 389–413.
5. Biosafety in Microbiological and Biomedical Laboratories. (2009) Fifth Edition, Department of Health and Human Services, Centers for Disease Control and Prevention, National Institutes of Health. HHS Publication No. (CDC) 21–1112.
6. Cost-Effective Methods for Design and Construction of Laboratories and Technical Facilities (2007). LCS Constructors, Inc. www.lcslab.com.
7. Primary Containment for Biohazards: Selection, Installation and Use of Biological Safety Cabinets. (2009) *Appendix A* in Biosafety in Microbiological and Biomedical Laboratories. Fifth Edition, Department of Health and Human Services, Centers for Disease Control and Prevention, National Institutes of Health. HHS Publication No. (CDC) 21–1112.

Chapter 2

Stem Cell Banks: Preserving Cell Lines, Maintaining Genetic Integrity, and Advancing Research

Lyn Healy, Lesley Young, and Glyn N. Stacey

Abstract

The ability to cryopreserve and successfully recover cell lines has been critical to the conservation of all cell lines, especially the preservation of pristine early-stage cultures and the preparation of well-characterized cell banks. Indeed, the systematic storage and establishment of cryopreserved banks of cells for the stem cell research community is fundamental to the promotion of standardisation in stem cell research and their use in clinical applications. In spite of the significant potential for the use of stem cells in research and therapy, they are challenging to maintain and have been shown to be unstable after prolonged culture that often results in permanent alterations in their genetic make-up, which ultimately alters the phenotype of the culture. This chapter will review the principles of cell bank production, techniques for the scale-up of human pluripotent stem cells, quality control, and characterisation methods for banked cell lines.

Key words: stem cell banking, master cell bank, working cell bank, stem cell characterisation, pluripotent stem cell characterisation, pluripotent stem cell quality control methods

1. Introduction

The establishment of large cryopreserved stocks or "banks" of microorganisms and cell cultures has been key to the reliability of industrial processes and to the manufacture of products based on the use of these cultures. Working in a similar way, cell culture collections, such as the UK Stem Cell Bank (UKSCB), that supply human stem cell cultures to researchers around the globe have promoted standardisation of methods that ensure that these cell lines remain available for decades. When setting out to establish a "Stem Cell Bank," it is important to identify its core function so that subsequent development can be directed appropriately. The cost associated with establishing and operating a cell bank is influenced by

Philip H. Schwartz and Robin L. Wesselschmidt (eds.), *Human Pluripotent Stem Cells: Methods and Protocols*, Methods in Molecular Biology, vol. 767, DOI 10.1007/978-1-61779-201-4_2, © Springer Science+Business Media, LLC 2011

the amount and type of quality control measures employed by the bank. Depending on the goal of the cell bank, these control measures may encompass mandatory quality standards including good manufacturing practices for therapeutic use in humans (1–5).

There is still much progress to be made in the optimisation of methodology in the areas of culture, preservation and characterisation of human stem cell lines. It is fundamental that those responsible for supplying stem cell lines to the stem cell community for research and clinical purposes work closely with leading stem cell researchers in order to remain in the forefront of these evolving methods. Projects funded by an international consortium of research funding bodies called the International Stem Cell Forum (www.stem-cell-forum.net/ISCF) have already promoted such developments (6, 7). Other organisations such as the International Society for Stem Cell Research (www.isscr.org/) have initiated high level guidance (*Guidelines for the Conduct of Human Embryonic Stem Cell Research, 2006, Guidelines for the Clinical Translation of Stem Cells, 2008*) as, has the US Academy of Sciences (*Guidelines for Human Stem Cell Research, 2005*). Key developments required to improve the standardisation of research and suitability of human pluripotent stem cells (hPSCs) for clinical applications will include development of robust methods for feeder-free culture (see Chapter 9) and passage of stem cells as single-cell suspensions (see Chapter 10) that do not result in genetic instability of the stem cell lines.

Characterisation of stem cell lines will no doubt be further enhanced through the application of proteomics, genomics and transcriptomics. Proteomic analysis has already been used to screen for phenotypic consistency in the analysis of human embryonic stem cell (hESC) cultures (8, 9), and has also been proposed for the quality control of cell lines (10). A major issue yet to be addressed is the development of robust and standardized in vitro and in vivo methods for evaluation of pluripotency in hPSC lines that can be used routinely in stem cell research laboratories.

This chapter summarises the various activities and issues involved in the preparation of master- and working-cell banks and quality control testing of lot-batches of human pluripotent stem cell (hPSC) lines that ensure the preservation of genetic integrity and increase experimental reproducibility.

2. The Principle of Master- and Working-Cell Banks

It is a commonly held belief that the banking of stem cell lines is simply a matter of growing up an extremely large batch of cells, aliquoting these cells into cryogenic vials, and freezing down one large lot that can then be used for experimental work and sent to collaborators. For specific projects, this may provide a short-term solution. However, in a research lab, with ongoing research programmes, it is important to be able to recover early passage stocks periodically in order to avoid the potentially detrimental effects on research data of experimenting with cells that have been in culture for prolonged periods of time. Irreversible genetic changes have been observed in hPSC cultures upon continuous culture over an extended period of time. Furthermore, resource centres that supply researchers with stem cell lines, as well as those supplying cells for clinical use, are expected to supply cells of consistent quality for many months and years with regards to characteristics and passage level.

In the biotechnology industry, where microorganisms and cell cultures have been used for manufacturing purposes for many decades, the establishment of a well-characterises, cryopreserved, seed stock, the master cell bank, as the source for all future work, is considered fundamental good practice. Individual vials of the master cell bank are then used to generate large "working" cell banks from which individual vials are used to initiate cultures for each production run or period of experimentation. This master/working bank system has been key to assuring long-term provision of high-quality cells for both research and industry and should be considered best practice for any stem cell culture laboratory (11, 12). The physical process of cryopreservation is not always well understood by stem cell workers and this can lead to failure to recover cells and even loss of cell lines. Furthermore, where preservation is suboptimal certain abnormal cells may be selected or induced giving rise to an altered cell culture. Reviews of preservation techniques and protocols used for pluripotent stem cell lines have been published respectively by (13, 14).

2.1. Feeder-Cell Banks

In order to maintain and expand undifferentiated hPSCs in culture these cells are usually co-cultured on inactivated fibroblasts of human or mouse origin known as feeder cells (see Chapter 8). It is not generally useful to extend the master- and working-stock principle to primary mouse embryonic fibroblasts (MEFs) because they have limited lifespan in culture and increased passaging affects their performance as feeder cells. MEFs are usually inactivated between passages 3 and 5 as they become senescent, depending on the mouse strain from which they were harvested, around passages 5–7. However, the preparation of large, pooled stocks of feeder cells at a consistent passage can improve the reliability of these cells in supporting the undifferentiated growth of PSC cultures. It is also important to obtain these primary cell cultures from animal colonies maintained under stringent animal husbandry standards and screened for a panel of infectious agents.

Human diploid fibroblasts, on the other hand, can be carried in culture much longer and passaged many more times than their murine counterparts, whilst retaining their capacity to support undifferentiated growth of hPSC lines. They therefore lend themselves to the production of master- and working-cell banks while providing a xenogenic-free alternative to MEFs.

Following preparation, feeder-cell banks should be subjected to functional assessment and to microbial quality control (QC) tests prior to their use as feeder-cells in PSC cultures. This will help prevent the contamination of the PSC line with bacteria, fungi, mycoplasma carried by the feeder-cells.

2.2. Scale-Up Techniques

Currently, the preferred method of hPSC culture requires co-culture with feeder cells and manual passaging methods that do not lend themselves readily to culture scale-up and production of large banks of cells. At the UKSCB, undifferentiated hPSC banks are generally prepared using a 6-well plate format where the hPSCs are in co-culture with feeder cells and subcultured manually. In order to produce a bank of high quality hPSCs, cells from the same cell line are harvested and pooled from several plates prior to cryo-preservation in an attempt to create a homogeneous bank of cells. Enhancing the "homogeneity" of cryopreserved cells promotes vial-to-vial consistency of these seed stocks, thus minimising variation that can arise in a single-well of any given plate.

Reduction of colonies to small clumps or single-cell suspensions using enzymes will facilitate the scale-up of cultures and the preparation of large banks of cells (15–18). While enzymatic passaging is preferable to manual passaging, especially when scaling-up cultures, chromosomal abnormalities (19–23) have been associated more commonly with cultures that are passaged using enzymatic methods. Until systems that sustain genetically stable cultures are developed and well-qualified, this may well remain the most important barrier to scale-up.

Scale-up of hPSCs is further complicated by the use of feeder-cell co-culture systems, which are still the most common method of hPSC culture. Effective scale-up of undifferentiated cells will only be achieved following the optimisation of culture conditions that do not use feeder cells. It is a challenge, however, to develop feeder-free systems that are capable of fully replacing the complex and as yet unknown mixture of components provided by feeder-cells. Currently, commonly-used systems require the coating of culture surfaces with extracellular matrices derived from animal sources, such as Matrigel™ or recombinant, xeno-free cell adhesion protein preparations of laminin (24) or vitronectin (25). However, these feeder-free systems will require significant development and validation for routine scale-up of hPSC cultures.

Despite these constraints, a number of scale-up methods for hPSC culture have been described (26–32). The ability to produce robust, reproducible protocols for scale-up is a fundamental requirement for the production of these cells, in both their undifferentiated and differentiated states, for use in cellular therapies.

Scale-up systems commonly used for non-stem cell cultures include roller bottles that are continuously rotated and can be supplied with ridged surfaces to increase the culture surface area (available from numerous suppliers). Stacked, static-flask systems are also available (e.g. Cell Cube™ [NUNC]) and there is a range of well-established scale-up systems for the culture of cell lines that grow as cell suspensions. Such systems are not readily amenable to the standard methods of culture and passage of hPSCs as enzymatic recovery of cells without altering their karyotype and potency needs to be developed for such systems. Most animal cell culture, scale-up systems are focused, maintaining growing cultures of homogenous cell suspensions and use an internal impellor device within the suspension of cells (e.g. spinner flasks), growth of cells in porous membrane compartments (e.g. "miniPerm"™ vessels, dialysis tubing systems, hollow fibre systems), or the culture of cells in agitated flexible culture bags (e.g. "Wave" bioreactors [GE Healthcare]). Despite the difficulty in growing human stem cell lines as single-cell suspensions, adherent cells, such as hPSCs, could be grown on "microcarriers", which could enable their growth in bioreactor formats normally used for suspension cell cultures.

A novel scale-up approach is adopted in low sheer-stress systems, sometimes called "microgravity" culture systems. Here, a gas-permeable membrane replaces the usual direct air-medium/gas–liquid interface and the cells are rotated in suspension to prevent gravitational settling (33). hPSCs grown in such systems form embryoid body-like structures. Other systems that enable metabolic activity to be measured without turbulence in the growth medium allow the cells to be grown in "orbital" or "free-fall" modes, or a combination of these approaches, as the culture develops (34). Examples of such systems that are commercially available include the Rotating Wall Vessel™ (Cellon) and Nova Pod™ (Novathera, now Medcell).

Automated systems for cell culture are now beginning to be applied to the culture of stem cell lines (35). Currently available systems focused on the automation of the cell culture process are primarily designed for enzyme-mediated passage of cultures (e.g. "CompacT", "SelecT" and "Cello" [Technology Automation Partnership]); however, some are designed to passage cultures, such as hPSCs, that grow as colonies (e.g. "Cellhost" and the BioLevitator™ [Hamilton]).

3. Quality Control and Characterisation

There are a number of central issues for the key quality control of all cell cultures since they are important for reliable research data and the quality and safety of products derived from such cultures.

Of these, the most critical characteristics are:

- Viability
- Identity (the cells are what they are purported to be)

- Purity (freedom from microbiological contamination)
- Stability on growth or passage in vitro

3.1. Viability

The ability of a cell culture to recover from the cryopreserved state is often determined using a dye-exclusion test such as trypan blue-exclusion. Whilst this is a useful indicator of the viability of cells, any one parameter will only give a narrow dataset on the overall status of a cell culture (36). If time and resources allow, it may be helpful to add additional parameters of viability, such as early markers of apoptosis (for example annexin IV expression). It is critical for the validity of any research programme that the culture recovered from a cell bank is representative of the original stock. In the case of hPSCs, this is indicated by demonstration of the typical phenotypic markers of the stem cell type, and importantly, the functional potential for pluripotency (see characterisation below).

3.2. Identity and Authenticity of Cell Lines

There is a long-established tradition of passing cell lines from one laboratory to another. Unfortunately, due to variation in local culture procedures and reagents, genetic alteration may occur following extended passaging, resulting in significant and permanent changes in the characteristics of the cell line. Furthermore, accidental cross-contamination, or mislabeling of cultures, can lead to the generation and publication of invalid data. Such events may go unrecognized for many years and the consequences may be far-reaching and cumulative, thus, resolution of such problems may be very time-consuming. In the history of cell culture, many examples of cell line cross-contamination have been identified (37, 38). Unfortunately, new generations of researchers do not appear to have learnt the lessons from these early publications since the problem of cross-contaminated cell lines, in particular, is still a significant issue (39–42). For human stem cell lines, and hPSCs in particular, the morphology and surface marker phenotype of cultures is very similar. Thus, identification and discrimination of different hPSC lines based on morphology and phenotype alone is almost impossible.

To ensure the correct identity and authenticity of a cell line, two primary factors must be addressed: firstly, cell line provenance (i.e., a traceable and documented origin, starting from the laboratory in which they were derived), and secondly, direct characterisation of the cells that can confirm cell line authenticity. A range of specialised tests may be employed to determine cell line identity including, HLA typing (43), isoenzyme analysis and karyology (44), but the level of specificity achieved using these methods is not very high (31). Genetic profiling of cell lines using multiplex PCR DNA Short Tandem Repeat (STR) profiling is the preferred method as it allows identification of specific cell lines with a high degree of specificity. This method has been utilised for a range of human tumor cell lines (45, 46) and is recommended as part of best practice in the quality control of human stem cell lines (7). Numerous companies now provide inexpensive DNA profiling services for human cells, making this technology available to all stem cell laboratories. It is important to note that, whilst these methods are human-specific, and therefore there is no cross-reactivity with the DNA derived from mouse feeder cells, it is possible that, when hPSCs are cultured with human feeder cells, the cell line DNA profile could be contaminated with the human feeder cell DNA. Thus, the specificity of the authentication method needs to be taken into account or evaluated. It is also important to remember that, whilst a DNA profile provides a highly specific DNA "bar code" that discriminates between different cell lines, this technique is only valuable for the formal authentication of a cell line when it can be compared with other material from the original donor/s and/or profiles from other qualified sources of the same cell line. Sharing genotypic data of this type between stem cell banks is therefore to be encouraged (4); however, national laws on release of individual, specific, genetic data may need to be considered before open publication of such information.

3.3. Microbial Contamination

Bacteria and fungi are common environmental contaminants that can infect and destroy cell cultures. A standard sterility-testing method using bacteriological broth cultures, such as outlined by Stacey (44), can be used on a regular basis to give assurance that general environmental contamination is not occurring. However, such methods do not have the capability to identify all possible bacterial and fungal contaminants that may arise; therefore it is best practice not to culture cells with antibiotics and to adhere to this principle most stringently when preparing cultures for banking.

The most common organisms known to cause unrecognized contamination are *Mycoplasma* and *Acholeplasma spp*. These organisms require special isolation media and growth conditions as described in Stacey (44). Details of standard methodologies for the above referenced testing methods can be obtained from the US and European Pharmacopoeia (47–50).

Any cell line can also harbour viral contamination arising from the original tissue or biological cell culture reagents. As a precaution, all cell cultures should be treated as potentially infectious, with appropriate containment and disposal according to local and national safety rules. Numerous viruses have been reported to establish persistent infection in cell lines (31). Viral contamination is clearly significant in that it may not only represent a hazard to laboratory workers, but also, where cells are persistently infected, it is likely to alter cell biology of the host culture (e.g., altered or deregulated biochemical pathways, transformation) and may cross-infect other cell lines in the laboratory.

It is not feasible to test cell banks for all potential viral contaminants but, where there is a very low risk of such contamination based on a risk assessment (51, 52), standard good laboratory practices, including containment of cell cultures (sealed culture vessels, use of biological safety cabinets, etc.), should provide adequate protection from contamination of cell cultures (11). Best practice guidance from the International Stem Cell Banking Initiative indicates that centres distributing stem cell lines to researchers, should aim to test for the most prevalent harmful human blood-borne viral pathogens (7).

It is suggested that all cell cultures suspected of harbouring microbial contamination be discarded in order to avoid health risks to laboratory workers and to assure the quality of the research. However, if the culture is irreplaceable, there may be a case for its maintenance in the laboratory under appropriate containment and isolation procedures prescribed to the infectious nature of the contaminating organism.

3.4. Characterisation and Stability Testing

Characterisation of each bank of cells will depend on the cell type and key phenotypic and genotypic markers for the particular stem cell line. *The master cell bank receives detailed characterisation with fundamental quality controls performed on working banks.* An international consensus, testing regime for master stocks of human stem cell lines has been established by the International Stem Cell Banking Initiative (7) and is shown in Table 1. Additional characterisation of the working cell bank and/or cultures passaged to anticipated maximum passage levels may be required to confirm the characteristics of the line (for example, assessing the chromosomal integrity of the line).

3.4.1. Phenotypic Characterisation

Early work on characterisation of hPSCs by flow cytometry and immunocytochemistry, using fluorescent antibodies often raised against embryonal carcinoma cells, established that hPSCs expressed certain phenotypic markers including OCT-4, Nanog, SSEA-3, SSEA-4, TRA-1-81, TRA 160 and alkaline phosphatase, and unlike mouse embryonic stem cells, were shown to be negative for SSEA-1 (53–55). A recent international study to characterise a large number of hPSC lines from numerous laboratories around the world established the consistent expression of such markers in hPSC cultures and provided a consensus for the characterisation of hPSCs including profiles of both surface markers and RNA expression (3). A selection of such antibody markers should be used to characterise banks of human stem cell lines. It is important to remember that these markers identify an undifferentiated state in the culture; if differentiation occurs, the presence of certain antigens is altered, with decreased expression of OCT-4 (which is associated with the ability of a hPSC cell to self-renew) and increased expression of SSEA-1. Those who have

Table 1
Outline of a typical testing regime for master cell bank of a human embryonic stem cell line

Test specification	Examples of test methods
Identity matches parent cell line	Short tandem repeat (STR) DNA profile Human leukocyte antigen (HLA) genotype
Bacteria/fungi	Inoculation of microbiological culture media to detect growth of bacteria and fungi
Mycoplasma	Direct culture in broth and agar and indirect test using indicator culture/DNA stain
Karyotype	Giemsa-band analysis Fluorescent in situ hybridization
Recovery	Post-thaw, trypan blue dye-exclusion Viable colonies recovered (qualified efficiency of recovery of each bank/lot should be given)
Pluripotency	Formation of teratomas in immune-deficient (SCID) mice Formation of embryoid bodies, in vitro "directed" differentiation
Growth characteristics	Doubling time Growth rate
Cell antigen expression	Flow cytometry or immunocytochemistry for a range of hPSC markers
Cell gene expression	Gene-expression profiling using DNA microarray or Q-PCR analysis
Genetic stability	Single nucleotide polymorphisms. Comparative genome hybridization by DNA microarray methodology (see also karyotype)

Reproduced from [7] with permission from Springer (Adapted from, ISCBI 2009)

worked with hPSC cultures will recognize that hPSC cultures often contain a population of differentiated cells that will vary to some degree from one time-point in culture to another and that it is not feasible or realistic to set absolute, stringent criteria for the expression of such markers.

3.4.2. Pluripotency

Key to the scientific and therapeutic potentials of human stem cell lines is their capacity to generate cells representative of all three human germ layers (endoderm, ectoderm and mesoderm), and thus, their potential to generate all the tissues of the human body – pluripotency. This is clearly a critical characteristic for cell banks to address; any human "stem cell line" claimed to be pluripotent must demonstrate this characteristic using accepted techniques. The current "gold-standard" technique is the formation of teratomas from hPSCs injected into immune-compromised (typically SCID) mice (see Chapter 17). The teratomas should reveal cells and tissue-like structures representative of all three germ layers. Other options to determine the pluripotency of hPSCs include the preparation of embryoid bodies showing evidence of markers from all three germ layers, and in vitro, directed differentiation into cell populations associated with each germ layer. It is important to remember that all of the described experimental procedures provide evidence for pluripotency but do not categorically confirm the ability of stem cells to generate all the cells and tissues of the human body. Even for the standard teratoma method, it is known that mouse strain and age and the method of preparation of the inoculated cells can influence the outcome of the assay. The lack of a robust, qualified laboratory assay for pluripotency is a significant challenge to researchers and stem cell banks alike. Establishing robust standardized assays that can be used for characterising cell banks remains a fundamental goal in regenerative medicine.

3.4.3. Genetic Characterisation

Another key requirement for hPSCs is a diploid karyotype (46:XY for male and 46:XX for female). In vitro culture, however, appears to promote the generation of clones of genetically abnormal cells that can take over a culture and rapidly replace the original diploid cells. Banks of cells should therefore routinely be assessed for karyotype. The guidance published by the International Stem Cell Banking Initiative gives recommended approaches for such karyological studies of hPSC lines (7). A number of other karyotyping methods are currently being used in clinical and research work but remain to be qualified for routine quality control of banks of human stem cell lines.

3.4.4. Stability Testing via Extended Culture

A desirable adjunct to the cell banking process is to passage cells beyond their anticipated limit of use in order to further evaluate the stability of the line. For hPSCs, this may include how well they maintain the undifferentiated phenotype, the consistency of the undifferentiated growth when compared with cells from the master bank and the confirmation of pluripotency. Ideally, this will involve establishing "extended cell banks" at intervals of around ten passages and comparing the characteristics of these banks to determine what, if any, drift has occurred in the culture over time. However, this would involve considerable time and effort, which would have to be balanced against the benefits. It might be better for banks to recommend that recipients of their lines should periodically check their cultures for chromosomal integrity.

Acknowledgements

The authors would like to thank the Medical Research Council (MRC), the Biotechnology and Biological Sciences Research Council (BBSRC), and the International Stem Cell Forum for funding and supporting the activities of the UK Stem Cell Bank.

References

1. EU Commission of the European Communities (1998) The Rules Governing Medical Products in the European Community, Good Manufacturing Practice for Medicinal Products, Volume IV, April 1998, ISBN 9289–20297, with additional Annexes and update of legal references, July 2004.
2. European Union (2004) Directive 2004/23/EC of the European Parliament and the Council of 31 March 2004 on setting standards of quality and safety for the donation, processing, preservation, storage and distribution of human tissues and cells. Official Journal of the European Union, L102:48–58.
3. US Food and Drugs Administration (2004a) 21 CFR Parts 16, 1270 and 1271. Current Good Tissue Practice for Human Cell, Tissue, and Cellular and Tissue-Based Products Establishments; Inspection and Enforcement; Final Rule. Department of Health and Human Services, November 24.
4. US Food and Drug Administration (2004b) Title 21, Code of Federal Regulations, Part 210, Current Good Manufacturing Practice in Manufacturing, Processing, Packing, or Holding of Drugs; General, US FDA.
5. US Food and Drug Administration (2004c) Title 21, Code of Federal Regulations, Part 211, Current Good Manufacturing Practice for Finished Pharmaceuticals, US FDA. US.
6. Adewumi O, Aflatoonian B, Ahrlund-Richter L, Amit M, Andrews PW, Beighton G, Bello PA, Benvenisty N, Berry LS, Bevan S, Blum B, Brooking J, Chen KG, Choo AB, Churchill GA, Corbel M, Damjanov I, Draper JS, Dvorak P, Emanuelsson K, Fleck RA, Ford A, Gertow K, Gertsenstein M, Gokhale PJ, Hamilton RS, Hampl A, Healy LE, Hovatta O, Hyllner J, Imreh MP, Itskovitz-Eldor J, Jackson J, Johnson JL, Jones M, Kee K, King BL, Knowles BB, Lako M, Lebrin F, Mallon BS, Manning D, Mayshar Y, McKay RD, Michalska AE, Mikkola M, Mileikovsky M, Minger SL, Moore HD, Mummery CL, Nagy A, Nakatsuji N, O'Brien CM, Oh SK, Olsson C, Otonkoski T, Park KY, Passier R, Patel H, Patel M, Pedersen R, Pera MF, Piekarczyk MS, Pera RA, Reubinoff BE, Robins AJ, Rossant J, Rugg-Gunn P, Schulz TC, Semb H, Sherrer ES, Siemen H, Stacey GN, Stojkovic M, Suemori H, Szatkiewicz J, Turetsky T, Tuuri T, van den Brink S, Vintersten K, Vuoristo S, Ward D, Weaver TA, Young LA, Zhang W (2007) Characterization of human embryonic stem cell lines by the International Stem Cell Initiative. *Nat Biotechnol.* **25**:803–16.
7. International Stem Cell Banking Initiative. International Stem Cell banking Initiative (2009) Consensus guidance for banking and supply of human embryonic stem cell lines for research purposes. *Stem Cell Rev.* **5**:301–14.
8. Prowse AB, McQuade LR, Bryant KJ, Van Dyk DD, Tuch BE, Gray PP (2005) A proteome analysis of conditioned media from human neonatal fibroblasts used in the maintenance of human embryonic stem cells. *Proteomics.* **5**:978–89.
9. Baharvand H, Hajheidari M, Ashtiani SK, Salekdeh GH (2006) Proteomic signature of human embryonic stem cells. *Proteomics.* **6**:3544–9.
10. Wagg SK, Lee LE (2005) A proteomics approach to identifying fish cell lines. *Proteomics.* **5**:4236–44.
11. Coecke S, Balls M, Bowe G, Davis J, Gstraunthaler G, Hartung T, Hay R, Merten O.-W, Price A, Shechtman L, Stacey G, Strokes W (2005) Guidance on Good Cell Culture Practice: a report of the second ECVAM task force on good cell culture practice. *ATLA*, **33**:261–287.
12. Knezevic I, Stacey G, Petricciani J, Sheets R, WHO Study Group on Cell Substrates (2010) Evaluation of cell substrates for the production of biologicals: Revision of WHO recommendations. Report of the WHO Study Group on

Cell Substrates for the Production of Biologicals, 22–23 April 2009, Bethesda, USA. *Biologicals*. 38(1):162–9.

13. Day DG, Stacey GN (2007) Cryopreservation and freezedrying methods. Humana Press, Totowa, USA.
14. Hunt CJ (2011) Cryopreservation of human stem cells for clinical application: a review. *Transfus Med Hemother*. 38(2):107–23.
15. Ellerström C, Strehl R, Noaksson K, Hyllner J, Semb H (2007) Facilitated expansion of human embryonic stem cells by single-cell enzymatic dissociation. *Stem Cells*. **25**:1690–6.
16. Bajpai R, Lesperance J, Kim M, Terskikh AV (2008) Efficient propagation of single cells Accutase-dissociated human embryonic stem cells. *Mol Reprod Dev*. **75**:818–27.
17. Thomson A, Wojtacha D, Hewitt Z, Priddle H, Sottile V, Di Domenico A, Fletcher J, Waterfall M, Corrales NL, Ansell R, McWhir J (2008) Human embryonic stem cells passaged using enzymatic methods retain a normal karyotype and express CD30. *Cloning Stem Cells*. **10**:89–106.
18. Ellerström C, Hyllner J, Strehl R (2010) Single cell enzymatic dissociation of human embryonic stem cells: a straightforward, robust, and standardized culture method. *Methods Mol Biol*. **584**:121–34.
19. Imreh MP, Gertow K, Cedervall J, Unger C, Holmberg K, Szöke K, Csöregh L, Fried G, Dilber S, Blennow E, Ahrlund-Richter L (2006) In vitro culture conditions favoring selection of chromosomal abnormalities in human ES cells. *J Cell Biochem*. **99**:508–16.
20. Baker DE, Harrison NJ, Maltby E, Smith K, Moore HD, Shaw PJ, Heath PR, Holden H, Andrews PW (2007) Adaptation to culture of human embryonic stem cells and oncogenesis in vivo. *Nat Biotechnol*. **25**:207–15.
21. Catalina P, Montes R, Ligero G, Sanchez L, de la Cueva T, Bueno C, Leone PE, Menendez P (2008) Human ESCs predisposition to karyotypic instability: Is a matter of culture adaptation or differential vulnerability among hESC lines due to inherent properties? *Mol Cancer*. **3**;7:76.
22. Lefort N, Perrier AL, Laâbi Y, Varela C, Peschanski M (2009) Human embryonic stem cells and genomic instability. Regen Med. **4**:899–909.
23. Närvä E, Autio R, Rahkonen N, Kong L, Harrison N, Kitsberg D, Borghese L, Itskovitz-Eldor J, Rasool O, Dvorak P, Hovatta O, Otonkoski T, Tuuri T, Cui W, Brüstle O, Baker D, Maltby E, Moore HD, Benvenisty N, Andrews PW, Yli-Harja O, Lahesmaa R (2010) High-resolution DNA analysis of human embryonic stem cell lines reveals culture-induced copy number changes and loss of heterozygosity. *Nat Biotechnol*. **28**:371–7.
24. Vuoristo S, Virtanen I, Takkunen M, Palgi J, Kikkawa Y, Rousselle P, Sekiguchi K, Tuuri T, Otonkoski T (2009) Laminin isoforms in human embryonic stem cells: synthesis, receptor usage and growth support. *J Cell Mol Med*. **13**:2622–33.
25. Braam SR, Zeinstra L, Litjens S, Ward-van Oostwaard D, van den Brink S, van Laake L, Lebrin F, Kats P, Hochstenbach R, Passier R, Sonnenberg A, Mummery CL (2008) Recombinant vitronectin is a functionally defined substrate that supports human embryonic stem cell self-renewal via alphavbeta5 integrin. *Stem Cells*. **26**:2257–65.
26. Azarin SM, Palecek SP (2010) Development of Scalable Culture Systems for Human Embryonic Stem Cells. *Biochem Eng J*. 2010 **48**:378–384.
27. Steiner D, Khaner H, Cohen M, Even-Ram S, Gil Y, Itsykson P, Turetsky T, Idelson M, Aizenman E, Ram R, Berman-Zaken Y, Reubinoff B (2010) Derivation, propagation and controlled differentiation of human embryonic stem cells in suspension. *Nat Biotechnol*. **28**:361–4.
28. Singh H, Mok P, Balakrishnan T, Rahmat SN, Zweigerdt R (2010) Up-scaling single cell-inoculated suspension culture of human embryonic stem cells. *Stem Cell Res*. Mar 12. [Epub ahead of print].
29. Fernandes AM, Marinho PA, Sartore RC, Paulsen BS, Mariante RM, Castilho LR, Rehen SK (2009) Successful scale-up of human embryonic stem cell production in a stirred microcarrier culture system. *Braz J Med Biol Res*. **42**:515–22.
30. Krawetz R, Taiani JT, Liu S, Meng G, Li X, Kallos MS, Rancourt D (2009) Large-Scale Expansion of Pluripotent Human Embryonic Stem Cells in Stirred Suspension Bioreactors. *Tissue Eng Part C Methods*. 2009 Sep 8. [Epub ahead of print].
31. Stacey GN, Davis J (2007) Medicines from Animal Cell Cultures, Wiley & Sons, Chichester, UK.
32. Oh SK, Choo AB (2006) Human embryonic stem cell technology: large scale cell amplification and differentiation.Cytotechnology. **50**:181–90.
33. Martin Y, Vermette P (2005) Bioreactors for tissue mass culture: design, characterization, and recent advances. *Biomaterials*. **26**: 7481–7503.

34. Kodidis T, Lenz A, Buoblik J, Akhyari P, Wachsmann B, Mueller-Stahl K, Hofmann M, Haverich A (2003) Pulsatile perfusion and cardiomyocyte viability in a solid three-dimensional matrix. *Biomaterials.* **24**:5009–5014.
35. Thomas RJ, Anderson D, Chandra A, Smith NM, Young LE, Williams D, Denning C (2009) Automated, scalable culture of human embryonic stem cells in feeder-free conditions. *Biotechnol Bioeng.* **102**:1636–44.
36. Stacey GN, Auerbach JM (2007) Quality Control of Stem Cell Lines. In: Culture of Stem Cells. (Eds.: Freshney IR, Stacey GN, and Auerbach JM, John Wiley & Sons). pp. 1–22.
37. Gartler SM (1967) Genetic markers as tracers in cell culture. *Natl Cancer Inst Monogr.* **26**:167–95.
38. Nelson-Rees WA, Daniels DW, Flandermeyer RR (1981) Cross-contamination of cells in culture. *Science.* **1212**:446–52.
39. MacLeod RA, Dirks WG, Matsuo Y, Kaufmann M, Milch H, Drexler HG (1999) Widespread intraspecies cross-contamination of human tumor cell lines arising at source. *Int J Cancer.* **83**:555–63.
40. Drexler HG, Dirks WG, Matsuo Y, MacLeod RA (2003) False leukemia-lymphoma cell lines: an update on over 500 cell lines. *Leukemia.* **17**:416–26.
41. Melcher R, Maisch S, Koehler S, Bauer M, Steinlein C, Schmid M, Kudlich T, Schauber J Luehrs H, Menzel T, Scheppach W (2005) SKY and genetic fingerprinting reveal a cross-contamination of the putative normal colon epithelial cell line NCOL-1. *Cancer Genet Cytogenet.* **158**:84–7.
42. Capes-Davis A, Theodosopoulos G, Atkin I, Drexler HG, Kohara A, Macleod RA, Masters JR, Nakamura Y, Reid YA, Reddel RR, Freshney RI (2010) Check your cultures! A list of cross-contaminated or misidentified cell lines. *Int J Cancer.* Feb 8. [Epub ahead of print].
43. Christensen B, Hansen C, Debiek-Rychter M, Kieler J, Ottensen S, Schmidt J (1993) Identity of tumourigenic uroepithelial cell lines and "spontaneously" transformed sublines. *Br J Can.* **68**:879–884.
44. Stacey GN. Quality Control of Human Stem Cell Lines. (2007b) In: Human Embryonic Stem Cells (Human Cell Culture - Volume 6) (Eds.: Masters J, Palsson B, Thomson J), Springer (Tokyo and new York), pp. 1–22.
45. Masters JRW, Thompson J, Daly-Burns B, Reid YA, Dirks WG, Packer PI, Toji LH, Ohno T, Tanabe TH, Arlett CF, Kelland LR, Harrison M, Virmani A, Ward TH, Ayres KL, Debenham PG (2001) Short tandem repeat profiling provides international reference standards for human cell lines. *PNAS (USA)*, **98**:8012–8017.
46. Parson W, Kirchebner R, Muhlmann R, Renner K, Kofler A, Schmidt S, Kofler R (2005) Cancer cell line identification by short tandem repeat profiling: power and limitations. *FASEB J.* **19**:434–436.
47. European Pharmacopoeia (2006a) European Pharmacopoeia section 2.6.1 (Sterility), 5th Edition, Maisonneuve SA, Sainte Ruffine (www.pheur.org).
48. European Pharmacopoeia (2006b) European Pharmacopoeia section 2.6.7 (Mycoplasma), 5th Edition, Maisonneuve SA, Sainte Ruffine.
49. US Food and Drugs Administration (2005a) Title 21, Code of Federal Regulations, Volume 7, revised April 2005, CFR610.12 (Sterility), FDA, Department of Health and Human Services.
50. US Food and Drugs Administration (2005b) Title 21, Code of Federal Regulations, Volume 7, revised April 2005, CFR610.30 (Test for Mycoplasma), FDA, Department of Health and Human Services.
51. Stacey GN, Tyrrel DA, Doyle A (1998) Source materials. In: Safety in Cell and Tissues Culture (Eds.: Stacey GN, Hambleton PJ, Doyle A), Kluwer Academic, Dordrecht, The Netherlands, pp. 1–25.
52. Stacey GN (2007c) Risk assessment of cell culture procedures. In: Medicines from Animal Cell Cultures, Wiley & Sons, Chichester, UK pp. 569–599.
53. Pera M, Reubinoff B, Thomson JA (2000) Human embryonic stem cells. *J Cell Sci.* **113**: 5–10.
54. Andrews PW, Przyborski SA, Thomson JA (2001) Embryonal carcinoma cells as embryonic stem cells. In: Stem Cell Biology (Eds.: Marshak DR, Gardner RL, Gottlieb D), Cold Spring Harbor Monograph Series, Cold Spring Harbor, New York, pp. 231–265.
55. Draper JS, Pigott C, Thomson JA, Andrews PA (2002) Surface antigens of human embryonic stem cells: changes upon differentiation in culture. *J Anat.* **200**:249–258.

Part II

Derivation

Chapter 3

Embryonic Stem Cell Derivation from Human Embryos

Paul Lerou

Abstract

Human embryonic stem cells (hESCs) are self-renewing, pluripotent cells that serve as a valuable research tool and hold promise for use in regenerative medicine. Most hESC lines are derived from cryopreserved human embryos that were created during in vitro fertilization (IVF) and are in excess of clinical need. Embryos that are discarded during the IVF procedure are also a viable source of hESCs. hESCs can be derived from pre-blastocyst embryos and even from single blastomeres. However, hESC line derivation efficiency is greatest using embryos that have reached the blastocyst stage. This chapter describes a protocol for the derivation of pluripotent stem cell lines from human embryos.

Key words: human embryonic stem cells, hES cells, hESC

1. Introduction

Human embryonic stem cells (hESCs) are self-renewing, pluripotent cells that serve as a valuable research tool and hold promise for use in regenerative medicine (1). Most hESC lines are derived from cryopreserved human embryos that were created during in vitro fertilization (IVF) and are in excess of clinical need. Embryos that are discarded during the IVF procedure are also a viable source of hESCs (2).Although hESC lines have been derived from pre-blastocyst stage embryos and even from single blastomeres, derivation efficiency is greatest using embryos that have reached the blastocyst stage (2, 3). Recently, Chen et al. reported a hESC derivation efficiency of 50% by timing the inner cell mass isolation to day 6 postfertilization (4). In this protocol, we describe how to derive novel hESC lines from human embryos. It is important to note that this protocol can yield various types of colonies and cells that represent both pluripotent and nonpluripotent cells, as is seen when somatic

Philip H. Schwartz and Robin L. Wesselschmidt (eds.), *Human Pluripotent Stem Cells: Methods and Protocols*, Methods in Molecular Biology, vol. 767, DOI 10.1007/978-1-61779-201-4_3,

cells are reprogrammed to generate induced pluripotent stem cells (iPSCs). It is important to establish that derived lines are bona fide hESCs. This is done by performing standard tests such as gene expression, promoter methylation, and teratoma formation analyses. Moreover, the microbiological safety, genomic integrity, and genetic identity of hESC lines must be confirmed soon after derivation and confirmed at regular passage intervals. Validated lines should be banked, for example, by creating frozen premaster, master, and distribution/usage stocks as described in (Chapter 2) of this volume.

2. Materials

2.1. Human Embryo Culture

1. G2.3 embryo medium (Vitrolife, cat. no. 10092).
2. 4-Well culture dish.
3. Acidic Tyrode's solution.

2.2. Human Embryonic Stem Cell Culture

1. *hESC derivation medium*: Dulbecco's modified Eagle's medium/ Nutrient Mixture F12 (DMEM/F12, Stem Cell Technologies, #36254) supplemented with 17% knockout serum replacement (KOSR, Invitrogen, 10828-028), 3% FBS, 2 mM L-glutamine, 0.1 mM MEM-NEAA, 100 μM beta-mercaptoethanol (2-ME), and 10 ng/ml human recombinant basic fibroblast growth factor (bFGF, dissolved in 0.1% BSA/PBS and stored in working aliquots at −20°C, Gemini Bio-Products, 400-432P).
2. *Standard hESC medium*: Dulbecco's modified Eagle's medium/ Nutrient Mixture F-12 (DMEM/F12, Stem Cell Technologies, #36254) supplemented with 20% knockout serum replacement (KOSR, Invitrogen, 10828-028), 2 mM L-glutamine, 0.1 mM MEM-NEAA, 100 μM beta-mercaptoethanol (2-ME), and 10 ng/ml human recombinant bFGF (dissolved in 0.1% BSA/ PBS and stored in working aliquots at −20°C, Gemini Bio-Products, 400-432P).
3. Irradiated CF-1 mouse embryonic fibroblasts (MEFs).
4. *MEF medium*: DMEM (high glucose, no pyruvate) supplemented with 10% FBS, 2 mM L-glutamine, 0.1 mM MEM-NEAA, 1 mM sodium pyruvate.
5. Nunclon multiwell dishes: 6-well, 12-well.
6. 5-ml Glass pipettes.

3. Methods

3.1. Human Embryo Culture

1. Use embryo culture and freeze/thaw conditions that match those used at the IVF laboratory from where the embryos have been received. Transport fresh embryos to the research laboratory in HEPES buffered medium in a portable 37°C incubator (see Notes 1 and 2).
2. To maximize the yield of hESC lines, culture the embryo to the blastocyst stage (day 5–6 postfertilization) in 0.5 ml of G2.3 medium, individually, in 1 well of a 4-well dish from day 3–5 at 37°C under low oxygen concentration in 5% O_2/5% CO_2/90% N_2 atmosphere.

3.2. Human Embryonic Stem Cell Line Derivation

1. 1–3 days prior to plating the embryo for derivation, thaw MEFs onto a gelatin-coated 4-well dish at 40,000 cells per well in 0.5 ml of MEF medium.
2. On the day of embryo plating, change MEF medium to 0.5 ml of hESC derivation medium.
3. Transfer embryo from the culture medium into 0.5 ml pre-warmed Acidic Tyrode's solution in a well of 4-well dish. Observe the embryo continually under stereo-microscope until the zona pellucida is completely dissolved (1–2 min). Then wash the embryo twice in 0.5 ml hESC derivation medium (see Note 3).
4. Plate embryos onto prepared MEF feeder cell dish from step 2 of Subheading 3.2 and incubate embryos on MEFs in 37°C, 5% O_2/5% CO_2/90% N_2 for 72 h, without disturbing the plates, allowing the embryos to attach to the culture dish. By days 5–7 of co-culture with MEF feeders, embryo outgrowths should appear (Fig. 1). However, attachment may take as long as 10 days (see Note 4).
5. When the embryo outgrowth reaches 5–10 mm in diameter, it should be passaged mechanically. Using either a flame-pulled Pasteur pipette or a bent 27–30 G needle attached to 1-ml syringe, "cut" the outgrowth into 2–4 pieces and dislodge the pieces from the bottom of the well. Separate the trophectodermal outgrowth (large, flat cells, along the periphery) away from the small, tightly packed cells in the centers. Using a sterile 20-μl pipette tip, transfer the small clumps composed of small, tightly packed cells to 0.5 ml hESC derivation medium on freshly plated MEFs in a well of 4-well dish.
6. Incubate in 37°C, 5% O_2/5% CO_2/90% N_2. Leave undisturbed for 48 h after passage, then feed every 24 h with hESC medium (the serum is no longer required and may induce differentiation).

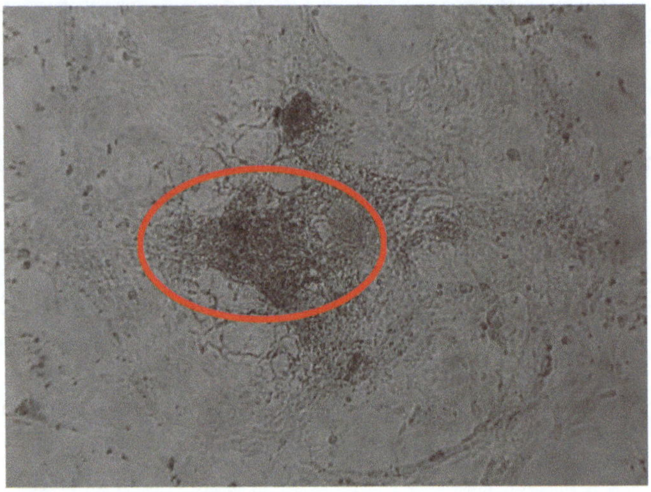

Fig. 1. Embryo outgrowth at day 5. *Red circle*: central clump of small tightly packed cells that should be passaged from the outgrowth to establish the hES cell line.

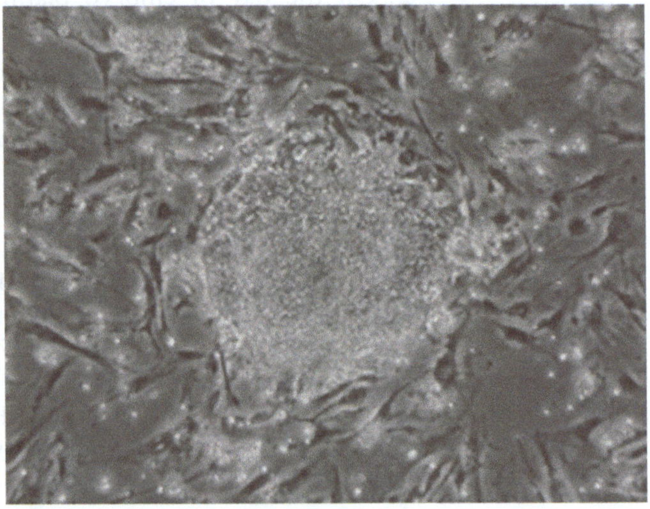

Fig. 2. Passage 8 hES cell colony. Flat, refractile, and composed of tightly packed small cells with a high nuclear to cytoplasmic ratio.

7. If a hESC line is successfully derived, the clumps of cells will attach by 48 h and over the next few days, the clumps will grow into colonies that resemble hESCs in morphology – flat, refractile, and composed of tightly packed small cells with a high nucleus-to-cytoplasm ratio (Fig. 2).
8. Continue to disperse hESC colonies in the manner described above for every 5–10 days depending on the number of colonies, their distribution in the well, and the degree of differentiation. Be sure to only passage colonies with hESC morphology and phenotype mentioned above (see Chapter 8 for hPSC morphology).

9. As the line is expanded, it is important to freeze premaster and master vials at early passages. Cells must also be set aside to perform standard testing to confirm that the newly derived line is a bona fide hESC line, by performing standard tests such as gene expression, promoter methylation, and teratoma formation analyses as described in other chapters of this book (Chapters 17, 19 and 20). Moreover, the microbiological safety, genomic integrity, and genetic identity of hESC lines must be confirmed soon after derivation and confirmed at regular passage intervals (see Chapter 2).

4. Notes

1. It is important to keep the embryos in the same medium as used by the IVF clinic. Consult with them prior to receiving the embryos for culture in your laboratory.
2. Embryos are very sensitive to changes in pH and temperature.

   The use of a portable incubator such as the "LEC960 Portable Incubator/Cell Carrier" can be useful in keeping them at standard temperature and pH during transport between the clinic and the laboratory.

   http://www.biogenics.com/html/products/p_05_02.html.
3. To improve derivation efficiency, the ICM can be isolated from the embryo either by immunosurgery or by mechanical or laser dissection (1, 4, 5).
4. The medium is not changed until the newly derived hESC colonies have reached sufficient size to be subcultured. This may take as long as 10 days of culture.

References

1. Thomson, J.A. et al. (1998) Embryonic stem cell lines derived from human blastocysts. *Science* **282**: 1145–1147.
2. Lerou, P.H. et al. (2008) Human embryonic stem cell derivation from poor-quality embryos. *Nat Biotechnol* **26**: 212–214.
3. Chung, Y. et al. (2008) Human Embryonic Stem Cell Lines Generated without Embryo Destruction. *Cell Stem Cell* **2**: 113–117.
4. Chen, A.E. et al. (2009) Optimal timing of inner cell mass isolation increases the efficiency of human embryonic stem cell derivation and allows generation of sibling cell lines. *Cell Stem Cell* **4**: 103–106.
5. Solter, D. & Knowles, B.B. (1975) Immunosurgery of mouse blastocyst. *Proc Natl Acad Sci USA* **72**: 5099–5102.

Chapter 4

Derivation of Human Parthenogenetic Stem Cell Lines

Nikolay Turovets, Andrey Semechkin, Leonid Kuzmichev, Jeffrey Janus, Larissa Agapova, and Elena Revazova

Abstract

Pluripotent stem cells (PSCs) derived from parthenogenetically activated human oocytes demonstrate the typical characteristics displayed by human embryonic stem cells (hESCs) including infinite division and in vitro and in vivo differentiation into cells of all germ lineages. Different activation techniques allow the creation of either human leukocyte antigen (HLA) heterozygous human parthenogenetic stem cell (hpSC) lines, which are HLA-matched/histocompatible with the oocyte donor, or HLA-homozygous hpSC lines, which may be histocompatible to significant segments of the human population. This immune-matching advantage, combined with the advantage of derivation from nonviable human embryos that originate from unfertilized parthenogenetically activated oocytes, makes hpSCs a promising source of PSCs for cell-based transplantation therapy. This chapter describes two approaches for the parthenogenetic activation of human oocytes, their cultivation to the blastocyst stage, and the subsequent derivation of PSC lines.

Key words: human parthenogenetic stem cells, embryonic stem cells, parthenogenetic activation, stem cell line derivation, human oocytes, parthenogenic

1. Introduction

Embryonic stem cells (ESCs) are derived from the inner cell mass (ICM) of blastocyst stage embryos, which have developed from fertilized oocytes (1). ESCs are capable of infinite division and differentiation into cells of all tissues types (2). The ESC is a potentially limitless source of pluripotent stem cells (PSCs) for transplant-based cell therapies because ESCs are likely to face the same immune-compatibility issues that limit other allogeneic transplants.

The risk of transplant rejection is proportional to the degree of cell-surface antigen disparity between the donor cells and that

Philip H. Schwartz and Robin L. Wesselschmidt (eds.), *Human Pluripotent Stem Cells: Methods and Protocols*, Methods in Molecular Biology, vol. 767, DOI 10.1007/978-1-61779-201-4_4, © Springer Science+Business Media, LLC 2011

of the recipient. In the ideal transplant, an autogenic transplant, donor tissue is identical to the recipient. However, autogeneic transplants are not usually practical. So donor tissues are screened for cell-surface antigens in order to determine the degree of histocompatibility with the recipient at the major histocompatibility complex (MHC). The human leukocyte antigen (HLA) system is the nomenclature designating the human MHC and represents antigens important for transplantation. Matching donor and recipient tissue for HLA antigens reduces the chance of a cytotoxic T-cell response in the recipient and thus greatly increases the likelihood of transplant survival.

Parthenogenetic activation of human oocytes may be one way to produce histocompatible/HLA-matched PSCs for cell-based therapy. Parthenogenesis is a form of asexual reproduction in which females produce eggs that develop without fertilization, in other words – without the participation of spermatozoa. Parthenogenesis in mammalian oocytes can be induced or "activated" by electrical or chemical stimuli that mimic spermatozoon penetration. Spontaneous parthenogenetic activation can also occur in human oocytes. Such activated oocytes have the ability to develop into parthenogenetic embryos that are capable of reaching the blastocyst stage. The ICM of such a parthenogenetic blastocyst can give rise to a parthenogenetic stem cell [human parthenogenetic stem cell (hpSC)] line. The resulting hpSCs do not contain male genetic material; therefore, they carry only the oocyte donor's HLA genes and are thus matched to the donor.

Using various activation protocols, it is possible to create hpSCs with different HLA status. Parthenogenetic activation of metaphase II (MII) oocytes with a combination of the activating agents ionomycin and 6-dimethylaminopurine (6-DMAP) blocks the extrusion of the second polar body; therefore, activated oocytes retain all of their genetic material. The HLA genotypes of hpSCs derived from these oocytes are heterozygous and genetically matched to the donors (3). Since the stem cells are HLA-matched to the oocyte donor, hpSCs created in this manner may be therapeutically useful for only the donor herself and a limited number of other people. The HLA diversity in the population combined with diversity of heterozygous individuals makes the chance of finding a donor–recipient match between one heterozygous individual and another heterozygous individual in the range of one in a thousand to one in several million. In addition, these heterozygous hpSCs may only benefit healthy women of reproductive age who are able to donate "healthy" oocytes. Finally, this method could not be used to create stem cells for the treatment of genetically caused diseases because the differentiated ("therapeutic") cells derived from these stem cells would carry the genetic defect.

Parthenogenetic activation of MII oocytes with a combination of the activating agents ionomycin and puromycin allows extrusion of the second polar body. The activated oocyte, therefore, contains only half of a set of metaphase II chromosomes. In spite of this, these activated oocytes lead to the formation of diploid stem cells with homozygous HLA genotypes (i.e., exhibiting identical alleles for each antigen-presenting protein) that contain a duplicated set of half of the alleles derived from the oocyte donor (4). The HLA status of these parthenogenetic stem cells may allow their application in allogenetic cell-based therapies for millions of people provided that the set of homozygous alleles contributed by the oocyte donor (called the "haplotype") is common in the population. It has been suggested that a panel of only ten HLA-homozygous human PSC lines selected from common genotypes can provide a complete HLA match in A, B, and DR loci for 37.7% of UK recipients, and a beneficial match for 67.4% (5). Using the US population, calculations suggest that there are close to 200 common haplotypes per racial group (6).

There are at least two approaches for deriving human HLA-homozygous hpSC lines: the ionomycin/puromycin as described above and ionomycin/6-DMAP which can be used to create activated oocytes from HLA-homozygous donors (4). However, this method is limited by the rarity of HLA-homozygous donors within the population.

The first intentional creation of hpSC lines was described by Revazova et al. (3), in which six pluripotent HLA heterozygous hpSC lines were derived from chemically activated human oocytes and were characterized as being PSCs. One-half year later, two more scientific groups claimed derivation of hpSC lines: two hpSC lines were produced by the combination of chemical and electrical oocyte activation (7) and one line was derived from spontaneously activated oocyte (8). A publication in 2008 reported the derivation or four HLA-homozygous hpSC lines from chemically activated oocytes (4). Interestingly, it has been suggested that parthenogenesis led to the unintentional creation of another hpSC line (SCNT-hES-1) reported to be of somatic cell nuclear transfer origin (9).

The generation of hpSC lines can be broken into three stages: (1) the process of obtaining donated oocytes, called the "oocyte pick-up stage"; (2) parthenogenetic activation of the oocyte and their cultivation to parthenogenetic blastocyst, called the "embryology stage"; and (3) the derivation of hpSC lines from the parthenogenetic blastocysts, called the "stem cell line derivation stage." In this chapter, we describe in detail the embryology stage and the hpSC line derivation stage. The oocyte pick-up stage, which is conducted by competent reproductive physicians working with oocyte donors under conditions that conform with government regulations and under the review of hospital administration and research approval committees, will be only briefly described.

2. Materials and Equipment

2.1. Equipment

1. K-MINC, embryology incubator (COOK).
2. STRIPPER®CC, pipette for work with gametes and embryos (MidAtlantic Diagnostics).
3. STRIPPER®, pipette for work with gametes and embryos (MidAtlantic Diagnostics).
4. Tips for STRIPPER® CC: 1,000 μm (MidAtlantic Diagnostics).
5. Tips for STRIPPER®: 200 μm (MidAtlantic Diagnostics).
6. Tips for STRIPPER®: 135 μm (MidAtlantic Diagnostics).
7. In vitro fertilization (IVF) workstation with integrated warming plate in workspace (e.g., K-Systems, MidAtlantic Diagnostics).
8. Warming block for tubes and 4-well dishes (e.g., K-systems, MidAtlantic Diagnostics).
9. Stereomicroscope with heated stage.
10. Dry bath incubator/test tube heater for heating tubes and aspirated follicular fluid (e.g., COOK, K-systems).
11. Thermometer for measurement of the heated stage temperature, ThermoDisc/surface thermometer or the same (Research Instruments).
12. Portable incubator.

2.2. Culture Supplies

1. Flushing medium (MediCult).
2. Universal IVF medium (MediCult).
3. SynVitro Hyadase solution (MediCult).
4. EmbryoAssist medium (Medicult).
5. BlastAssist medium (Medicult).
6. VitroHES medium (VitroLife).
7. DMEM, Dulbecco's modified Eagle medium (containing high glucose, L-glutamine, and sodium pyruvate).
8. 0.05% Trypsin/EDTA solution.
9. DPBS 1×, Dulbecco's phosphate buffered saline without Ca^{2+}, without Mg^{2+}.
10. Penicillin–streptomycin stock solution: 100×.

11. Ionomycin calcium salt.
12. Dimethylsulfoxide (DMSO).
13. 6-Dimethylaminopurine (6-DMAP, Sigma).
14. Puromycin.
15. Mitomycin C.
16. Pronase.
17. Human leukemia inhibitory factor (hu-LIF, Chemicon).
18. Recombinant human FGF-basic (PeproTech).
19. 4-Well IVF dish.

2.3. Solutions and Media

Ionomycin stock solution, 5 mM (1,000×)

- 1 mg Ionomycin calcium salt.
- 268 μL DMSO.

Prepare stock solution in dimmed light (avoid direct light). Divide stock solution into aliquots, store at −20°C, and protect from light. Prevent multiple freeze–thaw cycles. Usually, we use one thawed aliquot per day.

Puromycin stock solution, 10 mg/mL (1,000×).

- 10 mg Puromycin.
- 1 mL H_2O.

Divide stock solution on aliquots, store at −20°C. Prevent multiple freeze–thaw cycles. Usually, we use one thawed aliquot per day.

6-DMAP stock solution, 0.1 M (100×).

- 10 mg 6-DMAP.
- 613 μL DPBS

Divide stock solution on aliquots, store at −20°C. Prevent multiple freeze–thaw cycles. Usually, we use one thawed aliquot per day.

Neonatal skin fibroblast (NSF) culture medium

- DMEM (containing high glucose, L-glutamine, and sodium pyruvate).
- 1×, Penicillin–streptomycin.
- 10% Human umbilical cord blood serum (for derivation protocol, see text below).

Storage conditions +2°C to +8°C.

hpSC culture medium

- VitroHES medium.
- 10% Human umbilical cord blood serum (for derivation protocol, see text below).

- 4 ng/mL FGF-basic.
- 5 ng/mL LIF.

Prepare hpSC culture medium directly before use, do not store.

Mitomycin C stock solution, 1 mg/mL.

- 10 mg Mitomycin C.
- 10 mL DPBS.

Sterilize through 0.22 μm filter. Divide stock solution into aliquots, store at –20°C. Prevent multiple freeze–thaw cycles.

0.5% Pronase solution

- 0.5 g Pronase.
- 100 mL DPBS.

Sterilize through 0.22 μm filter. Divide stock solution into aliquots, store at –20°C.

3. Methods

3.1. Oocyte Pick-Up Stage: Donor Selection, Informed Consent, Super-Ovulation

1. *Donor selection.* We recruit donors from a pool of women who first present to the clinic with the intent of artificial reproduction via IVF and were found to be eligible for an IVF procedure according to clinical guidelines. Research eligibility was determined according to the FDA's Eligibility Determination for Donors of Human Cells, Tissues, and Cellular and Tissue-Based Products (10), as well as Order No. 67 (February 26, 2003) of the Russian Public Health Ministry. This included a thorough medical examination with chest X-ray, blood chemistry (including liver function tests), and urine analysis. Screening was also performed for *Chlamydia trachomatis*, *Neisseria gonorrhea*, syphilis, HIV, HBV, and HCV. In this protocol, the priority for oocyte harvest was a successful IVF procedure. The best fully developed mature cumulus oocyte complexes were selected for IVF. If the total number of oocytes harvested was less than 11, the woman was automatically excluded from donating for research purposes.
2. *Informed consent.* Each potential donor was informed and counseled by her doctor about the purpose of donating her oocytes. If the donor chose to participate, the donor was presented with a comprehensive informed consent document, which outlined the purpose of the study and the procedures. If the potential donor had questions, a medical doctor was made available. Only potential donors who signed the

informed consent participated in the study. Donors voluntarily donated their oocytes without financial compensation. The signed informed consent stated that all donated material was to be used for research and not for reproductive purposes, namely, the development of methods to derive hpSCs and their differentiated progeny.

3. *Super-ovulation*. Each donor underwent ovarian stimulation utilizing FSH (Gonal-F, Lab. Serono, Switzerland) from the 3rd to the 13th day of the menstrual cycle. A total of 1,500 IU was given. From the 10th to the 14th day of the donor's menstrual cycle, the gonadoliberin antagonist Orgalutran (Organon, Holland) was injected at 0.25 mg/day. From the 12th to the 14th day of the donor's menstrual cycle, a daily injection of 75 IU FSH+75 IU LH (Menopur, Ferring GmbH, Germany) was given. If an ultrasound examination displayed follicles between 18 and 20 mm in diameter, a single 8,000 IU dose of hCG (Choragon, Ferring GmbH, Germany) was administered on the 14th day of the donor's menstrual cycle. Ultrasound-guided transvaginal aspiration of follicular fluid from antral follicles was performed 35 h after hCG injection.

3.2. Embryology Stage: Oocyte Retrieval, Denudation, Activation, and Embryo Cultivation

High-quality oocytes are critical to maximize the occurrence of cleavage after activation. We activate only mature MII stage fresh oocytes that exhibit a well visible first polar body. The efficient and successful derivation of hpSC lines requires high-quality culture techniques that limit temperature and pH fluctuations, and culture under low oxygen tension. We strongly recommend maintaining the same high level culture conditions and embryo manipulation techniques that are routinely used in IVF procedures. Some of the key technical culture considerations are covered in Note 1.

Culture at this stage is performed in a K-MINC incubator with an atmosphere of: 5% CO_2+5% O_2+90% N_2.

1. *Oocyte retrieval, Day 0*.

   For oocyte retrieval procedure, two 4-well IVF dishes are prepared (Fig. 1). One dish is filled with Flushing medium, another dish is filled with Universal IVF medium. Two milliliters of medium should be put in the center well and 0.5 mL of medium should be put in the other four wells. Allow the medium in the dish to warm and equilibrate overnight in the K-MINC incubator.

   (a) Transfer the tubes containing the aspirated follicular fluid obtained in the surgical suite to the embryology laboratory under the appropriate conditions (see Note 2).

   (b) Decant the follicular fluid into a previously warmed (37°C) 100-mm Petri dish.

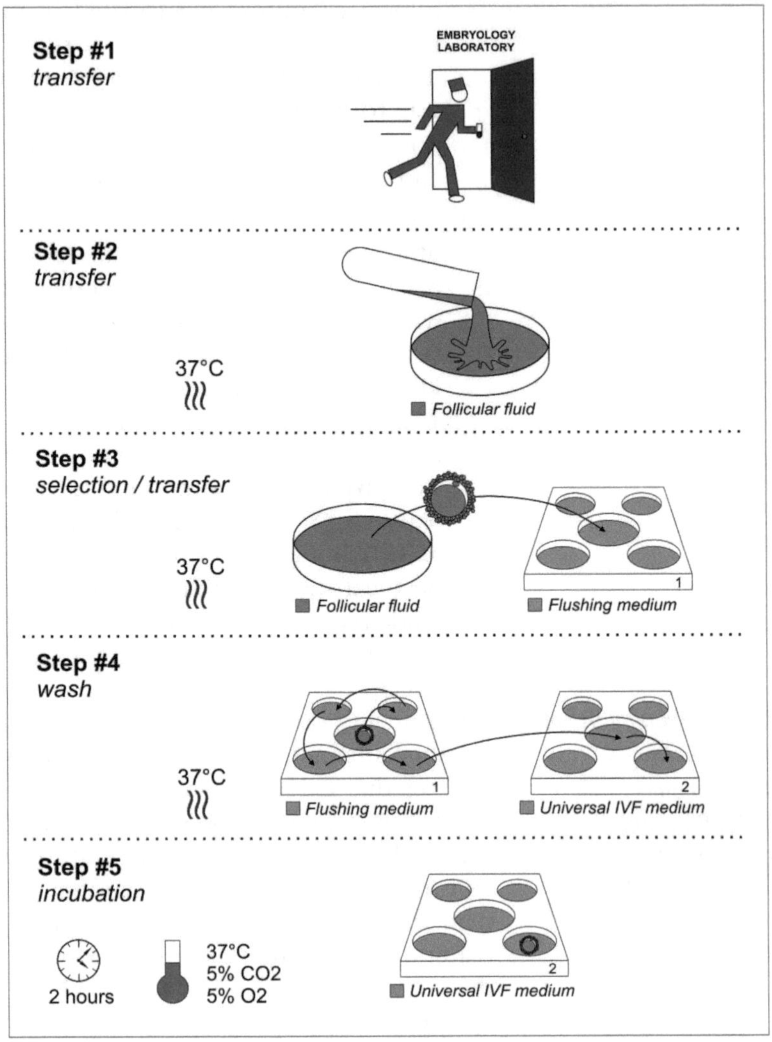

Fig. 1. Day 0. Oocyte retrieval.

(c) While observing follicular fluid under a stereomicroscope, select cumulus–oocyte complexes (COCs) and transfer them by STRIPPER®CC, with a 1,000 μm tip size, into the center well of the previously equilibrated 4-well dish containing Flushing medium.

(d) When all COCs have been picked up from the follicular fluid, wash the COCs, in series, in fresh Flushing medium (0.5 mL wells of the 4-well dish) and then transfer to the other 4-well dish containing Universal IVF medium. Do not leave more than five COCs in a well.

(e) Incubate COCs in K-MINC incubator for 2 h.

2. *Oocyte denudation, Day 0.*

Denudation requires the preparation of a 4-well dish as follows: fill one well with 0.5 mL SynVitro Hyadase; fill the other three wells with 0.5 mL Universal IVF medium; fill center well of the dish with 2 mL of Universal IVF medium. Allow the media in the dish to warm and equilibrate correctly. Use one dish for the denudation of up to five COCs; if you have more than five COCs you should prepare additional denudation dishes (see Fig. 2).

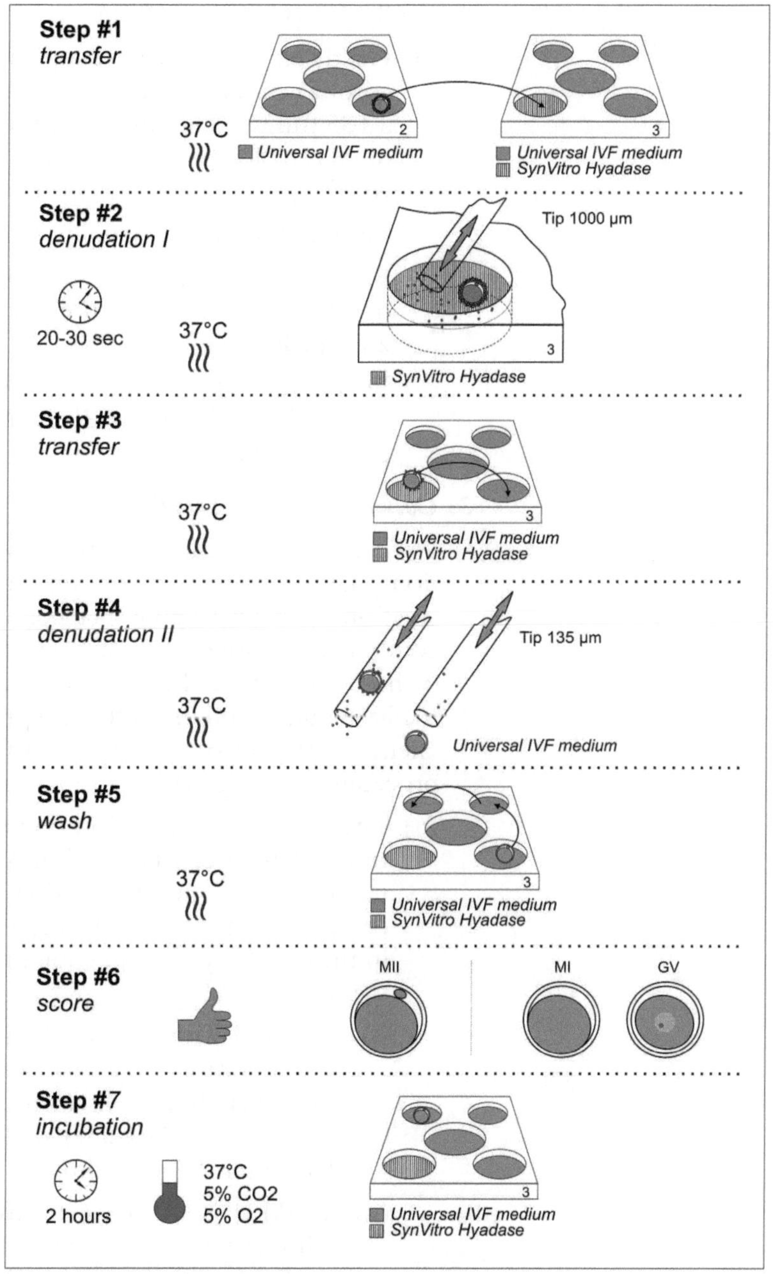

Fig. 2. Day 0. Oocyte denudation.

(a) Using a stereomicroscope with a warming stage, transfer COCs into the SynVitro Hyadase using STRIPPER®CC with a 1,000 μm tip (see Note 3).

(b) Pipette the SynVitro Hyadase up and down for around 20–30 s until most of the cumulus cells have been loosened. Do not aspirate oocytes in this step.

(c) Using a stereomicroscope, transfer the oocytes into Universal IVF medium using STRIPPER®CC with a 1,000 μm tip. Avoid transferring an excess volume of the SynVitro Hyadase solution with the oocytes.

(d) Gently pipette the oocytes up and down until most of the corona cells have been loosened, using STRIPPER® with a 135 μm tip. The subsequent manipulations should also use STRIPPER® with a 135 μm tip (see Note 4).

(e) Wash the oocytes thoroughly by transferring them serially through the wells containing Universal IVF medium.

(f) Score the oocytes for the proper metaphase stage of development. Select mature oocytes (MII) with an intact zona pellucida.

(g) Leave oocytes in fresh Universal IVF medium in the K-MINC incubator for 30 min before proceeding with parthenogenetic activation.

3. *Oocyte activation, Day 0.*

We use two different combinations of chemicals for activation. The combination of ionomycin and 6-DMAP allows the creation of heterozygous parthenogenetic stem cells. The combination of ionomycin and puromycin allows the creation of homozygous parthenogenetic stem cells. All dishes with media should be prepared the day prior activation procedure; chemicals should be added to wells 30 min prior activation procedure (see Fig. 3).

(a) The day before activation prepare the three 4-well culture dishes required for activation: an activation dish containing ionomycin; an activation dish containing 6-DMAP or puromycin; and a cultivation dish. Allow the medium in the dish to warm and equilibrate correctly.

(i) Ionomycin activation dish is prepared as follows: fill one well with 0.5 mL Universal IVF medium and add ionomycin to a final concentration of 5 μM. Fill the other three wells with 0.5 mL of Universal IVF medium. Fill the center well of the dish with 2 mL of Universal IVF medium.

(ii) 6-DMAP or puromycin activation dish is prepared as follows: fill one well with 0.5 mL of Universal IVF medium and add 6-DMAP to a final concentration

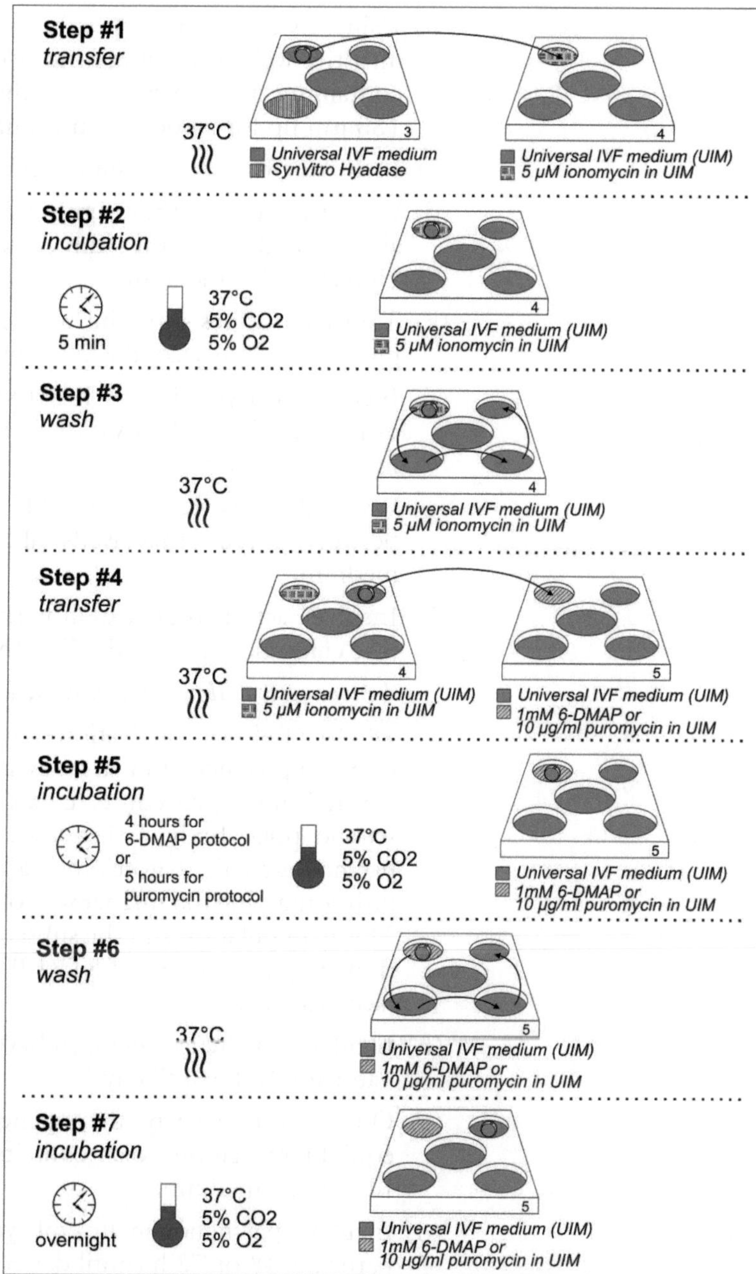

Fig. 3. Day 0. Oocyte activation.

of 1 mM or puromycin to a final concentration of 10 μg/mL. Fill the other three wells with 0.5 mL of Universal IVF medium. Fill the center well of the dish with 2 mL of Universal IVF medium.

(iii) The cultivation dish is prepared as follows: fill the whole dish with Universal IVF medium, 0.5 mL per each of the four wells, and 2 mL in the center well.

(b) Using a stereomicroscope, transfer the denuded oocytes into the well containing ionomycin (the activation dish containing ionomycin). Use the STRIPPER® with a 135 μm tip for all oocyte manipulations.

(c) Incubate oocytes for 5 min in the K-MINC incubator.

(d) Wash the oocytes thoroughly by transferring them serially through the other three wells that are filled with Universal IVF medium.

(e) Transfer oocytes into the well containing 6-DMAP or puromycin (the second activation dish).

(f) Incubate oocytes in the K-MINC incubator for 4 h if using the 6-DMAP activation or for 5 h if using the puromycin activation.

(g) Wash the oocytes thoroughly by transferring them between three other wells filled with Universal IVF medium.

(h) Incubate activated oocytes in fresh Universal IVF medium (the cultivation dish) in the K-MINC incubator overnight.

4. *Parthenogenetic embryo cultivation, Days 1–6* (Fig. 4).

(a) 18–20 h following activation (day 1) check for the formation of a pronucleus in all cases, and if using the puromycin activation procedure, check for the extrusion of a second polar body. Using a stereomicroscope, transfer parthenogenetic zygotes into a pre-equilibrated culture dish containing 0.5 mL per well of EmbryoAssist medium. This manipulation and the subsequent embryo manipulations require the use of STRIPPER® with a 200 μm tip (see Note 5).

(b) Incubate parthenogenetic embryos in the K-MINC incubator for 48 h (until day 3).

(c) On day 3, transfer parthenogenetic embryos to the pre-equilibrated culture dish containing 0.5 mL per well of BlastAssist medium.

(d) Incubate parthenogenetic embryos in the K-MINC incubator for 48 or 72 h (until day 5 or 6).

(e) On day 5 or 6, the parthenogenetic embryos that have achieved the blastocyst stage are ready for the derivation of parthenogenetic stem cells (see Note 6).

3.3. Parthenogenetic Stem Cell Line Derivation

1. *Human umbilical cord blood serum (hUCBS) preparation.*

During the cultivation of the parthenogenetic ICMs and during the early passages of hpSC culture, we use hUCBS in the hpSC culture medium. We have found that hUCBS supports ICM outgrowth and makes a positive contribution to the

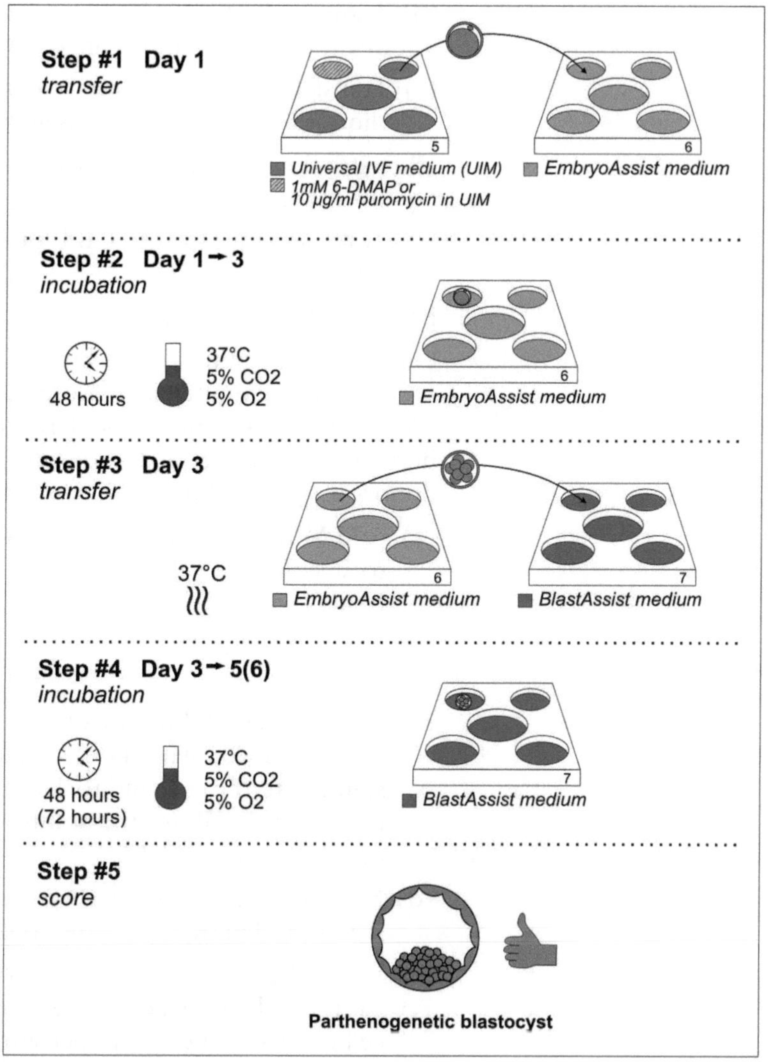

Fig. 4. Days 1–6. Parthenogenetic embryo cultivation.

derivation of hpSC lines. We also utilize hUCBS to support the growth of human NSFs used as a feeder-cell layer. Insure that cord blood is obtained under proper IRB (Internal Review Board) approval and that all donors have signed proper informed consent documents.

(a) Collect cord blood into a sterile vessel without using any anticoagulants (we have found that 50-mL tubes work well).

(b) Leave the cord blood for approximately 5–7 h at room temperature to allow clot formation, followed by overnight storage at 4°C.

(c) Centrifuge the specimen 10 min at 600 × *g*.

(d) Transfer the supernatant into new tubes and centrifuge one more time for 10 min at 600 × *g*.

(e) Transfer the supernatant into a new tube. Remove an aliquot to test for infectious agents.

(f) Filter the hUCBS through a 0.2-μm pore size filter. Store at −20°C.

(g) Test the aliquot of hUCBS for syphilis, HIV, HBV, and HCV.

2. Feeder-cell preparation

Our culture system utilized human NSFs (4). Suitable NSF are available commercially from Lifeline Cell Technology (Walkersville, MD) or can be derived from tissue. Insure that human tissue is obtained under proper IRB approval and all donors have signed proper informed consent documents.

(a) Add Mitomycin C to a final concentration of 10 μg/mL to a culture of NSF that are 80–90% confluent (see Note 7).

(b) Incubate NSF with Mitomycin C for 3 h in a 5% CO_2 incubator.

(c) Aspirate NSF culture medium from the Petri dish. Wash the cells twice with DPBS. Add 0.05% trypsin/EDTA solution to the cells and rock the Petri dish back and forth to evenly distribute the solution. Incubate for approximately 1 min at 37°C.

(d) Neutralize the trypsin/EDTA solution with NSF culture medium. Pipette repeatedly to break up clumps of cells.

(e) Pellet the cells by centrifugation for 3 min at 300 × *g*. Resuspend the pellet in NSF culture medium. Perform cell counts of the cell suspension.

(f) Plate mitotically inactivated NSF in NSF culture medium at a seeding density of 40,000 cells/cm^2. Allow the NSF to attach to the dish overnight.

(g) The day before plating hpSCs, exchange NSF medium with hpSC medium and allow the dish to equilibrate in the incubator.

3. Derivation of hpSC lines.

There are several methods for the isolation and derivation of PSCs from the blastocyst's ICM. We use a mechanical method described below.

(a) On a 100-mm dish that has been previously heated to 37°C, add a drop (~50 μL) of the preheated 0.5% pronase solution and several individual 50 μL drops of preheated BlastAssist medium.

(b) Using a stereomicroscope, transfer one parthenogenetic blastocyst to the pronase solution. Observe the parthenogenetic blastocyst under the stereomicroscope. Use STRIPPER® with a 200 μm tip for this and the subsequent manipulations (see Note 8).

(c) Incubate the parthenogenetic blastocyst in the pronase solution until the zona pellucida starts to disintegrate.

(d) Transfer the parthenogenetic blastocyst into a drop of BlastAssisit medium; gently pipette it up and down until the zona pellucida is totally disintegrated.

(e) Wash the parthenogenetic blastocyst thoroughly by transferring it between several drops of fresh BlastAssisit medium (see Note 9).

(f) Transfer the parthenogenetic blastocyst into the well of a 4-well dish previously plated with a feeder layer of mitotically inactivated NSF containing hpSC culture medium (see Note 10).

(g) Repeat steps 1–6 for each parthenogenetic blastocyst. Always use fresh drops of pronase solution and BlastAssist medium for each parthenogenetic blastocyst.

(h) Incubate the dishes containing parthenogenetic blastocysts in a 5% CO_2 incubator until they attach, usually 24–48 h.

(i) After the parthenogenetic blastocyst has attached to the feeder layer, change 2/3 of the medium volume, daily, using hpSC culture medium.

(j) After 3–4 days of culture, isolate the ICM by mechanically cutting the ICM away from the trophectoderm using a finely drawn glass pipette.

(k) Transfer the isolated ICM onto a fresh feeder-cell layer and culture for an additional 5–6 days. Change 2/3 of the medium volume daily. The first colony of hpSCs will be derived from this ICM.

(l) The first colony of hpSCs should be mechanically dissected and the bits replated after 5 days of culture. Usually, subsequent passages are performed every 5–7 days. For early passages, colonies are mechanically divided into clumps and replated. Once established and low passages are banked in cryogenic storage, subculture is performed with collagenase IV treatment as is common with human PSC culture. Once established a hpSC line can be propagated, including freezing–thawing in accordance with common techniques used in human PSC culture.

3.4. Characterization of hpSC Lines

hpSC lines are evaluated for the expression of common PSC markers such as SSEA-3, SSEA-4, TRA-1-60, TRA-1-81, OCT4,

NANOG, and are tested for alkaline phosphatase and telomerase activity. Results showed human hpSC lines have the same expression patterns found in other human PSC lines. The differentiation capacity of the hpSC was determined in vitro using the formation of embryoid bodies in suspension culture and in vivo though the formation of teratomas after injection into immunodeficient mice. hpSCs in both in vitro and in vivo tests gave rise to differentiated derivatives of all three embryonic germ layers. The determination of genetic status was performed by conducting typing studies of HLA loci classes I and II, SNP analysis, and G-banding karyotyping. All methods referenced above can be found in other chapters of this book.

4. Notes

1. For detailed information concerning clinical embryology techniques and tips see IVF manuals, for example, see ref. 11.

   Power of success: clinical embryology tips:

   (a) All dishes with medium that will be in contact with oocytes or parthenogenetic embryos must be precisely heated to 37°C. All manipulations of oocytes and parthenogenetic embryos must be performed only in dishes that are located on heated surfaces (including laboratory tables and microscope stages). The temperature of the surface of working areas must be equilibrated to the temperature of the surface of the dish (37°C). Special surface temperature thermometers are useful for this purpose. We do not recommend the use of air flow in the hood during any manipulation of oocytes or parthenogenetic embryos – air flow can lower the temperature of the culture dish. Safety requires the use of a mask during manipulation.

   (b) Dishes with medium should be prepared the day before they are to be used for oocyte or embryo culture in order to allow sufficient time for the medium to equilibrate in the CO_2 atmosphere. At least 6 h is required for proper equilibration of the oil-covered medium.

   (c) Minimize the changes in pH and temperature of the culture dishes by handling the medium and manipulating the oocytes and parthenogenetic embryos quickly. Minimize the time that culture dishes spend outside of the incubator. Before dishes containing oocytes or parthenogenetic embryos are removed from incubator, be sure that you are prepared. Have all the necessary materials (pipettes, tips, and so on) ready for manipulation.

(d) Use only sterile, nontoxic, powder-free gloves. Powder from gloves can be hazardous to oocytes and parthenogenetic embryos.

(e) Do not use toxic compounds in the embryology laboratory, including chemicals used for cleaning purposes.

(f) Do not use UV light in the embryology laboratory before work, to avoid creation of active radicals.

(g) All plastic consumables (dishes, tubes, pipette tips, and so on) should be "IVF grade" (tested for embryo-toxicity) or at least "cell culture grade."

2. Keep tubes with aspirated follicular fluid at 37°C during follicular fluid aspiration and transportation to the embryology laboratory. Use a dry bath incubator/test tube heater during the oocyte aspiration procedure to maintain a constant temperature of the tube and follicular fluid. Follicular fluid should be examined as soon as possible – avoid excessive incubation in follicular fluid.
3. Always wet the tip of the pipette with media before contact with parthenogenetic embryos or oocytes, the center well of the 4-well IVF dish is used for this purpose.
4. Avoid excessive or rough handling of oocytes during the denuding process as this can damage the oocytes.
5. The parthenogenetic embryos may be cultured singly or in multiples to a maximum of five per well.
6. Usually, the quality of parthenogenetic blastocysts is lower than the quality of IVF-embryos, but hpSC lines can be derived from low-quality parthenogenetic blastocysts even if they do not display a visible ICM.
7. For the derivation of hpSC lines, we used NSF in passages 6–10.
8. For transportation of parthenogenetic blastocysts from the embryology laboratory to the stem cell laboratory, a portable incubator is recommended; however, if very short distance one may use preheated warming blocks.
9. Blastocysts should be washed thoroughly after pronase treatment. Residual pronase can cause the detachment of feeder cells.
10. Wells of the 4-well dish should be plated with NSF feeder cells 1 or 2 days before use. NSF culture medium should be replaced with hpSC culture medium 2 h before plating the parthenogenetic blastocysts onto the NSF feeders. Do not use dishes with NSF feeder layers that are more than 48 h old.

References

1. Thomson, J.A., Itskovitz-Eldor, J., Shapiro, S.S., Waknitz, M.A, Swiergiel, J.J., Marshall, V.S., Jones, J.M. (1998) Embryonic stem cell lines derived from human blastocysts *Science* **282**, 1145–47.
2. Reubinoff, B.E., Pera, M.F., Fong, C.Y., Ttrounson A., Bongso A. (2000) Embryonic stem cell lines from human blastocysts: somatic differentiation in vitro *Nat Biotechnol*, **18**, 399–404.
3. Revazova, E.S., Turovets, N.A., Kochetkova, O.D., Kindarova, L.B., Kuzmichev, L.N., Janus, J.D., Pryzhkova, M.V. (2007) Patient-specific stem cell lines derived from human parthenogenetic blastocysts *Cloning and Stem Cells* **9**, 432–49.
4. Revazova, E.S., Turovets, N.A., Kochetkova, O.D., Agapova, L.S., Sebastian, J.L., Pryzhkova, M.V., Smolnikova, V.I., Kuzmichev, L.N., Janus, J.D. (2008) HLA homozygous stem cell lines derived from human parthenogenetic blastocysts *Cloning Stem Cells* **10**, 11–24.
5. Taylor, C.J., Bolton, E.M., Pocock, S., Sharples, L.D., Pedersen, R.A., Bradley, J.A. (2005) Banking on human embryonic stem cells: estimating the number of donor cell lines needed for HLA matching *Lancet* **366**, 2019–25.
6. Mori, M., Beatty, P.G., Graves, M., Boucher, K.M., Milford, E.L. (1997) HLA gene and HLA haplotype frequencies in the North American population: the National Marrow Donor Program Donor Registry *Transplantation* **64**, 1017–27.
7. Mai, Q., Yu, Y., Li, T., Wang, L., Chen, M.J., Huang, S.Z., Zhou, C., Zhou, Q. (2007) Derivation of human embryonic stem cell lines from parthenogenetic blastocysts *Cell Research* **17**, 1008–19.
8. Lin, G., OuYang, Q., Zhou, X., Gu, Y., Yuan, D., Li, W., Liu, G., Liu, T., Lu, G. (2007) A highly homozygous and parthenogenetic human embryonic stem cell line derived from a one-pronuclear oocyte following in vitro fertilization procedure *Cell Research* **17**, 999–1007.
9. Kim, K., Ng, K., Rugg-Gunn, P. J., Shieh, J.-H., Kirak, O., Jaenisch, R., Wakayama, T., Moore, M. A., Pedersen, R. A., Daley, G. Q. (2007) Recombination signatures distinguish embryonic stem cells derived by parthenogenesis and somatic cell nuclear transfer *Cell Stem Cell* **1**, 346–52.
10. Food and Drug Administration. (Draft) Guidance for Industry: Eligibility Determination for Donors of Human Cells, Tissues, and Cellular and Tissue-Based Products (HCT/Ps) dated May 2004.
11. Gardner, D., Lane, M., Watson, A. (2004) A Laboratory Guide to the Mammalian Embryo *Oxford University Press* p. 1–394.

Chapter 5

Generation of Induced Pluripotent Stem Cell Lines from Human Fibroblasts via Retroviral Gene Transfer

Justine D. Miller and Thorsten M. Schlaeger

Abstract

This chapter describes a protocol for deriving induced pluripotent stem cells (iPSCs) from human fibroblasts. Human fibroblasts, cultured in fibroblast medium, are infected with a cocktail of retroviral vectors expressing the transcription factors OCT4, SOX2, KLF4, and MYC. The culture conditions are then switched to conditions that support human embryonic stem cell growth and emerging iPSC colonies that morphologically resemble human embryonic stem cell (hESC) colonies and have silenced the retroviral vectors (as evidenced by downregulation of retroviral GFP expression) that are mechanically isolated and subsequently cultured in identical fashion to hESCs. Putative iPSC lines are validated to be bona fide human iPSC lines by analyzing them for the expression of pluripotency markers and by differentiation in vitro and in vivo.

Key words: human induced pluripotent stem cells, iPSC, reprogramming, pluripotency, retroviral transduction, human embryonic stem cells, fibroblasts

1. Introduction

Recently, forced expression of reprogramming genes (e.g., the transcription factors OCT4, SOX2, KLF4, MYC) has been used to directly convert adult and neonatal fibroblasts into induced pluripotent stem cells (iPSCs) (1–3). Like human embryonic stem cells (hESCs), human iPSCs have the properties of self-renewal and pluripotency that make them a promising resource for the study of human ontogeny, disease modeling, drug screening, and cell-based therapies (4). Here, we present a protocol for reprogramming human fibroblast cell lines through permanent genetic modification using amphotropic retroviral particles pseudotyped with the vesicular stomatitis virus G (VSV-G) envelope protein. It is important to note that this protocol, in addition to iPSC colonies, can yield various other types of colonies and cells that

Philip H. Schwartz and Robin L. Wesselschmidt (eds.), *Human Pluripotent Stem Cells: Methods and Protocols*, Methods in Molecular Biology, vol. 767, DOI 10.1007/978-1-61779-201-4_5,

represent aberrantly or incompletely reprogrammed cell states. It is therefore important to establish that the induced cell lines are bona fide iPSCs, by performing standard analyses such as gene expression, promoter methylation, and teratoma formation. Similarly, as with hESCs, it is important to confirm the microbiological safety and genomic integrity of newly derived iPSC lines, as well as to confirm their genetic identity with the original somatic cell line used for their generation. Independent iPSC lines derived from the same somatic cell line can be distinguished by genomic Southern blot analysis using retrovirus derived probes (1). Validated lines should be banked, by creating frozen pre-master- (≥4 vials), master- (≥10 vials), and working-stocks (≥20 vials). Validation should be repeated for each set of stocks and at the end of subsequent experiments, to ensure cell line identity and quality.

2. Materials

2.1. Human Fibroblast Cell Culture and Passaging

1. Human skin fibroblasts: fetal (e.g., Detroit 551; ATCC, Manassas, VA, CCL-110), newborn (e.g., BJ1; ATCC, PCS-201-010), or disease specific (e.g., ADA-deficiency; (Coriell, Camden, NJ, GM01390).
2. Human fibroblast medium: minimum essential medium alpha (MEM-alpha, Invitrogen, cat. no. 12561–072) supplemented with 10% heat-inactivated fetal bovine serum (FBS).
3. Dulbecco's phosphate-buffered saline (DPBS without calcium and magnesium).
4. Trypsin, 0.05% with ethylenediamine tetraacetic acid (Trypsin/EDTA).
5. Nunclon 10-cm sterile cell culture dish.
6. 0.1% Gelatin solution (Millipore, ES-006-B).

2.2. VSV-G Pseudotyped Retrovirus Preparation and Transfection

1. 293 T cells (ATCC, cat. no. CRL11268).
2. 293 T cell medium: Dulbecco's modified Eagle's medium (DMEM, Invitrogen, cat. no. 11965–118) supplemented with 10% FBS, 2 mM L-glutamine, 0.1 mM minimum essential medium nonessential amino acid solution (MEM-NEAA), 1 mM sodium pyruvate.
3. Nunclon 15-cm sterile cell culture dish.
4. 5-mL Falcon round-bottom polystyrene tubes.
5. Fugene 6 (Roche Applied Science, cat. no. 11814443001).
6. pMIG-OCT4 (Addgene, cat. no. 17225).
7. pMIG-SOX2 (Addgene, cat. no. 17226).

8. pMIG-KLF4 (Addgene, cat. no. 17227).
9. pMIG-MYC (Addgene, cat. no. 18119).
10. pCMV-VSV-G (Addgene, cat. no. 8454).
11. pUMVC (Gag-Pol) (Addgene, cat. no. 8449).
12. 250-mL 0.45-μm sterile vacuum filter unit.
13. Beckman Polyallomer tubes (Beckman Coulter, cat. no. 326823).
14. Nalgene cryogenic vials.
15. Protamine sulfate (Sigma, cat. no. P3369) is dissolved at 10 mg/mL in water, stored in working aliquots at −80°C, and used at a final concentration of 5 μg/mL (1:2,000).
16. 16% Paraformaldehyde (PFA, Electron Microscopy Sciences, cat. no. 15711) is diluted 1:4 in DPBS.
17. Hoechst 33342 DNA stain (10 mg/mL solution, Invitrogen, cat. no. H3570).

2.3. iPSC Culture and Passaging

1. Standard hESC medium: Dulbecco's modified Eagle's medium/F12 (DMEM/F12, cat. no. 36254, StemCell Technologies, Vancouver, BC) supplemented with 20% knock-out serum replacement (KOSR, Invitrogen, cat. no. 10828–028), 1 mM L-glutamine, 0.1 mM MEM-NEAA, 100 μM beta-mercaptoethanol, and 10 ng/mL human recombinant basic fibroblast growth factor (bFGF, dissolved in 0.1% bovine serum albumin (BSA)/DPBS and stored in working aliquots at −20°C) (Gemini Bio-Products, cat. no. 400-432P).
2. Irradiated CF-1 mouse embryonic fibroblasts (MEFs; e.g., GlobalStem, cat. no. GSC-6001G).
3. MEF medium: DMEM (Invitrogen, 11965–118) supplemented with 10% FBS, 2 mM L-glutamine, 0.1 mM MEM-NEAA, 1 mM sodium pyruvate.
4. Nunclon 6-well, 12-well, and 24-well sterile cell culture plates.
5. Collagenase IV (Invitrogen, cat. no. 17104–019) is dissolved in DMEM/F12 at 10 mg/mL, stored in working aliquots at −20°C, and used at 1 mg/mL in DMEM/F12.
6. Corning Cell lifter (Fisher Scientific, 07-200-364).
7. 5-mL Glass pipettes.

2.4. Immunostaining

1. Dulbecco's phosphate-buffered saline, containing calcium and magnesium ($DPBS^{+/+}$).
2. 96-Well tissue culture microplate (Matrix, cat. no. 4940).
3. 16% Paraformaldehyde (PFA, Electron Microscopy Sciences, cat. no.15711).
4. 7.5% BSA (Invitrogen, cat. no. 15260–037).

5. Donkey serum (Sigma, cat. no. D9663).
6. Triton X-100 (Sigma, cat. no.234729), is diluted at 1:10 in DPBS (without calcium and magnesium) and stored at room temperature.
7. OCT4, rabbit anti-human (Abcam, cat. no. ab19857) is optimally used at 1:1,400.
8. NANOG, rabbit anti-human (Abcam, cat. no. ab21624) is optimally used at 1:400.
9. OCT4/NANOG Secondary: Alexa Fluor® 555 donkey anti-rabbit IgG (Invitrogen, cat. no. A31572) is optimally used at 1:1,000.
10. TRA-1-60 Antigen Alexa Fluor® 647 (BD, cat. no.560122) is optimally used at 1:100.
11. SSEA-4 Alexa Fluor® 555 (BD, cat. no. 560218) is optimally used at 1:100.
12. Hoechst 33342 DNA stain (10 mg/mL solution, Invitrogen, cat. no. H3570) is optimally used at 1:10,000.

3. Methods

3.1. Culturing Human Fibroblasts (see Note 1)

1. Thaw human postnatal fibroblasts onto a gelatin-coated 10-cm dish in 12 mL human fibroblast medium.
2. *Feeding*: Complete medium change every 2–3 days.
3. *Passaging*: When dish reaches confluency (approximately every 3–4 days), harvest fibroblasts by washing each dish with 5 mL DPBS. Replace DPBS with 6 mL 0.05% trypsin/EDTA solution per 10-cm dish and incubate for 3–5 min at 37°C. Following incubation, inactivate trypsin with an equal volume of fibroblast medium. Collect cells and spin down at 200 × *g* for 4 min. Re-plate fibroblasts at a split ratio of 1:2–1:6 onto gelatin-coated 10-cm cell culture dishes in 12 mL human fibroblast medium per dish.

3.2. Preparation of Oct4, Sox2, Klf4, and Myc Retroviruses (see Note 2)

1. One day prior to transfection, split 70% confluent 293 T cells at ~1:4 onto 4 × 15 cm dishes per virus in 20 mL 293 T cell medium/dish.
2. The following day, add 3 mL DMEM/F12 into each of four 5-mL polystyrene tubes. Add 200 μL Fugene-6 per tube into the middle of the medium (without touching the well bottom or walls). Mix by pipetting up and down. Incubate for 15 min at room temperature.
3. For each of the four constructs, combine 25 μg pMIG vector, 22.5 μg pUMVC, and 3 μg pCMV-VSV-G vector. Add the

mixtures to each of the four virus tubes, mix by pipetting, and incubate for 15 min at room temperature.

4. During incubation, carefully replace old medium on 293 T cells with 22 mL new medium per dish (293 T cells do not adhere well).
5. Add 1 mL 293 T cell medium to each plasmid mixture, mix, and distribute 1 mL to each dish drop wise. Transfer plates back to tissue culture incubator. Carefully move plates back/forth and left/right to aid mixing and even distribution of transfection complexes. The cells and everything that may have been in contact with them or the supernatant should now be treated as BSL2+ hazardous biomaterial.

3.3. Concentration of Retroviruses

1. Three days following transfection, collect the viral supernatant and filter using a 250-mL 0.45-μm vacuum filter flask.
2. Transfer each viral supernatant to tubes for ultracentrifugation. Weigh and balance tubes using medium.
3. Centrifuge the supernatant at 4°C for 90 min at >70,000 ×*g* (e.g., using a Beckman Optima L-90K equipped with a SW 28 or SW 32 Ti rotor).
4. Pour off or aspirate the supernatant and add 1 mL DMEM/F12 (do not resuspend the pellet). Seal with parafilm and store tubes at 4°C overnight.
5. The next day, carefully flick the bottom of each tube, then aliquot 100 μL of virus into cryovials and transfer to −80°C for long-term storage.

3.4. Testing for Absence of Replication-Competent Retroviral Particles (see Note 3)

1. To test for the absence of replication-competent virus particles that may have formed inadvertently during virus generation, culture 100,000 293 T cells in a single well of a 6-well cell culture plate with 2 mL 293 T cell medium, 10 μL virus, and 2,000× protamine sulfate.
2. Forty-eight hours later, wash three times with DPBS and add fresh 293 T cell medium.
3. Forty-eight hours later, collect supernatant, supplement with 2,000× protamine sulfate, and add to freshly plated 293 T cells (100,000 cells per well).
4. Repeat steps 2 and 3 for a total of five cycles. In the last cycle, distribute the collected supernatant (supplemented with 2,000× protamine) to two identical test wells of freshly plated 293 T cells. Add 10 μL of fresh virus to one of the two wells (positive control). A third well of 293 T cells is required as a negative control.
5. Forty-eight hours later, carefully wash each well three times with DPBS$^{+/+}$, then add 1 mL DPBS$^{+/+}$ with Hoechst (1:10,000 dilution).

6. Using a fluorescence microscope, assess the number of Hoechst+ cells that are GFP+ or GFP− in each well. The test well without added virus should not contain any GFP+ cells (=no replication-competent retroviral particles were present). If GFP+ cells are detected in this well (replication-competent retroviral particles likely are present), the entire virus preparation and all transduced cells must be inactivated and discarded.

3.5. Retroviral Titering

1. To estimate the titer of each virus, culture 100,000 293 T cells per well of a 6-well cell culture plate with 2 mL 293 T cell medium, 2,000× protamine sulfate, and virus (to wells 1–3, add 0.1, 1, and 10 μL virus; to well 4, add no virus; *optional*: to wells 5–6, add 1,000 and 10,000 viral particles from an existing virus preparation with known titer).
2. Incubate at 37°C. Forty-eight hours later, carefully wash each well three times with DPBS to remove virus.
3. Add 500 μL trypsin/EDTA per well. When cells begin to lift off (carefully tap the plate), add 500 μL FBS. Mix to produce a single cell suspension. Transfer cells suspension to a tube containing 1 mL DPBS and mix.
4. Using a hemocytometer on an epifluorescence microscope, count the total number of cells and the number of GFP-positive cells (calculate the % GFP+ cells).
5. Determine the viral titer (expect $1.0–5.0 \times 10^3/\mu L$):

$$\text{Titer(infective particles per } \mu\text{L)} = \frac{100{,}000}{\mu\text{L virus}} \times \%\text{GFP}^+ \text{ cells.}$$

3.6. Retroviral Transfection of Fibroblasts (see Note 4)

1. Day −1: Passage exponentially growing human fibroblasts using 0.05% trypsin/EDTA solution as above. Plate 50,000–100,000 cells per well of a gelatin-coated 6-well plate in 2 mL/well human fibroblast medium.
2. Day 0: Change medium and add 1,000× protamine sulfate plus appropriate volumes of Oct4, Sox2, Klf4, and c-Myc retroviruses according to the titer and the number of cells plated (MOI = 2.5 viral particles per cell, per virus). *Note*: Leave one well uninfected or infect with an empty pMIG vector as a negative control.

3.7. Reprogramming (see Note 5)

1. Day 1: Wash cells two times with 1 mL DPBS and add 2 mL human fibroblast medium.
2. Day 4: Change medium. A large percentage of cells should be GFP+ by now.
3. Day 6: Change medium on transduced fibroblasts and prepare MEF plates for co-culture (1–6 wells per line). Thaw MEFs

onto a gelatin-coated 6-well plate at 7,000 cells per cm^2 using 2 mL MEF medium per well.

4. Day 7: Transfer transduced human fibroblasts to co-culture with MEFs in standard hESC medium. Replace the MEF medium on the 6-well plate with 2 mL of standard hESC medium. Lift the transduced human fibroblasts and negative control well with 1 mL/well 0.05% trypsin/EDTA solution for 5 min at 37°C. All cells should detach from the plate. Add 1 mL human fibroblast medium per well to inactivate the trypsin. Collect the cells and spin down at 200×g for 4 min. Seed at 50,000–100,000 cells per well in hESC medium in the prepared 6-well MEF plate.
5. Day 8 onward: Monitor reprogramming wells for the emergence of hiPSC colonies. Feed every 48 h (every 24 h once colonies begin to appear). Fully reprogrammed colonies can appear as early as 14 days and as late as 8 weeks postinfection.

3.8. Picking Induced Pluripotent Stem Cell Clones

1. Under a dissection microscope, mechanically pick hiPSC colonies based on hESC- of like morphology and GFP negativity. Fully reprogrammed hiPSC colonies have well-defined borders and are comprised of tightly packed, small cells that contain nuclei with prominent nucleoli (Fig. 1). Use a 27-gauge needle, bent at a 45° angle, to create borders that isolate the colony from the rest of the culture, thus preserving its clonality, and to divide the colony into small pieces (Fig. 2). If the fibroblasts start to come off in a single layer, discontinue the use of the needle and move on to the next step.
2. Using a P200 pipetteman, lift the fragments/colony off of the culture plate, starting at the outer edge and moving toward the center of the colony.
3. Directly transfer the fragments to a single well of a gelatin- and MEF-coated 24-well plate containing 500 μL standard hESC medium per well (see Note 6).

3.9. Expansion of Induced Pluripotent Stem Cell Clones

1. Monitor 24-well plates containing the picked iPSC colonies carefully. Feed daily. If significant growth of non-iPSCs or differentiation of iPSCs occurs, remove the corresponding areas by aspiration or scraping with a needle or P200.
2. Manually passage iPSCs using a needle and P200 pipetteman when colonies cover 30–70% of the well area (depending on the number of colonies and their distribution), or when colonies begin to touch. Be sure to only passage colonies with the morphology and phenotype mentioned above. It is possible that non-fully reprogrammed cells have contaminated the culture in the previous passage.
3. Collect the colony fragments into a 15-mL conical tube and spin down at 200×g for 2 min.

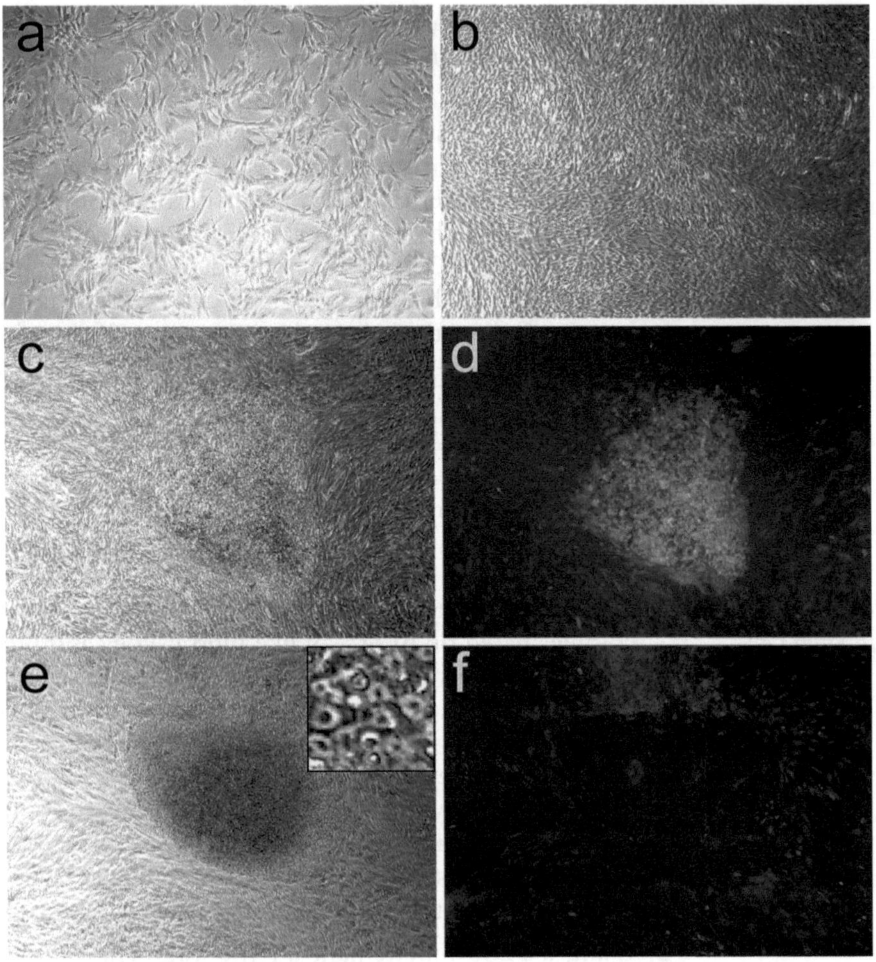

Fig. 1. (**a**) Virus-infected primary fibroblasts on day 4. (**b**) Fibroblasts on day 27 (uninfected control cells). (**c**) Incompletely reprogrammed colony on day 27. (**d**) GFP expression of the same colony. (**e**) Putative human iPSC colony (*inset*=20× Phase image showing tightly packed nuclei with prominent nucleoli). (**f**) Lack of GFP expression of the colony shown in (**e**).

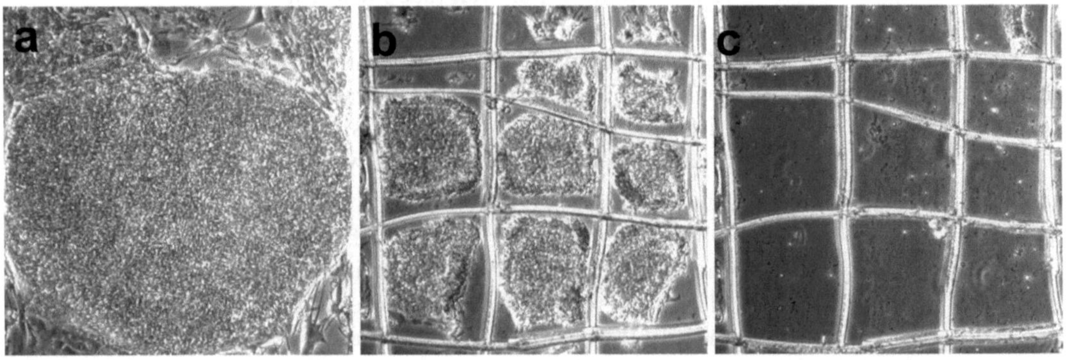

Fig. 2. (**a**) A hiPSC colony that is ready to be picked. (**b**) The same colony, after it has been scored several times with a bent needle. (**c**) The same area, after the colony pieces have been pushed off, picked up, and transferred to a new well with a P200 pipette.

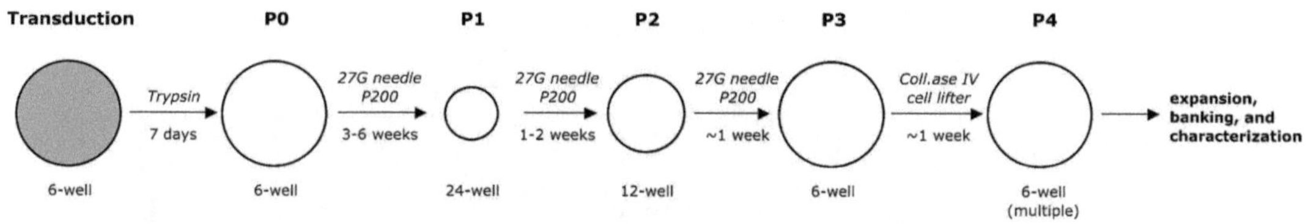

Fig. 3. Summary of steps and timeline of human fibroblast reprogramming. Gray indicates fibroblast culture conditions; white indicates human pluripotent stem cell (hESC/hiPSC) culture conditions. Individual candidate hiPSC colonies are mechanically picked into 24-well plate wells (P1) and expanded to the 6-well stage at which point cultures are passaged using collagenase IV.

4. Re-plate fragments in 1 mL of hESC medium onto a single well of a 12-well plate coated with gelatin and MEFs.
5. Repeat passaging according to the schematic in Fig. 3.
6. Once the iPSC clone has expanded to a 6-well, passage colonies enzymatically: wash wells with DMEM/F12, then incubate at 37°C for 5 min with 1 mL collagenase IV per well (see Note 7). Remove enzyme and add 1 mL DMEM/F12 per well. Use a cell lifter to scrape clean the entire well with gentle pressure. Collect fragments and wash well one time with additional DMEM/F12. Spin for 2 min at 200 × *g*. Carefully resuspend colony fragments in hESC medium (triturate with a P1000 as required if fragments are too large) and seed at a split ratio of 1:4–1:8 into 6-well plates coated with gelatin and MEFs.

3.10. Confirmation of Induced Pluripotent Clones by Immunostaining

1. Plate triturated colony fragments into a 96-well plate, coated with gelatin and MEFs, in 70 μL of hESC medium. Several wells containing iPSCs are required in order to stain with multiple antibodies. Parallel wells with human fibroblasts are recommended as negative controls.
2. Forty-eight hours later, begin feeding the wells daily until colonies are large enough to be fixed and stained (see Note 8).
3. Wash each well one time with 100 μL DPBS$^{+/+}$ and fix with 100 μL 4% PFA for 20 min at room temperature.
4. Wash three times with 100 μL DPBS$^{+/+}$.
5. Permeabilize the cells and block with 100 μL/well 3.5% BSA/5% Donkey serum/0.2% Triton X-100 in DPBS for 3 h at 4°C.
6. Stain with primary antibodies at appropriate concentrations in 40 μL 3.5% BSA/5% Donkey serum/0.2% Triton X/DPBS. Incubate overnight at 4°C.
7. Wash three times with 100 μL DPBS$^{+/+}$.
8. Stain with secondary/directly conjugated antibodies and Hoechst at their appropriate concentrations in 50 μL 3.5%

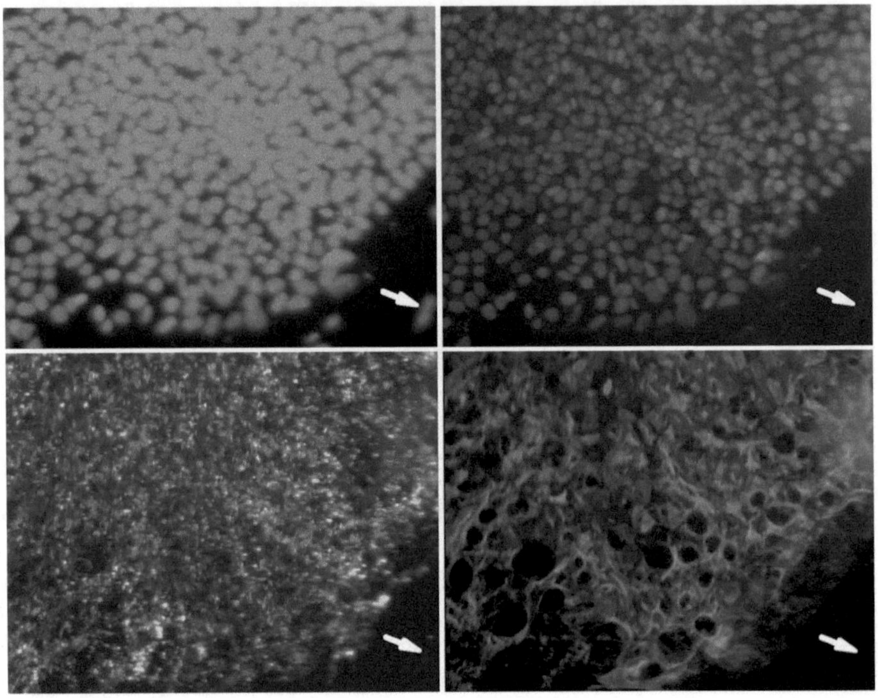

Fig. 4. Immunostaining of an induced pluripotent stem cell colony. Clockwise from *top-left*: Hoechst, OCT4, TRA-1-81, SSEA-4. Note the absence of stem cell marker expression by fibroblasts in the *bottom-right* corner of each panel (*arrow*).

BSA 5% Donkey serum/DPBS. Cover plate with aluminum foil and incubate for 3 h at room temperature.

9. Wash three times with 100 μL DPBS$^{+/+}$.
10. Observe marker expression in 50 μL DPBS$^{+/+}$ on an inverted microscope with fluorescence capability (Fig. 4).

4. Notes

1. All tissue culture procedures should be carried out in a Class II biological safety cabinet.
2. All work with retrovirus should be performed under BSL2+ biosafety conditions using appropriate safety precautions according to your institute's biosafety regulations.
3. Virus preparations containing replication-competent virus should immediately be inactivated and discarded.
4. Incubations are to be carried out in a humidified 37°C, 5% CO_2 incubator unless specified otherwise.
5. All medium should be warmed to 37°C before use.
6. Wash MEF-coated wells one time with DMEM/F12 before changing to standard hESC medium.

7. Following incubation with Collagenase IV, use a 5-mL glass pipette to collect fragments to prevent loss due to stickiness.
8. Once cells have been fixed, immunostaining may be performed on the bench top.

Acknowledgments

The authors would like to thank In-Hyun Park, Elayne Chan, Sutheera Ratanasirintrawoot, and Iris Huo for their advice and contributions.

References

1. Park, I.H., Zhao, R., West, J.A., Yabuuchi, A., Huo, H., Ince, T.A., Lerou, P.H., Lensch, M.W., and Daley, G.Q. (2008) Reprogramming of human somatic cells to pluripotency with defined factors. *Nature* **451**, 141–6.
2. Lowry, W.E., Richter, L., Yachechko, R., Pyle, A.D., Tchieu, J., Sridharan, R., Clark, A.T., and Plath, K. (2008) Generation of human induced pluripotent stem cells from dermal fibroblasts. *Proc Natl Acad Sci USA* **105**, 2883–8.
3. Takahashi, K., Tanabe, K., Ohnuki, M., Narita, M., Ichisaka, T., Tomoda, K., and Yamanaka, S. (2007) Induction of pluripotent stem cells from adult human fibroblasts by defined factors. *Cell* **131**, 861–72.
4. Park, I.H., Arora, N., Huo, H., Maherali, N., Ahfeldt, T., Shimamura, A., Lensch, M.W., Cowan, C., Hochedlinger, K., and Daley, G.Q. (2008) Disease-specific induced pluripotent stem cells. *Cell* **134**, 877–86.

Chapter 6

Derivation of Induced Pluripotent Stem Cells by Lentiviral Transduction

Hubert E. Nethercott, David J. Brick, and Philip H. Schwartz

Abstract

This chapter provides a method for reprogramming human dermal fibroblasts into induced pluripotent stem cells (iPSCs) using three lentiviruses containing cDNAs for OCT4 and SOX2, KLF4 and C-MYC, and NANOG and LIN28, respectively. Lentiviral vectors are based on the human immunodeficiency virus (HIV) and provide an effective means for the delivery, integration, and expression of exogenous genes in mammalian cells. Lentiviruses are attractive gene delivery vehicles as they are able to infect both proliferating and nonproliferating cells. Lentiviruses stably integrate into the genome without incurring cellular toxicity and can maintain sustained transgene expression during prolonged host cell proliferation and differentiation. In this protocol, we describe how to prepare lentiviruses, stably transduce human fibroblasts, and identify bona fide iPSC colonies based on morphological similarity to human embryonic stem cell (ESC) colonies and live-cell immunological staining using cell-surface markers of human PSCs such as Tra-1-60 and Tra-1-81.

Key words: induced pluripotent stem cells, lentivirus, live-cell imaging

1. Introduction

It is now possible to induce pluripotency in human somatic cells through the ectopic expression of a small number of transcription factors. The initial breakthrough was reported by the Yamanaka group at Kyoto University in 2006 when they demonstrated that induced expression of only four transcription factors, Oct3/4, Sox2, c-Myc, and Klf4, in mouse fibroblast cells resulted in the formation of embryonic stem cell (ESC)-like cells, termed induced pluripotent stem cells (iPSCs) (1). Subsequently, two independent groups demonstrated that human somatic cells can also be reprogrammed into iPSCs. Yamanaka's group successfully applied the same technique they used in mice. They used human dermal fibroblasts (HDFs) and two other human fibroblast populations from different

Philip H. Schwartz and Robin L. Wesselschmidt (eds.), *Human Pluripotent Stem Cells: Methods and Protocols*, Methods in Molecular Biology, vol. 767, DOI 10.1007/978-1-61779-201-4_6, © Springer Science+Business Media, LLC 2011

human donors and transduced them with retroviral vectors carrying human cDNAs for OCT4, SOX2, C-MYC, and KLF4 (2). The Thomson group at the University of Wisconsin discovered a new combination of factors sufficient for the generation of human iPSCs. They showed that, out of 14 genes, OCT4, SOX2, NANOG, and LIN28 represented a core set of genes able to reprogram human somatic cells with a mesenchymal phenotype to iPSCs following lentiviral (LV) transduction (3). These iPSCs exhibited the essential characteristics of human ESCs including their ability to differentiate into lineages of all three germ layers.

It is now evident that the combination of a small number of specific transcription factors can reprogram fully differentiated human somatic cells (4). The enforced expression of specific combinations of transcription factors can override and modulate existing gene networks and epigenetic marks. Indeed, an ability to induce pluripotency in somatic cells was previously demonstrated in elegant studies of the transfer of somatic cell nuclei into enucleated oocytes and the fusion of pluripotent stem cells with differentiated cells (5, 6).

Successful reprogramming requires expertise in a number of techniques from microbiology, molecular biology, and virology, to stem cell biology. Here, we present a protocol for (1) the production of rhabdoviral vesicular stomatitis virus G (VSV-G) envelope protein pseudotyped lentiviruses (LVs), (2) the transduction of HDFs, (3) a method for isolating iPSCs using live-cell immunocytochemistry, and (4) the culture and propagation of iPSC lines. This protocol emphasizes controlled steps for streamlining the iPSC derivation process.

The LV strategy for reprogramming detailed here provides a method to titer virus prior to transduction and to screen for the presence of human immunodeficiency virus (HIV) and replication-competent retroviruses (RCRs), both before and after transduction, using a single ELISA procedure.

Please note that Institutional Biosafety Committee approval is required before starting LV production and use.

2. Materials

2.1. Bacterial Cell Culture

1. Bacterial stabs:
   (a) pSIN4-EF2-O2S (Addgene, #21162). OCT4 and SOX2 genes.
   (b) pSIN4-EF2-N2L (Addgene, #21163). NANOG and LIN28 genes.
   (c) pSIN4-CMV-K2M (Addgene, #21164). KLF4 and C-MYC genes.

(d) pMD2.G (Addgene, #12259). Contains gene for VSV-G envelope protein.

(e) psPAX2 (Addgene, #12260). Lentiviral backbone, gag-pol genes.

2. 1,000× Ampicillin: 100 mg/mL Ampicillin in sterile water. Filter sterilize 1,000× Ampicillin solution through a 0.22 μm filter. Store at −20°C.
3. Luria Broth (LB) Medium: 2% (w:v) LB (Sigma, L3022) in distilled water. Adjust pH of LB Medium to 7.4 with 5M NaOH solution and/or 1N HCl solution. Autoclave for 30 min at 121°C. Store at room temperature.
4. LB Agar + Ampicillin plates: 3.5% (w:v) LB Agar (Sigma-Aldrich, L2897) in distilled water. Autoclave for 30 min at 121°C. Cool to ~55°C. Using a clean air bench or, alternatively, a laboratory bench top with an open flame, add Ampicillin to a final concentration of 1×. Transfer 20 mL of LB Agar + Ampicillin solution into sterile, untreated 100-mm Petri dishes. Allow the LB Agar + Ampicillin solution to form a gel at room temperature. Store the plates at 4°C, out of the light, and as dry as possible.
5. 50% Glycerol Solution: 1:1 (v:v) solution of 100% glycerol: distilled water. Autoclave for 30 min at 121°C. Store at room temperature.
6. Cryopreservation vials (Nalgene, 5000-1020).
7. Inoculating loop.
8. 37°C shaking incubator.
9. Dry-air incubator.

2.2. Isolation of Purified Plasmids

1. Endofree Plasmid Mega-Kit (Qiagen, 12381).
2. 100% Ethanol (Sigma, E7023).
3. 45 mm-Neck glass bottles (Pyrex, 1395-1L or equivalent).
4. Tris-EDTA (TE) buffer (included in Endofree Plasmid Mega-Kit).
5. UV spectrophotometer.

2.3. Lentivirus Production

1. 293FT cells (Invitrogen, R700-07).
2. *293FT medium*: Dulbecco's modified essential medium (DMEM, Invitrogen, 12430), 10% (v:v) defined fetal bovine serum (FBS, Hyclone SH30070.03), 4 mM L-glutamine, 1× minimum essential medium nonessential amino acids (NEAA, Hyclone, SH30238.01), 1 mM sodium pyruvate (GibcoBRL 11360-070), 50 μg/mL geneticin.
3. 100-mm TPP Tissue culture dishes.
4. 15-mL Polypropylene tubes.

5. Opti-Mem I reduced-serum medium (Invitrogen, 31985-070).
6. *Transfection medium:* Opti-Mem I reduced-serum medium (Invitrogen, 31985-070), 4% (v:v) defined FBS (Hyclone SH30070.03), 4 mM L-glutamine, 1× MEM non-essential amino acids (NEAA).
7. Lipofectamine 2000 (Invitrogen, 11668).
8. Purified pMD2.G.
9. Purified psPAX2.
10. Purified pSIN4-EF2-O2S.
11. Purified pSIN4-EF2-N2L.
12. Purified pSIN4-CMV-K2M.
13. 0.45-μm PVDF bottle filters (Millipore, SCHVU01RE).
14. Cryopreservation vials (Nalgene, 5000-1020).
15. Optiseal polyallomer ultracentrifuge tubes (Beckman Coulter, 361625).
16. XL-80 Ultracentrifuge.
17. Type 70 Ti Ultracentrifuge rotor.

2.4. Titering by ELISA

1. HIV-1 p24 Antigen Capture Assay (Advanced BioScience Laboratories, 5421).
2. MicroELISA plate reader, capable of absorbance readings at 450 nm.

2.5. Human Dermal Fibroblast Culture

1. Human dermal fibroblasts (HDFs).
2. *HDF medium*: Dulbecco's modified Eagle medium (DMEM, Invitrogen #12430054), 10% (v:v) defined FBS (Hyclone SH30070.03), 4 mM L-glutamine, 1× NEAA, 1 mM sodium pyruvate.
3. Recombinant human basic fibroblast growth factor (bFGF, Millipore, GF003).
4. Dulbecco's phosphate buffered saline without Ca^{2+} or Mg^{2+} (DPBS).
5. 0.05% Trypsin/EDTA solution.
6. Trypsin neutralizer solution (Cascade Biologicals, R-002-100).
7. 0.1% (w:v) Gelatin (Millipore, ES-006-B).
8. T-75 Tissue culture flasks.
9. 6-Well dishes.
10. Bright-Line Hemacytometer.

2.6. LV Transduction of HDFs

1. Lentiviral stocks (concentrated and titered).
2. Dulbecco's modified Eagle's medium (DMEM) (Invitrogen, 12430).

3. Dulbecco's phosphate buffered saline with Ca^{2+} and Mg^{2+} (DPBS +/+) (Hyclone, SH30264.01).
4. 0.1% (w:v) Gelatin (Millipore, ES-006-B).
5. 6-Well dishes (BD, 353046).
6. Bright-Line Hemacytometer.
7. Polybrene (Sigma, H9268).

2.7. Pluripotent Stem Cell Culture

1. *PSC medium*: DMEM/F12 (Invitrogen, 10565-018), 20% knockout serum replacement (KSR, Invitrogen, 10828-028), 2 mM GlutaMax-I (Invitrogen, 35050), 100 μM beta-mercaptoethanol (Invitrogen, 21985-023), 1× minimum essential medium NEAA (Hyclone, SH30238.01), 20 ng/mL bFGF (Millipore, GF003).
2. Mitotically inactive CF-1 mouse embryonic fibroblasts (MEFs) (see Chapter 8 for details).
3. *MEF medium*: Dulbecco's modified Eagle medium (DMEM, Invitrogen, 12430), 10% (v:v) defined FBS (Hyclone SH30070.03), 4 mM L-glutamine, 1× minimum essential medium NEAA (Hyclone, SH30238.01), 1 mM sodium pyruvate.
4. 0.1% (w:v) Gelatin (Millipore, ES-006-B).
5. 6-Well dishes (BD, 353046).
6. Y-27632 (ROCK inhibitor, Stemgent, 04-0012).
7. Optional: Microscope object marker (such as Nikon, MBW10020).

2.8. Live-Cell Imaging of iPSC Colonies

1. Dulbecco's phosphate buffered saline with Ca^{2+} and Mg^{2+} (DPBS +/+) (Hyclone, SH30264.01).
2. PSC medium.
3. PSC medium without phenol red: Use DMEM/F12 without phenol red (Invitrogen, 11039).
4. Tra-1-60-Alexa Fluor 647 antibody (BD Pharmingen, 560122).
5. Tra-1-81-Alexa Fluor 488 antibody (BD Pharmingen, 560174).
6. Hoechst 33342 (Invitrogen, H3570), see Note 1.
7. Fluorescence microscope in BSL2+ facility.
8. Optional: Microscope object marker (such as Nikon, MBW10020).

3. Methods

3.1. Lentiviral Vector Plasmids

The LV vectors to be used for the production of the iPSCs are self-inactivating (SIN), HIV-1-based vectors. These LV systems offer increased biosafety to users. The pSin4-promoter-gene vectors contain a 400-bp deletion in the U3 region (ΔU3) of the 3′ LTR, which results in the transcriptional inactivation of the 5′ LTR following reverse transcription and chromosomal integration. The inability to efficiently transcribe full-length vector RNA by these SIN-HIV-1 vectors in transduced target cells minimizes the possibility of the formation of RCRs. In addition, these vectors contain a central polypurine tract within the *pol* gene (cPPT) which facilitates nuclear translocation of the HIV preintegration complex and improves lentiviral transduction efficiency in several types of cells. Expression of the Oct-4 with Sox-2 and Nanog with Lin28 transgenes are driven by the human elongation factor 2 (EF2) promoters in plasmids pSin4-EF2-O2S and pSin4-EF2-N2L, respectively. Expression of KLF-4 with c-Myc is driven by the CMV promoter in pSIN4-CMV-K2M vector. Each of these three vectors will be individually packaged in 293FT cells by co-transfection with the pMD2.G-envelope plasmid encoding the VSV-G envelope glycoprotein and the psPAX2 2nd generation LV backbone packaging vector that is optimal for virus production.

3.2. Bacterial Cell Culture

Addgene ships plasmids as individual *Escherichia coli* (*E. coli*) bacterial stab cultures. A stab culture is made by inoculating (or transferring) bacteria into a vial containing LB agar with the appropriate antibiotic. Store bacterial stabs at 4°C for no longer than 2 weeks. The subsequent protocol describes how to streak bacterial stabs on LB agar plates, isolate a single colony from your stab culture, and create bacterial glycerol stocks for long-term storage. Please note that the bacteria containing psPAX2 and pMD2.G are "high-copy," while the bacteria containing pSIN4-EF2-O2S, pSIN4-EF2-N2L, and pSIN4-CMV-K2M are "low-copy" (see Note 2).

1. Over an open flame, sterilize an inoculating loop by holding its tip in the flame until it turns red. Wait for the loop to cool to prevent killing the bacterial sample.
2. For each bacterial stab obtained from Addgene, touch the end of the cooled inoculating loop to the bacterial sample you wish to transfer. Transfer bacterial cells to the LB Agar + Ampicillin plate by gently brushing the loop back-and-forth over the surface of the plate in a zig-zag fashion. Do not cross over previous streaks.
3. In separate dishes, repeat steps 1–2 for all five bacterial stabs obtained.
4. Cover the LB Agar + Ampicillin Petri plate, invert it, and incubate at 37°C in a dry-air incubator overnight.
5. The next morning, the plates should appear to have solid streaks of cells as well as isolated colonies. At this point, bacterial plates can be stored for several weeks at 4°C, if plates are inverted and sealed with parafilm wrap.

6. To make a LB liquid starter bacterial culture, aseptically transfer 6 mL of LB medium to a sterile 15-mL tube and add 6 μL of 1,000× Ampicillin stock. For multiple start-up cultures, the LB broth and antibiotic can be scaled up and aliquoted appropriately in 15-mL tubes.
7. Using a sterile inoculating loop, pick and transfer a single, well-isolated colony from a streaked bacterial plate to the tube containing LB medium + Ampicillin.
8. Close the top of the 15-mL tube and flame sterilize the inoculating loop before repeating the inoculations of the remaining four bacterial culture plates. These are your starter cultures.
9. Incubate at 37°C in a shaking incubator at 300 rpm for 12–16 h. These starter cultures will be used to inoculate larger volumes of LB for plasmid isolation and purification, and to create long-term glycerol stocks of *E. coli* harboring the respective plasmids.
10. For each high-copy bacterial culture, prepare 1 L of autoclaved LB Broth. For low-copy bacterial cultures, prepare 2.5 L of autoclaved LB Broth (see Note 2).
11. Just before inoculating LB with the respective *E. Coli* cultures, add fresh 1,000× Ampicillin solution to the LB to a final concentration of 1×. To ensure good aeration of the growing cultures, split the culture to a maximum volume of 500 mL contained within individual 2 L or larger clearly labeled Pyrex Flasks.
12. Inoculate each prelabeled LB + Ampicillin flask with 1 mL of its respective starter culture from step 9. Incubate at 37°C in a shaking incubator at ~300 rpm for 12–16 h.
13. With the remaining 5 mL of starter culture, make bacterial glycerol stocks. Under open-flame/bench-top sterile conditions, mix 500 μL of the overnight starter culture with 500 μL of 50% glycerol solution in a cryopreservation vial. Freeze and store the glycerol stock vial at −80°C. Glycerol stocks may be stored indefinitely. New LB Agar + Ampicillin plates may be streaked again from these glycerol stocks when needed.
14. Continuing from step 12, harvest the bacterial cell culture using centrifugation at 6,000 × g for 15 min at 4°C. Decant the supernatant and dispose of it in accordance with institutional guidelines. Bacterial pellets can be stored at this point, at −20°C.
15. To recover the plasmids, we recommend using the Qiagen EndoFree Plasmid Mega-Kit (Cat. no. 12381). This kit produces high-quality purified endotoxin-free plasmids, which are necessary to ensure high transfection efficiencies. Qiagen makes continuous improvements to this kit, so please refer to their most recent EndoFree Plasmid Purification Handbook for instructions.

16. Re-dissolve the air-dried purified plasmid DNA pellets in approximately 500–600 μL of endotoxin-free TE buffer (included in the kit). Calculate the concentration of the isolated plasmids using UV spectrophotometry at 260 nm. For reliable spectrophotometric DNA quantitation, A260nm readings should lie between 0.1 and 1.0. At a minimum, the recovered plasmid concentration should be 1 μg/mL. Concentrated preparations of plasmid aid in achieving high transduction efficiencies. If plasmid recovery is low, repeat the protocol and resuspend the DNA pellet in a smaller volume.

3.3. Lentivirus Production and Concentration

The following protocol is adapted from Invitrogen's ViraPower™ Lentiviral Expression System. Each LV is produced separately in 100-mm tissue culture dishes.

1. Day 1: Plate four 100-mm dishes with 5×10^6 293FT cells/dish in 10 mL 293FT medium. 293FT cells should be 90–95% confluent at the time of transfection. Record the time of plating. As antibiotics can negatively affect transfection efficiency, do not include any antibiotics in this medium.
2. Day 2: Approximately 24 h after plating the 293FT cells, aspirate the medium and replace with 5 mL transfection medium. As above, do not include any antibiotics in the medium.
3. For each 100-mm dish, prepare DNA–Lipofectamine 2000 complexes as follows:
   (a) Tube A (sterile): Combine the following: 1.5 mL of Opti-MEM I medium without serum, 7.5 μg psPAX2 plasmid, 2.5 μg pMD2.G, and 6 μg of one of the target clone vectors (pSIN4-EF2-O2S, pSIN4-EF2-N2L, or pSIN4-CMV-K2M). Mix together gently. Do not vortex.
   (b) In Tube B (sterile): Combine the following: 1.5 mL of Opti-MEM I medium without serum and 36 μL of Lipofectamine 2000. Gently mix the Lipofectamine 2000.
   (c) Incubate Tube B at RT for 5 min before proceeding.
   (d) After the 5-min incubation, combine the contents of Tube A with Tube B. Mix gently.
   (e) Incubate for 20 min at RT to allow the DNA–Lipofectamine 2000 complexes to form. The solution may appear cloudy, but this will not impede the transfection.
4. Drop-wise, add the DNA–Lipofectamine 2000 complexes to each plate of 293FT cells. Mix by gently rocking the plate back-and-forth. Incubate the cells overnight at 37°C in a humidified 5% CO_2 incubator. Record the time of transfection.

   Caution: Remember that you are working with potentially infectious virus following this stage. Therefore, all manipulations of virus and virus-transduced cells must be carried out with appropriate PPE under BSL-2+ conditions using a certified, Class II, biological safety cabinet (BSC) and all virus-contaminated media, serological pipettes, barrier pipette tips, and tissue cultureware must be deactivated in 10% (v:v) fresh sodium hypochlorite (household bleach) solution for at least 15 min prior to disposal. Follow the recommended NIH guidelines that are contained in the Biosafety in Microbiological and Biomedical Laboratories (BMBL) Manual and the NIH Guidelines for Research involving Recombinant DNA Molecules (http://www.cdc.gov/od/ohs/biosfty/bmbl4/bmbl4toc.htm, http://www.oba.od.nih.gov/oba/rac/guidelines_02/NIH_Guidelines_Apr_02.htm). In addition, consult your institution's policies and procedures and receive approval from your Institutional Biosafety Committee prior to initiating LV experiments.

Note: Expression of the VSV-G glycoprotein causes 293FT cells to fuse, resulting in the appearance of large, multinucleated cells known as syncytia. This morphological change is normal and does not affect LV production.

5. Day 3: Remove the medium containing the DNA–Lipofectamine complexes and replace with 10 mL 293FT medium. Do not include any antibiotics in the medium.
6. Day 4/5: Harvest LV-containing supernatants 48–72 h post-transfection. Differences in titer are minimal when LV is harvested at 48 or 72 h.
7. Pass the collected LV-containing supernatant through a 0.45-μm PVDF bottle filter (Millipore, SCHVU01RE) to clear cellular debris.
8. Transfer the LV-containing supernatant into a sterile Optiseal Beckman Coulter ultracentrifuge tube. Seal the ultracentrifuge tube in strict accordance with the manufacturer's instructions. In the BSC, load the ultracentrifuge rotor with the samples and seal the lid securely before removing from the BSC to the ultracentrifuge. Centrifuge at 70,000 × g for 90 min at 4°C.
9. Caution: Be sure the ultracentrifuge tube is completely filled with medium and make sure the cap and tube balancer is secured tightly.

   Note: If you do not have access to an ultracentrifuge or you do not wish to concentrate your viral stock, pipette 1 mL aliquots of LV supernatants into screw-capped cryovials, place in a secondary container, and store viral stocks at −80°C.
10. Remove supernatant by aspirating or decanting the supernatant. Aspirate away from the bottom of the tube so as not to aspirate the pellet. You may see a pellet, but LV pellets are often clear. For each 100 mm dish, add 200 μL of sterile

DMEM to the pellet. Scale accordingly. Do not triturate the pellet. Allow the pellet to slowly dissolve into the medium overnight at 4°C.

11. At this point, you may titer or freeze the virus, or transduce your HDFs directly. Viruses in general do not tolerate freeze–thaw well. We recommend freezing the prepared LV stocks in 20-μL aliquots in cryopreservation vials. In addition, freeze 1 vial with 10 μL of lentivirus. You will use this 10 μL vial for titering the virus by ELISA. Store viral stocks at −80°C.

3.4. Titering by ELISA

Titering the LV stock is helpful in producing consistent transduction results by using a consistent amount of active virus. Multiplicity of infection (MOI) is a measure of viral infectivity in a population of target cells. With lentivirus, the MOI is the ratio of transfer viral transducing particles to the target cells. A MOI of 5 indicates that there are five transducing units for every cell contained within a well. It is important to note that different cell types may require different MOIs for successful transduction. The following section will detail how to translate ELISA determined spectrophotometric readings to pg/mL p24 concentration and subsequently to transducing units (TU)/mL that are used to calculate the volume of LV required for a given MOI.

While it may not be necessary to determine the MOI for every application, it is necessary when predictably producing iPSCs for the following reasons:

1. Confirmation of the viability of the viral stock.
2. Determination of the maximum number of target cells that can be transduced by a given amount of virus.
3. Determination of the MOI-to-response ratio that produces the optimal induction profile.

Viral titer may be affected by any and all steps preceding viral titration. For instance, each freeze–thaw can reduce the functional titer (actual infection rate) by up to two- to four-fold. Likewise, low-quality cultures of 293FT will almost certainly lower the viral titer. Hence, titering the lentiviral stock serves as a useful quality control method. To more accurately gauge the performance of the lentivirus titer, we recommend freezing the viral stock before titering and/or titering your viral stock before each transduction.

The plasmid constructs used in this protocol do not contain a selectable marker gene (e.g., puromycin). Therefore, it is not possible to perform a colony-forming unit assay following LV transduction and drug selection.

We find it most convenient to titer lentiviral stocks by ELISA, using the HIV-1 p24 Antigen Capture Assay (Advanced BioScience Laboratories, 5421). The assay procedure is a double antibody sandwich enzyme immunoassay that is used to calculate the concentration of the amount of the HIV-1 p24 core antigen present in tissue culture samples. The assay's linear range is between 3.1 and 100 pg/mL. Since p24 is highly conserved among various HIV-1 isolates, this assay detects p24 from various isolates with comparable sensitivity. In addition to allowing the user to determine the titer of LV produced above, this assay also allows the user to screen research cell lines for HIV-1 contamination prior to LV transduction and iPSC generation and thus help eliminate the risk of generating RCR. Demonstration of p24-negativity in transduced cells is also necessary prior to downgrading the cells from BSL2+ to BSL2 status.

1. Before performing the p24 ELISA assay, make 10^{-6}, 10^{-7}, and 10^{-8} serial dilutions from 10 μL of your concentrated LV stocks. Make at least 250 μL per dilution for the assay. If you did not concentrate the LV, do not dilute the LV stock (see Note 3).
   (a) Note: From experience, this range of dilutions gives data that consistently fits the best-fit line of the p24 standards.
2. Follow the directions included in the HIV-1 p24 Antigen Capture Assay, using the serially diluted stocks of virus as the test samples.
3. After reading absorbance at 450 nm, ensure that the test is valid by checking absorbance values of the negative control and 100 pg/mL p24 standard. Two or more negative control absorbance values over 0.120 will invalidate the assay. Absorbance values of the 100 pg/mL p24 standard should be >1.200 and <2.200. Likewise, absorbance values outside of this range will invalidate the assay.
4. Calculate the mean absorbance for each test sample dilution, negative control, and p24 standard dilution. Subtract the mean of the negative control (background absorbance) from the means of each p24 standard dilution. Likewise, subtract the mean of the negative control (background absorbance) from the means of each test sample.
5. Determine the p24 concentration of each test sample dilution by extrapolating from a standard curve or by using linear regression analysis.
6. Find the mean p24 antigen concentration in pg/mL of each test sample: Multiply the mean of each test sample dilution (from step 4) by the reciprocal of its dilution. For instance, multiply your determined 10^{-6} value from step 4 by $1/10^{-6}$ (or 10^{6}).
7. Find the mean p24 antigen concentration in pg/mL of your LV Stocks: Calculate mean p24 antigen concentrations of your 10^{-6}, 10^{-7}, and 10^{-8} dilutions (from step 6).

8. Convert to TU/mL: 10 TU corresponds to approximately 1 pg of p24 (see Note 3). Multiply your mean concentration (from step 7) by the ratio 10 TU/pg.
9. Calculate the volume of virus needed for transduction. We recommend starting with an MOI of 5. Use the following equation to determine the volume of virus needed:

$$\text{Volume} = \frac{\text{Number of cells to be transduced} \times \text{MOI}}{(\text{TU/mL})}.$$

3.5. Transduction (Reprogramming) of HDFs

For optimal reprogramming of HDFs into iPSCs, fibroblast cultures should be healthy, relatively low passage (passage 6 or earlier), and rapidly dividing. For context, "rapidly dividing" means the following: If 2×10^5 fibroblasts are plated in a 35-mm dish, the culture should reach 70% confluence overnight. Such cultures are "rapidly dividing." Neonatal fibroblasts will generally fulfill this requirement. Adult fibroblasts usually do not. Addition of 4 ng/mL human bFGF will increase the cell division rate of adult fibroblast cultures and greatly improve their transduction efficiencies.

The following protocol describes the transduction of a HDF cell line.

1. Day 0: Coat two wells of a 6-well plate with 1 mL/well 0.1% (w:v) gelatin solution. One well will be infected with LV. The other will be mock-infected and serve as a negative (morphology) control. Incubate at 37°C for at least 1 h.
2. Detach HDFs from their culture vessel by trypsinization. Pipette the cells through a 10-mL serological pipette a couple of times to ensure a good dispersion of cells. Count HDF, either by Coulter counter or with a hemacytometer.
3. Remove the gelatin solution from the 6-well plate in step 1.
4. Plate 2×10^5 HDFs/well in the gelatin-coated well with 2 mL HDF medium/well.
5. Incubate overnight at 37°C in a humidified 5% CO_2 incubator.
6. Day 1 (Day of transduction): Prepare the transduction medium in plain, sterile DMEM. In a sterile tube, add the lentiviral stock(s) containing OCT4/SOX2, KLF4/c-MYC, and NANOG/LIN28 gene inserts. Bring the final volume up with DMEM to 600 μL/well/transduction. Include 4 μg/mL polybrene in transduction medium. (The control medium contains only DMEM and Polybrene.)

   Note: Polybrene is a polycation that reduces charge repulsion between the virus and the cellular membrane. Excessive exposure to polybrene (>24 h) can be toxic to cells. The optimum final concentration of polybrene may need to be determined empirically but generally falls within a range of 2–12 μg/mL.

7. Wash each well of HDFs twice with DPBS. Aspirate the final wash, and add the transduction medium, tilting the plate back-and-forth and side-to-side to disperse the virus.
8. Incubate at 37°C in a humidified 5% CO_2 incubator overnight. For the first hours after transduction, it is beneficial to rock the plate back-and-forth and side-to-side every 10–15 min to further disperse the virus.
9. Day 2: Remove virus-containing and control medium, wash each well three times with DPBS, and add 2 mL fresh HDF medium. Incubate at 37°C in a humidified 5% CO_2 incubator.
10. Day 3: Prepare 2 gelatin-coated 6-well plates: Coat all wells of a 6-well plate with 1 mL/well 0.1% (w:v) gelatin solution. Incubate at 37°C for at least 1 h.
11. After 1 h, thaw vials of irradiated MEFs (iMEFs), and plate iMEFs at a density of 2×10^4 cells/cm^2. For each gelatin-coated 6-well plate, plate $1\text{–}2 \times 10^6$ cells/dish (the total area of a 6-well plate is approximately 60 cm^2) in 2 mL MEF medium per well. Incubate at 37°C in a humidified 5% CO_2 incubator overnight.
12. Day 4: Lift the transduced HDFs and control with trypsin separately. Once the cells have sufficiently detached, inactivate the trypsin with trypsin inhibitor. Centrifuge at $200 \times g$ for 5 min. Aspirate the supernatant. Resuspend pellets in 12 mL HDF medium. Optional: We recommend supplementing this medium with 10 μM Y-27632 (ROCK inhibitor).
13. Aspirate the MEF medium from the 6-well plates containing iMEFs. Wash each well 1× with DPBS with Ca^{2+} and Mg^{2+} to clear out any debris.
14. Add 2 mL/well of resuspended transduced or control HDFs to the 6-well plate containing iMEFs. Incubate at 37°C in a humidified 5% CO_2 incubator.
15. Day 5: Aspirate the spent HDF medium, and replace with 2.5 mL/well fresh PSC medium. Replace medium daily, thereafter.

3.6. Live-Cell Staining and Live-Cell Imaging

Transduced fibroblasts are usually reprogrammed very quickly, if successful. Potential iPSC colonies are usually present in abundance at approximately day 14 posttransduction (see Fig. 1). Additional colonies may appear later. Many colonies, however, will be only partially transduced, but will have very similar morphology to bona fide iPSCs. These colonies may represent a stage of the spectrum of evolution toward bona fide iPSCs. Therefore, it is important to characterize the cultures by live staining during iPSC generation. Colonies that stain positive for Tra-1-60 and Tra-1-81 and stain dim for Hoechst 33342 are reliably bona fide iPSC colonies and can be passaged immediately after staining (see Note 1 and Fig. 2).

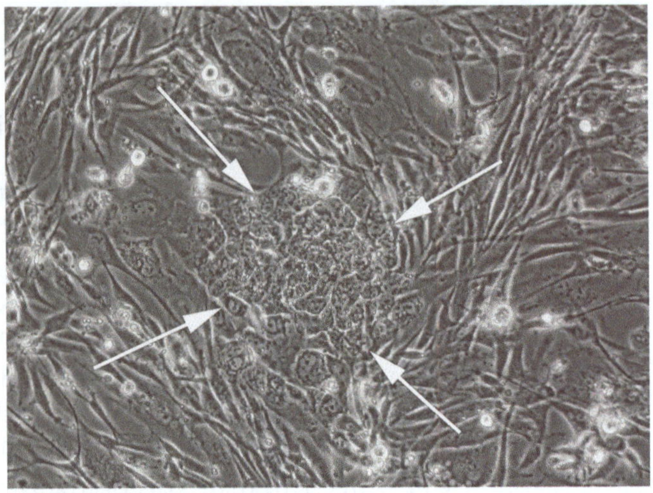

Fig. 1. Early morphology of transduced human dermal fibroblasts. Picture taken Day 9–10 post-transduction. Early in reprogramming, transduced human fibroblasts begin forming colonies, become small and rounded compared to nontransduced fibroblasts surrounding the colonies, but are still hyper-cytoplasmic compared to embryo-derived pluripotent stem cells. 10× objective.

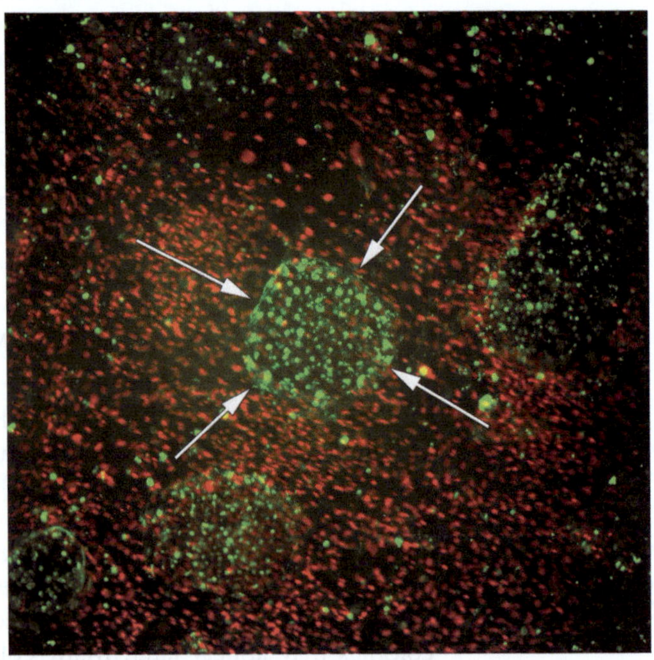

Fig. 2. Live staining immunocytochemistry of human iPSCs derived following lentiviral reprogramming of human dermal fibroblasts. Reprogrammed iPSCs stain positive for the pluripotent marker TRA-1-60 (*green*, colony marked by *arrows*). The surrounding iMEFs and undifferentiated HDFs brightly stain for Hoechst 33342 (*red*), while human iPSC colonies stain dim for Hoechst 33342. Of note is the discriminate Hoechst staining of the cells surrounding the iPSC colonies which stain brighter. 4× objective.

1. Make 1:100 dilutions of the conjugated antibodies in a sterile tube. For each well of a 6-well plate, add 6 μL of Tra-1-60-Alexa Fluor 647 and 6 μL of Tra-1-81-Alexa Fluor 488 to 600 μL of PSC medium.
2. Aspirate the medium from the well(s) to be stained. Wash each well three times with DPBS with Ca^{2+} and Mg^{2+} before adding the antibody-medium mixture.
3. After the final wash, apply the antibody-medium mixture (from step 1) to the well. Briefly, rock the plate back-and-forth and side-to-side gently to adequately disperse the antibody throughout the well.
4. Incubate at 37°C in a humidified 5% CO_2 incubator for 90 min.
5. In the meantime, prepare an excess of 2× solution of Hoechst 33342 in PSC medium (1.0 μg/mL). Prepare 1,000 μL for each well being stained. Filter the 2× solution using a 0.22-μm syringe filter.
6. After the 90 min incubation, add 600 μL of 2× Hoechst 33342 directly to the well. The final concentration of Hoechst is 0.5 μg/mL.
7. Incubate at 37°C in a humidified 5% CO_2 incubator for 30 min.
8. Wash each well three times with DMEM/F12 *without phenol red*. The subtraction of phenol red allows for better fluorescent visualization of the live-stained cultures.
9. Add 1 mL PSC medium *without phenol red*.
10. Visualize live-stained cultures under a fluorescence microscope. Take note of the morphology and staining patterns of iPSC colonies. The Hoechst is used primarily to train the eye (see Note 1).
11. Proceed to Subheading 3.7 on picking and passaging colonies.

3.7. Picking and Passaging iPSC Colonies

Colonies should be expanded depending on their number and size. As a point of reference, 2–4 colonies can be split into one well (~10 cm^2) of a 6-well plate but it is preferable to split only 1 colony per well to maintain clonality. Scale the reagents and tissue culture vessels appropriately. Once a colony is chosen for manual passaging, turn the objective wheel to the object marker (see Fig. 3), and mark the site of the colony on the underside of the plate. This product is relatively inexpensive and saves valuable time in picking the best colonies.

The following protocol can be used for both unstained and live-stained iPSC cultures.

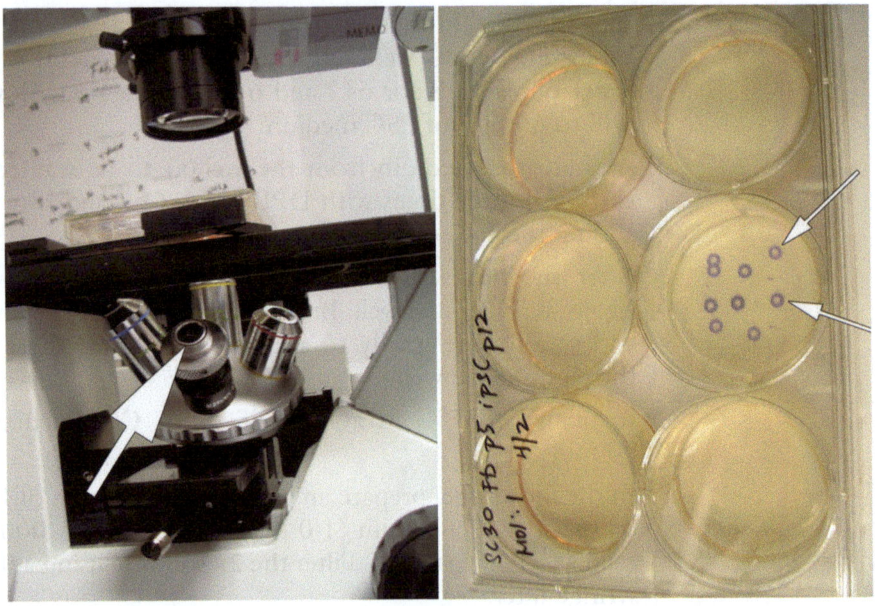

Fig. 3. Use of an object marker greatly simplifies identification of early iPSC colonies for picking and passaging. The marker (shown with the *arrow* in the *left panel*) simply takes the place of an objective. When a colony is identified (usually by using the 10× objective), the marker is rotated into place and the position of the colony is marked on the bottom of the plate (marked colonies are shown in the *right panel*). The plate can then be transferred to a BSC for picking and passaging of the marked colonies.

1. Day 0: For each colony to be passed, prepare 1 well of a 6-well tissue culture plate with gelatin. Add 1 mL of 0.1% (w:v) gelatin solution per well. Incubate at 37°C for at least 1 h.
2. After 1 h, thaw vials of irradiated MEFs (iMEFs), and plate iMEFs at a density of 2×10^4 cells/cm^2 in 2 mL MEF medium per well. Incubate at 37°C in a humidified 5% CO_2 incubator overnight.
3. Day 1: Aspirate MEF medium and wash plates twice with DPBS with Ca^{2+} and Mg^{2+}. Replace medium with 1 mL/well PSC medium. Optional: Supplement media with Y-27632 (ROCK inhibitor), of 10 μM at a final concentration.
4. Exchange the medium of iPSC cultures with 1 mL/well of fresh PSC medium.
5. Under a microscope equipped with a Microscope Object Marker, locate bona fide iPSC colonies (Tra-1-60/Tra-1-81-positive). Mark these colonies on the underside of the well.
6. In a Class II BSC, manually detach the marked iPSC colony with the end of a sterile P10 micropipette tip. Try to break this marked colony into the smallest pieces possible.
7. Pool together the manually detached iPSC colony pieces in a sterile tube.

8. Wash the well of the previously detached colony pieces with 1 mL/well of PSC medium. Pool this wash with the iPSC colony pieces. Repeat steps 4–8 for each colony to be passaged.
9. Evenly distribute the detached iPSC colony pieces over the 6-well plate containing iMEFs in PSC medium. Gently rock the plate back-and-forth and side-to-side to disperse the colony pieces evenly.
10. Incubate at 37°C in a humidified 5% CO_2 incubator. Leave newly passaged cultures undisturbed for 2–4 days. Feed daily thereafter. Passage as required. Cultures may be examined for the presence of markers of pluripotency by immunocytochemistry (see Chapter 15) after passage 10 (Fig. 4).

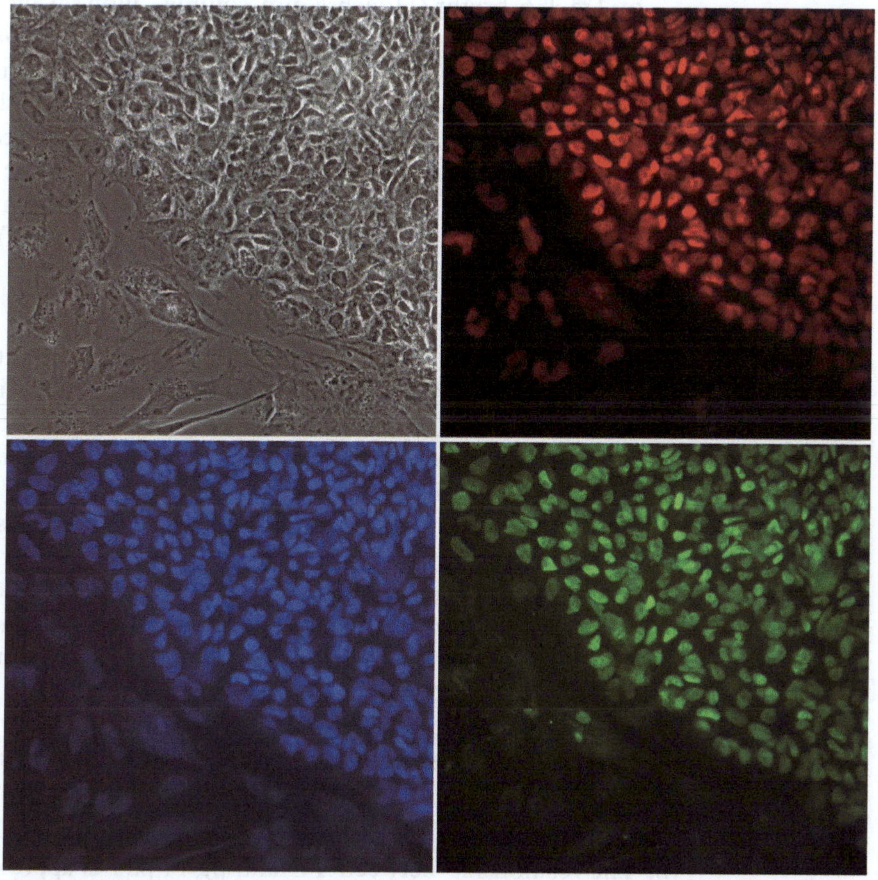

Fig. 4. Immunocytochemistry of lentivirus-derived human iPSC colonies propagated on iMEFs. iPSCs stain uniformly positive for the pluripotent markers OCT4 (*red*), SOX2 (*green*), and NANOG (*blue*). The phase contrast image illustrates the morphological similarity of the iPSC colony to classical human ESC colonies. 40× objective. Photos taken by Alexander Stover.

4. Notes

1. Hoechst 33342.

   The basis for the presence or absence of Hoechst staining after loading the cells with Hoechst is the presence or absence of the multidrug resistance transporter ABCG2. In the presence of ABCG2, which appears during the early phases of transduction, Hoechst is actively transported out of the cells, leading to cells that are Hoechst-dim compared to surrounding nontransduced cells that are Hoechst-bright (7). As the iPSCs mature, the ABCG2 transporter is downregulated and the cells remain Hoechst-bright after loading so it is of little use then. Since Tra-1-60 staining appears early in the transduction process and is not seen in nontransduced cells, the appearance of Tra-1-6-bright/Hoechst-dim colonies allows a nice, early indication of an iPSC colony that can be picked out of a mixed colony using fluorescence as your guide. Unfortunately, Hoechst is also mutagenic and thus cannot reasonably be used for routine picking of iPSCs. It is, therefore, omitted from routine use and is used solely to train the eye of the novice iPSC-deriver as to what a bona fide early iPSC colony looks like both by phase microscopy and by fluorescence with Tra-1-60 (with or without additional Tra-1-81).

2. High-copy versus low-copy plasmids.

   Plasmid copy number is perhaps the most important factor affecting plasmid DNA yield. Plasmids vary widely in their copy number depending on the origin of replication they contain, which determines whether they are under relaxed or stringent control, as well as the size of the plasmid and its associated insert. Some plasmids also have mutations which allow them to reach very high-copy numbers within the bacterial cell. Low-copy plasmids have approximately 10 copies/cell, while high-copy plasmids have up to 100's of copies/cell. Therefore, it is necessary to scale up the LB culture medium volume to obtain higher plasmid yield. It is best to use the suggested culture volumes. Using larger culture volumes will lead to an increase in biomass and can affect the efficiency of alkaline lysis during plasmid purification, leading to reduced yield and purity of the preparation.

3. The relationship between TU and pg of p24 is approximately 2,000 molecules of p24 per physical particle (PP) of HIV, which indicates there are approximately 10^4 PP/pg of p24 detected. It is also important to note that while freeze–thaw of the LV produced may not affect the quantity of p24 detected, this quantity is still used to estimate the MOI, and thus infectivity will likely be reduced. Thus, it is important to choose aliquot size carefully and, ideally, freeze–thaw no more than once.

Acknowledgments

This work has been funded by the National Institutes of Health (T15HL074286, R21MH087925, R01HD059967). The NCI Preclinical Repository supplied FGF-2. We also recognize Richard Pepple for proof-reading this chapter.

References

1. Takahashi, K. & Yamanaka, S. (2006) Induction of pluripotent stem cells from mouse embryonic and adult fibroblast cultures by defined factors. *Cell* **126**: 663–676.
2. Takahashi, K. et al. (2007) Induction of pluripotent stem cells from adult human fibroblasts by defined factors. *Cell* **131**: 861–872.
3. Yu, J. et al. (2007) Induced pluripotent stem cell lines derived from human somatic cells. *Science* **318**: 1917–1920.
4. Belmonte, J.C., Ellis, J., Hochedlinger, K., & Yamanaka, S. (2009) Induced pluripotent stem cells and reprogramming: seeing the science through the hype. *Nat. Rev. Genet.* **10**: 878–883.
5. Cowan, C.A., Atienza, J., Melton, D.A., & Eggan, K. (2005) Nuclear reprogramming of somatic cells after fusion with human embryonic stem cells. *Science* **309**: 1369–1373.
6. Stojkovic, M. et al. (2005) Derivation of a human blastocyst after heterologous nuclear transfer to donated oocytes. *Reprod. Biomed. Online.* **11**: 226–231.
7. Chan, E.M., Ratanasirintrawoot, S., Park, I-H., Manos, P.D., Loh, Y-H., Huo, H., Miller, J.D., Hartung, O., Rho, J., Ince, T.A., Daley, G.Q. & Schlaeger, T.M. (2009) Live cell imaging distinguishes bona fide human iPS cells from partially reprogrammed cells. Nat Biotechnol. Nov; 27(11): 1033–7.

Chapter 7

Transgene-Free Production of Pluripotent Stem Cells Using *piggyBac* Transposons

Knut Woltjen, Riikka Hämäläinen, Mark Kibschull, Maria Mileikovsky, and Andras Nagy

Abstract

Reprogramming of somatic cells to induced pluripotent stem cells (iPSCs) allows the derivation of personalized stem cells. Transposon transgenesis is a novel and viable alternative to viral transduction methods for the delivery of reprogramming factors (*Oct4*, *Sox2*, *Klf4*, *c-Myc*) to somatic cells. Since transposons can be introduced as naked DNA using common plasmid transfection protocols, they provide a safer alternative to viral methods. *piggyBac* transposons are host-factor independent and integrate stably into the target genome, yet benefit from the unique characteristic of seamless removal mediated by transient expression of *piggyBac* transposase. Thus, *piggyBac* transposition provides an effective means to generate human, transgene-free iPSCs. The protocol describes the production of iPSCs from human embryonic fibroblasts, delivering reprogramming factors via plasmid transfection and *piggyBac* transposition.

Key words: induced pluripotent stem cells, iPS cells, iPSC, reprogramming, transposon, transposition, *piggyBac*, transfection, transgene removal, transgene-free, factor-free, virus-free, nonviral, doxycycline regulated, dox

1. Introduction

Cellular reprogramming technology, epitomized by the ability to shift somatic cells from a state of limited potential to one of pluripotency that mimics that of embryonic stem cells (ESCs), has created a paradigm shift in how we conceive cell fate and determination (1). When applied to human biology, reprogramming somatic cells to induced pluripotent stem cells (iPSCs) (2) opens opportunities in the area of personalized medicine with new tools and perspectives in drug discovery, disease modeling, and cell therapy (3).

Philip H. Schwartz and Robin L. Wesselschmidt (eds.), *Human Pluripotent Stem Cells: Methods and Protocols*, Methods in Molecular Biology, vol. 767, DOI 10.1007/978-1-61779-201-4_7, © Springer Science+Business Media, LLC 2011

In order to overcome some of the issues associated with the use of viral vectors to deliver reprogramming factors to somatic cells, we have developed a method based on the *piggyBac* (PB) transposon/transposase system. Although retroviral systems may silence naturally (4) and lentiviruses can provide a high rate of transgene integration through repeated rounds of infection (5), the vectors themselves are permanently integrated at numerous locations throughout the genome raising the risk of insertional mutagenesis and the potential for re-activation of the reprogramming factors (6). Thus, viral-based reprogramming systems, including those that make use of Cre-mediated excision (5, 7), are not capable of retaining or fully reconstituting the genomic integrity of reprogrammed cells.

The ability to seamlessly remove PB transposon vectors (8) following stable genomic integration provides a novel aspect to iPSC production. Unique in the reprogramming field, the precise excision of the PB transposon from iPSC clones allows one to produce iPSCs that do not contain even a trace of the transgenic elements that mediated the reprogramming (9, 10). Coupled with the safety and simplicity of PB transposon vector preparation, the PB transposon system is potent alternative to viral methods for the production of iPSCs.

The PB reprogramming system described in this protocol (Fig. 1) makes use of doxycycline (dox) inducible *piggyBac* transposons (PB-TET) carrying the transcription factors *c-Myc*, *Klf4*, *Oct4*, and *Sox2* (11). The factors may be delivered independently on four separate transposons (PB-TET-*M*, -*K*, -*O*, and -*S*), or as one vector (PB-TET-*MKOS*), where all four are linked by 2A peptide sequences (12). Co-transfection with the *piggyBac* transposase (PBase) expression vector and a transposon constitutively expressing the dox-responsive rtTA transactivator (PB-CA-rtTA) results in stable genomic PB integration and reconstitutes dox-inducible regulation. Fully reprogrammed iPSC clones derived using this system may be treated with transient expression of PBase which excises the PB-TET transposons and permits the isolation of transgene-free iPSC subclones.

The theory and demonstration of this system has been described previously in Woltjen et al. (9). Reprogramming with the dox-regulated PB transposon system is similar to retroviral methods with regard to the kinetics of emergence and establishment of iPSCs (13). iPSC colonies are present 2–3 weeks after dox-induction and become fully reprogrammed and dox-independent over the next 1–3 weeks. After the iPSCs become dox-independent and the pluripotent state has stabilized, transient expression of PBase results in the mobilization and biased loss of the transposon elements. As part of the transposon removal process, the position of each transposon insertion is mapped, simplifying postremoval screening by conventional genomic PCR

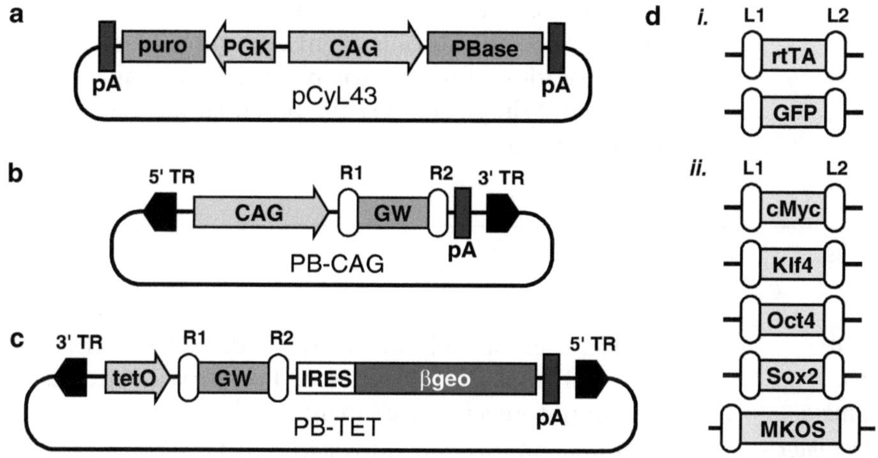

Fig. 1. *piggyBac* transposon vectors used to generate transgene-free human iPSCs. Three key vectors are utilized for the production of iPSCs from human fibroblasts. (**a**) pCyL43 is nonintegrating and constitutively expresses both *piggyBac* transposase (PBase) and puromycin resistance (puro) during transient transfection. Brief puromycin selection allows enrichment for PBase expressing cells. (**b**) PB-CAG is a Gateway destination vector that allows constitutive expression of a gene of interest inserted in place of the attR1/R2 flanked Gateway (GW) cassette. This vector can be used to express GFP as a reporter and is required to introduce the rtTA transactivator to mediate expression from tetO promoter-regulated transgenes. (**c**) PB-TET is used to achieve inducible expression of the reprogramming factors in the presence of the exogenous inducer, dox. The βgeo reporter allows indirect analysis of transgene induction (Fig. 2g) and could also be used to select for factor-containing cells by conferring resistance to G418 in the presence of dox. (**d**) Gateway entry vector cassettes used in the PB reprogramming protocol. Group *i* genes are expressed from PB-CAG, while the separate or linked reprogramming factors (c-Myc, Klf4, Oct4, Sox2) in Group *ii* are expressed from the PB-TET vector. Note that both vectors are designed to accept any series of Gateway compatible reprogramming gene candidates. *PGK* phosphoglycerate kinase promoter; *CAG* chicken β-actin/rabbit β-globin promoter; *puro* puromycin resistance gene; *PBase piggyBac* transposase gene; *pA* polyadenylation signal; *5′/3′ TR piggyBac* terminal repeats; *CAG* constitutive promoter; *R1/R2* Gateway attR1/R2 recombination sites; *GW* Gateway cloning cassette; *tetO* rtTA-responsive, dox-regulated inducible promoter; *IRES* internal ribosome entry site; β*geo* lacZ-neomycin fusion protein; *L1/L2* Gateway attL1/L2 recombination sites; *rtTA* reverse tetracycline transactivator; *GFP* green fluorescent protein; *MKOS* 2A-peptide linked reprogramming factors.

and sequencing. In order to produce completely transgene-free cell lines, it is necessary to remove both the rtTA- and PB-reprogramming transposons. The removal process is facilitated by using a *lacZ* reporter gene. Fully reprogrammed, factor-free, human iPSCs are nontransgenic and retain properties that are similar to human embryonic stem cells (hESCs).

The efficiency of reprogramming achieved with PB-transposons is near that of retroviral methods, although direct comparison between protocols is compromised by a prolific use of postinfection passage and culture expansion. PB reprogramming efficiency is directly related to the efficiency of transfection for any given cell type. Unlike viral vectors, where active entry into the cell is limited by viral titer and tropism, transposons are introduced into the cell by chemical or physical means as circular DNA. Thus, the permissiveness of a particular cell type to DNA transfection initially governs its ability to convert to iPSCs. For the purposes of this protocol, we will focus on the derivation of iPSCs from human embryonic fibroblasts (hEFs) using chemical transfection. The preparation of transposon plasmid DNA has obvious benefits of safety and simplicity over the potential biohazards of viral vector preparation.

2. Materials

2.1. piggyBac Transfection, Reprogramming, and iPSC Isolation

1. Mouse embryonic fibroblast feeder cells (MEFs). Prepared from 14.5 dpc DR4 (see Note 1) embryos, mitotically arrested with mitomycin-c treatment.
2. Human embryonic fibroblasts (hEFs).
3. *MEF culture medium*: DMEM (high glucose), 2 mM GlutaMAX, penicillin/streptomycin (50 U/mL, 50 μg/mL, Invitrogen), supplemented with 10% FBS (HyClone).
4. *hEF culture medium*: DMEM (high glucose), 0.1 mM MEM nonessential amino-acids, 2 mM GlutaMAX, 100 μM β-mercaptoethanol, penicillin/streptomycin (50 U/mL, 50 μg/mL, Invitrogen), supplemented with 15% Human Serum (Wisent) and 10 ng/mL bFGF (Peprotech).
5. *HEScGRO medium (Millipore/Merck)*: A fully defined, serum-free and animal-component-free medium used during the reprogramming phase.
6. *hES medium*: KO-DMEM, 0.1 mM MEM nonessential amino-acids, 2 mM Glutamax, 100 μM β-mercaptoethanol, and penicillin/streptomycin (50 U/mL/50 μg/mL, Invitrogen), supplemented with 20% KOSR (Invitrogen) and 10 ng/mL bFGF (Peprotech). This medium is used during iPSC maintenance culture.
7. Dulbecco's phosphate buffered saline (DPBS), Mg^{2+} and Ca^{2+} free.
8. TrypLE Select (Invitrogen).
9. Gelatin 0.1% (Millipore).
10. Trypan blue 0.4%.
11. Sterile ddH_2O or OptiMEM (Invitrogen).
12. FugeneHD transfection reagent (Roche).
13. Plasmid DNA Miniprep/Maxiprep kit (Qiagen) (see Note 2).
14. *piggyBac* PB-TET inducible reprogramming transposon plasmids (Fig. 1), transfection grade circular DNA, 500 ng/μL.
15. PB-CA-rtTA transactivator transposon plasmid, transfection grade circular DNA, 500 ng/μL.

16. pCyL43 PBase (transposase) plasmid, aka. pCyLo43 or pCyL43 (11), transfection grade circular DNA, 500 ng/μL (available through: http://www.sanger.ac.uk/technology/clonerequests).
17. Fluorescent reporter plasmid construct (optional). pCAG-eGFP or PB-CA-eGFP (see Note 3).
18. Doxycycline (dox), 1.5 mg/mL in water (1,000× stock), filter sterilized and protected from light (Sigma) (see Note 4).
19. 30 Gauge needles or pulled glass pipettes.
20. Hemocytometer or cell counter.
21. 15 or 50-mL sterile conical centrifuge tubes.
22. 10-cm Culture dishes.
23. 6-Well culture dishes.

2.2. Screening for Transgene Regulation, Pluripotency, and Transposon Integration Site

1. *Fixation solution*: 0.2% Glutaraldehyde (Sigma) in DPBS.
2. *lacZ Staining Solution*: 1 mg/mL X-Gal in *N,N*-dimethylformamide, 5 mM $K_4Fe(CN)_6$, 5 mM $K_3Fe(CN)_6$ (Invitrogen); protect from light, store in light-tight vessel. May be prepared in bulk and frozen. Prepare X-Gal stock solution in DMSO.
3. Alkaline Phosphatase Substrate Kit I (Vector Labs Inc.).
4. *Antibodies*: SSEA-3, SSEA-4, Tra-1-60, Tra-1-81 (Millipore/Merck), Nanog (ReproCell Inc.).
5. Quantitect Reverse Transcription Kit (Qiagen).
6. *Cell lysis buffer*: 100 mM Tris–HCl (pH 8.5), 5 mM EDTA, 0.2% SDS, 200 mM NaCl, 100 μg/mL ProteinaseK. Used for genomic DNA extraction.
7. 100% Ethanol.
8. 70% Ethanol.
9. Sterile ddH_2O.
10. *Southern blotting probes*:
    (a) Obtained by digestion: rtTA as a BsrGI fragment from PB-CA-rtTA.
    (b) PCR amplification:
        (i) neo-f (neomycin coding sequence with 5′ cloning site) CCGGAATTCATGGGATCGATGATTG;
        (ii) neo-r (neomycin coding sequence with 3′ cloning site) CCGCTCGAGTCAGAAGAACTCGTC.
11. Splinkerette primers (Operon):
    (a) Splinkerette Oligos: HMSpAa (top strand primer) CGAAGAGTAACCGTTGCTAGGAGAGACCGTGGC

TGAATGAGACTGGTGTCGACACTAGTGG; HMSpBb (lower strand primer/hairpin) gatcCCACTAGTGTCGACACCAGTCTCTAATTTTTTTTTCAAAAAAA.

(b) Primary PCR primers: PB-L-Sp1 (PB 5′TR sequence) GCGTGCTTGTCAATGCGGTAAGTGTCACTG; PB-R-Sp1 (PB 3′TR sequence) CCTCGATATACAGACCGATAAAACACATGC.

(c) Secondary nested PCR primers: PB-L-Sp2 (PB 5′TR sequence) GAGAGAGCAATATTTCAAGAATGCATGCGT; PB-R-Sp2 (PB 3′TR sequence) ACGCATGATTATCTTTAACGTACGTCACAA.

(d) Genomic amplicon sequencing primers: PB-L-Sq (PB 5′TR sequence) TCAAGAATGCATGCGTCAAT; PB-R-Sq (PB 3′TR sequence) CGTACGTCACAATATGATTATCTTTC

12. Sau3AI restriction enzyme (New England Biolabs).
13. Sau3AI restriction buffer (New England Biolabs).
14. Restriction buffer #2 (New England Biolabs).
15. 100× BSA, diluted to 10× working stock with sterile ddH_2O (New England Biolabs).
16. Qiagen Taq polymerase (Qiagen).
17. Q-solution (Qiagen).
18. dNTPs (Fermentas).
19. PCR buffer (containing Mg^{2+}) (Qiagen).
20. PCR thermocycler (Eppendorf used herein, others also suitable).
21. Spin Purification Columns (QIAquick PCR purification kit, Qiagen).
22. Genomic locus confirmation PCR primers (Operon):

    (a) Locus-specific primer sequence (see Note 5).

2.3. PB Transposon Removal

1. ROCK inhibitor Y27632 (Sigma). Dissolve in water at a concentration of 5 mM and filter sterilize (500× stock). Use at 10 μM.
2. Puromycin (Sigma). Dissolve in water at 10 mg/mL and filter sterilize (1,000× stock). Use at 0.5–1.0 μg/mL.
3. Southern blotting probes (see Subheading 2.2, item 10).
4. Genomic PCR screening oligos (Operon):

    (a) βgeo transgene (PB-TET vectors):

        (i) lacZ-f (*lacZ* sequence) ACGGTTTCCATATGGGGATT.

(ii) neo-r (neomycin sequence) AGTGACAACGTCG AGCACAG.

(b) PB-CA-rtTA:

(i) rtTA-f (pCAG promoter sequence) GCAACGTGCT GGTTATTGTG.

(ii) rtTA-r (rtTA sequence) AGAGCACAGCGGA ATGACTT.

3. Methods

The following protocol describes the production of iPSCs from hEFs, delivering reprogramming factors via plasmid transfection, and *piggyBac* transposition. The theory and demonstration of the system has been described previously in Woltjen et al. (9). Reprogramming with the dox-regulated PB transposon system is similar to retroviral methods with regard to the kinetics of emergence and the establishment of iPSC clones (13).

3.1. piggyBac Transfection, Reprogramming, and iPSC Isolation

1. Apply gelatin (0.1%) to the surface of a 6-well dish, aspirate, and allow dish to dry uncovered in the biosafety cabinet prior to hEF seeding.
2. Defrost low passage hEFs (see Note 6) in an appropriately sized dish in hEF medium. One 60 mm dish (total surface area ~20 cm^2) can yield about 1–1.5 × 10^6 hEFs, or enough hEFs to seed two 6-well dishes for reprogramming.
3. When the hEFs reach ~80–90% confluency, harvest them using enzymatic dissociation with TrypLE (5 min at 37°C), collect by centrifugation at 400 × *g* for 3.5 min, resuspend in fresh medium, and determine the total yield of viable cells using a hemocytometer and Trypan blue exclusion (50:50 mixture of Trypan blue and cell suspension), or an automated cell counter.
4. Seed hEFs at a density of 6.0–7.0 × 10^4 cells per 10 cm^2 surface area (i.e., one well of a 6-well dish) in 2 mL hEF medium for each induction condition. Culture overnight at 37°C, 5% CO_2 to allow the cells to adhere, and re-enter growth phase.
5. The next morning, change to fresh hEF medium (2 mL/well).
6. Set up FugeneHD transfection complexes (see Note 7). The amounts indicated below are optimized for a single well of a 6-well dish and may be scaled up according to the manufacturer's specifications:
   (a) Mix 500–800 ng of PB-TET-MKOS transposon plasmid (or 250 ng each of the four independent M, K, O, S vectors, for a total of 1 μg plasmid DNA) with 500 ng of

PB-CA-rtTA transposon plasmid and 500 ng pCyL43 plasmid (see Notes 7 and 8).

(b) 500 ng pCAG-eGFP plasmid (transient plasmid or PB-based) or an equivalent transient fluorescent reporter may be added to simultaneously determine relative transfection efficiencies (see Note 3).

(c) Dilute DNA to 100 μL with sterile ddH_2O or DMEM/OptiMEM (see Note 9). Mix with gentle vortexing.

(d) Add 6–8 μL of FugeneHD (FugeneHD:DNA ratio of 6 μL:1.5 μg or 8 μL:2 μg). Mix with gentle vortexing. Incubate at room temperature for 15–20 min.

(e) Following the incubation period, add FugeneHD:DNA transfection complexes directly to the hEF medium dropwise, and distribute by gentle swirling.

7. Allow the hEFs to grow undisturbed at 37°C in 5% CO_2 for 24 h, then add dox to a final concentration of 1.5 μg/mL (see Note 10).

8. Change the medium completely 48 h posttransfection from hEF medium to HEScGRO medium containing 1.5 μg/mL dox. Feed every second day with fresh dox-containing HEScGRO medium (see Note 11) and monitor the wells for the formation of foci of cells that have characteristic morphology of human ESCs. These foci typically become visible late in the second week (after ~10 days in culture) (see Notes 12 and 13).

9. Colonies suitable for picking should form as early as days 14–21; however, new clones will usually continue to arise through 21–28 days. Representative colony types observed in a standard induction are shown in Fig. 2a–f.

Fig. 2. Identification and screening of primary human PB-iPSC clones. Accurate identification of primary colony morphology and transgene regulation increases success rate. Like other integrative methods, a variety of colony types are to be expected. This variation is somewhat alleviated through the use of 2A-peptide linked transgenes (e.g., MKOS), presumably through the normalization of transgene stoichiometry. (**a**) A dense field of fibroblasts forms during extended culture. (**b**) Early colony formation may be detected at high magnification. Reprogrammed cells have distinct nuclei, a high nucleus:cytoplasm ratio, and display tight compact growth as compared to the surrounding fibroblasts. (**c**) Flat patches of adherent cells with epithelial properties may form, which do not give rise to densely packed iPSC colonies. (**d**) Primary PB-iPSC clones sometimes arise as a grape-like cluster of colonies, which may merge borders during the standard reprogramming period. These clusters are most likely the cause of posttransfection fibroblast migration and as they are clonal, may be isolated as such. (**e**) Compact, highly three-dimensional colonies that may arise are difficult to passage mechanically or enzymatically. These colony types are occasionally the result of prolonged dox exposure or failed reprogramming. (**f**) Typical hES/iPS cell colony morphology may arise spontaneously in primary colonies or will follow with passage. (**g**) Dox-regulated expression of the reprogramming factors expressed from PB-TET is revealed by staining for *lacZ* activity. A positive colorimetric reaction results in a strong blue stain (shown here in *black* and *white*). PB-iPSC lines that fail to stain blue in the absence of dox, yet maintain iPSC morphology and growth properties are desirable. A similar staining procedure may be used to screen for factor removal following the second round of PBase treatment. Scale bars shown are all 200 μm in length.

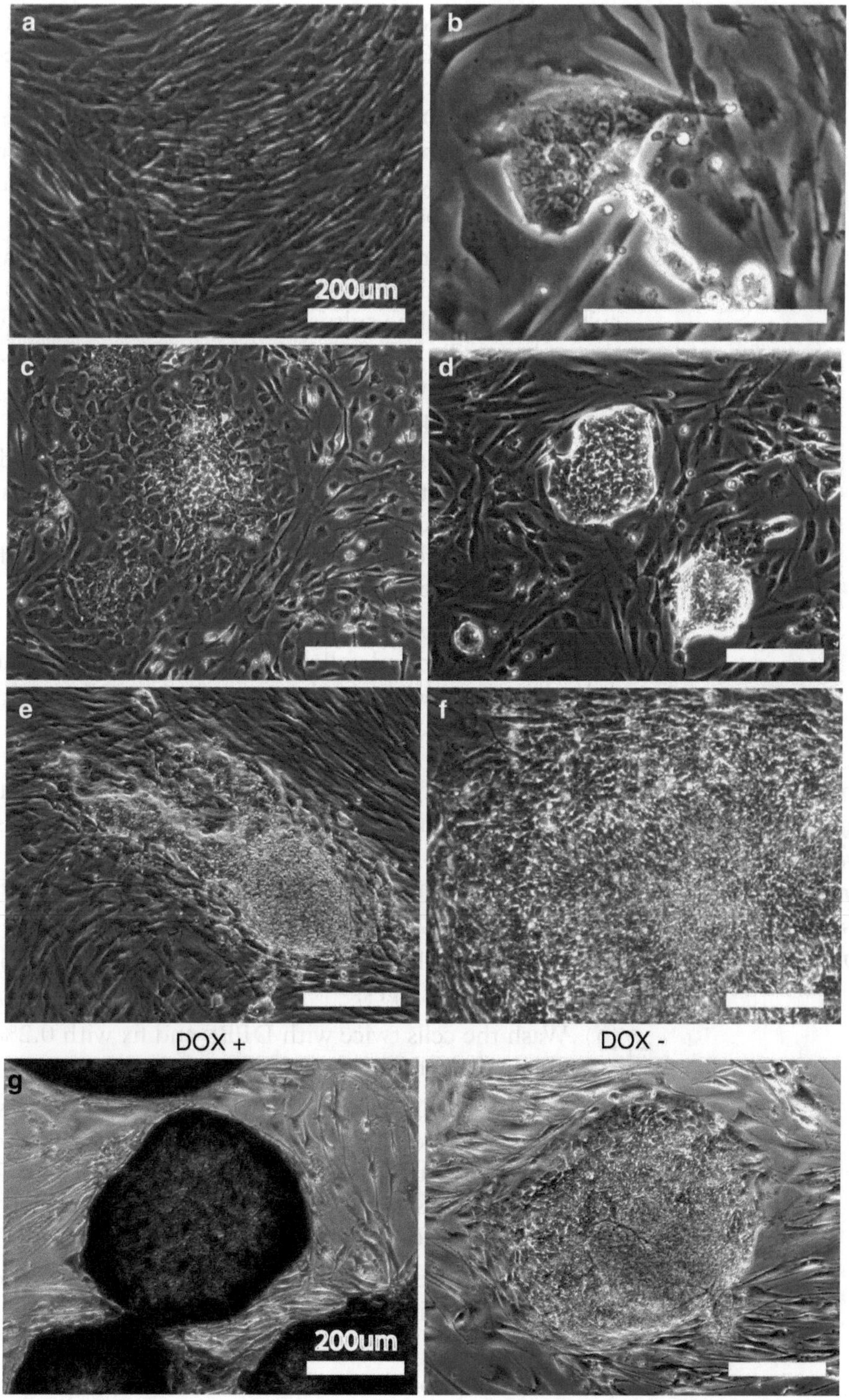
a
b
200um
c
d
e
f
DOX +
DOX -
g
200um

10. Pick colonies mechanically, dividing large colonies into small segments of ~100 cells with a pulled and fractured glass pipette, or 30 gauge needle. 24 h after picking, shift culture from HEScGRO medium to KOSR-based hESC medium (KO-DMEM supplemented with 20% KOSR and 10 ng/mL bFGF). Cells may be later adapted to TrypLE enzymatic passage (3 min at room temperature) (see Note 14).
11. Screen for dox-independence. After clones have established and grown sufficiently, split the culture 1:2 to set up replica cultures. One will be cultured in dox-containing medium and the other will be cultured in dox-free medium. Dox-independent growth parallels retroviral silencing and is as an indicator of complete reprogramming.
12. PB-iPSC clones should only be removed permanently from dox treatment if the iPSC cultures in the replica plates appear stable (with respect to growth rate, morphology, and lack of spontaneous differentiation) in dox-free conditions for 2–3 serial passages (see Note 15).
13. Additional expansion and cryopreservation of PB-iPSC clones should be performed following established hESC methods, culturing on mouse embryonic feeder cells with hESC medium (see Chapter 8).

3.2. Screening for Transgene Regulation, Pluripotency, and Transposon Integration Site

1. *lacZ* staining is applied as a preliminary screen to determine which PB-iPSC lines display a lack of transgene expression in the absence of dox (Fig. 2g) and are transgene factor-independent. When the colonies are of reasonable size, stain the plates for β-galactosidase (*lacZ*) activity as follows:
   (i) Grow the cells in the presence and absence of dox for at least two passages.
   (ii) Wash the cells twice with DPBS and fix with 0.2% glutaraldehyde in DPBS for 10 min at room temperature.
   (iii) Wash the cells three times with DPBS, add *lacZ* staining solution, and allow the enzymatic reaction and color development to occur in the dark at 37°C for 2–16 h. Choose lines that test negative for *lacZ* activity in the absence of dox for further studies.
2. Verification of reprogramming should be performed using established pluripotency assays (see Chapters 15, 17, 19 and 20)
   (i) Alkaline phosphatase can be used as a preliminary pluripotency screen. It is easy, low cost, and can be performed on unfixed iPSCs. However, it should not be considered a determinate marker of reprogramming.
   (ii) The cell surface markers SSEA-3, SSEA-4, Tra-1-60, Tra-1-81 (14), and the nuclear factor Nanog are excellent

pluripotency indicators, and can be detected by routine immunohistochemistry techniques (see Chapter 15).

(iii) Reverse transcriptase PCR of PB-iPSC RNA will indicate characteristic expression changes in genes such as *Dnmt3B* and *Rex1* (*Zfp42*), among others (15).

(iv) Considering the final application of the PB-iPSCs, it is likely important to exclude chromosomal abnormalities with karyotype analysis using either simple chromosome counting or more detailed methods such as G-banding or comparative genomic hybridization (CGH).

3. Functional validation of pluripotency, in vitro differentiation, and in vivo teratoma assay must be applied to each iPSC line in order to verify the pluripotency of the line.
4. Determine the number of transposon integration sites using genomic Southern blotting. Isolate genomic DNA using standard protocols. Suggested surface area minima for each application: Southern blotting (10–20 μg, 1×6-well), splinkerette (3–6 μg, 1×12-well), genomic PCR (1–2 μg, 1×48/24-well). MEF feeder depletion is not necessary in most cases, but may be performed for pure sample preparation.
5. In Southern blotting, PB-TET-based vectors (individual or 2A-linked factors) may be localized using a neomycin gene probe, amplified by PCR, and labeled with standard radioactive (^{32}P) or nonradioactive (DIG) methods. The PB-CA-rtTA transactivator-bearing transposons may be localized with an rtTA probe (restriction fragment) (see Note 16).
6. Integration sites are precisely located in the genome using a modified splinkerette protocol (9, 15).

(a) Digest genomic DNA with Sau3AI to generate average fragment sizes of 300 bp from a reaction containing: 10 U Sau3AI enzyme, 3 μL 10× restriction buffer, and 3 μL 10× BSA, bringing the volume up to 30 μL with sterile ddH_2O. Digest at 37°C for 2–16 h.

(b) Inactivate the restriction enzyme at 65°C for 20 min.

(c) Anneal 200 pmol of each of the splinkerette oligos (HMSpAa and HMSpBb) with 5 μL of NEB Buffer2 in a total volume of 50 μL (ddH_2O). Heat the mixture to 98°C for 2 min and allow it to cool slowly to less than 30°C (see Note 17).

(d) Ligate the HMSpAa/Bb splinkerette oligo mixture to the restricted genomic DNA in the following reaction (oligo:DNA ratio = 10:1 moles): 300 ng digested DNA, 20 pmol annealed splinkerette oligos, and 2 U T4 ligase, bringing the final ligation volume of 20 μL with ddH_2O. Ligate overnight at 16°C.

(e) Purify the ligation mixture through a column (Qiagen QIAquick PCR purification kit) to remove all salt and unannealed splinkerette oligos. Elute the DNA with a minimum volume of sterile ddH_2O.

(f) Perform a primary PCR on the purified ligation mix using the PBSp1 and HMSp1 primer set: 1 U Qiagen Taq polymerase, 5 μL 5× Q-solution, 12.5 pmol PBLSp1/PBRSp1, 12.5 pmol HMSp1, 250 μM dNTPs, 2.5 μL 10× PCR buffer, and 5 μL purified ligation, bringing the total volume to 25 μL with ddH_2O. Amplify the target fragments at: (94°C for 1 min, 68°C for 30 s) × 2 cycles, (94°C for 15 s, 65°C for 30 s, 72°C for 2 min) × 25 cycles.

(g) Perform a secondary nested PCR on the primary PCR mix using the PBSp2 and HMSp2 primer set as follows: 2 U Qiagen Taq polymerase, 10 μL 5× Q-solution, 25 pmol PBLSp2/PBRSp2, 25 pmol HMSp2, 250 μM dNTPs, 5 μL 10× PCR buffer, and 1 μL primary PCR, bringing the reaction volume to 50 μL with ddH_2O. Amplify the target fragments at: (94°C for 15 s, 60°C for 30 s, 72°C for 1 min) × 35 cycles.

(h) Gel-purify the resulting products (see Note 18). Fragments may be sequenced directly with the PBLSq or PBRSq primers, or TA-cloned and sequenced with standard vector primers (M13Froward/Reverse or T3/T7).

(i) Locate the PB sequence-tag in the human genome using human genome BLAST (http://www.ncbi.nlm.nih.gov/genome/seq/BlastGen/BlastGen.cgi?taxid=9606).

7. Design genomic primers to flank the PB-insertion site that may be utilized for locus-specific PCR either in combination with PBLSq or PBRSq to confirm integration sites, or postexcision locus amplification and sequencing (see Note 19).

3.3. PB Transposon Removal

1. Harvest iPSCs with TrypLE and plate on growth-arrested DR4 MEFs in hESC (KOSR) culture medium at a density of 2.5×10^5 cells per well on a 6-well plate.
2. Transfect the cells while still in suspension with 1 μg of pCyL43 plasmid and 4 μL of FugeneHD, establishing complexes as described in Subheading 3.1, step 6.
3. Allow the cells to attach and grow undisturbed in the medium containing the transfection complexes. After 48 h, initiate transient puromycin selection (0.7 μg/mL) to enrich for transfected cells coincidentally expressing PBase.
4. After 2 days of growth in puromycin containing medium, withdraw puromycin selection and resume daily feeding in hESC culture medium. Allow the surviving cells to recover and form colonies for 5–6 days.

5. Harvest PB-iPSCs with TrypLE and plate at clonal density (see Note 20). To enhance cell survival under these conditions, treat the cells with 10 μM ROCK inhibitor for 1 h prior to harvest. Dissociate cells carefully, making sure to achieve a single-cell suspension. Count viable cells, and plate at a density of 500–5,000 cells per 10 cm^2 surface area (see Note 21) on growth-arrested MEFs in hESC medium supplemented with 10 μM ROCK inhibitor. Change the medium the next day and withdraw the ROCK inhibitor.
6. Observe cultures and feed daily with hESC medium.
7. Allow the PBase-exposed iPSCs to grow for 10–12 days and wait for sizeable colonies to form. Pick the clones mechanically and transfer to fresh growth-arrested MEFs on 24-well plates (see Note 22).
8. After a few days of growth, harvest the cells with TrypLE and plate them on fresh feeders at a 1:1 or 1:2 ratio, depending on how well the cells have established (see Note 14).
9. Generate replica plates with and without dox. Screen the dox-treated culture via *lacZ* staining (Subheading 3.2, step 1) to determine which clones have lost either the PB-CA-rtTA or PB-TET-MKOS (see Note 23), or both (see Note 24).
10. Expand the cells further and extract DNA using standard protocols as described above (Subheading 3.2, step 3).
11. Confirm PB excision using molecular criteria such as Southern analysis or genomic PCR – for example, lacZ-f/neo-r to amplify all the M, K, O, S, and MKOS transposons, or rtTA-f, rtTA-r to amplify PB-CA-rtTA transposons – to ensure transgene loss without re-insertion, and locus-specific PCR with amplicon sequencing to determine the fidelity of removal required before cells may be considered modification-free.
12. Re-confirm the karyotype and pluripotency of the transgene-free iPSC lines by immunostaining and RT-PCR for ESC-markers and finally by teratoma assays.

4. Notes

1. MEFs should be resistant to puromycin. DR4 MEFs are prepared from DR4 mice, which are resistant to G418, 6-thioguanine, puromycin, and hygromycin. Tg(DR4)1Jae/J (Jackson Laboratory Stock number 003208).
2. Endotoxin-free preparation of plasmid DNA has not been found to be critical for reprogramming applications, although high-quality DNA (as determined minimally by $OD_{260/280}$) is recommended.

3. Transient transfection of a fluorescent reporter construct permits rapid determination of the transfection efficiency under any given condition without reporter gene integration into the genome of the resulting iPSC lines. The inclusion of a PB-based reporter construct has a high probability of labeling the resulting PB-iPSC clones because of the high levels of co-insertion during co-transfection. We have observed that about 60% of the iPSC lines co-transfected with PB-MKOS and PB-GFP are positive for GFP. This may be useful if the cells will be used in vivo (e.g., mouse transplantation studies).
4. Although dox is stable in stock solutions and culture medium at 4°C for at least 2 weeks, it is important to protect solutions by wrapping vessels in tinfoil and minimizing exposure to light as dox is light-sensitive.
5. The primer should be placed outside the splinkerette amplicon end to be certain of correct positioning, and paired with the appropriate PB TR internal primer: PB-L-Sq (PB 5′TR sequence), PB-R-Sq (PB 3′TR sequence).
6. hEF reprogramming efficiencies decrease significantly with subsequent passages. It is suggested that fibroblasts are prepared with minimal expansion and used for reprogramming within five passages following isolation from frozen stock or grown fresh from tissue.
7. Various transfection methodologies and reagents may be applied to deliver the transposon and transposase plasmid DNA to the target cells. We suggest that the delivery method be chosen and optimized appropriately for the particular cell type in question. Here, FugeneHD has been selected due to its broad-range transfection capability and low cytotoxicity over periods of long-term incubation. Human fibroblasts are amenable to lipofection, electroporation, and nucleofection. Other cell types may require different delivery methods.
8. If the intended goal is to remove all factors following reprogramming, the PB-TET-MKOS plasmid is recommended to minimize copy number. If transgene removal is not necessary, or various combinations of removed and remaining factors are desired, then the factors should be introduced individually in separate transposons.
9. No major difference has been noted in transfection efficiencies when either ddH_2O or DMEM/OptiMEM is used as a diluent. However, if DMEM is used, no supplements (antibiotics, FBS, etc.) should be added.
10. Due to the low toxicity (but see Note 7), it is not necessary to remove the FugeneHD complexes during the 48 h incubation period.

11. Some human fibroblast lines respond to culture in HEScGRO medium with a severe reduction in their proliferation. Although this response can be exploited to control the overall fibroblast density in an induction culture later in the induction process, early suppression of cell proliferation can have a negative impact on PB-iPSC colony formation. Considering this, it may be beneficial to shift from hEF medium to HEScGRO gradually (using stepwise mixtures of 1:0, 1:3, 1:1, 3:1, 0:1 hEF:HEScGRO) over the period of the first week.
12. If the fibroblast density becomes too high during induction, iPSC colony formation may be hindered. Periodic selection with G418 (for a 1–2 days at a time) can be used to suppress nontransfected fibroblast growth and reduce the fibroblast number. However, do not completely eliminate nonreprogrammed parental fibroblasts, since they act as feeders during co-culture. This approach has been applied with varying success and depends upon the particular growth properties of the fibroblast cell population.
13. Unlike most published protocols, induction with PB as described herein is performed *without* passage of the initial induction plates, ensuring a higher level of clonality and eliminating redundant clone isolation. To increase colony frequency, it is possible to expand the reprogramming cell culture onto a mouse fibroblast feeder layer following the initial 5–7 days of induction; however, clonality may be lost.
14. It can be beneficial to apply ROCK inhibitor during the first enzymatic passage of the cells, although at this point it is not necessary to dissociate the colonies beyond small clumps.
15. In some instances, reduction of dox concentration or complete removal of dox from PB-iPSC culture actually improves growth and morphology.
16. If the factors are to be removed in their entirety, lines displaying the least number of integration events are the most valuable.
17. Slow cooling suitable for annealing is achieved by simply switching off the boiling temperature block and monitoring the temperature with a thermometer.
18. As transposon copy number affects the number of PCR products amplified for a given PB-iPSC line, gel purification may be necessary to improve sequencing resolution.
19. Apply nested PCR during splinkerette to ensure specific fragment amplification, or during sequencing to avoid highly repetitive regions which impede analysis or primer design.
20. Following transient PBase expression, a pool of three different cell types exist – those in which: (1) no mobilization has

occurred, and the transgenes remain in their original sites; (2) the transgenes have jumped from one genomic site into another; and (3) the transgenes were excised from their original site but not inserted elsewhere, resulting in restoration of the genome.

21. The optimal cell number required to obtain a sufficient number of isolated colonies varies between different iPSC lines and is dependent on the growth and survival of the cells. In general, we have used from 3 to 30×10^3 cells on a 10-cm dish (~90 cm^2). For best results, the optimal cell density may be determined separately beforehand for each line. Alternatively, plate a tenfold dilution series over 3 orders of magnitude and choose the most appropriate dilution for picking.
22. As the frequency of excision varies considerably depending on the original PB integration site (observed to be ~5–15% in human), pick 48–96 subclones.
23. Loss of either the rtTA or MKOS transgenes will lead to a lack of *lacZ* reporter expression in the presence of dox.
24. Depending upon the intended application, it may be useful to maintain the PB-CA-rtTA transgene in the cells. The rtTA transactivator protein has no known side effects and may be used later to regulate a second set of transgenic constructs introduced into factor-free iPSC.

Acknowledgments

The authors would like to thank Ian Rogers for discussion and critical reading of the manuscript. This work was supported by the Canadian Stem Cell Network (K.W. and A.N.).

References

1. Nagy, A., and Nagy, K. (2010) The mysteries of induced pluripotency: where will they lead?, *Nature methods* 7, 22–24.
2. Takahashi, K., Tanabe, K., Ohnuki, M., Narita, M., Ichisaka, T., Tomoda, K., and Yamanaka, S. (2007) Induction of pluripotent stem cells from adult human fibroblasts by defined factors, *Cell* **131**, 861–872.
3. Lee, G., and Studer, L. (2010) Induced pluripotent stem cell technology for the study of human disease, *Nature methods* 7, 25–27.
4. Hotta, A., and Ellis, J. (2008) Retroviral vector silencing during iPS cell induction: an epigenetic beacon that signals distinct pluripotent states, *J Cell Biochem* **105**, 940–948.
5. Soldner, F., Hockemeyer, D., Beard, C., Gao, Q., Bell, G. W., Cook, E. G., Hargus, G., Blak, A., Cooper, O., Mitalipova, M., Isacson, O., and Jaenisch, R. (2009) Parkinson's disease patient-derived induced pluripotent stem cells free of viral reprogramming factors, *Cell* **136**, 964–977.
6. Markoulaki, S., Hanna, J., Beard, C., Carey, B. W., Cheng, A. W., Lengner, C. J., Dausman, J. A., Fu, D., Gao, Q., Wu, S., Cassady, J. P., and Jaenisch, R. (2009) Transgenic mice with defined combinations of drug-inducible reprogramming factors, *Nat Biotechnol* **27**, 169–171.
7. Sommer, C., Gianotti Sommer, A., Longmire, T., Christodoulou, C., Thomas, D., Gostissa,

M., Alt, F., Murphy, G., Kotton, D., and Mostoslavsky, G. (2009) Excision of Reprogramming Transgenes Improves the Differentiation Potential of iPS Cells Generated with a Single Excisable Vector, *Stem Cells* **28**, 64–74.

8. Ivics, Z., Li, M. A., Mátés, L., Boeke, J. D., Nagy, A., Bradley, A., and Izsvák, Z. (2009) Transposon-mediated genome manipulation in vertebrates, *Nat Meth* **6**, 415–422.
9. Woltjen, K., Michael, I. P., Mohseni, P., Desai, R., Mileikovsky, M., Hämäläinen, R., Cowling, R., Wang, W., Liu, P., Gertsenstein, M., Kaji, K., Sung, H.-K., and Nagy, A. (2009) piggyBac transposition reprograms fibroblasts to induced pluripotent stem cells, *Nature* **458**, 766–770.
10. Yusa, K., Rad, R., Takeda, J., and Bradley, A. (2009) Generation of transgene-free induced pluripotent mouse stem cells by the piggyBac transposon, *Nat Meth* **6**, 363–369.
11. Takahashi, K., and Yamanaka, S. (2006) Induction of pluripotent stem cells from mouse embryonic and adult fibroblast cultures by defined factors, *Cell* **126**, 663–676.
12. Szymczak, A. L., Workman, C. J., Wang, Y., Vignali, K. M., Dilioglou, S., Vanin, E. F., and Vignali, D. A. (2004) Correction of multi-gene deficiency in vivo using a single "self-cleaving" 2A peptide-based retroviral vector, *Nat Biotechnol* **22**, 589–594.
13. Hotta, A., Cheung, A. Y. L., Farra, N., Garcha, K., Chang, W. Y., Pasceri, P., Stanford, W. L., and Ellis, J. (2009) EOS lentiviral vector selection system for human induced pluripotent stem cells, *Nature protocols* **4**, 1828–1844.
14. Chan, E., Ratanasirintrawoot, S., Park, I., Manos, P., Loh, Y., Huo, H., Miller, J., Hartung, O., Rho, J., Ince, T., Daley, G., and Schlaeger, T. (2009) Live cell imaging distinguishes bona fide human iPS cells from partially reprogrammed cells, *Nat Biotechnol* **27**, 1033–1037.
15. Horn, C., Hansen, J., Schnütgen, F., Seisenberger, C., Floss, T., Irgang, M., De-Zolt, S., Wurst, W., von Melchner, H., and Noppinger, P. R. (2007) Splinkerette PCR for more efficient characterization of gene trap events, *Nature genetics* **39**, 933–934.

Part III

Growth, Maintenance, and Expansion

Chapter 8

Traditional Human Embryonic Stem Cell Culture

Philip H. Schwartz, David J. Brick, Hubert E. Nethercott, and Alexander E. Stover

Abstract

Culturing human embryonic stem cells (hESCs) requires a significant commitment of time and resources. It takes weeks to establish a culture, and the cultures require daily attention. Once hESC cultures are established, they can, with skill and the methods described, be kept in continuous culture for many years.

hESC lines were originally derived using very similar culture medium and conditions as those developed for the derivation and culture of mouse ESC lines. However, these methods were suboptimal for hESCs and have evolved considerably in the years since the first hESC lines were derived. Compared with mouse ESCs, hESCs are very difficult to culture – they grow slowly, and most importantly, since we have no equivalent assays for germline competence, we cannot assume that the cells that we have in our culture dishes are either stable or pluripotent. This makes it far more critical to assay the cells frequently using the characterization methods, such as karyotyping, immunocytochemistry, gene expression analysis, and flow cytometry, provided in this manual.

Key words: ESC, iPSC, PSC hESC culture, feeder cells, mechanical passaging, human pluripotent stem cells

1. Introduction

Following fertilization, the cells of the human embryo undergo several rounds of cell division and rearrange to form a hollow sphere of cells termed the blastocyst. The cells of the blastocyst segregate into an outer layer called the trophectoderm and an inner cell mass (ICM). The trophectoderm becomes the fetal contribution to the placenta while the cells of the ICM give rise to the embryo proper. The isolation and culture of a unique population of cells, embryonic stem cells (hESCs), from the ICM of human blastocysts was first accomplished in 1994 by Ariff Bongso's group (1). In 1998, James Thomson

Philip H. Schwartz and Robin L. Wesselschmidt (eds.), *Human Pluripotent Stem Cells: Methods and Protocols*, Methods in Molecular Biology, vol. 767, DOI 10.1007/978-1-61779-201-4_8, © Springer Science+Business Media, LLC 2011

and colleagues described the derivation of the first hESC lines that were propagateable and cryopreservable cell cultures (2). Although hESCs are typically produced from the ICM, they can also be produced from earlier stage embryos, including the morula and cleavage stages.

hESCs display two unique properties: the ability to self-renew indefinitely and the potential to give rise to all cell types of the human body (2). Thus, hESCs are pluripotent stem cells (PSCs) because they are able to form cell lineages of all of the three germ layers – endoderm, ectoderm, and mesoderm. These features of hESCs have made them extremely attractive tools not only for the study of development and cancer but also for their potential use in regenerative medicine strategies to repair or replace damaged tissues.

hESCs in culture (under the conditions described in this chapter) grow as tightly compact colonies of cells with high nucleus-to-cytoplasm ratios. On a molecular level, hESCs in culture express characteristic and specific (1) surface antigens such as the stage-specific embryonic antigen SSEA-4 and the teratocarcinoma recognition antigens TRA-1-60 and TRA-1-81 and (2) pluripotency-specific transcription factors such as Oct-4 and Nanog (3).

In this chapter, we describe basic culture methods for hESC (PSC) culture. These methods may also be used for the propagation of PSCs produced by cellular reprogramming (induced PSCs or iPSCs as described in Chapters 5–7). While these methods have been used successfully for years by many laboratories, and many publications cite them, it should be noted that more modern methods exist, and these are covered in Chapters 9 and 10. In addition, we also cover cryopreservation methods. Importantly, since PSC cultures are often kept in continuous culture for months, even years, it is critical that genetic and developmental drift (4) be monitored in the cultures (see Notes 1 and **2**). The best way to control for drift is to generate a large bank of frozen cells as soon as possible after the cultures are first expanded as described in Chapter 3. The importance of this cannot be overemphasized – the value of discoveries based on PSCs depends on the reproducibility of results.

2. Materials

2.1. Reagents

1. Dulbecco's modified Eagle medium: Ham's F12 (DMEM/F12 + GlutaMAX. Invitrogen, #10565).
2. Dulbecco's phosphate buffered saline (DPBS^{--}) without Mg^{2+} and Ca^{2+}.

3. Dulbecco's phosphate buffered saline (DPBS++) with Mg^{2+} and Ca^{2+}.
4. KnockOut™ Serum Replacement (KSR) (Invitrogen, #108280-028).
5. 2-Mercaptoethanol, 55 mM (1,000×) in PBS (such as Invitrogen, #21985-023).
6. GlutaMax (100×) (Invitrogen, #35050).
7. Human basic fibroblast growth factor (bFGF) (Stemgent, #03-0002).
8. MEM-Nonessential amino acids (NEAA) 100× (10 mM) (Hyclone, #SH30238.01).
9. Hybri-Max dimethyl sulfoxide (Sigma-Aldrich, #D2650).
10. Water for embryo transfer (Sigma, #W1503, see Note 3).
11. Fetal bovine serum (FBS) (Hyclone, #SH30070.03).
12. Pen-strep 100× (optional) (Invitrogen, #15070-063).
13. TrypeLE-Express (Invitrogen, #12604).
14. Collagenase IV (20,000 U, Invitrogen, #17104-019).
15. CF-1 mouse embryonic fibroblasts (ATCC, #SCRC-1040).
16. Nikon Object Marker, catalog # MBW10020 (optional).
17. 6-Well vacuum gas plasma-treated tissue culture dishes (such as BD Falcon, #353046).
18. Sterile nylon membrane syringe filter (Pall Life Sciences, #PN 4433).
19. Nalgene freezing container (containing isopropanol).
20. 20 μL pipette tips (Eppendorf and others).
21. 150 mm Tissue culture dishes (TPP, #93150).

2.2. Media and Stock Solutions

2.2.1. Human Basic FGF (bFGF) (10 μg/mL, 1 mL, See Note 4)

1. Dissolve 10 μg of human bFGF in 1 mL KSR.
2. Aliquot in 50 μL samples.
3. Store thawed aliquots at 4°C for up to 2 weeks.
4. Store frozen aliquots at −20°C or −80°C for 6 months.

2.2.2. Collagenase IV (200 U/mL, 100 mL, See Note 5)

1. Dissolve 20,000 U of collagenase IV in 100 mL of DMEM/F12 + GlutaMax. This is usually ~1 mg/mL final concentration.
2. Add to a 250 mL 0.2 μm filter unit and filter sterilize.
3. Aliquot in 5–10 mL in sterile tubes and store at −20°C until use.

2.2.3. KnockOut™ Serum Replacement

Aliquot as follows:

1. Thaw 500 mL bottle at 4°C and aliquot into sterile 50 mL tubes and store at –20°C.
2. Mix thoroughly when thawed, both for the initial aliquotting and when for use in media preparation.

2.2.4. MEF Medium (500 mL)

1. Combine 440 mL DMEM/F12 + GultaMax, 50 mL FBS, 5 mL GlutaMax, 5 mL NEAA.
2. Sterile filter 2 μm.
3. Store at 4°C.
4. Warm to room temperature in the hood before use, discard unused medium after 2 weeks.

2.2.5. Human PSC Medium (100 mL)

1. Combine, in order, 78.8 mL DMEM/F12 + GlutaMax with 20 mL KSR, 1.0 mL 100× NEAA, 100 μL bFGF stock solution, and 100 μL of 1,000× 2-mercaptoethanol.
2. Filter using 2 μm PES filter.
3. Store at 4°C when not in use and discard any unused medium after 2 weeks.

2.2.6. Human PSC Cryopreservation Medium (10 mL, 2×)

1. Combine 2 mL of human PSC medium, 6 mL of FBS, and 2 mL of DMSO.
2. Mix thoroughly.
3. Sterile filter using a syringe filter approved for use with DMSO (e.g., nylon membrane).
4. Keep cold and use immediately. This is a 2× solution.

3. Methods

These methods assume that all PSC culture is carried out in 6-well plates (see Notes 6–8 for helpful general suggestions).

3.1. Preparation of Feeder Cell Stocks

The traditional feeder cells are mitotically inactivated, low-passage mouse embryonic fibroblasts, usually from CF-1 strain mice (5). These MEFs are seeded at a wide range of densities depending on the different PSC cell line being grown. For example, the original "H" series lines from WiCell were grown in the presence of MEFs seeded at 75,000 cells/cm^2. However, these lines and others have also been successfully grown on denser feeder layers, as have many others. It will take some trial and error in your laboratory to determine the optimum density. Human-derived fibroblasts of various origins have also been successfully used as a feeder cell layer. Any distinct advantages of human versus mouse feeder cells have not been agreed upon. As with MEFs, if you choose to use human feeders, it will require some trial and error to determine the best concentration.

1. Thaw a cryopreserved vial (such as the ATCC product cited) of mouse embryonic fibroblasts quickly in a 37°C water bath (without submerging the cap), and wash with 70% alcohol before moving it to the tissue culture hood. Carefully move the contents into a 15 mL conical tube. Slowly and dropwise, add 10 mL of warm MEF medium, while gently shaking the tube.
2. Centrifuge at 200×*g* for 5 min, aspirate supernatant, and resuspend the pellet in 5 mL of MEF medium.
3. Seed onto a 0.1% gelatin-coated 150 mm TPP tissue culture dish and add an additional 15 mL of MEF medium. Place in incubator and gently move the plate back-and-forth and then side-to-side, so as to evenly distribute the cells.
4. Monitor the cells daily. They should divide extremely quickly but not require daily feeding before reaching confluence. This often takes only 24 h.
5. When the cells become confluent, split at a 1:2 ratio using TrypLE-Express. Aspirate the medium, rinse the plate with 5 mL DPBS^{--}, and add 10 mL of RT or 37°C TrypLE-Express. When the cells start to lift off the plate, inactivate the enzyme by adding 10 mL of warm MEF medium. Collect the cells in a sterile conical tube and centrifuge them at 200×*g* for 5 min. Aspirate the supernatant, resuspend in a 40 mL of medium, and reseed into 2–150 mm dishes. This is considered passage 1.
6. Continue monitoring and splitting the MEFs until passage 5 is reached. At this point, lift the cells, and irradiate them with 3,000 rads to inactivate them (see Note 9).
7. Freeze the cells in DMEM with 30% FBS and 10% DMSO. The freezing density will depend on the PSC line with which you are working. We recommend that you freeze enough feeder cells in one vial (~3.5×10^6/vial) to seed an entire 6-well plate.

3.2. Preparation of a Feeder Layer

1. 24 h before a plate of feeder cells is needed for PSC culture, coat the plate with 0.1% gelatin for 2–24 h before plating MEFs to help the PSCs attach.
2. Thaw a vial of prepared irradiated/inactivated MEFs as described above, and seed onto the plate in MEF medium.
3. After the MEFs have attached overnight, aspirate the MEF medium, rinse with DPBS^{++}, and add 1 mL/well PSC medium. Allow the MEFs to condition this medium for at least an hour before seeding the PSCs (1 mL/well, as described in Subheading 3.4.1, step 9).

3.3. Thawing of Cryopreserved PSCs

Quite frequently, there is a growth lag after thawing and plating PSCs – it may take several days to see colonies (6). It is advisable to observe the cultures under 4× magnification 24 h after thawing, but not to exchange the medium for at least 48 h. There may be considerable floating debris and dead cells upon thawing the cells – this is normal.

1. Gently but quickly thaw the vial of cells by shaking it in a 37°C water bath until the last sliver of ice has melted (about 60 s). Spray the tube with 70% alcohol and dry with a Kimwipe.
2. In the biosafety cabinet, aseptically remove the vial contents and place them in 15 mL conical tube. Slowly, with gentle tapping, add 10 mL of room temperature PSC medium.
3. Spin at 200 × *g* for 5 min.
4. Aspirate the supernatant.
5. Add 3 mL of PSC medium to the tube, triturate gently, and transfer the contents to one well of a 6-well dish that has been prepared with inactivated MEFs as described in Subheading 3.2.
6. Place plate into the incubator.
7. Allow 3–7 days for the cells to attach. During this time, replace half of the medium every other day being careful not to aspirate the cells.
8. The medium should be replaced daily starting 4–7 days after thawing the cells, or when the cells appear to have attached (Fig. 1 shows the appearance of traditionally cultured PSCs).

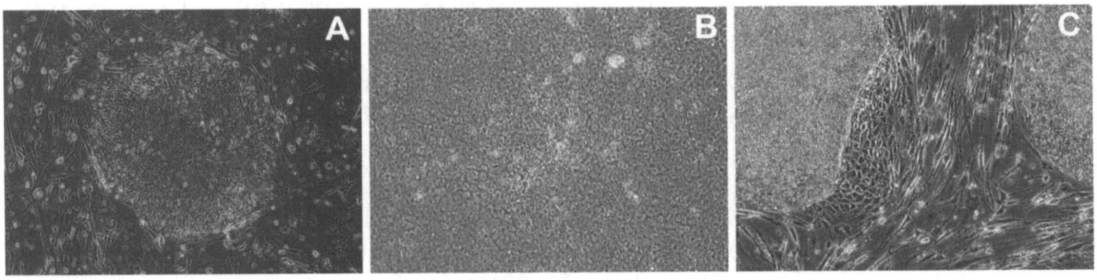

Fig. 1. Phase-contrast images of human PSC colonies in culture. (**a**) A PSC colony shown at 20× a few days after passaging. Note the distinct border between the PSCs and the feeder layer and the apparent heterogeneity of the PSCs, comparing the outside of the colony to the inside. This heterogeneity is typical in a recently passaged culture. (**b**) A closeup (40×) of a larger colony, indicating the classic cell morphology of PSCs. Note the high nucleus-to-cytoplasm ratio as well as the very prominent nucleoli. (**c**) A more mature culture at 10× showing both the relative homogeneity of cell morphology throughout the colony as well as an area of very noticeable differentiation (the area of larger *phase-dark cells*) on the *lower-right border* of an otherwise undifferentiated colony. Notice the difference between these differentiated phase-dark cells and the loose undifferentiated cells near the periphery of the early stage colony in (**a**).

3.4. Passaging PSCs

Human PSCs have traditionally not survived well when dissociated to single cells. Thus, the most reliable method for passaging undifferentiated PSC cultures has been manual dissection of the colonies. This method may seem tedious but it is virtually foolproof and we recommend that novices use this method until they have familiarity with the cells and can easily recognize differentiation in the cultures. We also recommend manual passaging for producing cell banks of low-passage PSCs (see Note 10).

3.4.1. Mechanical Dissociation

The choice of tool for mechanical passaging is an individual preference, but we have found that needles and pipette tips are the most common choice. They are inexpensive to obtain and provide consistency (see Note 11).

1. Evaluate the culture daily under 4× or 10× phase-contrast optics.
2. The cells can be split among 3–6 plates of the same size as the original culture, depending on the density of the original culture. If you wish to put the cells in different-sized plates or dishes, calculate the volume to add based on the surface area.
3. Mark (or remove) overtly differentiated colonies so as not to disturb these during the dissociation process.
4. Remove the medium from the dish and replace with fresh PSC medium.
5. Dissect the colonies by hand, either under a low-power dissecting microscope (in a horizontal flow hood) or without a microscope in the tissue culture hood (see Fig. 2).

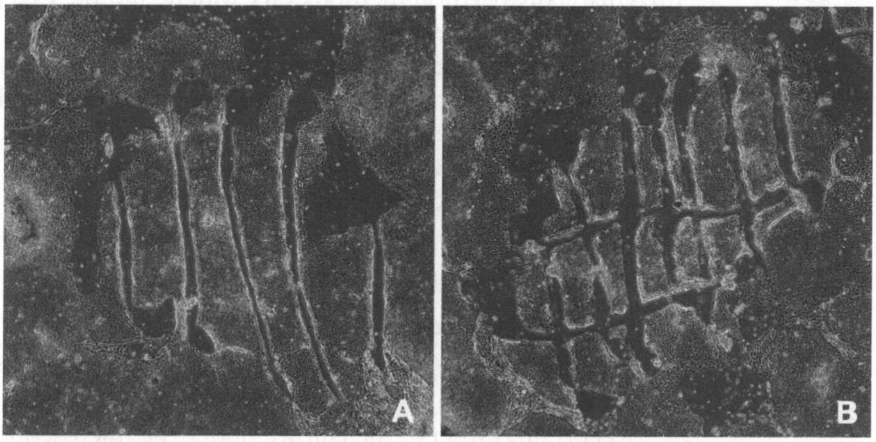

Fig. 2. Manual passaging of human PSCs in culture showing the sterile needle or pipette method used for slicing the colonies into about 100 pieces. The colony is cut into strips (**a**), and then into squares (**b**). Each piece of the colony has a few hundred cells (4× phase-contrast).

6. Break up each colony and move it into suspension by moving the tip around and across each colony in a crosshatch or a spiral motion. Pipette tips are a much better tool for this than needles due to their larger bore. Since the colonies are large at the time of passage, it is relatively easy to see individual colonies on the plate and, with practice, one can quickly passage an entire plate in less than 20 min (see Note 12).
7. After all of the colonies are dissected (from an entire 6-well plate, for example), use a 5 mL pipette to transfer the culture medium containing the dissected colonies to a 15 or 50 mL conical tube. Rinse the plate with 1 mL PSC medium, moving the medium sequentially from well to well before adding it to the same 15 mL tube.
8. Using PSC medium, bring the volume of medium and cells in the tube to the appropriate amount for seeding new plates. For example, if you have just passaged one well of a 6-well plate, and are passaging 1:6, you should bring the final volume to 12 or 2 mL for each new well that will be seeded (but see the slightly different procedure if you are cryopreserving the cells, Subheading 3.5).
9. Gently triturate the cell clumps using a sterile 10 mL pipette and divide the cell suspension into the prepared culture dishes on feeder layer. Do not overtriturate; triturate gently, trying to achieve a relatively uniform distribution of the cell clumps without creating single cells.
10. Place the newly seeded plates in the incubator. Briskly move the plate(s) back-and-forth, side-to-side, and forwards-and-backwards to ensure even dispersion, while being careful not to splash any medium onto the cover of the culture dish.

3.4.2. Collagenase Dissociation

Enzymatic dissociation methods vary widely, and the exact conditions need to be developed for each laboratory. Most importantly, cultures that have been maintained by manual passaging cannot be passaged by enzymatic dissociation unless exceptional care is taken to adapt the cells to this new set of conditions. When done properly though, enzymatic passaging can provide the investigator with a convenient and efficient way of maintaining PSC culture stocks (see Note 5).

1. Remove the culture medium.
2. Rinse the culture with DPBS^{++}.
3. Treat with 2 mL/well 200 U/mL of collagenase IV solution for 5–10 min at 37°C or until the edges of the colonies start to curl up. Observe the culture under the microscope.
4. Remove the collagenase and replace with 2 mL/well of PSC medium.

5. Using a 5 mL pipette, gently dislodge the "good" colonies from the plate and transfer them to a 15 mL conical tube. Alternatively, one could remove the differentiated colonies prior to treating the culture dish with collagenase.
6. Gently triturate the cell clumps using a sterile 10 mL pipette and plate on a feeder layer of MEFs. Do not make a single-cell suspension but try to achieve a relatively uniform suspension of cell clumps containing several hundred cells each.
7. The cells can be divided among 3–6 dishes of the same size as the original culture, depending on density of the original culture. If you wish to put the cells in different-sized dishes, calculate the dilution based on surface area.

3.5. Cryopreservation

Cryopreservation is used to stabilize cultures with specific genetic characteristics at specific points in time. Without the ability to cryopreserve our cell lines, we are forced to continuously subculture them, during which time the cells may accumulate genetic changes and become heterogeneous. Using validated stock vials to initiate new experiments maximizes the long-term usefulness of a cell line and minimizes experimental variation.

For many years, the traditional method of cryopreserving PSCs has involved freezing the cells in large clusters with a medium containing FBS and DMSO. PSCs have very poor survival with this cryopreservation method, however (6). As a result, the time from thawing the vial to having cultures suitable for experimentation can be weeks to months. Vastly more efficient techniques have been recently developed in conjunction with alternative culture systems, and these are outlined in Chapters 9 and 10. Nonetheless, for researchers interested in pursuing traditional PSC culture, the historical method is presented in this chapter.

1. Prepare cells for cryopreservation when they have reached the same stage at which you would normally passage them.
2. Change the culture medium just before harvesting the cells.
3. Label 1.8 mL cryogenic vials with cell line name, date, and passage number.
4. Prepare 2× stock cryopreservation medium (see Subheading 2.2) and keep on ice.
5. Dislodge the colonies from the plate, mechanically, using a sterile pipette tip or treat with 2 mL/well 200 U/mL of collagenase IV in DMEM/F12 + Glutamax for 5–10 min at 37°C.
6. Remove collagenase and replace with PSC medium (3 mL for each well of a 6-well dish).
7. For each well of a 6-well dish, collect the cells in 3 mL of PSC medium and transfer to a 15 mL conical tube.

8. Centrifuge 5 min at 200×*g*. Aspirate supernatant, leaving a small amount of medium covering the pellet.
9. Gently resuspend the pellet in conditioned PSC medium (usually 1.5 mL for each well of a 6-well dish or one half of the final freezing volume). Use a 5 mL pipette to gently triturate the clumps.
10. Dropwise, add an equivalent volume of ice-cold 2× cryopreservation medium, mixing constantly by tapping the tube (see Note 13).
11. Place 1.0 mL of cell mixture in each cryogenic vial (i.e., about three vials per well).
12. Rapidly transfer the vials to a precooled (4°C) Nalgene freezing container (containing isopropanol), and place immediately in a freezer at –70°C to –80°C. The next day, transfer cells to liquid nitrogen for long-term storage.

4. Notes

1. We know that PSCs acquire chromosomal abnormalities over long periods of culture, so karyotyping or other genetic analysis methods must be performed on a regular basis. For detailed information about how to monitor genetic drift, see Chapters 13 and 14. Keep in mind that changes during the time the cells are cultured in your laboratory can only be detected if you first analyze the cells very soon after you obtain them.
2. PSCs can also drift toward a more differentiated state over periods of extended culture. Since there is no assay for pluripotence equivalent to the germline transmission assay for mouse PSCs, surrogate markers, such as antibody markers, should be routinely checked, especially if the morphology of the cells seems to be different from the earlier cultures. The gold standard for measuring the pluripotency of a PSC line is to transplant it to an immune-deficient mouse to form a teratoma (Chapter 17). Keep in mind that it will require histological expertise to identify cell types and tissues in the tumors. In vitro differentiation of PSCs using embryoid body culture (Chapter 28) will allow at least a cursory analysis of PSC differentiation potential. However, embryoid bodies never achieve the maturity of cells that develop in teratomas, and since the methods used to assess differentiation in vitro usually involve a small number of markers assayed by immunocytochemistry (Chapter 15), it is more difficult to judge the full range of pluripotence using this method. The best

approach to monitoring developmental drift is to pick a particular method and differentiated cell type to check periodically (see Chapter 30 on embryoid body and neuroepithelial differentiation, as well as the specific chapters on neuronal, cardiac, and hematopoietic cells, Chapters 29–33).

3. An important consideration for all cell cultures, but most especially for PSC culture, is that all reagents, including the water used to make them up, be qualified for use in the particular culture at hand. Water quality, even that purified by double distillation and/or "MilliQ" water, can vary significantly from geographic area to geographic area and from season to season. With this in mind, our laboratory tends not to buy media powders that we must ourselves reconstitute but rather the fully diluted media. In addition, for all dilutions of reagents that will come into contact with live cells, directly or indirectly, we use water that has been qualified for embryo culture, such as the Sigma product we cite. Although one might save pennies doing everything oneself, the likelihood, in this research area, of losing dollars by doing so is too high to justify it.
4. For all stock or small quantity growth factor solutions, prewet all pipette tips, tubes, and filters with DPBS with 0.2% BSA to lessen the loss of the growth factor.
5. The type of enzyme used for dissociation is critical. For example, passaging with trypsin appears to put more selective pressure on the cultures than other methods, resulting in a higher incidence of drift of PSC lines toward aneuploidy. But some PSC lines have been derived using trypsin from the outset; thus some lines can be routinely passaged using whatever enzymatic technique is used by the supplier. Microbial collagenase is preferred by many laboratories, perhaps because of the way in which it is used. Collagenase is used to loosen the PSC colonies from the dishes, not to dissociate them to single cells, but the cell clumps have to be further dissociated by trituration. Collagenase is isolated from Clostridium histolyticum. Type IV is selected because of its low tryptic activity, and is recommended for isolation of pancreatic islets. This is a crude product, so expect a wide lot-to-lot variation. EDTA inhibits this enzyme's activity. Accutase is also becoming more popular because of its unique ability to dissociate PSCs into single cells while maintaining viability (7); however, this requires the use of defined medium and a Matrigel substrate (see Chapter 10). The potent apoptosis-blocking ROCK Inhibitor Y27632 is frequently used alongside Accutase and other enzymatic and mechanical passaging techniques to further help maintain viability (8). Another enzymatic reagent which has been used is Invitrogen's TrypLE-Express. TrypLE

is a fungally derived Trypsin-like enzyme that dissociates cells extremely quickly. It has not been widely used for PSC passaging, but may be useful for breaking apart colonies for use in flow cytometry or cytogenetics. Nonenzymatic cell dissociation buffers are also available. The latter are Ca- and Mg-free saline solutions containing EDTA or EGTA. They have not been as widely used for PSC dissociation as the methods described above, however, they should offer advantages for assays that require intact cell surface proteins such as flow cytometry. Commercial formulations are available, such as Cell Dissociation Buffer (Invitrogen catalog no. 13150016) which contains glycerol as well as a proprietary mixture of salts and chelators. Keep in mind that enzymes are not highly purified recombinant products, and they may contain animal products. Trypsin is generally prepared from porcine tissue, and collagenase is a crude microbial product.

6. PSCs are usually cultured without antibiotics; with good culture technique, bacterial and fungal contamination should not be a problem. However, we recommend that antibiotics be used while new investigators are being trained in the techniques. Antibiotics such as amphotericin, penicillin, and streptomycin, however, do not have any effect on mycoplasma. Mycoplasma is highly infectious and commonly occurs when new cells are introduced into laboratories. The ATCC (American Type Culture Collection) estimates that 16% of cell cultures are contaminated by mycoplasma. This bacterium can also come from tissue culture reagents such as serum and media supplements as well as from laboratory staff. Mycoplasma is a serious problem in laboratories that culture multiple cell lines or have inadequately trained personnel. Cultures must be monitored for mycoplasma on a regular basis, and contaminated cultures must be destroyed. The best defense against mycoplasma contamination is good aseptic technique; the laboratory should not allow inexperienced or careless workers to share cell lines, solutions, or tissue culture equipment. As a precaution, all cultured cell lines should be tested at least four times a year. Testing for mycoplasma can be done by enzymatic, polymerase chain reaction (PCR), fluorescent staining, or culture methods.
7. It is important to keep in mind the actual source of the materials and reagents used in the culture and maintenance of PSCs. Since many reagents are derived from animal sources, there is inherent lot-to-lot variability. While vendors make every effort to control the variability by setting production specifications, these are usually ranges and as long as the product falls within the approved range, the product passes inspection and is distributed. Ideally, you should have your

own quality control methods to test new lots of products. At the very least, record the lot numbers of reagents used; if an experimental result cannot be replicated, or a cell line fails to thrive, you will save considerable time if the problem is traceable to a suboptimal reagent.

8. The cells should be passaged at about 1:3–1:6 every 5–7 days. Prepare the feeder layer or extracellular matrix (ECM) substrata the day before passaging. Depending on the cell line, passaging on a Friday afternoon may be a good routine. The cells are usually left undisturbed for 2–3 days following passaging, which allows them to settle down on the substratum, attach and begin dividing before the medium is changed. There will be considerable variation in the size of colonies in a single dish. Human PSCs do not substantially pile up on each other, and their colonies can grow to a large diameter while remaining undifferentiated. Culture conditions affect the flatness of the colonies, but as an approximation, they are ready to split when the diameter fills the 10× field when observed under the microscope. As shown in Fig. 3, a colony about half the diameter of the 10× field contains about 4,400 cells. A colony filling the field would contain about 15,000 cells. When the colonies grow very large and start to merge into one another (see Fig. 4), they must be passaged to avoid differentiation and/or starvation of the culture.

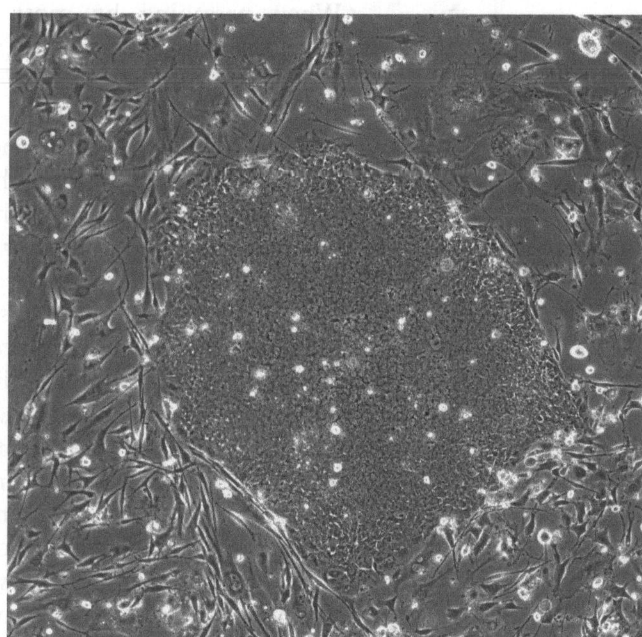

Fig. 3. A colony about half the diameter of the 10× field contains about 4,400 cells; a colony filling the field contains about 15,000 cells.

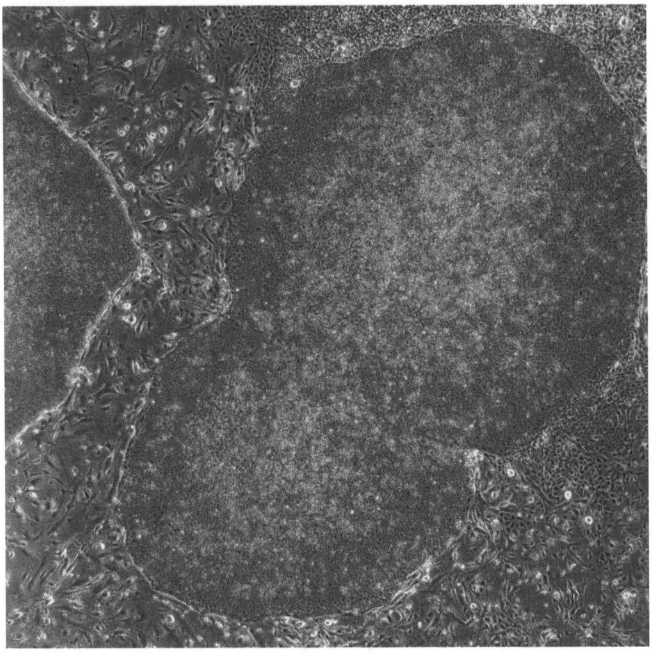

Fig. 4. Example of a high-density culture on mouse embryonic fibroblasts. Note how two large colonies have merged together and there is some noticeable differentiation around the edges of the colonies (4× phase-contrast). This culture would need to be passaged right away.

When passaging by most methods, do not make a single-cell suspension; dissociate the colonies into smaller colonies of a few hundred cells. Examine the culture daily for colony morphology under the phase-contrast or dissecting microscope. With experience, one can get a good overview of colony morphology by holding the dish up to a light and looking at the bottom of the dish. The differentiated colonies will have ragged edges and hollow or dark centers. On the bottom of the dish, mark colonies that are badly differentiated or parts of the colony that you do not wish to transfer to a new culture dish. This can be easily accomplished with a special microscope attachment sold by Nikon (see Subheading 2.1 "Reagents"). To be certain that the colonies selected are undifferentiated, it is advisable to dissect the colonies while viewing the dish under a dissecting microscope with illumination from the base. But this is not absolutely necessary, and some prefer passaging cells without magnification.

The single most important skill in successful culturing of PSCs may be the ability to recognize the morphology of undifferentiated cells under a variety of conditions (see Fig. 1).

(a) Feed cells every day, except for 1 or 2 days following passage.

(b) Examine the cultures every day under 4× and 10× phase-contrast. This will allow you to become familiar with the morphologies of undifferentiated and differentiated cells and colonies.

(c) When they are cultured on feeder layers, PSCs tend to undergo spontaneous differentiation in the centers of the colonies. When passaging, take care to avoid passaging these differentiated "centers" to the new culture.

(d) Most PSC lines double every 31–35 h.

(e) Store medium at 4°C, protect from light, and discard any unused medium after 10 days. Best results are achieved when medium is prepared in small batches once a week.

9. Mitomycin C can be used in the place of irradiation for inactivation but it is a cytotoxic antitumor agent and must be handled carefully; it works by cross-linking the DNA, which blocks cell division. Follow your institution's rules for safe handling and disposal. Handlers should wear latex or nitrile protective gloves and work in a biological safety or fume hood. One effective method is to inactivate the mitomycin C with an equal volume of household bleach. Inactivation is rapid.

   (a) Remove the feeder cell medium.

   (b) Add 10 mL/75 cm^2 of mitomycin C medium. Make sure the entire flask is covered with mitomycin C medium so that the inactivation is complete and all cells are exposed for the entire incubation time.

   (c) Incubate for 3 h at 37°C in 5% CO_2.

   (d) Remove mitomycin C, neutralizing it with bleach or other recommended procedure.

   (e) Wash inactivated feeder layer three times with 10 mL each of $DPBS^{++}$.

   (f) Trypsinize the cells to remove them from the dish.

   (g) Use the cells immediately for plating and/or cryopreserve them for later use.

10. An unfortunate historical accident has been using the number of passages as a measure of the age (or of the number of cell divisions) of a PSC line. Because of the inconsistencies in PSC culture procedures in different labs, cells are passaged at different time intervals, ranging from 4 to 7 days. Therefore, the number of passages for one line might not be representative of another, even though the cells have been in culture for exactly the same amount of time. Passage ratios vary from 1:2 to 1:80 among different laboratories. A better measure would be the number of doublings, but to count the number of cells in a culture is difficult since the cells form tight clusters and are not passaged as single cells, but as clumps.

11. Fire-drawn glass pasteur pipettes were the choice for early PSC labs, but they are labor-intensive to prepare, and no two are alike. Needles have the advantage of greater precision; however, for the novice user, they have a tendency to scrape ribbons of plastic off of the plate and introduce them into the medium. Needles, however, are very useful in instances where one may need to isolate a very small colony or where there is a small patch of undifferentiated cells surrounded by areas of spontaneous differentiation. Pipette tips, on the other hand, have the advantage of greater efficiency for more confluent plates with larger, less-differentiated colonies. This is due to their larger bore, and the ability to use them like a scoop to shovel colony chunks into suspension. At the same time, relatively precise cuts can be made using the edge of the tip. We recommend individually wrapped 20 μL pipette tips from Eppendorf since this eliminates the possibility of another technician accidentally contaminating a shared box of autoclaved pipette tips.
12. As the basal media used for most common human cell cultures is bicarbonate-based, do not keep a culture outside the incubator for more than about 15 min at a time. Any time the medium is outside the CO_2 environment, it loses CO_2 and its pH rises, going above pH 8 in about 30 min. Thus, if your dissections take longer than about 20 min, put the cultures back into the incubator for about 10 min before continuing.
13. Do not leave the cells in DMSO at room temperature for long periods of time as DMSO is toxic to the cells and is, under a variety of conditions, also known to induce differentiation.

Acknowledgments

This work has been funded by the National Institutes of Health (T15HL074286, R21MH087925, R01HD059967). The NCI Preclinical Repository supplied FGF-2. We would also like to recognize the contributions of Chris Stubban, Rodolfo Gonzalez, Richard Pepple, Heather Maxwell, and Nicole Sheridan in the production of these protocols.

References

1. Bongso, A., Fong, C.Y., Ng, S.C., and Ratnam, S. (1994) Isolation and culture of inner cell mass cells from human blastocysts. *Hum Reprod.* **9**, 2110–7.
2. Thomson, J.A., Itskovitz-Eldor, J., Shapiro, S.S., Waknitz, M.A., Swiergiel, J.J., and Marshall, V.S., et al. (1998) Embryonic stem cell lines derived from human blastocysts. *Science.* **282**, 1145–7.
3. Müller, F.J., Laurent, L.C., Kostka, D., Ulitsky, I., Williams, R., Lu, C., Park, I.H., Rao, M.S., Shamir, R., Schwartz, P.H., Schmidt, N.O.,

and Loring, J.F. (2008) Regulatory networks define phenotypic classes of human stem cell lines. *Nature*. **455**, 401–5.

4. Mitalipova, M.M., Rao, R.R., Hoyer, D., Jones, K., Johnson, J., Dalton, S., Meisner, L., and S.L. Stice. (2005) Preserving the genetic integrity of human embryonic stem cell lines. *Nature Biotechnology*. **23**, 19–20.
5. Amit, M., Margulets, V., Segev, H., Shariki, K., Laevsky, I., Coleman, R., and Itskovitz-Eldor, J. (2003) Human Feeder Layers for Human Embryonic Stem Cells. *Biology of Reproduction*. **68**, 2150–6.
6. Heng, B.C., Ye, C.P., Liu, H., Toh, W.S., Rufaihah, A.J., Yang, Z., Bay, B.H., Ge, Z., Ouyang, H.W., Lee, E.H., and Cao, T. (2006) Loss of viability during freeze-thaw of intact and adherent human embryonic stem cells with conventional slow-cooling protocols is predominantly due to apoptosis rather than cellular necrosis. *Journal of Biomedical Science*. **13**, 433–45.
7. Bajpai, R., Lesperance, J., Kim, M., and Terskikh, A.V. (2007) Efficient propagation of single cells accutase-dissociated human embryonic stem cells. *Molecular Reproduction and Development*. **75**, 818–27.
8. Watanabe, K., Ueno, M., Kamiya, D., Nishiyama, A. Matsumura, M., Wataya, T., Takahashi, J.B., Nishikawa, S., Nishikawa, S., Muguruma, K., and Sasai, Y. (2007) A ROCK inhibitor permits survival of dissociated human embryonic stem cells. *Nature Biotechnology* **25**, 681–6.

Chapter 9

Xeno-Free Culture of Human Pluripotent Stem Cells

Rosita Bergström, Susanne Ström, Frida Holm, Anis Feki, and Outi Hovatta

Abstract

Stem cell culture systems that rely on undefined animal-derived components introduce variability to the cultures and complicate their therapeutic use. The derivation of human embryonic stem cells and the development of methods to produce induced pluripotent stem cells combined with their potential to treat human diseases have accelerated the drive to develop xenogenic-free, chemically defined culture systems that support pluripotent self-renewal and directed differentiation. In this chapter, we describe four xeno-free culture systems that have been successful in supporting undifferentiated growth of hPSCs as well as methods for xeno-free subculture and cryopreservation of hPSCs. Each culture system consists of a xeno-free growth medium and xeno-free substratum: (1) TeSR2™ with human recombinant laminin (LN-511); (2) NutriStem™ with LN-511; (3) RegES™ with human foreskin fibroblasts (hFFs); (4) KO-SR Xeno-Free™/GF cocktail with CELLstart™ matrix.

Key words: xeno-free culture of hPSC, human embryonic stem cells, induced pluripotent stem cells, extracellular matrix, recombinant human laminin, LN511, defined culture medium, human foreskin fibroblasts

1. Introduction

Human embryonic stem cells (hESCs), induced human pluripotent stem cells (iPSC) were first cultured in medium containing fetal bovine serum (FBS) and in co-culture with mouse embryonic fibroblasts (MEFs) (1, 2). This system, which proved highly effective for maintaining mouse embryonic stem cells (ESCs), was subsequently shown to be inadequate for maintaining human ESC cultures and has been abandoned by most laboratories in favor of a somewhat more defined system that consists of a growth medium containing KnockOut™ serum replacer (KO-SR) and beta fibroblast growth factor (bFGF) instead of FBS,

Philip H. Schwartz and Robin L. Wesselschmidt (eds.), *Human Pluripotent Stem Cells: Methods and Protocols*, Methods in Molecular Biology, vol. 767, DOI 10.1007/978-1-61779-201-4_9, © Springer Science+Business Media, LLC 2011

but continues the co-culture with MEFs (3). This KO-SR containing medium has proven reliable in maintaining existing hESC lines and for the derivation of new hESC lines (Fig. 1) (4). The use of KO-SR instead of FBS in culture medium helped lower the variability of cultures and helped limit the differentiation, but did not eliminate animal components from the growth medium.

As researchers moved toward developing xenogenic-free (xeno-free) culture conditions, human foreskin fibroblasts (hFFs) were found to be supportive of long-term culture of both preexisting ESC lines and for the derivation of new lines (5, 6) and are routinely used in many laboratories instead of MEFs. hFFs have been immortalized and transduced by bFGF gene to over-produce bFGF, which makes them easier to use and even more standardized (7). These same fibroblasts were transduced using the Nanog, Oct-4, Sox2, and Lin28 vectors (8). The obtained induced pluripotent stem cells (iPSCs) grew well on the immortalized hFFs (7). For xeno-free culture, hFFs can be cultured or derived from fresh skin biopsies and cultured in a humanized fibroblast culture medium (9, 10).

Additional improvements in defining the culture system have been made by eliminating the use of feeder cells all together. Feeder-free cultures on either Matrigel™ or xeno-free matrices have been developed. As an extracellular matrix (ECM), Matrigel™ has been most widely used (11) in conjunction with MEF- or hFF- (12) conditioned medium. Of the newer defined media, mTeSR1™ (StemCell Technologies) and StemPro™ (Invitrogen, see Chapter 10) can be used in feeder-free culture on Matrigel. However, Matrigel is a product derived from a mouse sarcoma line and therefore is xenogenic. mTeSR1™ also contains animal proteins.

Efforts to culture hPSCs under totally xeno-free conditions have recently proven successful. We have cultured human ESC and iPSCs in either TeSR2™ or NutriStem™ XF/FF media on LN511 without any particular adaptation. It is also possible to culture hPSCs on CELLstart™ in xeno-free KO-SR™ medium supplemented with a growth factor (GF) cocktail mixture, but the cells need long adaptation for this medium to work (see also Chapter 10).

Survival of hPSCs during passaging and subculture can be improved through the addition of a rho-kinase inhibitor (Y-27632) (13) to the growth medium prior to passage. It also increases the survival of cells when used in the freezing medium (14). The enzymes collagenase type IV, dispase, and trypsin are widely used to dissociate cells, but are derived from animal sources. The recombinant trypsin, TrypLE™ Select, offers a xeno-free alternative that is well tolerated by hESCs (15).

Over the years, there have been reports of various culture media and matrices, and combinations of matrices and feeder cells,

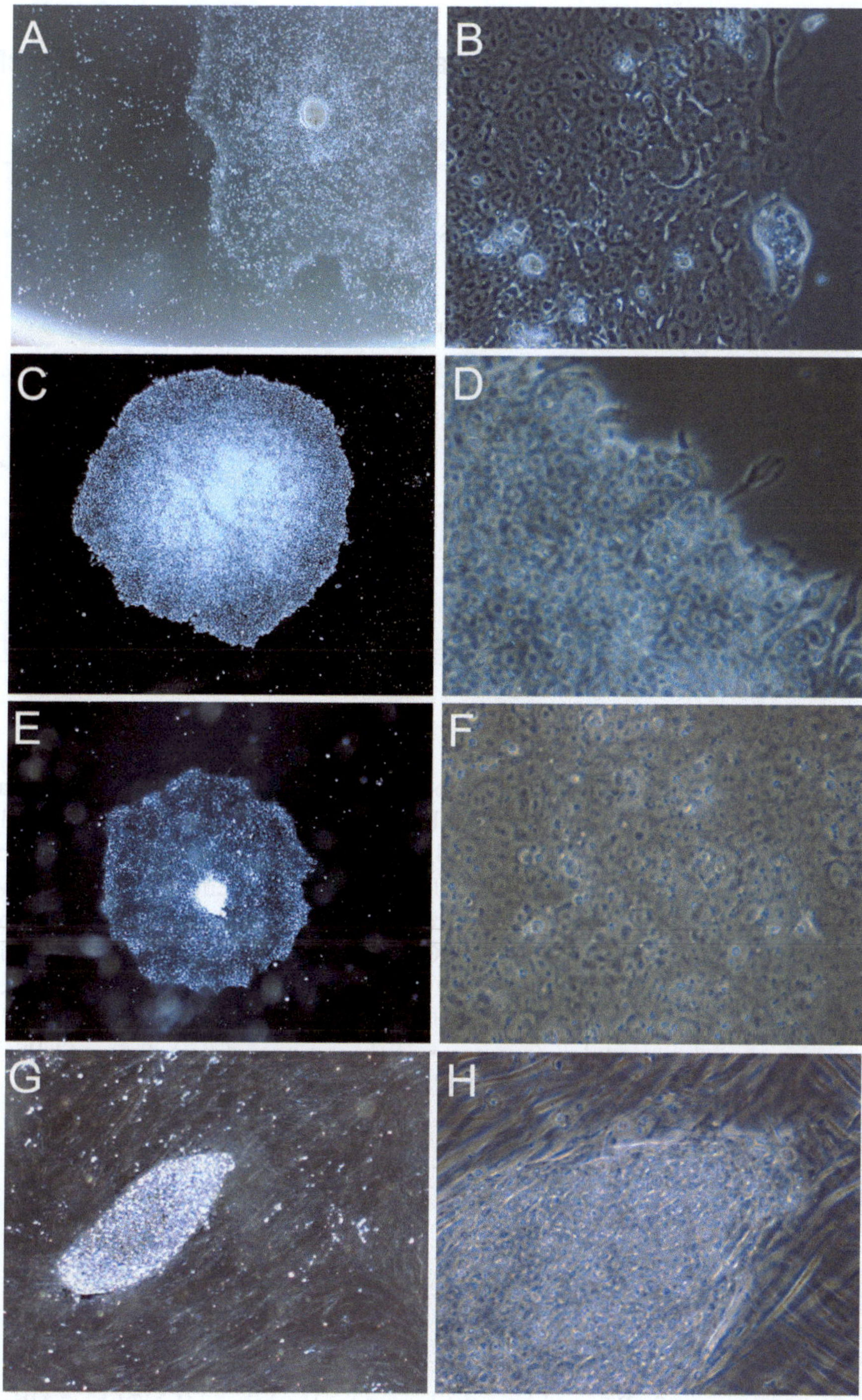

Fig. 1. hESC line HS207 growing on human recombinant laminin LN-511, passage 6 in a medium with the same composition as mTeSR1, but in which the bovine protein has been replaced by dialyzed human albumin (corresponds to commercial TeRS2). (**a**) Original magnification ×40, (**b**) original magnification ×400. (**c** and **d**) Line HS 346, passage 5 in TeSR2 medium, (**c**) original magnification ×40, (**d**) original magnification ×400. (**e** and **f**) hESC line 346, passage 5 in NutriStem XF/FF medium, (**e**) original magnification ×40, (**f**) original magnification ×400. (**g** and **h**) The hESC line HS181, passage 4, in RegES medium on hFF feeders, (**g**) original magnification ×40, (**h**) original magnification ×400.

Table 1
Current commercially available culture systems for hPSCs

Medium	Substrate
F12/DMEM + FBS	MEF or hFF
KO-DMEM + bFGF + SR	MEF or hFF
HesGro	hFF
MEF- or hFF-conditioned medium	Matrigel
mTeSR1	Matrigel or LN-511
StemPro	Matrigel
TeSR2, xeno-free	Matrigel or LN-511
Serum replacement, Xeno-free	hFF
Serum replacement, Xeno-free + GF cocktail	CELLstart
NutriStem XF/FF	Matrigel or LN-511
RegES, Xeno-free	hFF

which have been reported to support hESC cultures (Table 1). Many of them have been reviewed by Rajala et al. (16, 17) and Unger et al. (9), and many have been tested in the International Stem Cell Initiative (ISCI2) (18). Current commercially available, xeno-free culture systems that we have shown to support the long-term culture of hPSCs are: TeSR2™ with LN511, NutriStem™ XF/FF with LN511, RegES™ with hFFs, and KO-SR Xeno-Free™ supplemented with GF Cocktail with CELLstart™. This protocol reviews state-of-the-art xeno-free culture of hPSCs using commercially available media and substrata.

2. Materials

2.1. Cell Culture Materials and Supplies

1. Tissue culture-treated dishes: 35 mm (11.78 cm^2) and 60 mm (23.65 cm^2).
2. hFFs (ATCC, #CRL-2429).
3. Recombinant human laminin (LN511) (Biolamina, Sweden).
4. CELLstart™ (Invitrogen, #A10142).
5. PromoCell™ Human Fibroblast Growth Medium (#23010, Promocell).

6. TeSR2™ Medium Kit (StemCell Technologies, #05860).
7. NutriStem™ XF/FF medium (Stemgent, Inc., #01-0005).
8. RegES™ xeno-free medium (Vitrolife, Sweden, will be made available in 2010).
9. Xeno-Free KnockOut-SR™ (Invitrogen).
10. Growth factor (GF) cocktail (Invitrogen).
11. Knockout DMEM (Invitrogen).
12. 200 mM L-GlutaMax™ (Invitrogen).
13. 10 mM Nonessential amino acids (NEAA, 100×).
14. Dulbecco's phosphate buffered saline without Mg^{2+} and Ca^{2+} (DPBS).
15. Dulbecco's phosphate buffered saline with Mg^{2+} and Ca^{2+} (DPBS+/+).
16. Rho-kinase inhibitor (ROCK) 5 mM solution (Calbiochem, #688001 InSolution™ Y-27632), this is light sensitive, store frozen in aliquots at −20°C.
17. TrypLE™ Select Animal-Origin-Free (Invitrogen, #12563-011) (see Note 1).
18. StemCell Banker (Xenoaq, Japan).
19. Cell Lotion (Xenoaq).

2.2. Culture Media

2.2.1. TeSR2™ Xeno-Free hPSC Medium

1. Thaw TeSR2 5× and 250× supplements at room temperature (15–25°C) or 2–8°C overnight (see Note 2). Aliquots can be made and stored at −20°C for 6 months.
2. To make complete TeSR2 medium, add 100 mL of 5× and 2 mL of 250× supplements to 400 mL of TeSR2 basal medium, mix well. Complete TeSR2 medium is stable for 2 weeks at 2–8°C or can be aliquotted and frozen at −20°C for up to 6 months.

2.2.2. NutriStem™ XF/FF hPSC Medium

This is a complete medium, no need to add supplements. Thaw one bottle of complete NutriStem™ XF/FF culture medium at 4°C. Thawed medium is stable for 2 weeks when stored at 4°C. Warm to 37°C prior to use.

2.2.3. KO-SR Xeno-Free hPSC Medium

1. Thaw KO-SR Xeno-Free at 4°C, make aliquots of 7.5 mL and store in −20°C.
2. To make complete medium, add 7.5 mL KO-SR Xeno-Free, 0.5 mL GlutaMAX-1, 0.5 mL NEAA to 41.4 mL KO-DMEM and just prior to use, add bFGF to final concentration of 8 ng/mL.
3. For feeder-free culture, add 1 mL of GF cocktail to the medium above.

2.2.4. RegES™ Xeno-Free hPSC Medium

1. Thaw one bottle of complete RegES medium in a 37°C water bath. Minimize dwell time.
2. Aliquots can be made and re-frozen at −20°C. Aliquots are stable for 3 months at −20°C and for 1 week at 4°C.
3. Pre-equilibrate RegES medium to 37°C prior to use.

2.2.5. Xeno-Free hFF Medium

This is a serum-free medium that consists of a supplemented basal medium.

1. Thaw the vial containing the supplement mix at room temperature.
2. Add the supplement mix to basal medium and mix. Complete medium is stable if kept at 4°C for 6 weeks. Do not freeze the complete medium.
3. Prewarm the desired volume of medium before use at 37°C.

3. Methods

3.1. Preparation of hFF Feeder-Cell Plates

1. Equilibrate the human fibroblast feeder cells medium in the incubator (5% CO_2 in air, at 37°C) for 1 h prior to plating or passaging hFFs (10 mL/10 cm dish).
2. To passage using TrypLE Select, wash hFFs two times with DPBS.
3. Add 2 mL TrypLE Select to each 10-cm dish or 75-cm^2 culture flask.
4. Incubate at 37°C for 5–7 min, or until the cells have detached from the dish.
5. Transfer the cells to a 50-mL centrifuge tube and rinse the culture vessel, two times, with 10 mL of hFF culture medium.
6. Spin the cells at 200 × *g* for 7 min.
7. Discard the supernatant and re-suspend the cell pellet in 10–15 mL of growth medium.
8. Count the cells to determine the concentration.
9. 0.8–1.0 × 10^6 nonirradiated hFF cells are seeded/10-cm culture dish for further expansion up to passage 16.

3.1.1. Inactivate hFFs Prior to Use as Feeder Cells

1. If hFFs are to be used as feeder-cell layer, they are irradiated with 40 Gray.
2. Count the cells and plate them in the desired culture dish, for example:
   (a) A confluent feeder-cell layer in a 2.89-cm^2 dish plate: 0.10 × 10^6 cells/0.7 mL hFF medium.

(b) A confluent feeder-cell layer in a 11.78-cm^2 dish plate: 0.3×10^6 cells/ 2 mL medium is needed.

3. Allow the cells to attach to the culture vessel overnight in the incubator.
4. The next day, change the hFF medium (see Note 3).
5. Prior to seeding hPSCs, replace hFF medium with hPSC medium such as RegES medium.

3.2. Preparation of ECM-Coated Plates

3.2.1. Laminin 511-Coated Plates

1. Thaw recombinant laminin slowly at 4°C. After thawing, dilute the solution with DPBS to get desirable amount needed per surface area (see Table 2).
2. Coat plated with 5 μg/cm^2 of laminin in a large enough volume to avoid complete evaporation before use. It is recommended to add a small volume of sterile DPBS or ddH$_2$O in the space between the wells to slow evaporation.
3. Cover with parafilm and store at 4°C. Plates are ready to use after overnight incubation at 4°C, but can be stored up to 3 weeks at 4°C.
4. Before use, incubate the laminin-coated plates at 37°C for 1 h.
5. Wash each well two times with prewarmed culture medium.
6. Add prewarmed cultured medium supplemented with all suitable growth factors to each well.

3.2.2. CELLstart™-Coated Plates

According to the manufacturer's instructions:

1. Dilute CELLstart 1:50 in DPBS+/+. Pipet gently to mix. Do not vortex.
2. Add diluted CELLstart to cell culture dish at a final volume of 0.078 mL/cm^2 (Table 3, see Note 4).
3. Incubate in a 37°C, 5% CO_2 incubator for 1–2 h.
4. Aspirate CELLstart from the culture vessel. The culture dish is now ready for the addition of cells. It is not necessary to rinse the culture dish after removal of CELLstart. *The bottom of the coated culture dish should have a clear and wet appearance.*

Table 2
Laminin 511 coating regimen

Plate type	Surface area (cm^2)/well	Laminin needed (μg)	Total volume (μL)
24-Well	2.0	10	300
12-Well	3.8	19	450
6-Well	9.6	48	1,000

Table 3
CELLstart coating regimen

Plate type	Surface area (cm²)/well	Total volume (μL per well)
6-Well	9.6	750
12-Well	3.2	250
24-Well	2.0	160

3.3. Adapting hPSCs to Feeder-Free Culture

Feeder-free culture is initiated by manually dissecting hPSC colonies using a sterile scalpel, lifting undifferentiated bits of colonies, and placing them onto matrix-coated dishes (see Notes 5–7).

1. Using a sterile scalpel, divide an undifferentiated, medium-sized colony into 6–9 pieces, avoiding differentiated parts of colonies and culture. Gently "lift" small pieces of PSC colonies off the feeder layer.
2. Transfer the bits of colonies using a 20-μL pipette to the prepared plates. 10–15 pieces can be put on a 2.89-cm² plate and 20–30 pieces can be put on a 11.89-cm² plate.
3. Carefully transfer the culture dishes into the incubator.
4. Incubate overnight at 37°C.
5. The next day, observe the cultures under phase-contrast microscopy and feed.
   (a) No medium is removed, but 50% of total volume of fresh medium is carefully added to each well or culture vessel.
6. Cultures are fed daily thereafter, with a complete medium change using freshly prepared, prewarmed culture medium (1 h at 37°C, see Note 8).
7. Cultures should adapt to feeder-free culture and be ready for subculture passaging every 4–6 days.
8. Carefully monitor the culture for differentiation. This is especially important during the first three passages while the cells are adapting to feeder-free growth.
9. Mechanically remove differentiated cells from the culture prior to subculture as described below.

3.4. Xeno-Free Subculture

Xeno-free subculture is accomplished by using one of the two methods described below.

3.4.1. Mechanical Splitting

1. The culture plates are supplemented with stem cell culture medium 30 min prior to use and placed in the incubator to equilibrate.
2. A surgical scalpel is used to "lift" small pieces of hPSC colonies from the feeder layer. One medium size colony can be split into 6–9 pieces. Differentiated parts of the colonies are avoided.
3. The small pieces are transferred with a 20-μL pipette to the prepared fresh feeder plates. 10–15 pieces can be put on a 2.89-cm^2 plate and 20–30 pieces can be put on a 11.89-cm^2 plate.
4. Carefully transfer the culture dish into the incubator.

3.4.2. Enzymatic Passaging Using TrypLE™ Select

The cells are routinely passaged once in 6–7 days by exposure to TrypLE Select for 1.5 min at room temperature. Wash cells two times with prewarmed medium, gently scrape with a pipette tip to break into small pieces, and plate 1:2 or 1:3.

3.4.3. Single-Cell Passaging Using ROCK Inhibitor

1. ROCK inhibitor is diluted 1:500 in PSC culture medium.
2. Culture medium is replaced with ROCK inhibitor-containing medium and incubated for 1 h.
3. Remove medium, wash the plate once with DPBS.
4. To one 35 mm (11.78 cm^2) dish, add 0.6 mL TrypLE Select and incubate at 37°C for 5–7 min.
5. Add 1.2 mL medium and make single-cell suspension by gently triturating the culture with a micropipette. Transfer the suspension to a 14-mL centrifuge tube.
6. Centrifuge at 200 × *g* for 5 min.
7. Use a 5-mL pipette to remove the supernatant.
8. Add 2 mL PSC culture medium supplemented with ROCK (1:500) and re-suspend the cells to single cells.
9. Count the cells.
10. About 10,000 cells/mL of culture medium containing ROCK inhibitor and 8 ng/mL bFGF can be added to one 11.78-cm^2 dish.
11. The next day, 0.8 mL PSC culture media is added.

3.5. Xeno-Free Cryopreservation of hPSCs

StemCell Banker™ is a commercially available, xeno-free cryopreservation medium that is highly efficient in the cryopreservation of hPSC (see Note 9).

This is an optimized cryoprotectant that has been used for both hESCs and iPSCs with viability of 90–96% and without any impact of proliferation and differentiation capacity. Standard slow freezing in DMSO has a much lower viability (46%) (19).

3.5.1. Cryopreservation of hPSC Colonies

1. Mechanically remove undifferentiated colonies using a surgical scalpel (10–30 colonies).
2. Transfer cell aggregates to a cryogenic vial.
3. When cells have settled to the bottom, remove all surplus medium.
4. Add cold StemCell Banker (500 μL) to cryogenic vial and immediately place tubes in −80°C freezer.
5. After 24 h, move cryogenic vials to cryogenic freezer, below −130°C, for long-term storage.

3.5.2. Cryopreservation of hPSCs as Single Cells

1. Add ROCK inhibitor (10 μM) to cell cultures 1 h prior to freezing.
2. Dissociate the cells into single cells by treating the cultures with TrypLE Select for 5 min at 37°C and flush the colonies until they detach from feeder layer.
3. Move cell suspension to a centrifuge tube and spin at 200 × *g*.
4. Re-suspend the cells in cold StemCell Banker and transfer to a cryogenic vial.
5. Place vials immediately in −80°C freezer.
6. After 24 h move frozen vials to liquid nitrogen for long-term storage.

3.6. Thawing Human Embryonic and iPSC Colonies and Dissociated Cells Using Cell Lotion

1. Remove cryogenic vials from liquid nitrogen freezer and place in a 37°C water bath.
2. When a small ice crystal remains, remove cells and add to a centrifuge tube and dilute with Cell Lotion (7 mL).
3. Centrifuge at 200 × *g* for 5 min.
4. Re-suspend cells with warm (37°C) culture medium and seed onto fresh hFFs or prepared, recombinant ECM.
5. If cells were frozen as a single-cell suspension with ROCK inhibitor-containing medium, add ROCK (10 μM) inhibitor to culture medium when seeding.

4. Notes

1. As described by the manufacturer, TrypLE™ Select is free of animal- and human-derived components. TrypLE™ Select can be directly substituted for 0.25% trypsin EDTA in your current protocol. Inactivation with trypsin inhibitors is not required.

2. Ensure that the lot numbers of TeSR2 components end with the same letter (Basal medium, 5× supplement, and 250× supplement).
3. The new hFF feeder plates are used no earlier than day 3 after irradiation.
4. CELLstart™: The manufacturer recommends coating the culture dishes on the day of use or the day before. If precoating is done on the day before, the culture dish must be stored at 2–8°C wrapped with parafilm to avoid drying, after the incubation at 37°C.
5. Colonies in TeSR2 are passaged when they begin to merge. The centers of the colonies should be dense and phase-bright under low magnification, phase-contrast illumination. There is a 24-h window for passaging. Cultures are usually passaged 4–5 days postseeding. Colonies will appear transparent and loosely packed up to 3 days postseeding.
6. Colonies passaged too frequently can have attachment problems.
7. Colonies grown in Xeno-Free KO-SR medium suffer from a high percentage of differentiation in the beginning. Sequential adaptation to the new culture medium from 25 to 100% of the new medium has to be used to overcome this problem and, if needed, the adaptation period can be increased. Adaptation is not necessary when RegES or TeSR2 are used.
8. The volume of medium required for the daily medium change is placed in the incubator, for pre-equilibration of temperature and pH. Recommended amounts are 2–2.5 mL per well of a 6-well plate.
9. StemCell Banker is an effective serum- and xeno-free chemically defined freezing procedure for both hESCs and iPSCs. The earlier established slow-freezing protocols have resulted in low viability and thawed cells have had a high tendency to differentiate. StemCell Banker is completely free of serum and animal substances and it contains dimethylsulfoxide, anhydrous dextrose, and a polymer as cryoprotectants.

Acknowledgments

Our hESC and iPSC culture optimization studies have been supported by EU: ESTOOLS, ISCI2, the Swedish Research Council, and the R&D Funds (ALF) of Stockholm County and Karolinska Institute and KID funds of Karolinska Institute. Special Note: Outi Hovatta is a patent holder of the RegES medium.

References

1. Thomson JA, Itskovitz-Eldor J, Shapiro SS, et al. (1998) Embryonic stem cell lines derived from human blastocysts. Science; **282**:1145–7.
2. Reubinoff BE, Pera MF, Fong CY, Trounson A, Bongso A. (2000) Embryonic stem cell lines from human blastocysts: somatic differentiation in vitro. Nat Biotechnol; **18**:399–404.
3. Koivisto H, Hyvarinen M, Stromberg AM, et al. (2004) Cultures of human embryonic stem cells: serum replacement medium or serum-containing media and the effect of basic fibroblast growth factor. Reprod Biomed Online;**9**:330–7.
4. Inzunza J, Gertow K, Stromberg MA, et al. (2005) Derivation of human embryonic stem cell lines in serum replacement medium using postnatal human fibroblasts as feeder cells. Stem Cells;**23**:544–9.
5. Richards M, Fong CY, Chan WK, Wong PC, Bongso A. (2002) Human feeders support prolonged undifferentiated growth of human inner cell masses and embryonic stem cells. Nat Biotechnol;**20**:933–6.
6. Hovatta O, Mikkola M, Gertow K, et al. (2003) A culture system using human foreskin fibroblasts as feeder cells allows production of human embryonic stem cells. Hum Reprod;**18**:1404–9.
7. Unger C, Gao S, Cohen M, et al. (2009) Immortalized human skin fibroblast feeder cells support growth and maintenance of both human embryonic and induced pluripotent stem cells. Hum Reprod. Oct;24(**10**):2567–81. Epub 2009 Jun 25.
8. Yu J, Vodyanik MA, Smuga-Otto K, et al. (2007) Induced pluripotent stem cell lines derived from human somatic cells. Science;**318**:1917–20.
9. Unger C, Felldin U, Nordenskjold A, Dilber MS, Hovatta O. (2008) Derivation of human skin fibroblast lines for feeder cells of human embryonic stem cells. Curr Protoc Stem Cell Biol;**Chapter 1**:Unit 1 C 7.
10. Unger C, Skottman H, Blomberg P, Dilber MS, Hovatta O. (2008) Good manufacturing practice and clinical-grade human embryonic stem cell lines. Hum Mol Genet;**17**:R48-53.
11. Carpenter MK, Rosler E, Rao MS. (2003) Characterization and differentiation of human embryonic stem cells. Cloning Stem Cells;**5**:79–88.
12. Montes R, Ligero G, Sanchez L, et al. (2009) Feeder-free maintenance of hESCs in mesenchymal stem cell-conditioned media: distinct requirements for TGF-beta and IGF-II. Cell Res;**19**:698–709.
13. Watanabe K, Ueno M, Kamiya D, et al. (2007) A ROCK inhibitor permits survival of dissociated human embryonic stem cells. Nat Biotechnol;**25**:681–6.
14. Martin-Ibanez R, Unger C, Stromberg A, Baker D, Canals JM, Hovatta O. (2008) Novel cryopreservation method for dissociated human embryonic stem cells in the presence of a ROCK inhibitor. Hum Reprod;**23**: 2744–54.
15. Ellerstrom C, Strehl R, Noaksson K, Hyllner J, Semb H. (2007) Facilitated expansion of human embryonic stem cells by single-cell enzymatic dissociation. Stem Cells;**25**: 1690–6.
16. Rajala K, Hakala H, Panula S, et al. (2007) Testing of nine different xeno-free culture media for human embryonic stem cell cultures. Hum Reprod;**22**:1231–8.
17. Rajala K, Lindroos B, Hussein SM, et al. (2010) A defined and xeno-free culture method enabling the establishment of clinical-grade human embryonic, induced pluripotent and adipose stem cells. PLoS One;**5**:e10246.
18. Akopian V, Andrews PW, Beil S, et al. (2010) Comparison of defined culture systems for feeder cell free propagation of human embryonic stem cells. In Vitro Cell Dev Biol Anim;**46**:247–58.
19. Holm F, Strom S, Inzunza J, et al. (2010) An effective serum- and xeno-free chemically defined freezing procedure for human embryonic and induced pluripotent stem cells. Hum Reprod;**25**:1271–9.

Chapter 10

Adaptation of Human Pluripotent Stem Cells to Feeder-Free Conditions in Chemically Defined Medium with Enzymatic Single-Cell Passaging

Alexander E. Stover and Philip H. Schwartz

Abstract

This protocol describes the culture of human pluripotent stem cells (PSCs) under feeder-free conditions in a commercially available, chemically defined, growth medium, using Matrigel as a substrate and the enzyme solution Accutase for single-cell passaging. This system is strikingly different from traditional PSC culture, where the cells are co-cultured with feeder cells and in medium containing serum replacement. PSCs cultured in this new system have a different morphology than those cultured on feeder cells but retain their characteristic pluripotency. This feeder-free PSC culture system is conceptually similar to feeder-free systems that use mouse embryonic fibroblast (MEF)-conditioned medium (MEF-CM) and Matrigel substratum. Instead of MEF-CM, a very complex and undefined medium, this new system uses StemPro SFM, a chemically defined medium that permits enzymatic passaging with Accutase to disaggregate the colonies into single cells. Accutase passaging has been used in conjunction with Stempro in our hands for 20+ passages without detectable karyotypic abnormalities. We will also review techniques for adapting cultures previously grown on MEFs, routine passaging of the cells, and cryopreservation.

Key words: feeder-free, stemPro, matrigel, single-cell passaging, accutase, cryopreservation, adapting to feeder-free culture

1. Introduction

Feeder cells support the growth of pluripotent stem cells (PSCs) in culture by contributing an as yet undefined and complex mixture of extracellular matrix components and growth factors. Feeder cells used for the co-culture of PSCs are usually fibroblasts and are usually mitotically inactivated so that they remain viable but cannot replicate and overgrow the PSC culture. Mouse embryonic fibroblasts (MEFs) are the most commonly used

Philip H. Schwartz and Robin L. Wesselschmidt (eds.), *Human Pluripotent Stem Cells: Methods and Protocols*, Methods in Molecular Biology, vol. 767, DOI 10.1007/978-1-61779-201-4_10, © Springer Science+Business Media, LLC 2011

feeder-cell type and have reliably served as feeder cells for co-culture with mouse embryonic stem cells (1, 2) and human embryonic stem cells (3). The search for methods to develop nonxenogenic culture systems has led to the use of human-derived feeder cells, extracellular matrix components, and growth factors. The development of fully defined, nonxenogenic, culture systems is an important milestone for the PSC field and will greatly improve the usefulness of PSCs in basic science programs and, over the longer term, in human therapeutic applications. The use of defined culture systems will eliminate much of the inherent variability in culture media whose components are sourced from animals and are likely to improve our ability to predictably and reliably direct differentiation. One such defined medium is StemPro SFM, sold by Invitrogen. Its exact formulation is proprietary, but it is based on a system known to contain IGF1, heregulin1, and activin A, which act through different tyrosine kinase pathways to maintain pluripotency (4). Like traditional PSC media and other defined media, it also contains FGF2 (bFGF).

Simultaneously, a need has arisen for efficient expansion of PSCs using single-cell passaging. The traditional mechanical passaging used to separate clumps of PSCs from feeder layers is laborious and slow, and does not easily result in the large number of cells that would be desired for extensive experiments or on-demand clinical use. Not only is mechanical passaging a highly qualitative process as the cells cannot be counted when they are in clump form, dissociating the clumps into single cells under traditional culture conditions results in extensive cell death. Enzymatic passaging methods employing 0.05% trypsin, TrypLE Express, and nonenzymatic cell dissociation buffer have been used by some laboratories, but their use has not been for single-cell dissociation and has been frequently shown to generate populations of PSCs with abnormal karyotypes (5). A recent report has shown that Accutase can be used to passage human embryonic stem cells (6) in very small clumps and, therefore, we have begun using Accutase to support single-cell passaging of PSCs. When used in conjunction with defined media, single-cell passaging forms the basis of a modernized PSC culture system.

2. Materials

2.1. Reagents and Supplies

1. Matrigel, reduced growth factor (BD Biosciences, #354231).
2. Dulbecco's phosphate buffered saline without Mg_2 and Ca_2 (DPBS).

3. StemPro SFM kit (includes 50× supplement, DMEM-F12 with GlutaMax, and 25% BSA Solution) (Invitrogen, #A1000701).
4. 2-Mercaptoethanol (2-ME, Invitrogen, #21985-023).
5. Accutase (Millipore, #SCR005). Store in frozen at −20°C in 10 mL aliquots. Once thawed at 4°C, use within 7 days. Take care when thawing, as the enzyme can separate and settle at the bottom of the bottle or tube; mix the freshly thawed Accutase thoroughly prior to aliquotting or adding to cells (see Note 1).
6. Human bFGF/FGF2 (such as Stemgent, #03-0002). Dilute in DPBS, 1% BSA to a concentration of 20 μg/mL, or 1,000×, and freeze in small aliquots at −20°C.
7. Hybri-Max dimethyl sulfoxide (DMSO, Sigma-Aldrich, #D2650).
8. 6-Well vacuum gas plasma-treated tissue culture dishes (such as BD Falcon, #353046).
9. Nylon membrane syringe filter (Pall Life Sciences, #PN 4433).
10. Mr. Frosty Nunc #5100-0001.

2.2. Media

1. StemPro basal medium: Add 25% BSA solution to DMEM/F12/GlutaMax to make a final BSA concentration of 1.8%. Add 2-ME to a final concentration of 55 μM. Sterile filter, and store at 4°C. Discard unused medium after 2 weeks.
2. StemPro complete medium: Add 50× cytokine supplement to StemPro basal medium to a final concentration of 1× (e.g., 1 mL of supplement to 49 mL of basal medium), followed by bFGF to a final concentration of 20 ng/mL. This complete medium is only stable for 24 h, so it is recommended to prepare small volumes and use it immediately.
3. StemPro cryopreservation medium: Combine equal volumes of fresh complete StemPro medium, and 1-day-old "conditioned" medium from the cells themselves. Add DMSO to a concentration of 10%, and sterile filter with a syringe filter approved for use with DMSO (e.g., nylon membrane). Keep cold and use immediately.

3. Methods

3.1. Preparing Matrigel

1. Thaw a 10 mL bottle of Matrigel at 4°C overnight. Do not place it in a water bath or attempt to thaw it rapidly (see Note 2).

2. Prepare 1.5-mL screw-top tubes for receiving 200 μL aliquots of Matrigel. Have the caps loose and easily opened.
3. Chill several 2 mL serological pipets in a −20°C freezer for a few minutes.
4. Remove the Matrigel bottle from 4°C, and quickly spray the entire bottle with 70% alcohol. The rubber stopper caps that BD uses for its bottles are easily contaminated, so make sure the bottle is fully bathed in 70% alcohol.
5. Using a chilled 2 mL serological pipette, distribute 0.2 mL (200 μL) of Matrigel in each screw cap tube. Change pipettes if the Matrigel starts to gel inside the pipette. Matrigel gels at 15°C, so this process must be done quickly. If necessary, take a break, and rechill the pipettes and the stock Matrigel bottle. A repeater pipette may also be used for fast and efficient aliquotting.
6. Store the aliquots of Matrigel at −20°C. 200 μL, when diluted 1:30 into 6 mL of cold medium, is sufficient for one 6-well dish. Different-sized aliquots can be made according to each user's individual needs.

3.2. Plating Matrigel

This protocol describes a 1:30 Matrigel dilution. Less concentrated solutions can also be used, but the optimal concentration depends on the cell type.

1. Take a 200 μL aliquot of stock concentration Matrigel (in a 1.5-mL screw-top tube), and thaw overnight at 4°C.
2. In the morning, add 6 mL of ice-cold DMEM to a 15-mL conical tube.
3. Remove the Matrigel from the 4°C fridge, and immediately mix the 200 μL of Matrigel with the 6 mL of DMEM. If not done immediately after removal from fridge, the Matrigel will polymerize too quickly, and a glob of Matrigel will form, which will be useless for coating plates. An extremely effective way to quickly and safely do this is to take 1 mL of the ice-cold DMEM, and use a 2-mL pipet to rinse all the Matrigel out of the tube. In addition, if the Matrigel prematurely solidifies in solution, the tube may be stored at 4°C until the pellet is no longer visible (i.e., it depolymerizes), and then used for coating.
4. Immediately coat your plate with the 6 mL of working concentration Matrigel that you have just created. If not done immediately, the Matrigel will coat the inside of the conical tube.
5. Place the Matrigel-coated plate at 37°C for at least 30 min to an hour. Longer incubations increase cell attachment. An overnight incubation is preferred for hard-to-stick cell types.
6. Aspirate excess Matrigel before adding cells. There is no need to rinse the plate.

3.3. Transitioning PSCs from Co-culture with Feeder Cells to StemPro/Matrigel Culture

It is recommended that cultures be adapted to feeder-free conditions with defined medium before routinely subculturing by single-cell passaging. Although a sudden, complete transition is certainly possible (we have done it several times in our laboratories), some cell lines do not tolerate it well, so it is not recommended. A more gradual shift minimizes the shock to the cells and provides the best possible chance of a smooth transition with little cell death or differentiation.

1. Start feeding a feeder-cell-grown culture with a mixture of 50% StemPro complete medium/50% standard growth medium daily, 2–3 days prior to the day when the culture would normally be passaged.
2. 24 h Prior to passaging, feed the PSC culture with 100% StemPro complete medium.
3. On the day of passaging, exchange the medium with fresh StemPro complete medium and mechanically passage colonies onto a fresh Matrigel plate. Simply pipette the lifted colonies onto the receiving Matrigel-coated plate (see Note 3).
4. Feed the cultures daily with StemPro complete medium until the colonies have grown so that an average colony on the plate completely fills a 10× objective view. Some moderate differentiation may appear during this adaptation phase. Remove differentiated cells and colonies mechanically (see Chapter 8).
5. When the colonies are large enough to be passaged, proceed with Accutase passaging.

3.4. Single-Cell Passaging of PSCs with Accutase

1. Aspirate the medium from the culture.
2. Rinse well using 1 mL of DPBS/well and aspirate.
3. Add 1–2 mL of 37°C Accutase (fresh from the water bath or incubator) to each well. Only very dense cultures should require 2 mL (see Note 4).
4. Immediately take the culture dish to a phase-contrast microscope, and observe the culture carefully. Watch for signs that that the individual cells are starting to loosen from the dish. This is characterized by both an increase in phase brightness and a more rounded appearance. At the periphery of colonies, where you can commonly see cells that have membrane stretched across the culture dish, you will see this membrane detach and the cells visibly loosen from the Matrigel-coated dish. After 1–2 min, the vast majority of the cells should appear very phase bright and rounded. There is no need to wait until the cells float into suspension, nor to tap or shake the culture dish to induce this effect.
5. Return the cells to the tissue culture hood. Using a 5-mL serological pipet, add 5 mL of DPBS for each milliliter of Accutase

to dilute the enzyme, and then forcefully pipet up and down to dislodge the cells from the well surface. Be careful do not splatter/splash any of the resulting cell suspension (the wells will be quite full). If passaging multiple wells, add the DPBS to each well before dislodging the cells from the individual wells. This helps guard against Accutase overexposure. In our experience, diluting 1 mL of Accutase with 5 mL of DPBS is more than sufficient to prevent Accutase overexposure.

6. Transfer the Accutase/PBS cell suspension to an appropriately sized conical tube.
7. Rinse the well with an additional 5 mL of DPBS to remove remaining cells, especially around the edges of the well.
8. Spin the cells at $100 \times g$ for 5 min at room temperature.
9. Aspirate the supernatant, and resuspend the cells in 2 mL of StemPro complete medium.
10. Count the cells using a hemacytometer, and determine the number of receiving wells or plates required. Cells should be plated at 5×10^4–1×10^5 cells/cm^2 (see Notes 5 and 6).

3.5. Cryopreservation of Accutase-Passaged Cells

With traditional PSC cryopreservation, initial viability following a thaw tends to be very low and it may take up to 2 weeks for even a single colony to appear on the plate. When freezing PSCs as a dense single-cell suspension using StemPro and conditioned medium (see below), thawing viability is greatly improved, although there is still some noticeable cell death. We have found that the use of the apoptosis-blocking Y27632 ROCK inhibitor is not necessary for successful cryopreservation using the method described below (see Note 7).

1. Save spent culture medium (conditioned medium, CM) by removing it with a serological pipette and placing it into a sterile conical tube for later use in the procedure.
2. Rinse the cells with DPBS, and lift with Accutase as described above in Subheading 3.4.
3. While the cells are in the centrifuge, prepare the freezing medium containing 45% conditioned medium, 45% fresh StemPro complete medium, and 10% DMSO. Chill at 4°C.
4. Aspirate the supernatant from the pelleted cells, and resuspend the cell pellet in 1 mL of freezing medium for each well lifted (~3×10^6 cells/mL).
5. Dispense 1 mL of the cell suspension to a cryopreservation vial (1 vial for each well of 6-well plate).
6. Place the vials in isopropanol-jacketed freezing containers (Mr. Frosty) and place at −80°C overnight.
7. The next day, transfer the vials to long-term storage at cryogenic temperatures below −130°C.

3.6. Thawing Cells in StemPro Complete Medium

1. Remove a vial from the liquid nitrogen.
2. Rapidly thaw in a 37°C water bath. Gently swirl the vial in the water, being careful not to submerge the cap or get the water near the cap threads. Thawing should take about 60 s. Any unnecessary time the cells spend in DMSO-containing medium at room temperature will result in increased cell death.
3. Spray the vial with 70% alcohol, wipe with a tissue, and allow to air dry briefly in the tissue culture hood.
4. Transfer the contents of the tube to a 15-mL conical tube.
5. Slowly and drop wise, add 10 mL of fresh StemPro complete medium to the cells while swirling the tube to promote even mixing. Do not add the medium along the side of the tube – this will result in a medium gradient forming. Instead, have the drops of media fall directly vertical into the tube. Gently triturate once or twice before capping the tube.
6. Centrifuge the cells at 150 × *g* for 5 min at room temperature, and aspirate the supernatant.
7. Resuspend the pellet in 2 mL of fresh StemPro complete medium and seed onto one well of a Matrigel-coated plate.
8. Feed and observe daily.
9. Begin passaging with Accutase when the cells reach confluence (see Notes 8 and 9).

4. Notes

1. Accutase: The exact formulation of Accutase is proprietary, but it is known to contain proteases and collagenases of non-mammalian and nonbacterial origin. It was originally developed by Innovative Cell Technologies of San Diego, CA and is distributed by several vendors, including Thermo Fisher-Hyclone, Sigma, and Millipore. We have found no functional differences between the different brands of Accutase.
2. Matrigel: Matrigel is an undefined mixture of extracellular matrix proteins isolated from the Engelbreth-Holm-Swarm (EHS) mouse sarcoma cell line. It is a solid when stored at −20°C and a liquid when thawed and stored at 4°C, but at 15°C it turns to a colloid that is difficult to manipulate. For this reason, when diluting, pipetting, and coating plates with Matrigel, it must be worked with very quickly and it must be kept cold.
3. Do not worry about any MEFs that are carried over during the initial transition – they will die off and disappear. Use a higher passaging ratio than you would normally use (i.e., if you normally passage cells 1 well to 6 wells, passage to only 2 or 3 wells).

4. Accutase passaging of PSCs on Matrigel: Although Accutase is relatively gentle to cells compared to harsher enzymes such as trypsin, it reacts with cells extremely quickly, and exposure to the enzyme must be kept to a minimum. When performed properly, lifting cells with Accutase can take as little as 1 min (this can vary with culture density). Second, although the manufacturers of Accutase claim that the enzyme(s) self-digest and do not require inactivation, we have found that Accutase still needs to be either diluted or completely removed after it has served its purpose. In other words, although Accutase passaging is easy, it must be done carefully and quickly.

   Temperature is another variable that affects Accutase performance. Some laboratories use room-temperature (25°C) Accutase to passage their cells. This has the overall effect of slowing down the reaction, meaning digestion can take up to 10 min. In addition, the cells will tend to come up in clusters rather than as single cells, perhaps due to incomplete digestion of the attachment proteins.

   A further consideration is the appearance of the cells once they have been passaged with Accutase. The cultures may appear differentiated to researchers used to standard culture morphologies – they typically have large amounts of membranous material surrounding them, and the individual cells will appear abnormally large and phase dark. However, close observation of the cells should show that the cells are still grouping together, and rarely appear alone even when recently seeded as single cells. They will attain a high nucleus-to-cytoplasm ratio as they expand, and the nuclei should have prominent nucleoli. Several days after passaging (if not the next day), it should be obvious at low magnification that the cells are organized as colonies, and not as single cells (see Fig. 1). We typically passage cells when they first form a monolayer that covers the entire surface area of the well/dish, or slightly beforehand.

5. Seeding density: The cells migrate to form colonies and, if the seeding density is low, they will have difficulty doing this. Under-seeded plates can eventually recover, but will experience more differentiation and apoptosis than normal. Over-seeded plates, on the other hand, will simply need to be passaged sooner.

6. Cell concentration: In our experience, one confluent well generates 3–6 million cells, so it is possible to perform routine 1:6 splits every 5–7 days. This should help predict how many Matrigel plates should be prepared in advance of passaging. In addition, it should help predict the rate of expansion, thus helping preparations for experiments involving large numbers of cells. For example, one plate initially containing 3×10^6

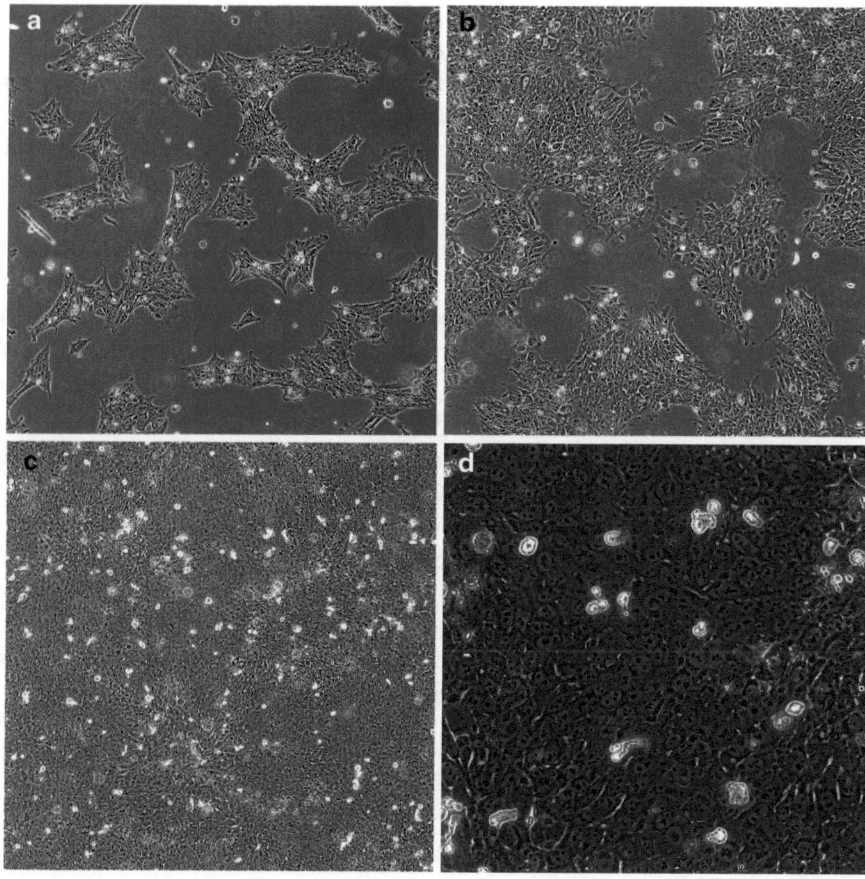

Fig. 1. Growth progression and morphology of hESCs when plated as single cells. (**a**) Despite being seeded as single cells, Accutase-passaged cells will quickly migrate to form small colonies with lots of obvious membranous material (shown with a 10× objective at day 1 after seeding). (**b**) After 2 days, the colonies have gradually grown larger and begun to fuse. (**c**) After 4–5 days, the small colonies have expanded so that the entire area of the cell culture well is essentially one large, monolayer colony. This is when the cells should be passaged to avoid cell death from underfeeding the dense culture. (**d**) At 40× magnification, dense Stempro cells appear very similar to traditional feeder-grown colonies. *Note*: The presence of small numbers of floating cells is normal. Larger numbers indicate an underfed culture.

cells, passaged weekly, will generate 3.9×10^9 cells in 1 month (1,296 plates)!

7. Cryopreservation: This frozen cell concentration is much greater than the concentration typically used with other cell types (about 1×10^6 cells). However, the high number of cells is necessary for PSC survival; these cells appear to depend on being able to group together shortly after being seeded, otherwise they differentiate or die. Also, it is not unusual to experience some cell death after any thaw. Increasing the frozen cell concentration tends to ensure a more successful thaw.
8. A note on aneuploidies: Although we have successfully grown cells in StemPro with accutase passaging for 20+ passages without seeing genetic abnormalities, we have observed some

higher passages picking up common PSC trisomies, such as 12 and 17. It needs to be noted, however, that all culture systems (including traditional mechanical passaging) have had reported (either in the literature or in anecdote) incidents of aneuploidies, and some of this may be related to individual laboratories' handling of their cells. For example, Accutase overexposure and underfeeding may contribute to stress on the cultures, and, over time, select for a subpopulation of cells with both an abnormal karyotype and a proliferative advantage. The exact mechanisms are not yet fully understood. Therefore, it is extremely important to regularly karyotype PSCs grown in any culture system, and to keep reserves of earlier passages banked.

9. You can easily seed Accutase-passaged cells onto a MEF layer in traditional PSC medium and obtain "normal"-looking colonies once again. They take on a "normal" appearance less than 24 h after seeding. Since the seeding densities prescribed in this chapter are for Matrigel-coated plates, we recommend reducing the density to take into account the surface area occupied by feeders. Otherwise, the culture will be very crowded soon after seeding.

Acknowledgments

This work has been funded by the National Institutes of Health (T15HL074286, R21MH087925, R01HD059967). The NCI Preclinical Repository supplied FGF-2.

References

1. Martin, G. R. (1981) Isolation of a pluripotent cell line from early mouse embryos cultured in medium conditioned by teratocarcinoma stem cells. Proc Natl Acad Sci, 78, 7634–8.
2. Evans, M. J., and Kaufman, M. H. (1981) Establishment in culture of pluripotential cells from mouse embryos. Nature, 292, 154–6.
3. Thomson, J. A., Itskovitz-Eldor, J., Shapiro, S. S., Waknitz, M. A., Swiergiel, J. J., Marshall, V. S., Jones, J. M. (1998)Embryonic stem cell lines derived from human blastocysts. Science, 282, 1145–7.
4. Wang, L., Schultz, T. C., Sherrer, E. S., Dauphin, D. S., Shin, S., Nelson, A. M., Ware, C. B., Zhan, M., Song, C. Z., Chen, X., Brimble, S. N., McLean, A., Galeano, M. J., Uhl, E. W., D'Amour, K. A., Chesnut, J. D., Rao, M. S., Blau, A., and Robins, A. J. (2007) Self-renewal of human embryonic stem cells requires insulin-like growth factor-1 receptor and ERBB2 receptor signaling. Blood 110, 4111–119.
5. Mitalipova, M. M., Rao, R. R., Hoyer, D., Jones, K., Johnson, J., Dalton, S., Meisner, L., and Stice, S. L. (2005) Preserving the genetic integrity of human embryonic stem cell lines. Nature Biotechnology. 23, 19–20.
6. Bajpai, R., Lesperance, J., Kim, M., Terskikh, A. V. (2008) Efficient propagation of single cells Accutase-dissociated human embryonic stem cells. Mol Reprod Dev., 75, 818–27.

Chapter 11

GMP Scale-Up and Banking of Pluripotent Stem Cells for Cellular Therapy Applications

Lara J. Ausubel, Patricia M. Lopez, and Larry A. Couture

Abstract

Human pluripotent stem cells (PSCs), which include human embryonic stem cells (ESCs) as well as induced pluripotent stem cells (iPSCs), represent an important source of cellular therapies in regenerative medicine and the study of early human development. As such, it is becoming increasingly important to develop methods for the large-scale banking of human PSC lines. There are several well-established methods for the propagation of human PSCs. The key to development of a good manufacturing practice (GMP) bank is to determine a manufacturing method that is amenable to large-scale production using materials that are fully documented. We have developed several banks of hESCs using animal feeder cells, animal-based matrices, or animal-free matrices. Protocols for growing hESCs on mouse embryonic fibroblasts (MEFs) are well established and are very helpful for producing research grade banks of cells. As most human ESCs cultured by research laboratories have been exposed to xenogeneic reagents, it is not imperative that all materials used in the production of a master cell bank be animal-free in origin. Nevertheless, as the field develops, it will no doubt become increasingly important to produce a bank of cells for clinical use without xenogeneic reagents, particularly nonhuman feeder cells which might harbor viruses with potential risk to human health or cell product integrity. Thus, even for cell lines previously exposed to xenogeneic reagents, it is important to minimize any subsequent exposure of the cell lines to additional adventitious agents. We have specifically described procedures for the growth of hESCs on Matrigel, an animal-matrix, and CELLstart, an animal-free matrix, and these can be used to produce hESCs as part of a clinical manufacturing process.

Key words: human embryonic stem cells, good manufacturing process, cell banking, matrigel, CELLstart

1. Introduction

Ever since the isolation of human embryonic stem cells (hESCs) was first described by Thomson and colleagues (1), individuals have been looking for ways to translate this basic research into

Philip H. Schwartz and Robin L. Wesselschmidt (eds.), *Human Pluripotent Stem Cells: Methods and Protocols*, Methods in Molecular Biology, vol. 767, DOI 10.1007/978-1-61779-201-4_11, © Springer Science+Business Media, LLC 2011

the clinic in order to treat human disease. Recently, several clinical-grade hESCs have been generated (2) and the FDA has started approving investigational new drug applications for the use of differentiated cells derived from hESC lines to treat humans. As such, it is becoming increasingly important to develop methods for the large-scale banking of human pluripotent stem cell (hPSC) lines that are suitable for eventual clinical use and thus follow good tissue practice (GTP) and good manufacturing practice (GMP) requirements (3).

There are many challenges to manufacturing stem cell-based products. Although currently there are no FDA guidelines specific to the generation of hPSCs for clinical use, one can turn to the general guidelines for manufacturing cell-based products and human gene therapy investigational new drug applications (4, 5). One must insure that the starting material, or PSC line, is qualified extensively. For human ESCs, this may involve genetic fingerprinting (STR and SNP), the determination of expression of intracellular and cell surface stem cell markers (FACS and RT-Q-PCR), the presence of a normal chromosomal array (karyotyping), as well as the epigenetic profile of the PSCs (6, 7). As few reagents used in the manufacturing process are listed as GMP grade, one must thoroughly investigate the fitness of the biomaterials to be used for clinical manufacturing. This includes demonstrating that the reagents work reliably, are free from contaminants or adventitious agents, and can produce a consistent product. In order to do this, Certificates of Analysis or Certificates of Origin for products must be scrutinized in order to determine the suitability of the origin of any animal-derived products. Often, custom manufacturing agreements need to be in place in order to obtain large quantities of identical lots of any reagents that may have lot-to-lot variability. In addition, in-house testing needs to be performed to show that consistent results are obtained and that products that meet quality standards can be produced repeatedly with minimum variability.

2. Materials

2.1. Banking on Matrigel

1. Matrigel™ hESC-qualified matrix (BD Bioscience, #354277) (see Note 1). Matrigel should be aliquoted according to manufacturer specifications for that particular lot and frozen. Aliquots of Matrigel™ can be stored at −80°C for up to 6 months.
2. Dispase (Invitrogen, #15105-041). Prepare a working solution of dispase at 1 mg/mL and filter through a 0.2-μm filter. Diluted dispase can be stored at 4°C for up to 2 weeks.

3. mTeSR Medium Kit (Stemcell Technologies, #05850) (see Note 2). Thaw mTeSR 5× supplement at room temperature or overnight at 2–8°C. Aseptically add the entire 100 mL of the thawed 5× supplement to the 400 mL basal medium for a total volume of 500 mL and mix well. The complete mTeSR medium is stable when stored at 2–8°C for up to 2 weeks (see Note 3). If desired, complete medium can be aliquoted and stored frozen at −20°C. Use aliquots within 3 months. Do not re-freeze aliquots after thawing.

2.2. Banking on Cellstart

1. CELLstart (Invitrogen, Carlsbad, CA, #A10142-01) (see Note 4). CELLstart should be aliquoted but never frozen. Aliquots of CELLstart can be stored at 4°C for up to 12 months, but should be protected from light. Dilute CELLstart 1:50 in PBS with Ca^{2+} and Mg^{2+} (Irvine Scientific #9236) right before use.
2. b-FGF solution is prepared by dissolving 10 μg b-FGF (Invitrogen, #13256-029) into 1 mL 0.1% BSA in DPBS with Ca^{2+} and Mg^{2+} that has been filtered through a 0.2-μm filter. Aliquot into 80 μL aliquots and store for up to 6 months at −20°C.
3. StemPro media kit (Invitrogen, #A1000701) (see Note 5). Thaw supplement in 37°C water bath. In order to make 500 mL of partially complete StemPro medium, add 454 mL DMEM-F12 (provided in kit), 10 mL StemPro growth supplement (provided in kit), 36 mL BSA 25% (provided in kit), and 400 μL b-FGF solution at 10 μg/mL (for a final concentration of 8 ng/mL). The partially complete StemPro medium is stable when stored at 2–8°C in the dark for up to 1 week. In order to make complete StemPro medium, supplement 500 mL of partially complete StemPro media with 909 μL 2-ME (Invitrogen, #21985-023). The 2-ME should be added daily, to the required daily volume of partially complete StemPro medium, just prior to use to make it complete.
4. TrypLE (Invitrogen, #12563-011) can be used for passaging the cells (see Note 6).

2.3. Cryopreservation

Freezing medium is composed of 30% FBS (Hyclone, #SH30070.03, see Note 7), 10% DMSO/Stemsol (Protide Pharmaceuticals, #PP1130), and 60% growth medium. A 2× solution of 60% FBS, 20% DMSO, and 20% growth medium should be prepared and mixed in a 1:1 ratio with cells in growth medium from cultures. Sterile filter the FBS and growth medium through a 0.2-μm filter. Add sterile DMSO to the filtered solution. It is highly recommended to use a sterile syringe and needle to withdraw the DMSO. Store at 4°C until ready for use (see Note 8).

Cells are frozen in 1.2-mL cryogenic vials.

3. Methods

A key factor to the development of GMP-grade banks of PSCs is the use of processes that can be scaled up consistently. Although many laboratories can successfully produce high-quality PSCs using manual passaging of cells, these types of methods are not easily replicated at large scale. More appropriate may be methods that employ some type of chemical passaging, although it is imperative to show that such methods do not cause chromosomal abnormalities in the cells or affect their pluripotency. In addition, if these enzymes are animal-based, the final product will have to be tested for adventitious agents associated with that animal species.

Protocols for growing human PSCs on MEFs are well established and are very helpful for producing research grade cell banks. Since most human PSCs that are cultured by research laboratories have been exposed to xenogeneic reagents, it is not imperative that all reagents used in the production of a master cell bank be animal-free in origin. In addition, many laboratories have successfully grown human PSCs using human feeder cells such as human fibroblasts (8) or using human fibronectin (9). With the goal of implementing scalable processes that are free from animal feeder cells, banks of human PSCs can be generated using feeder-free methods. One method relies on an animal-based matrix (Matrigel™) and therefore would require the final product to be tested for potential mouse-derived adventitious agents (10). Another, CELLstart, uses a xenogeneic-free matrix (11). Such methods should serve as a starting point with which to determine the best procedure for expansion of a specific PSC line.

3.1. Banking on Matrigel™

3.1.1. Coating Plates with Matrigel™ (see Note 9)

1. Thaw one aliquot of Matrigel™ at 4°C on ice. Keep Matrigel™ on ice before use and use precooled medium, pipettes, tips, and tubes.
2. Add one aliquot of Matrigel to 25 mL of cold DMEM-F12 and mix well. Immediately use the diluted Matrigel solution to coat 4× 6-well plates (1 mL/well). Adjust volume for other vessels depending on the surface area. Swirl the plate to spread the solution evenly (see Note 10).
3. Keep Matrigel™-coated plates at room temperature for at least 1 h before use. Coated plates can be sealed with parafilm and stored at 4°C overnight.

3.1.2. Initial Thaw of Vial of PSCs

1. Thaw the vial of PSCs in a water bath at 37°C for 30–60 s.
2. Spray the vial with 70% isopropanol and transfer to a biosafety cabinet (see Note 11).
3. Transfer the contents of the vial to a 15-mL conical tube. Add 4 mL of mTeSR medium drop wise while gently shaking the 15-mL tube.

Fig. 1. WA09 (H9) hESCs grown on feeder cells or matrices. WA09, p36 were passaged onto either MEFs (**a**) or Matrigel (**b**) coated 6-well plates. Medium was changed daily after the first 48 h and cells were imaged on day 6 under 4× power. WA09, p49 were passaged onto a CELLstart (**c**) coated 6-well plate. Medium was changed daily after the first 48 h and cells were imaged on day 4 under 4× power.

4. Wash emptied vial with one additional milliliter of medium and transfer to the same conical tube.
5. Pipette up and down once to mix.
6. Spin at 200 × *g* for 5 min at room temperature.
7. Just prior to seeding the PSCs, aspirate matrix solution from plates and discard.
8. Decant supernatant and resuspend pellet in 1 mL of mTeSR medium. Pipette slowly to avoid breaking cell clumps into single-cell suspension.
9. Cell aggregates can be seeded in 2–4 wells of a 4-well plate (1 mL/well) or a single well of a 6-well plate (2.5–4.0 mL/well) (see Note 12).
10. Aliquot the PSC aggregates drop wise into the Matrigel™ coated wells.
11. Incubate at 37°C, 5% CO_2.
12. Change medium daily after 24–48 h (see Note 13). An example of a hESC line grown on Matrigel™ is shown in Fig. 1.

3.1.3. Passaging of Cells for Cultures Not Highly Differentiated (<25% Differentiation)

1. Warm mTeSR, dispase solution, and DMEM-F12 to room temperature before use.
2. Aspirate spent medium from wells and discard.
3. Wash 1× with warm DMEM-F12 and discard wash.
4. Add 1 mL of Dispase solution per well. Incubate at 37°C for 5–10 min. Colonies should start to fold upward (see Note 14).
5. Add 2 mL/well of DMEM-F12 to dilute the enzyme and use a sterile glass 1-mL pipette to gently scrape the colonies off.
6. Collect the cell colony pieces into a conical tube.

7. Wash wells with 1 mL of DMEM-F12 and collect wash.
8. Spin colonies at $200 \times g$ for 5 min in a microcentrifuge at room temperature.
9. Decant supernatant and resuspend pellet in 1–2 mL of mTeSR media. To mix, pipette gently only a few times to prevent colonies from breaking into pieces that are too small.
10. Add enough mTeSR media to increase the volume as is needed to seed wells (4 mL/well).
11. Right before seeding, aspirate Matrigel™ from Matrigel™-coated plates and discard.
12. Wells are seeded to an appropriate confluence, with an appropriate number of colonies, or at a specific split ratio. If an appropriate concentration is not known, seed with a range of seed densities (1:3, 1:6, 1:10 dilution). A typical scale-up to a 100-vial bank generated on Matrigel™ is provided (Table 1).
13. Swirl plates to distribute cells evenly across the surface of the wells.
14. Change medium daily after 24–48 h.
15. Re-passage after cells become confluent (usually, 1–2 times per week) and scale up accordingly (see Note 15).

Table 1
Scale-up in production of a 100-vial bank of hESCs

Week	Activity	Scale
1 (p27)	Thaw H9 p27 on Matrigel coated plate	1 Well of 6-well plate seeded
2 (p28)	Enzymatic passage (dispase)	1:12 Split 2× 6-Well plates seeded (12 wells)
3 (p29)	Pick to remove followed by dispase (differentiation > 20%)	1:6 Split 4× 6-Well plates seeded (24 wells)
4 (p30)	Pick to keep without dispase (differentiation > 50%)	Approximately 6 colonies placed per well 2× 6-Well plates seeded (11 wells)
5 (p31)	Enzymatic passage (dispase) (differentiation < 10%)	1:10 Split 17× 6-Well plates seeded (102 wells)
6 (p32)	Enzymatic passage (dispase) (differentiation < 10%)	101 Vials frozen

WA09, p27 cells were thawed and grown on Matrigel™ coated plates. Cells were passaged once a week at the scale indicated until enough cells existed to freeze a bank of approximately 100 vials (2×10^6 cells/vial)

3.2. Banking on CELLstart

3.2.1. Coating Plates with CELLstart (see Note 10)

1. Add diluted CELLstart solution to 6-well plate (1 mL/well) or any other appropriately sized vessel. Swirl the plate to spread the solution evenly.
2. Incubate the CELLstart-coated plate at 37°C for at least 2 h before use. Coated plates can be sealed with parafilm and stored at 4°C overnight.

3.2.2. Initial Thaw of Vial of hESCs

Follow procedure for initial thaw of hESCs (Subheading 3.1.2) except replace mTeSR medium with StemPro complete medium. An example of an hESC line grown on CELLstart is shown in Fig. 1.

3.2.3. Passaging of Cells for Cultures Not Highly Differentiated (≤25% Differentiation)

1. Warm StemPro complete medium, TrypLE solution, and DMEM-F12 to room temperature before use. If coated plates were kept at 4°C, warm to room temperature.
2. Aspirate spent medium from wells and discard.
3. Wash once with warm DMEM-F12 and discard wash.
4. Add 1 mL of TrypLE solution per well. Incubate at 37°C for 30–60 s (see Note 14). Tap the plate against the palm of the hand a few times to detach the cells.
5. Add 2 mL/well of DMEM-F12 to dilute the enzyme and use a sterile glass 1-mL pipette to gently scrape the colonies off the well.
6. Collect the colony pieces into a conical tube.
7. Wash the wells with 1 mL of DMEM-F12 and collect wash.
8. Spin colonies at $200 \times g$ for 5 min at room temperature.
9. Decant supernatant and wash cell pellet with DMEM-F12. Add medium drop wise while shaking the tube to mix cells.
10. Spin colonies at $200 \times g$ for 5 min at room temperature.
11. Decant supernatant and resuspend the pellet in 1–2 mL of complete StemPro medium. To mix, pipette gently only a few times to prevent colonies from breaking into pieces that are too small.
12. Add enough complete StemPro medium to increase the volume as is needed to seed wells (4 mL/well).
13. Just before seeding, aspirate the CELLstart from the CELLstart-coated plates and discard.
14. Wells are seeded to an appropriate confluence, with an appropriate number of colonies, or at a specific split ratio. If an appropriate concentration is not known, seed with a range of seed densities (1:2, 1:4, 1:8 dilution) (see Note 16).
15. Swirl plates to distribute cells evenly across the surface of the wells.
16. Re-passage after cells become confluent (usually, 1–2 times per week) and scale up accordingly (see Note 15).

3.3. Methods to Clean Up Stem Cell Cultures (see Note 17)

3.3.1. Pick to Remove (When Differentiation Is 25–50%)

1. Wipe an inverted microscope with 70% IPA and place inside the biosafety cabinet.
2. Remove half of the spent medium from each well and discard.
3. Use a glass 1-mL pipette to gently scrape the differentiated colonies. Sterile pipette tips, such as a P200 tip, can be used as well. Discard the remaining spent medium with the differentiated colonies.
4. Add 1 mL of the appropriate enzyme per well and incubate for an appropriate period of time.
5. Add 2 mL of basal media to dilute the enzyme and use a sterile glass 1-mL pipette to gently wash/scrape off the undifferentiated colonies. Collect the cell aggregates in a conical tube.
6. Wash the wells with 1 mL of basal medium. Collect the wash.
7. Spin the cells at 200×g for 5 min in a centrifuge at room temperature.
8. Wash pellet with DMEM-F12, gently.
9. Spin the cells at 200×g for 5 min at room temperature.
10. Decant the supernatant and resuspend the pellet in 1–2 mL of growth medium. To mix, pipette gently only a few times to prevent colonies from breaking into pieces that are too small.
11. Add enough growth medium to bring up the volume as is needed to seed wells (up to 4 mL/well) of a 6-well plate.
12. Wells are seeded at an appropriate previously determined density.

3.3.2. Pick to Keep (When Differentiation Is >50%)

1. Wipe an inverted microscope with 70% IPA and place inside the biosafety cabinet.
2. Aspirate all spent medium from wells and discard.
3. Add 1 mL of fresh growth medium.
4. Collect undifferentiated colonies by gently scraping them off using a sterile 1-mL glass pipette. Sterile pipette tips, such as a P200 tip, can be used as well. Work one well at a time for no longer than 15 min. If longer time is needed, place the plate back in the incubator for 15–30 min to regain appropriate temperature and pH.
5. Collect the colony pieces and transfer into a conical tube.
6. Add enough growth medium to increase the volume as is needed to seed an appropriate number of new wells (4 mL/well) of a 6-well plate.

3.4. Cryopreservation and Cell Banking (see Notes 18 and 19)

1. When a sufficient number of cells have been produced, calculate the amount of freezing medium (50% complete growth medium and 50% 2× freeze medium) necessary. Label an appropriate number of cryogenic vials and chill on ice. Place the growth medium and 2× freeze medium on ice to chill (see Note 20).
2. Prepare 2× freezing medium just prior to freezing the cells.
3. Collect cells using the appropriate enzyme following the same protocol for that is used for passaging the cells.
4. Spin cells at 200 × *g* for 5 min at room temperature.
5. Wash pellet once with DMEM-F12.
6. Spin cells at 200 × *g* for 5 min at room temperature.
7. Transfer the prelabeled cold vials, cold growth medium, and cold freeze medium to an ice tray and place in the biosafety cabinet after a thorough cleaning with 70% isopropanol.
8. Loosen the caps on the cryogenic vials. Generally, cells are frozen in 1.2 mL cryogenic vials with 1.0 mL of solution per vial.
9. After the cells have been pelleted, remove and discard the supernatant.
10. Gently resuspend the pellet with prechilled complete growth medium in a volume equal to half of the total volume (i.e., if the total volume is 20 mL for 20 vials, then resuspend cells with 10 mL of growth medium).
11. Very slowly and while shaking the tube, add an equal volume of prechilled 2× freeze medium. To mix, pipette gently only a few times to prevent colonies from breaking into too many pieces.
12. Aliquot the cell solution into cryogenic vials, making sure to gently resuspend the solution during the procedure. This is particularly important when larger banks (i.e., >100 vial banks) are being produced. For optimal recovery, the cells should be kept on ice at all times once in DMSO.
13. Transfer filled vials on ice to a controlled-rate freezer.
14. Once frozen, the cells should be transferred within a 24-h period to a liquid nitrogen tank for long-term storage in the vapor phase below −130°C.
15. Vials should be thawed at various times (i.e., 1 week, 1 month, and 6 months) after the freeze (stability test) to determine the ability of the cells to be re-cultured, to be pluripotent, and to express an appropriate panel of stem cell markers (5, 0).

4. Notes

1. It is important to use the hESC-qualified version of Matrigel™. This product is marketed for research purposes only, is animal-based, and is not a defined substrate. Testing is now performed by the manufacturer to insure that it is free from the mouse virus LDEV (lactate dehydrogenase-elevating virus).
2. mTeSR is reported to be a complete, defined medium. Check to ensure that the lot numbers of both components, 5× Supplement and Basal Medium, end with the same letter. Stemcell technologies have indicated that mTeSR has been designed to specifically work well with Matrigel™ as a matrix. Thus, only lots of mTeSR that allow for the healthy growth of hESC lines on Matrigel™ are sold. This combination of products (Matrigel™ with mTeSR medium) should provide consistent results with little lot-to-lot variability.
3. It is important to note that antibiotics are not used in GMP manufacturing.
4. CELLstart is marketed to be a defined, humanized substrate.
5. StemPro is marketed to be a defined, xeno-free medium.
6. It is possible to use other enzymes with CELLstart. Invitrogen recommends the use of TrypLE or Accutase (Millipore #SCR005), but dispase can be used successfully as well.
7. It is important to check that the lot of FBS to be used is suitable for GMP manufacturing.
8. There are multiple freezing media that can be used. The freezing method listed is one that is reported to be used by the WiCell Research Institute to freeze hESCs. Some other examples include mFreSR (Stemcell Technologies), Xeno-FREEze (Millipore), and hESfreeze (Globalstem). It is important to test these reagents before banking your specific cell line to determine its stability in a variety of freezing media.
9. One of the key features to GMP banking of cells is to keep complete records of all procedures. All steps in the manufacturing process should be detailed in a protocol or standard operating procedure (SOP) and any deviations from that procedure should be noted. The specifics of all reagents and equipment used should be recorded. In addition, the operators performing the procedure (listed by name) should document when and where the manufacturing process took place.
10. Cells can be grown in a variety of size vessels including, but not limited to, 4-well plates, 6-well plates, 25 cm^2 flasks,

75 cm^2 flasks, and 150 cm^2 flasks. These can be Nunc, Falcon, or any other tissue culture-treated plate or flask.

11. A properly calibrated biosafety cabinet should be used for all manipulations. It is essential that all equipments be fully functioning and calibrated and that products are segregated to the best means possible to avoid any mixing of samples.
12. Overfeeding with 4 mL of medium is recommended early in the split schedule when skipping medium change for 24–48 h. Otherwise, feed with 2.5–3.0 mL/well when 6-well plates are used.
13. As a back-up during the first medium change on newly thawed PSCs, the spent medium (and possibly nonadhered colony pieces) can be plated onto a newly prepared matrix-coated well. Supplement the cells with 1–2 mL medium and continue to change medium daily.
14. Avoid overexposure to enzyme. Colonies should not completely lift off of matrix.
15. The size of the bank desired as well as the ability of the hPSC line to grow at high density will dictate the scale of production necessary for the generation of sufficient cells for banking. There are numerous options for vessels to use in which to grow the adherent PSCs. It is best to test the ability of your specific cell line to grow on dishes, plates, as well as larger single layer flasks. For example, a 100 vial hESC bank would require 17× 6-well plates if the cells grow to 2×10^6 cells/well and cells are frozen at 2×10^6 cells/vial (Table 1). It is best for sterility issues as well as technical issues to be able to grow cells in the largest flask size that still allows the cells sufficient cell to cell contact as well as the appropriate aeration. During later passages of the cells, just before freezing down the bank, when cells of high quality (limited differentiation) are obtained, multilayer flasks such as the Nunc T-500 flask (Nalgene, Nunc, Rochester, NY) can also be tried. At very large scale, larger vessels such as a Nunc single layer factory (630 cm^2) can be used. Often, the cells will divide at a slower pace with the larger size vessels and thus the passage schedule may change slightly as production scale increases. It is always helpful to perform preliminary experiments testing the growth of the specific hESC line in a variety of vessel sizes.
16. CELLstart cultures should be maintained at a high density. This may require that the cells be passaged twice per week. It is always best to begin to grow the cells at high density and then set up a few conditions where the cells are split to lower density to insure that the cells continue to grow well with minimal differentiation (less than 10%). We have found 1:4 to be a good split ratio.

17. It is best to minimize the need to manually manipulate cells during the last few passages of a scale-up procedure. In order to proceed to scale-up and production of a cell bank, the culture should be extremely healthy with minimal differentiation.
18. hESC lines to be banked should be karyotyped prior to scale-up. In addition, a full profile of the stem cell characteristics of the cells should be assessed. This could include expression of cell surface markers or intracellular markers by FACS, Q-PCR, or immunocytochemistry. It is helpful to perform in-process testing to monitor the expression level of these markers just before freezing (5, 6).
19. It is important to do a test before banking your specific cell line to determine the optimal density of the cryopreserved cells. $0.5–2 \times 10^6$ cells/vial has proven appropriate for several hESC lines in our hands and is a good starting point for a density assessment. Often one confluent well of a 6-well plate is usually sufficient for one vial.
20. It is imperative that the freezing of stem cells be carried out as quickly as possible with all reagents chilled prior to the start of the freezing process.

Acknowledgments

The authors would like to thank David Hsu for scientific advice, and Sylvana Couture and Christine Knoblauch for technical assistance. This work was supported by City of Hope and a grant from the California Institute of Regenerative Medicine (CIRM).

References

1. Thomson, J. A., Itskovitz-Eldor, J., Shapiro, S. S., Waknitz, M. A., Swiergiel, J. J., Marshall, V. S., and Jones, J. M. (1998) Embryonic stem cell lines derived from human blastocysts, *Science* **282**, 1145–1147.
2. Crook, J. M., Peura, T. T., Kravets, L., Bosman, A. G., Buzzard, J. J., Horne, R., Hentze, H., Dunn, N. R., Zweigerdt, R., Chua, F., Upshall, A., and Colman, A. (2007) The generation of six clinical-grade human embryonic stem cell lines, *Cell Stem Cell* **1**, 490–4.
3. Carpenter, M. K., Frey-Vasconcells, J., and Rao, M. S. (2009) Developing safe therapies from human pluripotent stem cells, *Nat Biotechnol* 27, 606–613.
4. FDA, Guidance for Industry (2009) Current good tissue practice (CGTP) and additional requirements for manufacturers of human cells, tissues, and cellular and tissue-based products (HCT/Ps).
5. FDA, Guidance for FDA reviewers and sponsors (2008) Content and review of chemistry, manufacturing, and control (CMC) information for human gene therapy investigational new drug applications (INDs).
6. Loring, J. F. and Rao, M. S. (2006) Establishing standards for the characterization of human embryonic stem cell lines, *Stem Cells* **24**, 145–150.
7. The international Stem Cell Initiative (2007) Characterization of human embryonic stem cell lines by the International Stem Cell Initiative, *Nat Biotech* **25**, 803–816.
8. Gonzalez, R., Loring, J. F., Snyder, E. Y. (2008) Preparation of autogenic human feeder cells for growth of human embryonic

stem cells in, *Current Protocols in Stem Cell Biology* **4**, 1 C.5.1-1 C.5.15.

9. Amit, M., and Itskovitz-Eldor, J. (2006) Feeder-free culture of human embryonic stem cells in, *Methods in Enzymology* **420**, 37–49.
10. Xu, C., Inokuma, M. S., Denham, J., Golds, K., Kundu, P., Gold, J. D., and Carpenter, M. K. (2001) Feeder-free growth of undifferentiated human embryonic stem cells, *Nat Biotechnol* **19**, 971–974.
11. Swistowski, A., Peng, J., Han, Y., Swistowska, A. M., Rao, M. S., and Zeng, X. (2009) Xeno-free defined conditions for culture of human embryonic stem cells, neural stem cells and dopaminergic neurons derived from them, *PLoS One* **4**, 6233.

Chapter 12

Culture of Human Pluripotent Stem Cells on Glass Slides for High-Resolution Imaging

Victoria Fox

Abstract

For certain applications, particularly experiments involving high-resolution imaging, it is necessary to culture cells on glass slides or cover glasses. This chapter describes techniques for successfully growing human embryonic stem cells (hESCs) on glass surfaces under three different conditions – serum-containing, serum-free, and following single-cell dissociation. It is anticipated that these techniques will extrapolate to other types of pluripotent stem cells such as induced pluripotent stem cells (iPSCs) and embryonic germ cells (EGCs).

Key words: pluripotent stem cells, high-resolution imaging, immunostaining, culture on glass slides

1. Introduction

Applications involving high-resolution imaging require samples to be prepared on glass slides or cover glasses. This has proved difficult for human pluripotent stem cells, such as human embryonic stem cells (hESCs) as they experience difficulties attaching to glass surfaces and readily undergo spontaneous differentiation, even when co-cultured with MEFs (1, 2). Despite these difficulties, it is possible to successfully culture hESCs on a number of different types of glass if the surface is properly coated with biomolecules and extra care is taken when handling the cells. Following the recent discovery that a RHO-associated Kinase (ROCK) inhibitor can improve the cloning efficiency of dissociated hESCs it has also become possible to plate single cells for high-resolution intracellular analysis (3). This chapter describes

Philip H. Schwartz and Robin L. Wesselschmidt (eds.), *Human Pluripotent Stem Cells: Methods and Protocols*, Methods in Molecular Biology, vol. 767, DOI 10.1007/978-1-61779-201-4_12, © Springer Science+Business Media, LLC 2011

techniques for culturing hESCs on cover glasses under a variety of culture conditions. It is anticipated these techniques will extrapolate to other types of pluripotent stem cells such as induced pluripotent stem cells (iPSCs) and embryonic germ cells (EGCs) (4, 5).

2. Materials

2.1. General Solutions

1. Dulbecco's Phosphate-Buffered Saline (DPBS), without magnesium or calcium.
2. 70% Alcohol (Harleco Alcohol, 70%, EMD Cat# 65350–85).

2.2. Cover Glass Preparation

1. Cover glasses: 12 mm diameter, #1 round, German glass cover glasses (Electron Microscopy Sciences Cat# 72196-12); 12 mm diameter #1 round (borosilicate) cover glasses (Electron Microscopy Sciences Cat# 72195-12).
2. 1 N Hydrochloric Acid (HCl).
3. pH indicator strips: pH range 3.8–5.5 and pH range 6.0–8.1.
4. 0.1% Gelatin, (can be purchased as a ready-made solution, Millipore Cat# SF008).
5. BD Matrigel™ (hESC qualified Cat# 354277) should be aliquoted on ice into cold tubes using precooled pipettes and stored at −70°C until use. Matrigel™ is diluted in ice-cold Dulbecco's Modified Eagle Medium (DMEM, high glucose, no pyruvate, no glutamine formulation) for coating.

2.3. MEF Culture and Preparation

1. MEF culture medium contains DMEM (high glucose, no glutamine, no pyruvate formulation) supplemented with 10% FBS and 1 mM L-glutamine.
2. Mitomycin C (Sigma-Aldrich Cat# M0503) is distributed as a lyophilized powder which should be stored at 4°C. Dissolve 2 mg of powder into 200 mL of MEF culture medium to make a 10 μg/mL working stock solution. Filter sterilize using a 0.2 μm filter. Store excess solution at 4°C for 1 week or −20°C for prolonged periods of up to 6 months.

2.4. PSC Culture Reagents

1. *Serum-free (SR) hESC medium* is composed of DMEM/F12 (Sigma, Cat# D6421) substituted with 20% KSR (Invitrogen Cat# 10828-028), 1% non-essential amino acids (diluted 1:100 from a ×100 stock), 1 mM L-glutamine; 4 ng/mL bFGF (diluted from a 10 μg/mL stock solution Peprotech Cat# 100-18B) (see Note 1) and 0.1 mM β-mercaptoethanol. Filter sterilize using a 0.2 μm filter and store at 4°C. Use within 2 weeks, warm aliquot to 37°C just prior to cell culture.

2. *Collagenase IV* (Invitrogen Cat# 17104-019). To make a working solution, dissolve collagenase powder into pre-warmed DMEM, at a concentration of 1–5 mg/mL. Filter sterilize using a 0.2 μm filter. Store at 4°C for up to 1 week. For optimal results make fresh collagenase weekly. Warm to 37°C prior to use, but avoid repeated warming.
3. *Serum-containing hESC medium* is composed of Knockout DMEM (Invitrogen Cat# 10829-018) supplemented with 20% FBS (Hyclone Cat# SH30071.02), 1% non-essential amino acids, 1 mMl-glutamine; 1% Insulin–Transferrin–Selenium (Invitrogen # 41400–045) and 0.1 mM β-mercaptoethanol. Filter sterilize using a 0.2 μm filter and store at 4°C. Use within 2 weeks, warm to 37°C prior to cell culture.

2.5. Culture of Dissociated PSCs

1. Rho-associated kinase (ROCK) inhibitor Y-27632 (Sigma-Aldrich Cat# Y0503): Reconstitute Y-27632 in 0.2 μM in filtered, autoclaved water to make a 5 mM stock solution. Store in aliquots at −20°C.
2. 0.05% Trypsin/EDTA solution.
3. 0.25% Trypsin/EDTA solution.

3. Methods

Human ESCs can be successfully cultured on borosilicate and German glass and to some extent on soda-lime glass though, the efficiency is more variable. "Tissue culture glasses" can be obtained in one of two formats; cover glasses or chamber slides (see Note 2). A comparison of the two systems is provided in Table 1.

This chapter will describe techniques for culturing hESCs on round borosilicate or German glass cover glasses and assumes the reader has basic experience working with these cells under standard co-culture conditions with MEF's in KSR and FBS -containing medium (1, 6, 7), described in Chapter 8. It is anticipated that many of the principles discussed in this chapter will extrapolate to other types of pluripotent stem cells such as iPSC and EG cells (see Note 3) (4, 5). Whenever necessary, techniques for using chamber slides are discussed in the notes section.

3.1. Preparing Cover Glasses for Cell Culture

A three-step process is required to prepare cover glasses for cell culture: the glass must be acid washed, sterilized, and coated with biomolecules.

3.1.1. Acid Washing

Cleansing with hydrochloric acid removes small particles and organic contaminants. It also renders the surface of the glass porous and increases hydrophilicity, which improves attachment and spreading.

Table 1
A comparison of cover glasses vs. chamber slides

	Coverslips	Chamber slides
Composition	Typically German or borosilicate glass.	Typically soda-lime glass but German and borosilicate chamberslides are available.
Durability	Brittle and Fragile. Break easily.	Durable, do not break easily.
Sterility	Sterilization with alcohol or UV light required prior to use. Incidence of contamination is rare following sterilization.	Sterile unless exposed to atmosphere during acid washing. Culture media often seeps into the lid and outside of the slide causing problems with contamination.
Convenience	Relatively inconvenient. Coverslips, require plating and sterilizing which is time consuming. They are also difficult to handle.	Relatively convenient. Chamberslides do not require plating or sterilizing (unless exposed to atmosphere outside of the tissue culture hood). They are relatively simple to handle.
Acid washing	Optional. 1 N HCl for 15 h at 60°C.	Optional. 1 N HCl for 15 h at 60°C.
Coating	Required (Matrigel or Gelatin).	Required (Matrigel or Gelatin).
Cell culture	Must be placed into a cell culture vessel. Each well is completely isolated enabling samples to be kept separate.	Complete with media chamber for cell culture. On occasion media from one chamber leaks into a neighboring chamber causing cross contamination of samples.
Staining	Staining can be carried out within the well of the plate. Cross contamination is not an issue since each well is isolated.	Staining should be carried out with the plastic chamber in place. Remove after staining to prevent cross contamination. On occasion staining solution leaks from one chamber to another.
Mounting	Mount onto glass slides. Low through put and tricky. Coverslips are easily dropped and broken during mounting.	Mounting carried out with a single elongated coverslip. Fast but relatively in efficient.

Spread the cover glasses out in a wide-bottomed glass container and add enough 1 N HCl to fully submerge the glass. Cover the container with foil and place on a hot block heated to 60°C for around 15 h. Wash the cover glasses thoroughly four to five times with distilled water then leave to soak in water for 1 h. Carry out a second round of washing and soaking this time with 0.2 μm filtered water. The pH of the water should then be checked using indicator paper to ensure the acid has been fully neutralized. Drain the cover glasses and wash one to two times with 70%

alcohol to remove residual water. Store at room temperature in 70% alcohol. Acid washing can be carried out on all types of cover glasses regardless of composition or thickness (see Note 4).

3.1.2. Plating and Sterilizing

1. Since cover glasses do not contain a media chamber they must be placed into a suitable vessel for cell culture. We recommend using round-shaped cover glasses as they can be inserted into the wells of 12- or 24-well plates. The cover glass should be slightly smaller than the well to ensure it can be removed. This chapter will assume the use of 12-mm diameter cover glasses and 24-well culture plates. Adherent cell culture-treated plates are also recommended as they permit the entire surface area of the well to be coated with MEFs.
2. Using ethanol-cleaned forceps, place a single 12-mm diameter cover glass into the wells of a 24-well plate. Fill each well to the top with 70% alcohol and soak for 30 min to 1 h to sterilize the glass (see Note 5). Remove the alcohol and wash two times with DPBS unless storing (see Note 6). Use immediately after washing with DPBS. Do not allow the DPBS to dry as it will form crystals on the surface of the glass.

3.1.3. Biocoating

1. Neither hESCs nor MEFs grow well on bare glass. It is essential to prime the surface using 0.1% gelatin or Matrigel™, if attachment to gelatin is poor (see Note 7). Coating should be carried out at 37°C for a minimum of 2 h.
2. To Matrigel™-coat cover glasses, thaw an aliquot of Matrigel™ at 4°C and dilute 1:50 to 1:30 with ice-cold DMEM. Add 0.5 mL per well of a 24-well plate (or 250 μL/cm^2) and incubate at 37°C for 2–3 h or overnight. Remove excess Matrigel™ prior to the addition of hESCs. Do not allow Matrigel™ (or gelatin)-coated cover glasses to dry out. Take care not disrupt the coating by scratching the surface of the glass.

3.2. Plating Mouse Embryonic Fibroblasts on Glass Surfaces

Mitotically inactivated Mouse Embryonic Fibroblasts (MEFs) can experience as many difficulties attaching to glass as PSCs, especially in serum-free medium. Although most MEF cells attach, they typically die at an accelerated rate and do not always spread out. It is critical to use high quality feeders that support hPSCs well under routine culture conditions on adherent cell culture-treated plastic (see Note 8). We do not recommend plating inactivated MEFs on bare glass even at high density in serum-containing medium. The surface must first be primed with gelatin or Matrigel™ (see Subheading 3.1 for instructions on preparing glass surfaces).

1. MEFs can be inactivated for culture on glass according to standard protocols, i.e., by irradiation (up to 5,500 rads) or Mitomycin C treatment (10 μg/mL for 2.5–3 h). See Chapter 8.

2. MEFs should be plated at the same density used for routine culture for that particular cell line, unless it is to be maintained in serum-containing medium (see Notes 9, 17 and Subheading 3.4).
3. Cultures that experience problems on glass (such as low growth rates or spontaneous differentiation) are rarely improved by increasing the number of MEFs (see Note 9). It is much more effective to use MEF-conditioned medium supplemented with bFGF. Instructions for the production of MEF-conditioned medium are provided in Subheading 3.3 of this protocol and Chapter 8.

3.3. Serum-Free Culture of PSCs on Glass

Successful transfer to growth on glass surfaces is highly dependent on the overall state of the hESC culture. It is absolutely essential to begin with high quality cells, i.e., cells that have been carefully tended to and have grown consistently well as a predominately undifferentiated stem cell population (>70%) for a minimum of 2 weeks. Do not be tempted to use poor quality, temperamental cultures or cell lines that grow poorly on cell culture-treated plastic, as few cells will attach and those that do are likely to differentiate. Indeed, the most important factor for successfully growing hPSCs on glass surfaces is to start with healthy undifferentiated stem cell populations (see Note 10).

3.3.1. Plating hESC Cultures on MEF-Coated Glass Surfaces

Co-culture with MEFs does not always provide an adequate level of support for maintaining hESCs on glass in serum-free medium (see Note 11). However, the quality of the culture can be improved if in addition to MEFs, the cells are maintained in MEF-conditioned medium supplemented with bFGF (Fig. 1a–c, photomicrographs). Preparation of extra cover glasses is strongly advised in order to generate enough good quality cultures. This protocol describes methods for lifting the PSC culture with either collagenase IV or by manual dissection.

1. *Prepare MEF-conditioned medium*: Use double the concentration of MEFs and half the volume of medium normally used during routine culture of your PSC line. Allow MEFs to condition the medium for 24 h before collecting. Add 4 ng/mL of bFGF and filter to remove dead cells (see Note 12). Store at 4°C and use within 10 days. For long-term storage, freeze at −70°C and substitute with fresh bFGF upon thawing. Plated MEFs can be used to condition medium for 1 week; apply fresh serum-free hESC medium daily. Use large flasks, such as 75–150 cm^2, when large volumes are required.
2. *Plate the PSCs on prepared glass slides:*
   (a) *Collagenase IV treatment*: The following protocol can be used to plate robust lines that respond well to glass. Treat an appropriate number of dishes containing predominately

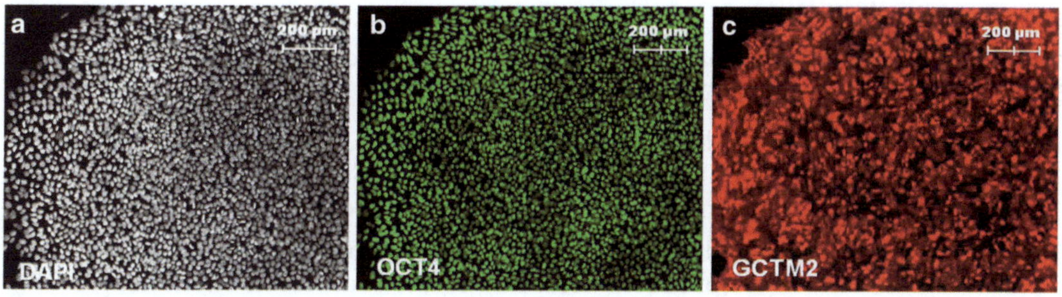

Fig. 1. hESC colonies maintained on glass with MEFs in serum-free MEF-conditioned medium. This colony was immunostained for the stem cell markers OCT-4 (**b**) and GCTM2 (**c**) and the nuclear dye DAPI (**a**).

undifferentiated stem cells with 1–5 mg collagenase for 10–15 min at 37°C. Wash once with serum-free hPSC growth medium then replace with enough MEF-conditioned medium to cover the dish. Dissociate the colonies to generate smaller fragments using a sterile P10/P20 pipette-tip by passing the tip over the surface of the dish in a tight circular motion. Observe the resulting fragments under phase-contrast microscopy. If they are larger than required, pipette up and down gently, one to two times using a P1000 pipette to dissociate further (see Note 13). Transfer the clumps of hESC colonies to a 15-mL conical tube and add enough conditioned medium to make a total volume of 5 mL. Leave the suspension for 30 s to 2 min or until large fragments containing differentiated material have settled to the bottom of the tube (see Note 14). Transfer the remaining suspension containing small/medium-sized fragments of undifferentiated stem cells to the prepared cover glasses and return to the incubator immediately (see Note 15). Feed the next day to remove dead cells and every day thereafter.

(b) *Manual dissection of colonies*: When cultivating lines that do not cope well on glass it is preferable to dissociate the colonies manually under a dissecting scope. This enables fragments of the desired size to be generated from the highest quality colonies, without the need for pipetting or fractionation. 1–200 µL pipette tips can be used as a tool for cutting the colonies. Starting at the edge of the colony, dig the tip into the cells and push away to release a fragment into suspension. It is important to scrape away from the colony rather than pulling the tip toward you as it enables a greater control over the size of the resulting fragments. Continue until the entire colony has been cut into small pieces by working inward from the edge. Transfer the fragments of the colony to the prepared cover glasses using a P200 pipette and return to the incubator immediately (see Note 16).

3.3.2. Feeder-Free Culture of HESCs on Glass Surfaces

The following factors are critical to successfully culturing hESCs on glass under "feeder-free" conditions in MEF-conditioned medium on Matrigel™.

1. The surface of the glass must be coated with a Matrigel™ (1:30 dilution) as described in Subheading 3.1.3. Gelatin will not support feeder-free culture of hESCs on glass.
2. The cells must be maintained in MEF-conditioned medium (produced according to the protocol in Subheading 3.3.1). We recommend using fresh rather than pre-made frozen medium for feeder-free culture on glass.
3. The cells may be lifted for transfer by manual dissociation or collagenase IV depending on the character of the line and its ability to grow on glass (see Subheading 3.3.1 for methodology).

3.4. Culture of hESCs on Glass in Serum-Containing Medium

Human PSCs can be cultured on glass surfaces in serum-containing medium for short periods (up to 5 days), even if they were previously grown under serum-free conditions (Fig. 2) (see Note 17).

1. The surface of the glass should be coated with Matrigel™ or gelatin depending on the attachment properties of the cell line in use (see Subheading 3.1.3).
2. The MEF feeder density must be low enough to enable the colonies to expand but higher than the densities typically used

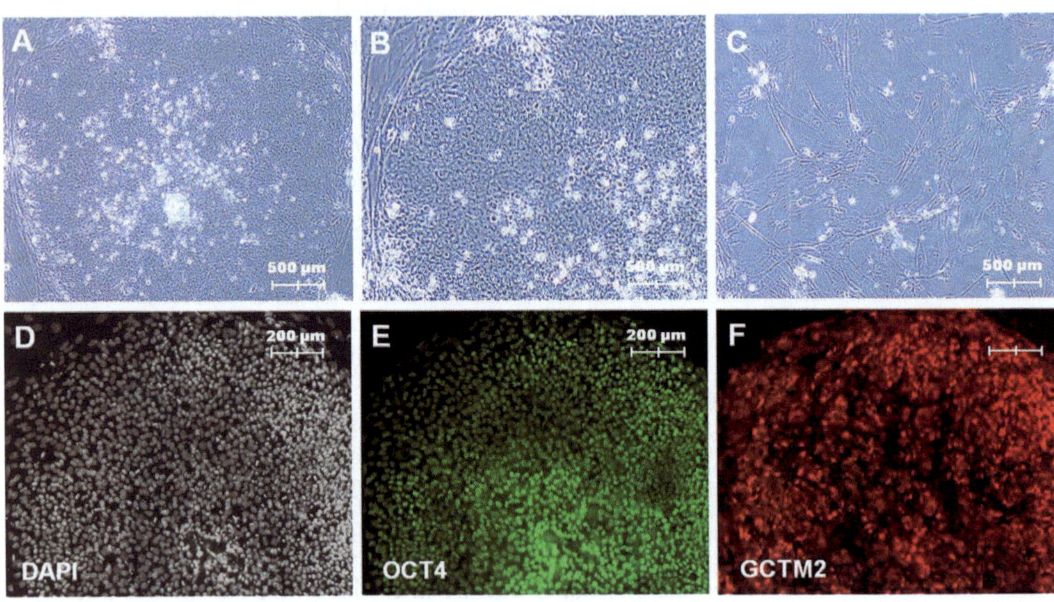

Fig. 2. Culture of hESCs in serum-containing medium as flat monolayers. (**a**, **b**) Phase-contrast image of a typical hESC colony grown on MEF feeders in FBS-containing medium. (**c**) Optimal feeder density for hESCs to grow as a monolayer in FBS-containing medium. (**d**, **e**) An hESC colony immunostained for the stem cell markers OCT-4 and GCTM2 and the nuclear dye DAPI (**f**).

for serum-free culture. The MEFs should form a thin layer covering up to 80% of the glass surface (Fig. 2c, photomicrograph). We recommend using 3–5 × 10^4 MEFs/cm^2.

3. Successful growth on glass in serum-containing medium is greatly dependent on the quality of the starting culture. It is critical to transfer healthy, undifferentiated, proliferating cells. We find that days 3–4 cultures contain the highest number of cycling stem cells. The colonies should be dissociated according to the manual dissection protocol outlined in Subheading 3.3.1. Transfer only pristine colonies that appear to be composed primarily of undifferentiated stem cells.
4. Preparing 25–50% extra cover glasses is strongly advised in order to generate enough cells for analysis.
5. *Experiments requiring undifferentiated cells should be completed within 4–5 days of transfer for optimal results.*

3.5. Plating Dissociated HESCs for High-Resolution Single-Cell Analysis

Human PSCs can be plated on glass as single cells for the purpose of high-resolution intracellular imaging (Fig. 3). We recommend conducting single-cell analysis under feeder-free conditions to prevent interference from the MEFs during imaging. Instructions for coating cover glasses with Matrigel™ and generating MEF-conditioned medium can be found in Subheadings 3.1 and 3.2.

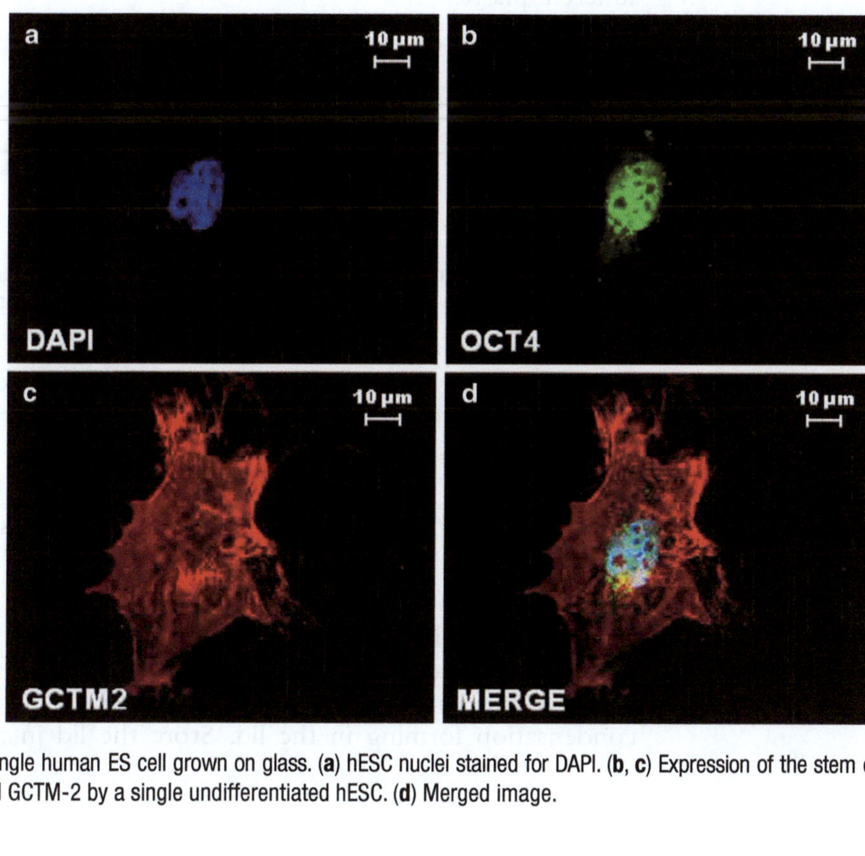

Fig. 3. A single human ES cell grown on glass. (**a**) hESC nuclei stained for DAPI. (**b**, **c**) Expression of the stem cell markers OCT-4 and GCTM-2 by a single undifferentiated hESC. (**d**) Merged image.

of this protocol, respectively, and Chapter 8 of this book. The propensity for cell death following dissociation is reduced using the Rho-associated kinase (ROCK) inhibitor Y-27632 (3).

1. Treat the cells for 1–2 h prior to plating with 10 μM Y-27632 diluted in serum-free hESC medium.
2. To generate single cells, wash the flask once in DPBS and treat with 0.05–0.25% trypsin for 3–5 min at 37°C (see Note 18). Pipette the cells up and down gently in the trypsin/EDTA solution one to two times using a P1000 pipette then transfer to a conical tube containing regular serum-free hESC medium without ROCK inhibitor.
3. Observe the cells under phase-contrast microscopy to ensure that they are fully dissociated. Mix thoroughly and remove a small aliquot for counting.
4. Centrifuge at 200 × *g* for 2 min. Meanwhile count the cells (see Note 19).
5. Remove the supernatant and resuspend in MEF-conditioned medium containing 10 μM ROCK inhibitor.
6. Plate onto prepared cover glasses at high density (around $2–5 \times 10^4$ cells/cm^2).
7. Add a sample of cells to a control well without cover glasses to ensure they are able to survive under normal conditions on adherent plastic.
8. Culture for 48–36 h depending on whether single cells or small clumps are required (see Note 20).

4. Notes

1. bFGF stock solution should be made in 0.1% BSA. Aliquot and store long term at −20°C. Store thawed bFGF at 4°C and use within 2 weeks.
2. Most chamber slides are composed of soda-lime glass. However, it is possible to buy slides made from German or borosilicate glass from some manufacturers.
3. Tumor-derived (pluripotent) stem cells (such as Embryonal Carcinoma (EC) cells) generally grow well on all types of bare glass.
4. Chamber slides are precleaned by the manufacturer to remove contaminants. However, they may be treated with acid to improve cell attachment if required. Add 1 N HCl to each chamber and cover the chamber slide with foil to prevent condensation forming in the lid. Store the lid in a sterile

container in the tissue culture hood until needed. Heat the slide to 60°C for 15 h on a heated hot block in the fume hood. Since the volume of the chamber is small it must be washed multiple times with water (i.e., 20 or more) to fully remove the acid. Soak for 1 h after every 10 washes.

5. Expose to UV light in the tissue culture hood for 30 min if recurrent contamination is observed.
6. To prepare sterilized glass coverslips for storage, remove the ethanol and allow to air dry in the tissue culture hood. Store in a sterile environment. Wash two times with DPBS prior to use.
7. Glass coverslips can also be coated with purified ECM molecules such as fibronectin, vitronectin or collagen. This is particularly encouraged during differentiation studies as culture on glass can alter differentiation trajectories and even hinder the development of certain lineages.
8. Some MEF strains do not settle well on glass or support hESC culture if they have been frozen postinactivation.
9. For high-resolution imaging, the MEF density must be low enough to permit hESC colonies to expand as a flat monolayer. hESC lines that typically grow on high-density MEFs in serum-free medium can be cultured on fewer feeder cells in conditioned medium. A concentration of 7.5×10^3 to 1×10^4 MEF cells/cm^2 is usually sufficient to support the growth of most hESC lines when combined with conditioned medium.
10. We do not recommend maintaining PSCs on glass for applications that involve monitoring culture quality. PSCs should only be maintained on glass for high-resolution image analysis.
11. It is not unusual for hESCs to generate differentiated cells that resemble fibroblasts when cultured on glass. This occurs most often when the cells are maintained under serum-free conditions on low-density MEFs. It is likely that these cells are produced in an effort to generate a more supportive microenvironment. The majority of the time, this can be overcome by co-culturing the cells with MEFs and MEF-conditioned medium.
12. This can be increased to 8 ng/mL for lines that are particularly prone to undergoing spontaneous differentiation.
13. The optimal colony fragment size for successfully plating hESCs on glass varies from line to line and should be determined empirically prior to the onset of sensitive experiments. Some lines are able to grow well on glass when cultured as small colonies whereas others must be plated as large clumps in order to prevent spontaneous differentiation.

14. Unlike ESCs colonies which break apart easily, differentiated material is typically difficult to dissociate and as a result forms large clumps in suspension. These clumps settle more quickly than smaller stem cell-containing fragments, thus permitting their exclusion.
15. The plating conditions for each hESC line grown under local conditions should be optimized prior to the onset of sensitive experiments. For instance, hESC attachment rates to glass vary greatly from line to line and also according to the quality of the culture. It may be necessary to compensate for cell loss by increasing the plating density 30–50%.
16. It may be necessary to treat some hESC lines with 1–5 mg/ml Collagenase IV solution for 5–10 min prior to dissociation to prevent the colony fragments from forming aggregates in suspension.
17. Some hESC lines (i.e., the ESI family of lines) are maintained routinely in serum on high-density MEFs as thick multilayered colonies that are passaged by manual dissection. To successfully recapitulate this culture system on glass, the entire surface of the glass must be coated with MEFs. It may be necessary to increase the density by 10–15% to ensure proper coverage. Typically $6–7.5 \times 10^4$ MEF/cm^2 is sufficient. To maximize the transfer of undifferentiated cells, the colony should be cut at the very outer edge to generate small fragments of around 0.5–0.75 cm^2. We recommend using days 4–5 cultures as the colonies are easier to assess morphologically and less depleted of OCT-4 positive cells. It is noteworthy that the ESI lines can also be cultured on low-density MEFs as flat monolayer colonies using the methodology outlined in Subheading 3.4.
18. Trypsin concentration may need to be varied according to the character of the line.
19. Dissociated hESCs should be kept on ice at all times.
20. It is not unusual for a large number of hESCs to die 1–2 days after plating. The plating density should be optimized to compensate for this.

Acknowledgments

I would like to acknowledge the technical staff at the USC Stem Cell Core for providing cell culture support.

References

1. Skottman H, Hovatta O. (2006) Culture conditions for human embryonic stem cells. *Reproduction* **132**:691–8.
2. Skottman H, Narkilahti S, Hovatta O. (2007) Challenges and approaches to the culture of pluripotent human embryonic stem cells. *Regen Med* **2**:265–73.
3. Watanabe K, Ueno M, Kamiya D, et al. (2007) A ROCK inhibitor permits survival of dissociated human embryonic stem cells. *Nat Biotechnol* **25**:681–6.
4. Shamblott MJ, Axelman J, Wang S, et al. (1998) Derivation of pluripotent stem cells from cultured human primordial germ cells. *Proc Natl Acad Sci USA* **95**:13726–31.
5. Takahashi K, Tanabe K, Ohnuki M, et al. (2007) Induction of pluripotent stem cells from adult human fibroblasts by defined factors. *Cell* **131**:861–72.
6. Reubinoff BE, Pera MF, Fong CY, Trounson A, Bongso A. (2000) Embryonic stem cell lines from human blastocysts: somatic differentiation in vitro. *Nat Biotechnol* **18**:399–404.
7. Thomson JA, Itskovitz-Eldor J, Shapiro SS, et al. (1998) Embryonic stem cell lines derived from human blastocysts. *Science* **282**:1145–7.

Part IV

Characterization

Chapter 13

Classical Cytogenetics: Karyotyping Techniques

Steven E. Bates

Abstract

Classical cytogenetics by karyotyping has been utilized in clinical research laboratories for more than 50 years and remains the key method used in the stem cell laboratory to assess the genetic stability of stem cell cultures. It is currently the most readily accessible method for detecting chromosomal abnormalities in pluripotent stem cell cultures. This chapter will describe (1) how to prepare a culture to maximize the number of metaphase cells, (2) how to prepare slides containing chromosome spreads (3) methods used to stain chromosomes, and (4) how to interpret the cytogenetic report.

Key words: human embryonic stem cell, karyotype, cytogenetic analysis, G-banding, euploid, aneuploid, chromosome preparation, pluripotent stem cell, karyotype

1. Introduction

Classical cytogenetics, karyotyping, is one of the most common and easily accessible assays used to determine the chromosomal complement of human pluripotent stem cell (PSC) cultures. One of the first things done to characterize novel PSC lines is to karyotype the line. This is to determine whether it is a normal diploid line with full a complement of chromosomes, including the sex chromosomes, or whether it has rearrangements, such as inversions, deletions, or is aneuploid, containing an abnormal number of chromosomes. Since one of the major advantages of working with PSC lines is the fact that they are capable of long-term proliferation while maintaining a normal diploid chromosome complement, karyotyping is performed frequently. An essential assay upon the derivation of new lines, karyotyping is performed every 15–20 passages during routine culture and maintenance, or when cell lines are expanded, banked, or are exposed to potentially

Philip H. Schwartz and Robin L. Wesselschmidt (eds.), *Human Pluripotent Stem Cells: Methods and Protocols*, Methods in Molecular Biology, vol. 767, DOI 10.1007/978-1-61779-201-4_13, © Springer Science+Business Media, LLC 2011

stressful experiments such as gene targeting or growth in non-traditional media. The maintenance of PSC cultures in a state that promotes a normal diploid karyotype is a major goal in the stem cell laboratory and assessment of the karyotype is a routine part of laboratory operations.

Human embryonic stem cell (hESC) lines have been maintained for extended periods of time in culture without developing detectible chromosomal abnormalities (1–3). However, even during routine culture PSC lines can become aneuploid, which may alter the characteristics of the cells affecting gene expression, cellular growth kinetics, and differentiation potential (4–6). Two commonly observed abnormalities found in hESC cultures are a gain of chromosome 17q or the presence of one or more isochromosomes of 12p (7, 8).

Several methods can be used to assess the chromosomal complement of PSC lines, including the classical cytogenetic method described in this chapter and molecular based techniques such as fluorescent in situ hybridization (FISH), spectral karyotyping (SKY) (9), single nucleotide polymorphism (SNP) and copy number polymorphism (CNP) mapping (10). These methods differ in resolution and the types of abnormalities they detect. The best resolution obtainable by karyotyping is estimated to be about 10 Mb, while SKY resolves at 1–2 Mb, and SNP and CNP mapping can yield 30 Kb resolution (11). However, more resolution is not necessarily better; for example, SNP genotyping cannot be used to detect translocations or inversions, and SKY cannot detect inversions, both of which are readily detected by classical cytogenetic techniques.

The basic cytogenetic method involves chromosome harvest, slide preparation, staining and banding of the chromosomes, analysis of chromosome numbers in the population, and analysis of the banding patterns. Chromosome preparation consists of arresting the cells, harvest and production of a single cell suspension, hypotonic treatment (controlled swelling of the cells) and fixation. After fixation and dehydration, the swollen metaphase cells are dropped onto glass slides, air-dried, and aged before staining or banding. Banding is a staining method used for visualizing chromosomes and results in a continuous series of longitudinal light and dark staining regions (12, 13). The process of arrest, hypotonic treatment, fixation and slide preparation profoundly affects the quality of the resultant banding. The G-banding method (trypsin treatment followed by Giemsa staining) is the most commonly used method for visualizing the banding pattern of the chromosomes, and can generate up to 1,000 bands per haploid human genome. Each band has been assigned a unique designation to indicate its location on each human chromosome. The nomenclature of band assignment and chromosome aberrations is summarized by the International System for Human Cytogenetic Nomenclature (ISCN) (14). For routine surveillance of PSC cultures, a chromosome preparation is required every 15–20 passages. To validate critical experiments, parallel cultures for chromosome preparations should be established. It is emphasized that cytogenetic aneuploidy (not 46 chromosomes) or pseudodiploidy (incorrect 46 chromosomes) are the last stages of departure from normality. The molecular causes of such gross rearrangements occur long before they are visible in the chromosome complement of dividing cells. This said, aneuploid ESCs and PSCs may continue to exhibit many, or all, of the classic stem cell antigens. Thus, cytogenetic analysis remains an essential tool in the evaluation of PSC cultures.

2. Materials

1. Water, use fresh from ultra filtration, or commercially available "for cell culture" (see Note 1).
2. Colcemid Solution (10 μg/mL) (see Note 2).
3. Hypotonic solution (0.075 M KCl): Dissolve 0.56 g KCl in 100 mL water. Warm to 37°C before use.
4. Carnoy's Fixative is methanol/acetic acid, 3:1. Make in small batches, as fixative is very hygroscopic. Chill before use to minimize evolution of heat when it contacts the aqueous hypotonic suspension.
5. Gurr's Buffer Solution, dissolve one tablet into 1,000 mL of water (see Note 3).
6. 0.9% NaCl, dissolve 0.9 g in 100 mL water.
7. 0.05% trypsin and 1 mM ethylenediamine tetra acetic acid (EDTA) in PBS.
8. Dulbecco's Modified Eagle's Medium (DMEM).
9. Dulbecco's Phosphate-Buffered Saline (DPBS) (without Mg^{++} and Ca^{++})
10. Giemsa Stain (5% solution): 2.5 mL Giemsa stain to 47.5 mL Gurr's buffer solution (see Note 4).
11. Lint-free cloths or tissues for cleaning slides.
12. Slide mounting medium with and without DAPI.
13. Microscope slides and coverslips.
14. Coplin jars.

3. Methods

The only way to obtain a sufficient number of high quality metaphases is to have many actively dividing cells. Chromosome preparation is best done the day before the cells would normally be passaged. During the metaphase of mitosis, the chromosomes reach their maximum level of condensation and become identifiable under the microscope. The chromosomes are less condensed at early metaphase and become more condensed as the cell progresses toward the end of metaphase. Since the goal of harvesting the cells is to obtain as many quality metaphase cells as possible in order to make an accurate analysis of the culture, colcemid is added to the cultures as it blocks the cells in metaphase. Longer treatment with colcemid will increase the mitotic index, but prolonged treatment will lead to higher fraction of cells with shortened chromosomes. The optimum length of time the cells are incubated with colcemid can be determined empirically in order to obtain *both* a good mitotic index and good chromosome length.

3.1. Metaphase Chromosome Preparation

1. Add 1/100 volume of colcemid stock solution to a culture with actively dividing cells.
2. Return the culture to the CO_2 incubator for 2–3 h.
3. Aspirate the culture medium.
4. Wash cells three times with DPBS.
5. Remove PSC colonies and make a single cell suspension of PSCs by incubating the colonies with trypsin/EDTA solution for 3–5 min (see Notes 5 and 6).
6. Transfer the cell suspension to a microfuge tube, take an aliquot, count the cells during centrifugation, and pellet at $200\times g$ for 5 min. The speed and force of all centrifugations should be just sufficient to form a pellet, and no more.
7. If there are sufficient cells ($>2\times10^6$), divide the harvest into three tubes. Aliquot cells into three tubes, labeled as follows: Tube #1 1:6 dilution, Tube #2 1:6.5 dilution, Tube #3 1:7 dilution.
8. Centrifuge at $200\times g$ for 5 min at room temperature.
9. Suspend the cells in a volume of DMEM to produce 1×10^7 cells/mL. If there are 1×10^6 cells per tube, resuspend these in 100 μL.
10. Centrifuge the cells at $200\times g$ for 5 min at room temperature. After spinning remove DMEM and dislodge the cell pellet by gently flicking the bottom of the tube.

11. Hypotonic Treatment. 0.075 M KCl is added, causing the cells to swell; add 600, 650, and 700 μL of the hypotonic solution to tubes #1 through #3, respectively. While gently flicking the tube containing the cell suspension, add dropwise, or down the wall of the tube, 10% of the total hypotonic solution. Continue agitating the suspension. After 3 s, add the remainder of the hypotonic solution as fast, but as evenly, as possible, without splashing.
12. Cap and invert several times.
13. Allow cells rest 20 min at room temperature.
14. Gently invert the tube several times in order to suspend the cells, and then add a few drops of cold Carnoy's Fixative. Mix by inverting the tube several times.
15. Add more cold fixative to the maximum capacity of the tube. Cap and invert gently.
16. Centrifuge at 200 × *g* for 5 min at room temperature. Aspirate the medium carefully, leaving about 50–100 μL and suspend the cell pellet by tapping the tube.
17. Drop wise, add cold Carnoy's Fixative, and centrifuge. Aspirate the medium carefully, leaving about 50–100 μL.
18. Suspend the pellet, add 200 μL of cold Carnoy's Fixative.
19. Cells are now fixed and can be stored at 4°C for up to 1 week before making slides. Excess suspension can be stored at 4°C for several weeks, but usually, the best slides are those made promptly.

3.2. Preparing Metaphase Slides

Making metaphase spreads requires a bit of trial and error. Several factors affect the quality of the spreads and practice slides allow one to tweak the slide making to achieve high quality slides that are easy to read and provide accurate assessment of the culture. If the cell density is low, but the preparation good, try making the next slides with two or three drops of suspension instead of a single drop. If the metaphases are not well spread, or if there is a significant of water remaining, further changes of fixative are indicated. If the chromosomes are under-spread, place some slides in a freezer to chill. Before dropping the suspension on the chilled slides, gently exhale on the slide to dampen it. Alternatively, precleaned slides can be soaked in methanol, and quickly dipped into water. Apply one or two drops of suspension immediately to the slide. Shake and drain at an angle.

1. Clean slides with a lint-free cloth and ethanol (see Note 7).
2. Using a Pasteur pipette or a P_{1000} tip, allow one drop of the suspension to fall in the middle third of the slide.
3. Shake, and dry slides at an angle in a fume hood (see Note 8).

4. Make a test slide for each cell preparation.
5. Evaluate the following criteria by viewing with a phase microscope at 400×.
   (a) Evaluate density (sparse slides are very labor intensive to read).
   (b) The extent of spreading and/or scattering of chromosomes.
   (c) Degree of contrast (high standing, bright chromosomes usually band poorly).
   (d) Overall chromosome morphology (too much/too long exposure to colcemid results in shrunken chromosomes).
6. Make more slides of the best preparation, diluting or applying 2–3 drops of suspension to correct for density. If metaphases are slightly under-spread, repeat steps 16–18, and make another slide. Some loosening or even scattering may occur with subsequent fixative changes. Scattered metaphases are useless. High standing (high contrast) chromosomes require more fixative changes to remove water.

3.3. Chromosome Staining

The simplest cytogenetic evaluation is the counting of chromosomes. Three simple methods are presented for staining chromosomes. It is not necessary to dry slides prior to staining. The slides can be stained immediately after making the spreads.

3.3.1. Giemsa Staining

1. Assemble four Coplin jars:
   (a) #1 Coplin jar: water
   (b) #2 Coplin jar: Giemsa Stain (5%)
   (c) #3 Coplin jar: water
   (d) #4 Coplin jar: water
2. The Giemsa stain may be dissolved in Gurr's buffer stock solution, or Mc Ilvaine's buffer, ph 6.8.
3. Stain for 20 min.
4. Rinse the slide twice in water.
5. Shake and dry at an angle.
6. Clean coverslips with ethanol.
7. Place 1 drop of mounting medium that does not contain DAPI on the slide.
8. Apply the coverslip at an angle to avoid entrapment of bubbles.
9. Invert the slide on a flat absorbent surface, and press to expel excess mounting medium.

10. Allow to dry, clean slide with xylene to remove excess mounting medium, necessary before inspecting on the microscope.
11. Use 100× oil immersion bright field to scan slide for well-spread, unscattered metaphases.
12. Record coordinates of metaphase spreads and make a drawing or take a photograph of spreads.
13. Round to oval metaphases having no touching or crossing, or scattered chromosomes are the only ones acceptable for data collection.

3.3.2. DAPI Staining

This is a simple method that one can use to stain DNA and count chromosomes, but will not allow identification of individual chromosomes.

1. Add a drop of mounting medium containing DAPI to the slide (see Note 9).
2. Seal the cover slip.
3. Count chromosomes under UV light using 100× oil immersion lens.

3.3.3. In Situ Giemsa Staining (see Figs. 1 and 2)

Gross aneuploidies can be detected by fixation and Giemsa staining of the intact stem cell colony.

1. Establish cultures on coverslips, or in small dishes.
2. One day before harvest, carefully rinse the culture three times with serum-free medium.
3. Incubate the stem cells in a 1:9 mixture of Carnoy's fixative in serum-free medium for 10 min.
4. Aspirate, and add full strength fixative for 10 min.
5. Repeat three times to fully dehydrate the cells.
6. Dry at an angle.
7. Stain with Giemsa as if for chromosome staining or banding. Rinse well.
8. Observe on a bright field microscope (see Figs. 1 and 2).

3.4. Chromosome Banding: Trypsin G-Banding

The drying process can affect how the chromosomes spread as well as the banding quality. In general, it is sufficient to dry the slides at ambient conditions. Store slides for 2–6 days in a cool, dry, dark, dust free condition. If urgent, fresh slides can be baked at 90°C for 30 min, but this introduces another variable in the process. As slides age, the chromosomes become progressively resistant to the trypsin digestion.

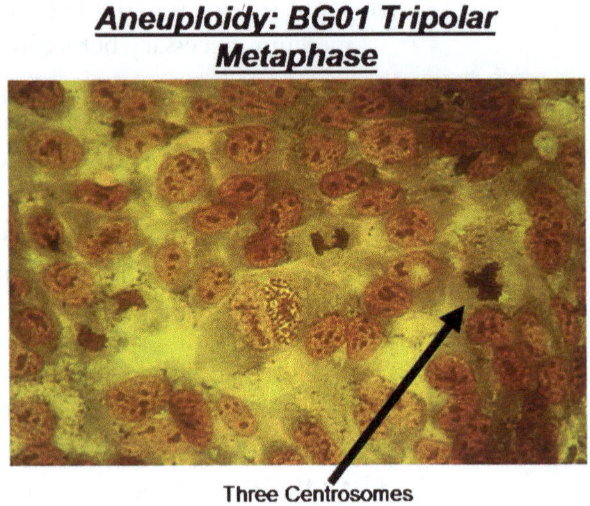

Fig. 1. Fixed and stained culture of BG01 HESC, showing a bipolar (presumably normal) anaphase, and a tripolar (inherently abnormal) metaphase, at *arrow*. Identification of individual chromosomes and accurate chromosomal numbers cannot be determined by this method.

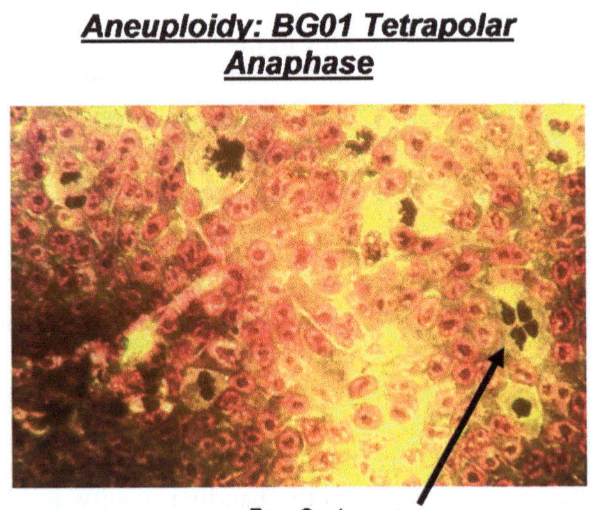

Fig. 2. Fixed and stained culture of BG01 HESC, showing bipolar anaphases, a polyploid metaphase, and a tetrapolar metaphase (*at arrow*).

1. Make a set of Coplin jars:

(a)	#1	Coplin jar:	DMEM
(b)	#2	Coplin jar	Gurr's Buffer
(c)	#3	Coplin jar:	Giemsa Stain (5 %)
(d)	#4	Coplin jar:	Water
(e)	#5	Coplin jar:	Water

2. Place 100 μL of ¼ concentration Trypsin-EDTA solution on the middle third of the first of three slides, quickly followed by application of a 22 × 30 mm coverslip.
3. After 5 s, deftly immerse the slide in Jar #1. Hold slide such that the cover slip falls down and away from the slide.
4. Dip slide twice and transfer to Jar #2. Treat the second slide for 8 s, the third for 12 s. Skim the meniscus of Jar #3, and place the slides in the stain for 20–25 min. Skim the meniscus again, and transfer slides sequentially to Jars #4 and 5. Rinse, dry at an angle, and inspect as above.

3.5. Interpreting Results

Clinical cytogeneticists examine at least 20 metaphases. Generally, 6 metaphases are analyzed by banding pattern and the other 14 metaphase spreads are counted to determine the number of chromosomes present. However, if an abnormal chromosome is found during a routine screen, the cytogeneticist will search the slide for more of these abnormalities and may evaluate more than 20 spreads to determine whether this particular abnormality represents a clonal expansion of an aneuploidy or is a random change that does not represent a significant karyotype in the culture. A typical report will contain a karyotype, and a formula which describes the population. Unlike mouse chromosomes, human chromosomes generally have distinct arms visible on both sides of the centromere. Chromosomes often overlap in the metaphase spread, so it can be difficult for untrained individuals to identify individual chromosomes.

3.5.1. Karyotype

A photograph is taken of a G-banded metaphase and individual chromosomes are cut out of the photograph and arranged in a standardized method by size, specific banding pattern, and centromere location (see Figs. 3 and 4).

3.5.2. Resolution

One of the variables in classical karyotyping by G-banding is the "resolution." The resolution of the karyotype is related to the number of bands that are visible and therefore the smallest segment of the genome that can be detected using this method. The most common method to determine the resolution of banded

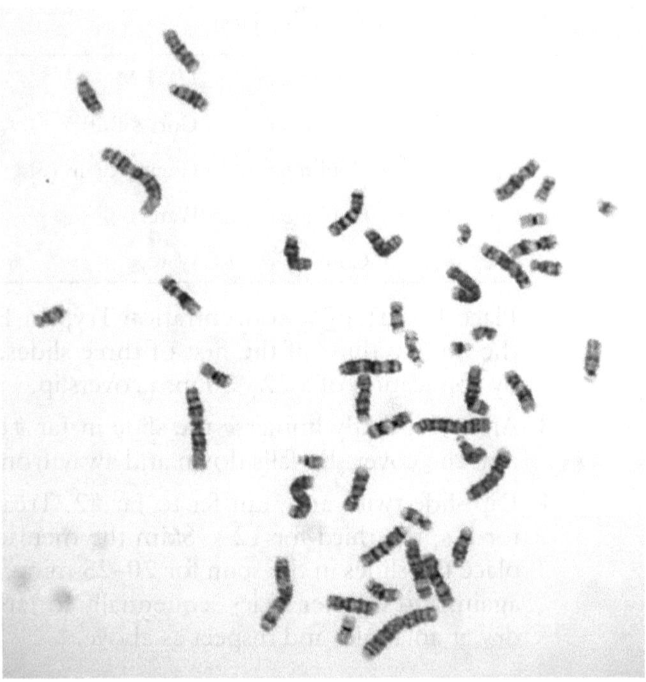

Fig. 3. BG03. Mitotic cells must be swollen sufficiently in three dimensions, so that the chromosomes will not overlap when flattened to a two-dimensional slide.

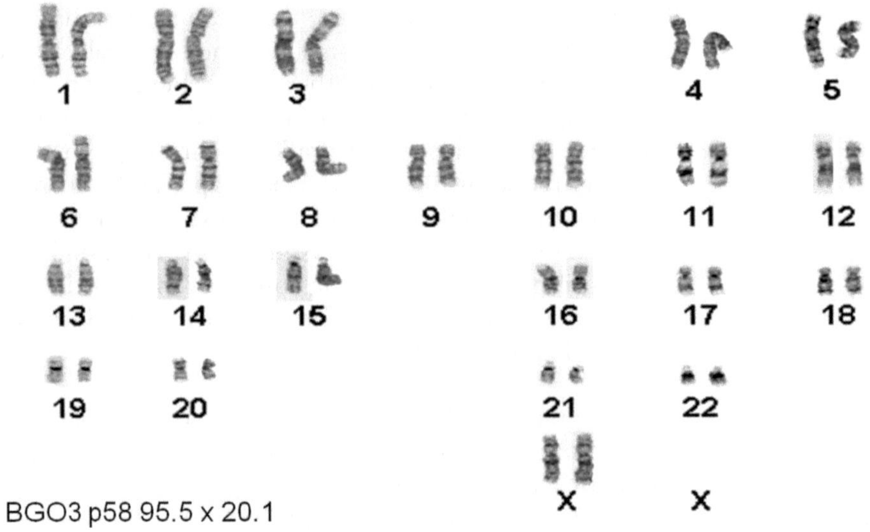

Fig. 4. G-Banded HESC Karyotype, from metaphase in Fig. 1.

chromosomes is to count the number of bands on chromosome 10 (Fig. 5). Then, using the chart in Table 1, the resolution of the karyotype can be determined. The highest resolutions are relevant only to synchronized lymphocyte preparations from which very early (pro-metaphase, prophase) cells can be obtained.

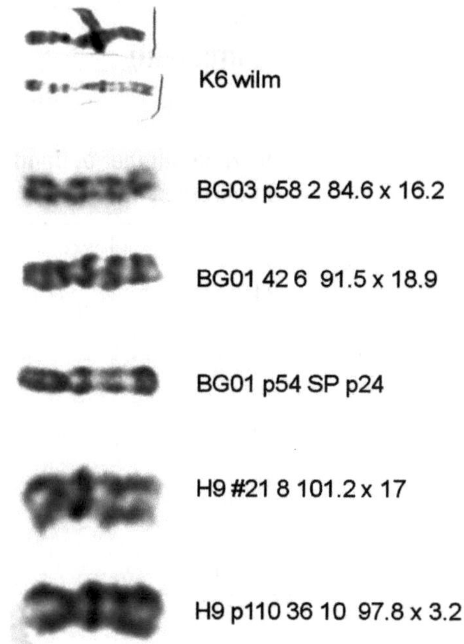

Fig. 5. Comparative Banding Resolution of Human #10 Chromosomes.

Table 1
Resolution of a karyotype is determined by counting the bands found on chromosome 10

Average number of bands on chromosome 10	Estimate of the total number of bands in one haploid set (=resolution)
12	375
13–14	400
15–16	425
17–18	450
19–21	475
22–23	500
24–25	525
26–28	550
29	575
30	600
31	625
32	650

(continued)

Table 1 (continued)

Average number of bands on chromosome 10	Estimate of the total number of bands in one haploid set (=resolution)
33	675
34	700
35	725
36	750
37	775
38	800
39	825
40–41	850

From hESC cultures, a resolution of 1,000 bands per haploid genome would be considered an excellent result.

Three examples of the shorthand used to cytogenetically describe cell populations are given:

46, XY

All cells in the population have 46 chromosomes and are male.

47, XY, +17 [26]/46, XY [4]

Thirty metaphases were counted, ([26]+[4]). Two populations occur in the culture, the larger having 26 metaphases, each bearing an extra chromosome 17. This aneuploid population has come to dominate the culture. The smaller population (4 of 30 metaphases) is normal human male. The development of trisomies is not uncommon in hESCs. Careful dissection and laborious culture of individual colonies followed by chromosome preparations may allow isolation and reestablishment of the normal population.

51,XXY,+8,+12,+14,+17[2]/51,XXY,+7,+12,
+14,+17[2] /50,XXY,+12,+14,+17[40]

This culture contains three cytogenetically distinct populations. Forty four cells were counted ([2]+[2]+[40]). The majority population has 50 chromosomes, with one extra copy each of X, 12, 14, and 17. The two minority populations each possess 51 chromosomes with one extra copy of X, 8, 12, 14, 17 or one with extra X, 7, 12, 14, 17. Such a culture is completely aneuploid, with no remnant of the original normal 46 XY population, therefore it cannot be rescued.

4. Notes

1. Water is very difficult to store. The higher the quality, the more quickly it deteriorates. Atmospheric carbon dioxide dissolves and reacts with it to form carbonic acid, and its derivatives.
2. Colchicine at $1–3 \times 10^{-6}$ g/mL can also be used to arrest cells.
3. McIlvaine's Buffer may be used instead of Gurr's Tablets. Make stocks of 0.1 M Citric Acid and 0.2 M Na_2PO_4. Add 45.5 mL of citric acid to 154.5 mL of sodium phosphate to give 200 mL of buffer at pH 6.8.
4. Giemsa Stain. Giemsa stock is a methanol solution of various aniline dyes such as Methylene Blue, Eosin, Azure Blue, etc. To avoid debris, never shake the stock solution, and always remove solution from the top of the meniscus, never from the bottom of the container. Make a 5% solution by combining 2.5 mL of stain with 47.5 mL of Gurr's or McIlvaine's buffer. Mix, and filter through a tissue before use. Always skim the meniscus of staining solutions with a tissue immediately prior to insertion or removal of slides to prevent the deposition of accumulated "scum."
5. If PSCs are co-cultured with feeder cells, differentially detach the colonies with collagenase IV or dispase. Once PSC colonies are detached, pellet the colonies and rinse with DPBS, suspend the cell pellet in trypsin/EDTA solution, incubate at 37°C for 2–5 min in order to produce a single cell suspension. Pellet cells in a microfuge tube.
6. A single cell suspension is an absolute prerequisite for proper swelling during the hypotonic treatment. Remove any irresolvable clumps by gravity settling before proceeding.
7. Slide cleaning. Use a clean tissue, or lint-free cloth, sprayed with ethanol. Wipe the slides in one uniform motion across the middle third. A clean surface will evaporate evenly and quickly, leaving no streaks. This is an important step to insure spreading of the metaphases.
8. Fixative Fumes are Toxic. Dry slides in a fume hood.
9. Any stain that will bind either DNA or protein can serve to identify nuclei and chromosomes in fixed cells.

References

1. Thompson JA, Itskovitz-Eldor J, Shapiro SS, Waknitz MA, Swiergiel JJ, Marshall VS, Jones JM (1998). Embryonic stem cell lines derived from human blastocysts. Science **282**: 1145–1147.
2. Reubinoff BE, Pera MF, Fong C-Y, Trounson A, Bongso A (2000). Embryonic stem cell lines from human blastocysts: somatic differentiation in vitro. *Nat Biotechnol* **18**: 399–404.
3. Rosler ES, Fisk GJ, Ares X, Irving J, Miura J, Rao MS, Carpenter MK (2003). Long-term culture of human embryonic stem cells in feeder-free conditions. *Devl Dyn* **229**: 259–274.

4. Holmquist GP and Ashley T (2006). Chromosome organization and chromatin modifications influence gene function and evolution. *Cytogenetic Research* **114**: 96–125
5. Yang S, Lin G, Tan YQ, Zhou D, Deng LY, Cheng DH, Lou SW, Liu TC, Zhou XY Sun Z, Xiang Y, Chen TJ, Wen JF, Lu GX (2008). Tumor progression of culture-adapted human embryonic stem cells during long-term culture. *Genes Chromosomes Cancer* **47**: 665–679.
6. Baker DE, Harrison NJ, Maltby E, Smith K, Moore HD, Shaw PJ, Heath PR, Holden H, Andrews PW (2007). Adaptation to culture of human embryonic stem cells and oncogenesis in vivo. *Nat Biotechnol.* **25**: 207–215.
7. Draper JS, Smith K, Gokhale P, Moore HD, Maltby E, Johnson J, Meisner L, Zwaka TP, Thomson JA, Adrews PW. (2004) Recurrent gain of chromosomes 17q and 12 in cultured human embryonic stem cells. *Nat Biotechnol.* **22**: 53–54.
8. Buzzard JJ, Gough, NM, Crook J, Colman A. (2004) Karyotype of human ES cells during extended culture. *Nat Biotechnol.* **22**: 381–382.
9. Peterson S, Rehen S, Westra W, Yung Y, Chun J (2007) Spectral karyotyping and fluorescent in situ hybridization. In *Human Stem Cell Manual: A Laboratory Guide.* Loring JF, Wesselschmidt RW, Schwartz PH (eds.). New York: Elsevier.
10. Gunderson K, Barker DL, Bibikova M, Fan JB (2007) Genotype and epigenotype by single nucleotide polymorphism (SNP) analysis. In *Human Stem Cell Manual: A Laboratory Guide.* Loring JF, Wesselschmidt RW, Schwartz PH (eds.). New York: Elsevier.
11. Wesselschmidt RL and Loring JF. (2007) Classical Cytogenetics: Karyotyping. In *Human Stem Cell Manual: A Laboratory Guide.* Loring JF, Wesselschmidt RW, Schwartz PH (eds.). New York: Elsevier.
12. Barch, M.J. (ed.) (1991) *The ACT Cytogenetics Laboratory Manual,* (2nd Edition), Raven Press, New York.
13. Hsieh, CL. (1998) Basic cytogenetic techniques: culturing, slide making, and G-banding. In *Cell Biology: A Laboratory Handbook,* (2nd edition.) Vol.3. Elsevier, Academic Press. New York.
14. ISCN (2005) "An International System for Human Cytogenetic Nomenclature". S. Karger, Basel/New York.

Chapter 14

FISH Analysis of Human Pluripotent Stem Cells

Suzanne E. Peterson, Jerold Chun, and Jeanne Loring

Abstract

Human pluripotent stem cells (PSCs) hold promise for treating a multitude of diseases. These fascinating cells are unique in their ability to both self-renew and differentiate into cells from all three germ layers. However, PSCs, as well as other cultured cells, are prone to genetic instability. Given the possibility that these cells may one day be used clinically, identifying, and perhaps preventing, genetic instability is of particular concern for human PSC researchers. One type of genetic alteration that has been observed in PSCs is aneuploidy. Aneuploidy is defined as any divergence from the normal diploid number of chromosomes. So for human cells, any cell with more or less than 46 chromosomes would be considered aneuploid. Interestingly, there is a tendency for human PSCs, regardless of culture conditions, to gain specific chromosomes. In particular, gains of chromosomes 12, 17, 1, and X have been reported from labs all over the world. Since gains of these specific chromosomes are by far the most common aneuploidy seen in human PSCs, it is relatively easy and inexpensive to screen for these using fluorescent in situ hybridization (FISH). Here we will describe a cytogenetic method for screening human PSCs using FISH.

Key words: ESCs, PSCs, karyotyping, FISH, aneuploidy, chromosomes, genetic instability

1. Introduction

Fluorescent in situ hybridization (FISH) is a molecular cytogenetic technique that uses fluorescently labeled chromosome- or locus-specific sequences to specifically label chromosomes or specific genetic loci. FISH probes hybridize to metaphase chromosomes and interphase nuclei allowing one to assay nondividing cells, which can be a distinct advantage over classical karyotyping by G-banding or spectral karyotyping (SKY).

Philip H. Schwartz and Robin L. Wesselschmidt (eds.), *Human Pluripotent Stem Cells: Methods and Protocols*, Methods in Molecular Biology, vol. 767, DOI 10.1007/978-1-61779-201-4_14,

2. Materials

2.1. Cell Preparation and Harvest

1. Colcemid solution (10 μg/mL).
2. 0.05% Trypsin/EDTA solution.
3. Dulbecco's Modified Eagle's Medium (DMEM) with 10% fetal bovine serum (FBS).
4. 0.075 M KCl warmed to 37°C.
5. Transfer pipettes.
6. Fixative: 3:1 Methanol: Glacial Acetic Acid. Make fresh, do not store.

2.2. Metaphase Chromosome Spread Preparation

1. 80°C water bath.
2. Metal plate ~2–4 mm thick that will fit on top of the water bath.
3. Precleaned glass slides.

2.3. Slide Pretreatment and Denaturation

1. *Pepsin solution*: Add 25 μL of 100 mg/mL pepsin (Sigma) to 50 mL of 0.01 M HCl solution that has been prewarmed to 37°C. This gives a final pepsin concentration of 50 μg/mL. Make immediately before use (see Note 1).
2. Dulbecco's Phosphate-Buffered Saline (DPBS, without Mg^{2+} or Ca^{2+}).
3. DPBS with 0.5 mM $MgCl_2$.
4. *Formaldehyde solution*: 1% formaldehyde in DPBS with 50 mM $MgCl_2$. (see Note 2) Make fresh daily in chemical fume hood.
5. *Ethanol series*: 70, 80, and 100% ethanol diluted in water. Make up fresh ethanol solutions. Discard any unused solutions after 7 days.
6. *20× SSC stock solution*: To make 1 L of 20× SSC, dissolve 175.3 g of NaCl and 88.2 g of sodium citrate with water to a final volume of 1 L. Dilute the 20× SSC stock appropriately to make 1× and 4× SSC.
7. *Denaturation solution*: 70% formamide in 2× SSC pH 7.0. Make 1 mL aliquots and store at −20°C for up to a year.
8. Coplin jars (see Note 3).
9. 75°C heating plate.
10. 24 × 50 mm coverslips.

2.4. Probe Hybridization

1. FISH probe(s) Store at –20°C and protected from light.
   (a) Premade probes may be purchased commercially from vendors such as:
      i. Vysis (http://www.abbottmolecular.com).
      ii. Cambio (http://www.cambio.co.uk).
   (b) Probes may also be made via nick translation of homemade DNA probes.
2. 22 × 22 mm coverslips.
3. Rubber cement.
4. 37°C hybridization oven.

2.5. Slide Washing and Mounting

1. *Formamide wash solution:* 50% Formamide in 2× SSC pH 7.0. Use Prepare in a chemical fume hood. Pre-warm the wash solution to 45°C. Discard unused formamide wash solution after 2 days.
2. *1× SSC pH 7.0.* Store at room temperature, discard unused solution after 6 months.
3. *Tween-20 wash solution*: 0.1% Tween-20 in 4× SSC. Store at room temperature, discard unused solution after 6 months. Ensure the Tween-20 is completely dissolved before use.
4. *DAPI (4′, 6-diamidino-2-phenylindole) stain solution*: Dilute 5 μL of 5 mg/mL DAPI into 50 mL of 4× SSC to make a 0.5 μg/mL solution. Protect solution from light. Discard unused solution after 2 days.
5. Vectashield (Vector Labs).

2.6. Viewing and Interpretation

1. Fluorescence microscope with filters that match the fluorophore used on probe(s) and DAPI.
2. Laboratory counter with at least 4 channels.

3. Methods

Molecular cytogenetic analysis by FISH involves 5 distinct steps. First, the PSCs are arrested in metaphase with colcemid. While FISH can be performed on intact, nonmitotic nuclei, the colcemid treatment is often useful. Next, cells are trypsinized to a single cell suspension, swollen in a hypotonic solution, and fixed. Fixed cells are then dropped onto slides to make metaphase chromosome spreads. Acceptable slides are then pretreated with a set of solutions that open the chromatin and denature it. The slide is then hybridized overnight with a fluorescently labeled, chromosome-specific FISH probe. The next day, nonhybridized probe is removed through a series of washes and the slide is visualized with a fluorescence microscope and levels of aneuploidy are quantified.

3.1. Cell Preparation and Harvest

1. Add colcemid to cells in their culture medium at a final concentration of 0.1 μg/mL and incubate at 37°C for 3–4 h (see Notes 4 and 5). Two 70% confluent wells of a six-well plate should provide plenty of cells for this analysis.
2. Trypsinize the cells with 0.05% trypsin/EDTA solution, transfer the cells to a 15 mL conical tube, and gently triturate the cells until you have a single cell suspension. Add an equal volume of DMEM, 10% FBS medium to inactivate the trypsin.
3. Centrifuge the cells at 200×*g* for 3 min and aspirate the supernatant. Flick the tube resuspend the cell pellet and add 10 mL of prewarmed 0.075 M KCl. Incubate at 37°C for 10–15 min.
4. Add 3 drops of fresh fixative to the cells and centrifuge at 200×*g* for 3 min (see Note 6).
5. Aspirate the supernatant and flick tube to resuspend the pellet. Using a transfer pipette, add fixative slowly in drops while vortexing the cells at the lowest speed possible (see Note 7). Add approximately 5–10 mL of fixative. Store fixed cells at 4°C, for up to 6 months.

3.2. Metaphase Chromosome Spread Preparation

1. Prepare an 80°C water bath with a thin metal plate across part of the top of the water bath. Ensure the water is only 1–2 in. below the metal plate (see Fig. 1).
2. Centrifuge the fixed cell suspension at 200×*g* for 3 min. Aspirate the supernatant and wash two times with 5 mL of fresh fixative. Resuspend the pellet in 1 mL of fixative. Depending on the number of cells, you may have to spin it down again and resuspend in a smaller or larger volume to achieve the best concentration.
3. Flick the tube to resuspend the cell pellet and take 20–30 μL of your cell suspension and load it onto the slide. Hold the slide level for 10–30 s or until the center of the slide becomes granular due to evaporation of the fixative.
4. Immediately, flip over the slide and expose it to the steam for 5–10 s (see Note 8).
5. Quickly place the slide, cell side up, on the metal plate and leave it there until the fixative solution has almost completely evaporated.

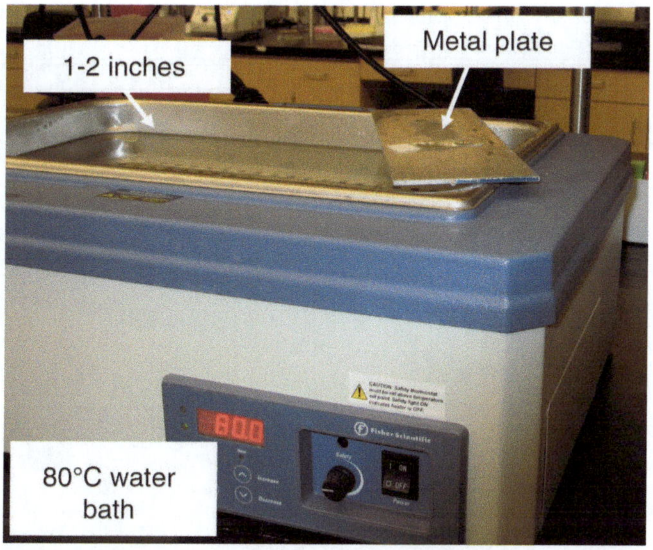

Fig. 1. Metaphase chromosome spread setup. Fill water bath to 1–2 below the top and set the temperature to 80°C. Set a metal plate across the top of the water bath to warm. After dropping cells onto the slide, expose the slide to steam and then place it on the hot metal plate just until all the liquid is evaporated.

6. Look at the slide under a light microscope and check the density of the metaphase spreads. Chromosome spreads should not be so dense that they are overlapping, but do get as many spreads on the slide as reasonably possible. If the cells are too sparse, spin your cell suspension down and resuspend in a smaller volume. If they are too dense, resuspend in a larger volume.
7. After the cell density is acceptable, check the morphology of the spreads. Chromosomes should not be over lapping but they should be contained within a reasonably tight circle. If the chromosome morphology is not acceptable, experiment with different fixative drying times and different steam exposure times (see Fig. 2a–d).
8. Also, check the color of the chromosomes in the spreads. Ideally, they should not be too bright nor too dark. Good chromosome spreads are typically light gray in color.
9. Make 5–10 slides. Pick the best one to hybridize with the FISH probes. Slides should be stored in the dark at room temperature for no more than a week before hybridization.

3.3. Slide Pretreatment and Denaturation

1. Wash slide in a coplin jar with room temperature 2× SSC for 5 min (see Note 9).
2. Incubate slide in prewarmed pepsin solution for 5 min.

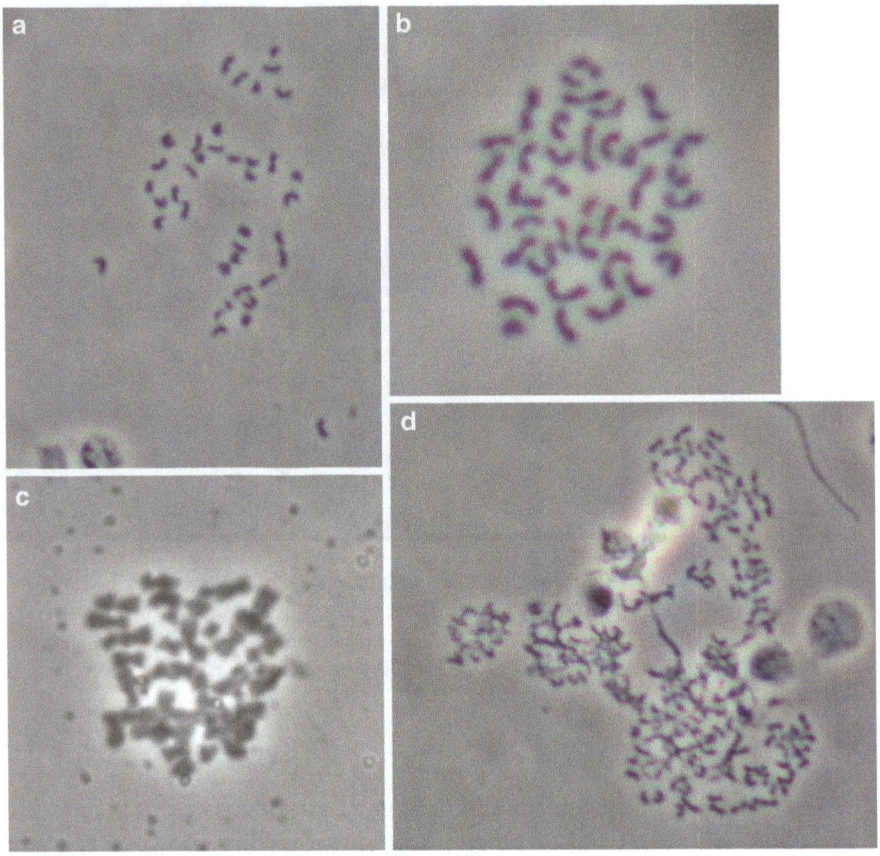

Fig. 2. Metaphase chromosome spread morphology. (**a**) The chromosome spread is too spread out. Chromosomes may be lost from the spread. (**b**) Perfect chromosome spread. Chromosomes are in a tight circle but are not overlapping. (**c**) Chromosomes in this spread are clumping and overlapping too much. (**d**) Too many chromosome spreads in the same area. It is too hard to tell which chromosomes belong to which spread.

3. Transfer the slide to room temperature DPBS for 5 min.
4. Wash the slide in DPBS with 50 mM $MgCl_2$ at room temperature for 5 min.
5. Incubate the slide in formaldehyde solution for 10 min at room temperature.
6. Dehydrate the slide in the 70, 80, and 100% ethanol series, 1 min in each solution.
7. Air-dry the slides. The slides can now be denatured and hybridized with the FISH probe or they can be stored in a desiccator at −20°C for at least 1 year.
8. Add 100 μL of denaturation solution to the slide and cover with a 24 × 50 mm coverslip (see Note 10).
9. Place the slide on a 75°C heating block for 1.5 min.

10. Carefully remove the coverslip and immediately dehydrate in the 70, 80, and 100% ethanol series, 1 min in each solution (see Note 11).
11. Air-dry the slide.

3.4. Probe Hybridization

1. Thaw the probe, if necessary, then vortex vigorously. Protect it from light.
2. Place the manufacturer recommended volume, typically 10 μL, into an PCR or microfuge tube.
3. Denature the probe for 10 min at 80°C, then at 37°C for 60 min in a thermocycler or water bath.
4. When the probe is ready, place slide and a 22 × 22 mm coverslip on a 37°C heating block.
5. Add the probe to the area of the slide that contains the spreads. Add the coverslip to the slide and quickly seal the edges with rubber cement (see Note 12).
6. Incubate overnight, in the dark, at 37°C.

3.5. Slide Washing and Mounting

1. The next day, carefully remove the rubber cement from the slide with a forceps and by rubbing across it with your fingertips.
2. Incubate the slide in 2× SSC until the coverslip lifts off by itself. Try to keep the slide protected from light for this and all subsequent steps.
3. Wash the slide in formamide wash solution for 5 min at 45°C.
4. Wash in 1× SSC for 5 min at 45°C.
5. Wash in Tween-20 solution for 5 min at 45°C.
6. Wash in DAPI stain solution for 5 min at room temperature.
7. Dehydrate the slide in a 70, 80, and 100% ethanol series, 1 min per solution.
8. Air-dry the slide.
9. Add 1 drop of vectashield to the slide and cover with a 24 × 50 mm coverslip.

3.6. Viewing and Interpretation

1. Observe the slide using a fluorescence microscope equipped with filters that are appropriate for the fluorophore used to label your probe as well as DAPI (see Note 13).
2. In diploid interphase nuclei, probes should show two signals (see Fig. 3a). In metaphase chromosome spreads, the signal may appear as 1 signal per chromosome (see Fig. 3b) or as two closely spaced signals per chromosome (see Fig. 3c). Two closely spaced signals on a chromosome spread should be counted as one signal.

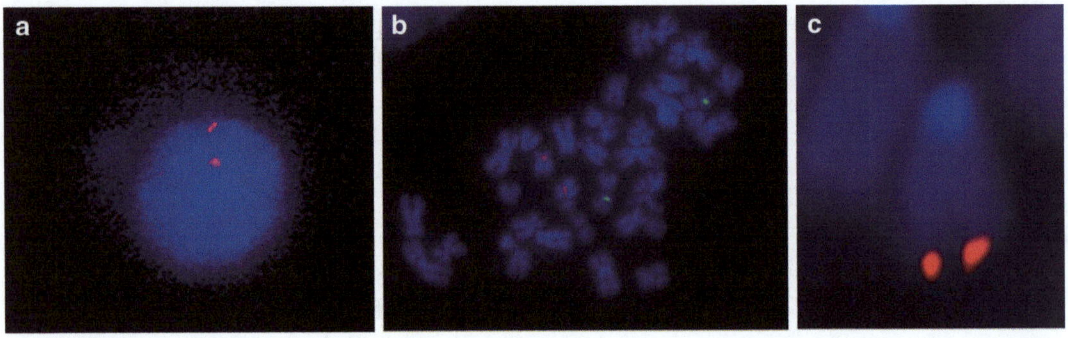

Fig. 3. Nuclei and chromosome spreads hybridized with FISH probes. (**a**) Interphase nuclei hybridized with chromosome 12 probe and stained with DAPI. Disomic cells should show two fluorescent spots. (**b**) Metaphase chromosome spread hybridized with chromosome 12 (*red*) and chromosome 17 (*green*) FISH probes. The probes show up as a single dot on the chromosome. This spread is disomic for both chromosome 12 and 17. (**c**) Example of a chromosome hybridized with a probe that shows up as two dots on a single chromosome. Either morphology – single dot or double dot per chromosome – can be observed.

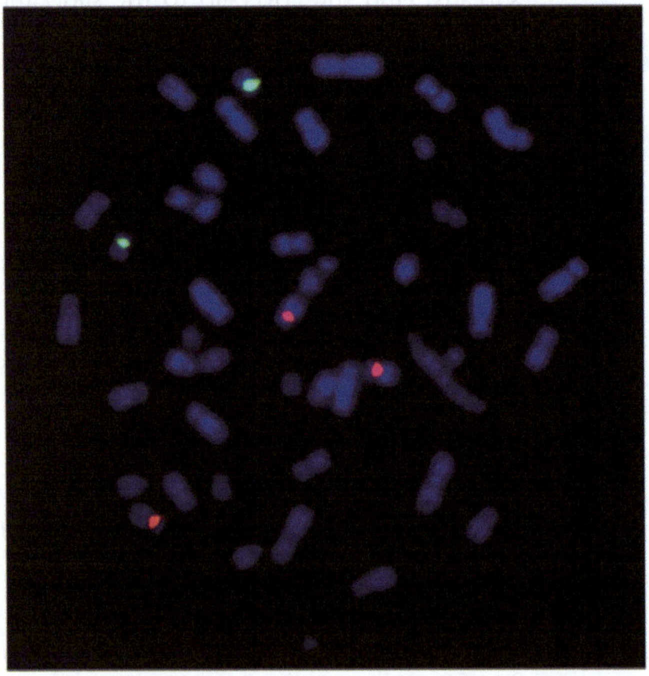

Fig. 4. Metaphase chromosome spread hybridized with probes for chromosome 12 (*red*) and 17 (*green*). Note that this cell is disomic for chromosome 17 but trisomic for chromosome 12 so this cell is considered aneuploid. Photograph courtesy of Dr. Zoltan Simandi.

3. Using a multichannel laboratory cell counter, count the number of signals from 200–500 nuclei or chromosome spreads. Record the number of monosomic and trisomic chromosome events in the culture.
4. Figure 4 shows a metaphase chromosome spread from human ESCs hybridized with probes for chromosome 12 in red and 17 in green. The cell is trisomic for chromosome 12 and disomic for chromosome 17, making the cell aneuploid.

4. Notes

1. Ensure the 25 μL of pepsin is thoroughly distributed in the 0.01 M HCl solution by swirling the coplin jar before adding the slide.
2. Stock formaldehyde solutions are 37%.
3. If coplin jars are not available, 50 mL conicals can be used.
4. If the PSCs were cultured on inactivated human fibroblasts, the fibroblasts will be present on your slide and indistinguishable from human PSCs that are not in metaphase. This typically is not a problem due to the much lower number of fibroblasts compared to PSCs in the culture. However, if this is a concern, only analyze metaphase chromosome spreads as these are derived from actively dividing cells and can only be PSCs, assuming the feeder layer is fully inactivated.
5. FISH can be done on intact, nonmitotic nuclei, so the colcemid treatment is optional, but it is often useful. Longer colcemid incubation times can be used but this will result in shorter, more condensed chromosomes. Incubations much longer than ~8 h can be toxic to the cells.
6. Treatment with hypotonic 0.075 M KCl solution leads to cell swelling. When the cells are centrifuged after treatment with 0.075 KCl, the cell pellet should be visibly larger.
7. It is very important that the cells are moving (slowly) and not in clumps when the fixative is added. If not, the cells will become "fixed" together in the clumps, rendering those cells uninformative when analyzed.
8. The quality of metaphase chromosome spreads is dependent on drying time. Steam is used to slow down the evaporation process. Depending on the atmospheric conditions in the lab on that particular day, it may or may not be necessary to slow the evaporation process with steam. One must empirically determine the best drying procedure for any particular day. In addition to manipulating the drying time with steam, the fixative can be altered so that it has more or less methanol or glacial acetic acid. A fixative with more methanol (e.g., 6:1 methanol to glacial acetic acid) would dry faster than one with more glacial acetic acid (e.g., 1:1 methanol to glacial acetic acid). Getting good metaphase chromosomes spreads can be very, very difficult some days but, fortunately, FISH can be done on intact, nonmitotic nuclei as well.
9. From this point on, slides must not dry out until step 7.
10. From this point on, slides must not dry out until step 11.

11. Often the easiest way to remove the coverslip is to turn the slide perpendicular to the floor (preferably over a sharps container) and quickly flick your wrist in a downward motion.
12. The rubber cement seal does not have to be pretty – just make sure that all the edges of the coverslip are sealed.
13. Typically, it is easiest to find the cells first in the DAPI channel then look at your probe in the appropriate channel.

References

1. Draper, J.S., Smith, K., Gokhale, P., Moore, H.D., Johnson, J., Meisner, L., Zwaka, T.P., Thomson, J.A., Andrews, P.A. (2004) Recurrent gain of chromosomes 17q and 12 in cultured human embryonic stem cells. *Nat Biotechnol* **22**, 53–54.
2. Buzzard, J.J., Gough, N.M., Crook, J.M. & Colman, A. (2004) Karyotype of human ES cells during extended culture. *Nat Biotechnol* **22**, 381–382; author reply 382.
3. Schrock, E., du Manoir, 2., Veldman, T., Schoell, B., Weinberg, J., Ferguson-Smith, M.A., Ning, Y., Ledbetter, D.H., Bar-Am, I., Soenksen, D., Garini, Y., Ried, T. (1996) Multicolor spectral karyotyping of human chromosomes. *Science* **273**, 494–497.
4. Peterson S., Rehen S., Westra W., Yung Y., Chun J. (2007) *Spectral Karyotyping and Fluorescent In Situ Hybridization.* Human Stem Cell Manual, A Laboratory Guide. First Edition, Eds. Jeanne F. Loring, Robin L. Wesselschmidt, Philip H. Schwartz, Elsevier Inc. ISBN: 978-0-12-370465-8

Chapter 15

Immunocytochemical Analysis of Human Pluripotent Stem Cells

Hubert E. Nethercott, David J. Brick, and Philip H. Schwartz

Abstract

This chapter will describe the most common immunocytochemical method utilized in the stem cell field – using fluorescently tagged secondary antibodies to detect a primary antibody that is bound to an epitope on a molecule of interest. Secondary antibodies recognize the heavy chain of the primary antibody's isotype. Generally, these methods employ an incubation period of the sample with the primary antibody, a series of washes to remove unbound primary antibody, a secondary incubation period of the sample with the fluorescently conjugated secondary antibody, followed by washes and preparation for microscopy.

Key words: immunocytochemistry, antibodies, fluorescent tags, immunofluorescence

1. Introduction

Immunocytochemistry, using antigen-specific antibodies, is a fast and easy way to determine whether a population of cells is homogeneous or heterogeneous with regard to a particular molecular marker. Immunocytochemistry allows for the visualization of individual cells within a colony or culture and thus provides an overall assessment of expression of a particular marker throughout the culture under specific culture conditions (1–4). Antibodies, in combination with specific stains/dyes such as the commonly utilized nuclear stain DAPI, can also reveal the subcellular localization of the particular antigen in question. In addition, translocation of signaling factors from one cellular location to another following signal transduction may be easily examined following staining under alternate conditions. Primary antibodies are raised against an antigen, which may be a protein, glycolipid (such as the SSEA-4 epitope), carbohydrate, small molecule, or DNA.

Philip H. Schwartz and Robin L. Wesselschmidt (eds.), *Human Pluripotent Stem Cells: Methods and Protocols*, Methods in Molecular Biology, vol. 767, DOI 10.1007/978-1-61779-201-4_15, © Springer Science+Business Media, LLC 2011

Antibodies (also known as immunoglobulins, Igs), first described by Paul Ehrlich in 1891, have proven to be one of the most useful research tools available. They are typically made of basic structural units – each with two large heavy chains and two small light chains – to form, for example, monomers with one unit, dimers with two units or pentamers with five units. Antibodies are produced by white blood cells called plasma cells. There are several different types of antibody heavy chains, and several different kinds of antibodies, which are grouped into different isotypes based on which heavy chain they possess.

Though the general structure of all antibodies is very similar, a small region at the tip of the protein is extremely variable, allowing millions of antibodies with slightly different tip structures, or antigen-binding sites, to exist. This region is known as the hypervariable region. Each of these variants can bind to a different target, known as an antigen. The unique part of the antigen recognized by an antibody is called an epitope.

Primary antibodies vary widely in their binding affinities and specificities and must be tested to determine whether they recognize the antigen when the specimen is prepared for immunocytochemistry. Antibodies bind to specific epitopes on antigens. Epitopes may consist of short stretches of amino acids in a protein, conformational characteristics such as an exposed alpha helix, or structural elements of a small molecule. Polyclonal antibodies contain multiple antibodies that usually recognize several different epitopes on a single molecule. In contrast, monoclonal antibodies are of a single defined antibody type and recognize a single epitope on a single molecule (2, 5–8).

Specimens are often described as "weakly positive" or "strongly positive." When using a new antibody or testing a new sample, it is usually a good idea to confirm the presence of the antigen using an alternate method, such as RT-PCR, if the antigen is a protein. In general, "weakly positive" samples must always be verified. If both mRNA and protein are present in your cells, then there is compelling evidence that the antigen you are examining is present. Other methods used for confirmation of antibody staining include the use of a second antibody that recognizes another epitope on the same molecule, and immunoblots ("Western blots"), in which predefined or predicted molecular weight determination adds confirmation of the identity of the antigen.

Immunocytochemistry for cultured cells uses an amplification technique to make submicroscopic molecules visible. Ideally, every experiment should include negative controls (such as no primary antibody) and positive controls (such as a cell type known to express the antigen) in order to assess the efficacy of staining.

2. Materials

2.1. Preparation of Samples

1. Chamber Culture slides, Lab-Tek II, (Thermo Fisher Nunc).
2. Extracellular Matrix Component such as Matrigel, laminin, or fibronectin.
3. Bromodeoxyuridine (BrdU, Sigma B-9285), 10 μM final concentration.

2.2. Fixation

1. Fume hood for working with paraformaldehyde.
2. 0.2 M sodium phosphate buffer, pH 7.4.
3. 4% Paraformaldehyde. In the fume hood: Add 40 g of paraformaldehyde to 500 mL of dH_2O, heat to 60°C (do not exceed this temperature), and stir. Add a few drops of 1 N NaOH until solution is clear (the solution will not completely clear without the addition of NaOH as the basicity is needed to depolymerize the paraformaldehyde). Filter (0.2 or 0.45 μm) and add 500 mL of 0.2 M sodium phosphate buffer, pH to 7.4 (recheck pH after cooling and adjust if necessary with phosphoric acid or sodium hydroxide). Store at 4°C up to 1 week or alternatively store aliquots at −20°C up to 6 months.

2.3. Immunostaining

1. Dulbecco's phosphate-buffered saline with Ca^{++} and Mg^{++} (DPBS).
2. *Blocking buffer*: DPBS, 0.3% (v/v) Triton X-100, 3% (v/v) serum from secondary antibody host species: rat, mouse, goat, donkey, etc.
3. *Antibody dilution buffer*: DPBS, 0.3% (v/v) Triton X-100, 1% (v/v) serum from secondary antibody host species: rat, mouse, goat, donkey, etc.
4. Antibodies, primary and secondary:
   (a) *Primary antibodies* can be purchased from various commercial vendors, such as BD Biosciences, Millipore, R&D Systems, Santa Cruz Biotechnology, Serotec, Sigma, or Developmental Studies Hybridoma Bank, or provided by colleagues.
   (b) *Secondary antibodies* (ex.: AMCA, Cy2, Cy5, RRX, AlexaFluors, DyLights) Jackson ImmunoResearch Laboratories, Invitrogen and other commercial sources.
5. ProLong Gold antifade reagent (Invitrogen, P-36934).
6. ProLong Gold antifade reagent with DAPI (Invitrogen, P-36934).
7. Cover slips, No. 1 thickness range for high magnification objectives (Thermo Fisher Scientific, 12-548-5P).

8. Nail polish "Clear" Top coat.
9. Sodium azide (Sigma-Aldrich, S8032).
10. Hoechst 33342 (Invitrogen, H3570).

2.4. Imaging

1. Fluorescence microscope.
2. Objectives: 10×, 20×, 40×, and perhaps 60× or 100×.
3. Filter cubes appropriate for secondary antibody fluorophores. It is important to make sure that the cubes will give maximal signal for one fluorophore but not allow bleed-through excitation of another fluorophore.
4. Digital Camera.
5. Image Pro 4.0 and AFA Plug-in (or other imaging software).
6. Adobe Photoshop.

3. Methods

The protocol described below, which has routinely produced high quality images for publication, is easy and can be performed by devoting only a short period of time each day. If rapid analysis is desired, the alternative protocol can be used, with timing indicated at the end of each section.

3.1. Preparation of Slides

3.1.1. Growth on Glass surface

Several days prior to staining, passage the cells to Lab-Tek glass chamber slides coated with extracellular matrix such as laminin or a feeder layer of cells, such that the cells will adhere strongly to the surface and not wash off during the staining process. Fluorescent antibody staining on plastic culture dishes is not advised. It is also advisable to incubate the slides in a large (165 mm) culture dish so that the slides do not need to be handled – handling increases the probability of breaking the seal between the wells. For a detailed description of pluripotent stem cell culture on glass slides, see Chapter 12.

3.1.2. Bromodeoxyuridine (BrdU) Labeling

1. BrdU (10 μM final concentration) should be incubated with the cells for 2–24 h prior to fixation (in some cases it will be desirable to remove the BrdU-containing medium and culture the cells in regular medium for a few days before fixation).
2. BrdU-labeled cells should be treated with HCl (1 N HCl for 20–30 min at 37°C) after fixation, but prior to blocking and antibody incubation.
3. Wash well with DPBS after HCl incubation.

3.2. Fixation

1. Carefully aspirate the growth medium and rinse cells one time with DPBS (see Note 1).
2. Fix cells for 10 min at room temperature with 4% paraformaldehyde in DPBS (see Note 2). Dispense the solution down the side of the well so that it slowly floods the well without disturbing the cell surface. Use this same gentle technique at all times while adding any solution to the wells.
3. Wash cells twice with DPBS, allowing the cells to incubate in the wash for approximately 5 min before aspirating the wash.
4. For best results, stain fixed cells within 24 h of fixation. Alternatively, store fixed cells at 4°C in DPBS, 0.05% (w/v) sodium azide.

3.3. Immunostaining

The method described is used for simultaneous staining with more than one antibody. Staining for more than one antigen involves use of multiple primary antibodies, each of a unique class or animal species, followed by use of multiple secondary antibodies, each specific for one of the primary antibodies and each carrying a unique fluorophore (see Note 3 for a summary of the entire procedure).

3.3.1. Day 1

1. Design a plan for each sample well as in Fig. 1. Make certain antibody isotypes do not overlap within a given well (see Notes 4 and 5).
2. Aliquot antibody dilution buffer (ADB) into single 0.65 mL micro-centrifuge tubes for each well. If using eight-well culture slides, you will need a final volume of 250 μL per well. For four-well culture slides, use 400 μL per well (adjust volume per well accordingly for wells that are other sizes). Add appropriate volume of primary antibody (or antibodies) to each tube with ADB and gently mix. We typically dilute primary antibodies 1:100. Note that secondary-only control wells (see Fig. 1) should be incubated in ADB alone (no primary antibody) or with a control Ig diluted in ADB.
3. Remove protein precipitates from the primary antibody solution by spinning at 16,000 × *g* for 5 min in a micro-centrifuge.
4. Gently remove primary antibodies to new tubes, leaving a small amount of liquid at the bottom where the sediment remains (if the hinge of the tube is placed toward the outside of the rotor, then the sediment, if any, will be directly under the hinge). Keep diluted antibodies on ice until added to cells.
5. Wash cells gently with DPBS. Note – incubate any BrdU-treated wells with HCl then rinse with DPBS (see notes on BrdU above in Subheading 3.1.2).

Well	Blue (AMCA, Hoechst, DyLight 405)	Green (Cy2, AF488, DyLight 488)	Red/Orange (Cy3, AF555, DyLight 549)	Far Red (Cy5, AF 647, DyLight 649)
1*	Oct3/4-Mouse IgG	Tra-1-60-Mouse IgM	Nanog-Rabbit	Sox2-Goat
2	Secondary only: Anti-mouse IgG	Secondary only: Anti-mouse IgM	Secondary only: Anti-rabbit	Secondary only: Anti-goat
3	Tra-1-60-Mouse IgM	Nanog-Rabbit	Ki-67-Mouse IgG(Note)	GATA4-Goat
4	Secondary only: Anti-mouse IgM	Secondary only: Anti-rabbit	Secondary only: Anti-mouse IgG	Secondary only: Anti-goat
5*	PAX6-Rabbit	GATA4-Goat	SSEA1-Mouse IgM	BrdU-Mouse IgG
6	Secondary only: Anti-rabbit	Secondary only: Anti-goat	Secondary only: Anti-mouse IgM	Secondary only: Anti-mouse IgG
7	DAPI (DNA staining)	SSEA4-Mouse IgG	Brachyury-Rabbit	SSEA1-Mouse IgM
8	No antibody	Secondary only: Anti-mouse IgG	Secondary only: Anti-rabbit	Secondary only: Anti-mouse IgM

Fig. 1. An example staining plan for an eight-well slide. Note: This well must be treated with HCl prior to applying primary antibody. See Subheading 3.1.2 above.

6. Remove DPBS and add approximately 250 μL of Blocking Buffer to each well. Incubate for 15 min at room temperature.
7. Wash cells gently with DPBS.
8. Remove DPBS and immediately add the diluted primary antibodies to the wells.
9. Remove the covers from the (eight-well) slides and place slides into a humidity controlled bin (i.e., covered Tupperware with damp Kimwipes). Condensation on the eight-well slide cover increases the probability of cross-contamination among the wells.
10. Recommended method: Incubate chamber slides overnight at 4°C.

    Alternate method: incubate slides 1–2 h at room temperature.

Table 1
Common fluorophores and their peak excitation and emission spectra

Fluorophore	Excitation peak (nm)	Emission peak (nm)
AMCA, Hoechst, DAPI	~350	~450
FITC, Cy2, Alexa488	~492	~520
TRITC, Cy3, Alexa555	~550	~570
Cy5, Alexa647	~650	~670

3.3.2. Day 2

1. Dilute secondary antibody (or antibodies, see Table 1 for fluorophore selection criteria) in ADB using the concentration recommended by vendor or determined empirically to give the best results. *We typically dilute secondary antibodies* 1:250, *Alexafluor* 1:1,000.
2. Remove the primary antibody from each well (see Note 6).
3. Wash cells twice with DPBS. Replace aspirator tips after each use.
4. Spin secondary antibodies at 16,000 × g for 5 min, to remove any protein precipitates (see steps 3 and 4 of Day 2 above).
5. Carefully add secondary antibodies to aspirated wells immediately after aspiration.
6. Recommended: Incubate slides overnight at 4°C in a humidity controlled bin (i.e., covered Tupperware with damp Kimwipes).
   (a) *Alternate method*: Incubate 1 h at room temperature.

3.3.3. Day 3

1. Wash wells three times with DPBS, incubating for 5 min during each wash (cells in the chamber slide can be visualized under the fluorescence microscope during this procedure to ensure that enough washes have been performed to adequately reduce background signal).
2. If nuclear counterstaining is desired, cells can be incubated for a short period with the counterstaining reagent following by washing and mounting of the slide. Alternatively, the use of a mounting medium which already contains DAPI may eliminate the need for a separate staining step. In the first method, the counterstaining reagent (Hoechst 33342 Invitrogen) at 1 mg/mL in DMSO (stored at 4°C in the dark) is diluted 1:500 in DPBS and incubated with cells for 1–5 min at room temperature, followed by washing excess stain away with DPBS prior to mounting. If a mounting

medium such as ProLong Gold Antifade reagent with DAPI (Invitrogen P36935) is used, excess moisture is removed from the slide by gently tapping the side of the slide or coverslip onto a clean Kimwipe prior to addition of the prewarmed-to-room-temperature reagent. It is often useful to have a cellular counterstain if it does not interfere with an antibody being detected by a fluorophore in the blue channel, such as 7-Amino-4-methylcoumarin-3-acetic acid (AMCA). A nuclear counterstain is also helpful when evaluating the nuclear localization of an antigen (particularly in stem cells that have a high nucleus-to-cytoplasm ratio).

3. Prepare mounting medium (used to minimize photobleaching of fluorescence) in accordance with manufacturer's instructions. Examples of mounting media are as follows: Vectashield (Vector Labs), Slow Fade (Invitrogen), and Prolong Gold Antifade Reagent (Invitrogen). It should be noted that certain antifade reagent solutions contain glycerol and may be incompatiable with certain applications, such as specimens that contain lipophilic plasma membrane stains such as DiI.
4. Aspirate wells.
5. Snap off plastic wells according to the manufacture's recommendations. Carefully use a razor on one of the short ends of the silicone gasket (if present; otherwise, skip this step). Using fine tweezers peel back the gasket slowly.
6. Pipette a bead of the mounting medium along the long end of the slide (approximately 300 μL). Be careful not to allow bubbles to form on the bead. Gently lower a rectangular cover slip at a 45° angle on the slide. Allow the mounting medium to spread.
7. Using two fingers very gently squeeze out the extra mounting medium and/or trapped air bubbles over a disposable paper towel. Pressing too hard will displace and/or damage cultures. Aspirate the extra medium off the slide.
8. Allow the slide to dry at room temperature in a dark, dry place overnight.
   (a) *Alternate method*: Allow samples to dry briefly then proceed to the steps in Day 4 below. Note that the coverslips will still move around and should be handled with care.

3.3.4. Day 4

1. Remove excess mounting medium by gently wiping the slide with 70% ethanol (use Kimwipes or cotton swab).
2. Seal slide edges with nail polish.
3. Allow to dry.

4. View slides on fluorescence microscope. Afterward, store slides at −20°C (with desiccant for best preservation). Storage at −20°C can preserve the signal for months (depending on the sample, antibody, choice of antifade reagent, etc.).
5. See Notes 7–12 for troubleshooting hints.

3.4. Imaging

After immunostaining, cells are usually viewed on a fluorescent microscope and images of the stained cells captured with a digital camera. There are a variety of cameras and image capturing software packages available; therefore, we won't go over the specific details of a particular program here (details about one program, ImagePro, can be found in the Appendix of this chapter). However, many scientists bring the captured images into Adobe Photoshop to create output for publications. Therefore, in the next section we will describe how to use several features in Photoshop and briefly introduce a program available for image quantization (NIH Image, also known as ImageJ).

3.4.1. Microscope Setup

1. Seat slide on microscope stage with the cover slip facing the objective lens.
2. Make sure the microscope shutter is closed. Turn on mercury lamp and incandescent lamp.
3. Using a phase-contrast 20× objective, bring the sample into the focal plane.
4. Turn off incandescent light and use mercury light (preferably through lower frequency filters). Bleaching of fluorochromes is accelerated during exposure to higher frequency light. We prefer an excitation of ~570 nm (Cy3 channel) to first evaluate staining. Open shutter and analyze cultures through the microscope's binocular eyepieces.
5. Scan through areas of interest while cycling through the other channels. Remember to limit the exposure of the slide to mercury light. Close shutter when not analyzing samples.

3.4.2. Adobe Photoshop

Photoshop can open a wide variety of image files captured from a microscope-mounted camera, including ".tiff" and ".jpg" formats, and provides a variety of means to manipulate images; however, it may not be able to open 16-bit files (if this is the case, be sure to save your files as 8-bit files). Here, we will briefly describe how to set the color mode, alter the image size, create scale bars for an image, adjust the image brightness and contrast, and create color overlays of images.

Setting the Color Mode

Color digital images can either use RGB (red, green, blue) or CMYK (cyan, magenta, yellow, black) for color encoding. RGB images are more compatible for computer monitors or projectors, since they use an additive light system (printers rely on a subtractive light system such as CMYK). Bright greens, reds, and blues cannot be reproduced as readily in print as they can on a monitor, so prints of an RGB image may not convey the bright colors or fine detail visible on the computer monitor. For print purposes (and therefore for most journal submissions), it is best to convert an RGB image to CMYK. To convert to CMYK for printing, go to "Image" → "Mode" and select CMYK.

Adjusting the Image Size

Images captured by image acquisition software programs can come in a variety of sizes and resolutions. To find the size of your image, go to "Image" → "Image Size." Images often are captured at 72 pixels/in. and are of fairly large dimensions (in terms of inches). It is often desirable to set the resolution to 300 pixels/in. but not change the overall size of the file so that the dimensions (in inches) of the image are more suitable for printing or incorporating into a figure. To do this, make sure that the checkbox next to "Resample Image" is unchecked (as in Fig. 2) then adjust the resolution [see the figure and note that the overall pixel dimensions (1.83 M, 1,600 × 1,200 pixels) are the same for both while the document sizes (width, height, resolution) are different].

Scale Bars

One way to generate scale bars for your images and to make size/length determinations is to use a scale micrometer. These are slides that have lines etched a particular distance apart from each other. The micrometer can be placed on the microscope and an image taken using each of the microscope objectives. As the images are captured at the same width (in terms of pixels); you can determine a conversion factor that will allow you to measure real distances on your images. As an example, if an image taken with a camera on a particular microscope using a 20× objective has a total width of 580 μm (from the scale micrometer) and 1,600 pixels; this means that 100 μm would equal ~276 pixels on that image. Note that these measurements will be specific to the objective, microscope, and camera used, so attention must be

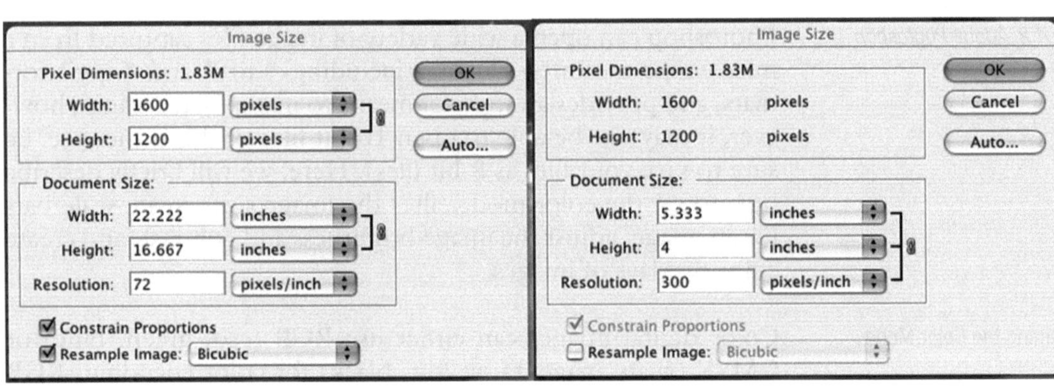

Fig. 2. Adjusting the image size in Adobe Photoshop.

paid to the conditions under which a particular image was captured in order to appropriately determine the scale. You can draw a line of a particular length (in pixels) in Photoshop by using the line tool (on the tool bar, which also contains the move tool, text tool, etc.) and watching the pixel location in the Navigator window ("Window"→"Navigator"; click on the "Info" tab in the Navigator window). The X and Y coordinates of the cursor location will be in pixels as long as the rulers for the image are set to "pixels" ("Preferences"→"Units and rulers"). Your image capture software, in most cases, also has the capacity for generating scale bars on your images.

Adjusting the Brightness/ Contrast of an Image

There are multiple ways to adjust images in Photoshop, and most are found under "Image"→"Adjustments." One straightforward way to adjust the brightness/contrast is to use the "Levels" option ("Image"→"Adjustments"→"Levels") and adjust the sliders under the histogram (see Fig. 3). The advantage of this option is that by viewing the histogram, you can more accurately adjust the intensity of the image without altering the data. It is IMPERATIVE when using any image adjustment for data images to be extremely careful not to alter the data with the adjustment. For example, decreasing the brightness should not remove signal and increasing the brightness should not create signal or expand the signals zone. See examples below for images that have been appropriately and inappropriately adjusted.

Changing Grayscale Images to Color and Overlaying Color Images

Cells or tissues are often double- or triple-labeled with different fluorescent molecules to allow visualization of multiple signals. Photoshop can be used to convert captured grayscale images to color and overlay the color images so that all fluorescent signals can be visualized simultaneously. In order to create a color overlay, the images of the different fluorescent channels are brought together into a single file. The separate images are maintained on individual layers and then assigned a different color. To begin, select the entire image ("Select"→"All") and copy ("Edit"→"Copy"). Make a new file ("File"→"New") and the size, resolution, etc., will be identical to what you just copied. In the window that opens and describes the new file, switch from "Grayscale" to "RGB" (or "CMYK" if the image is solely for print media). Once the new file is created, paste in the copied image ("Edit"→"Paste"). Select all and copy the other images to be overlaid then paste them into the new file. Each image will automatically be pasted into a different layer ("Window"→"Layers"). To change the color of an image in a layer, open the "Levels" option ("Image"→"Adjustments"→ "Levels") and use the tab marked "RGB" to select either the Red, Green, or Blue channel. Use the "Output Levels" to alter the color: for a Red image, make the Green and Blue output levels 0

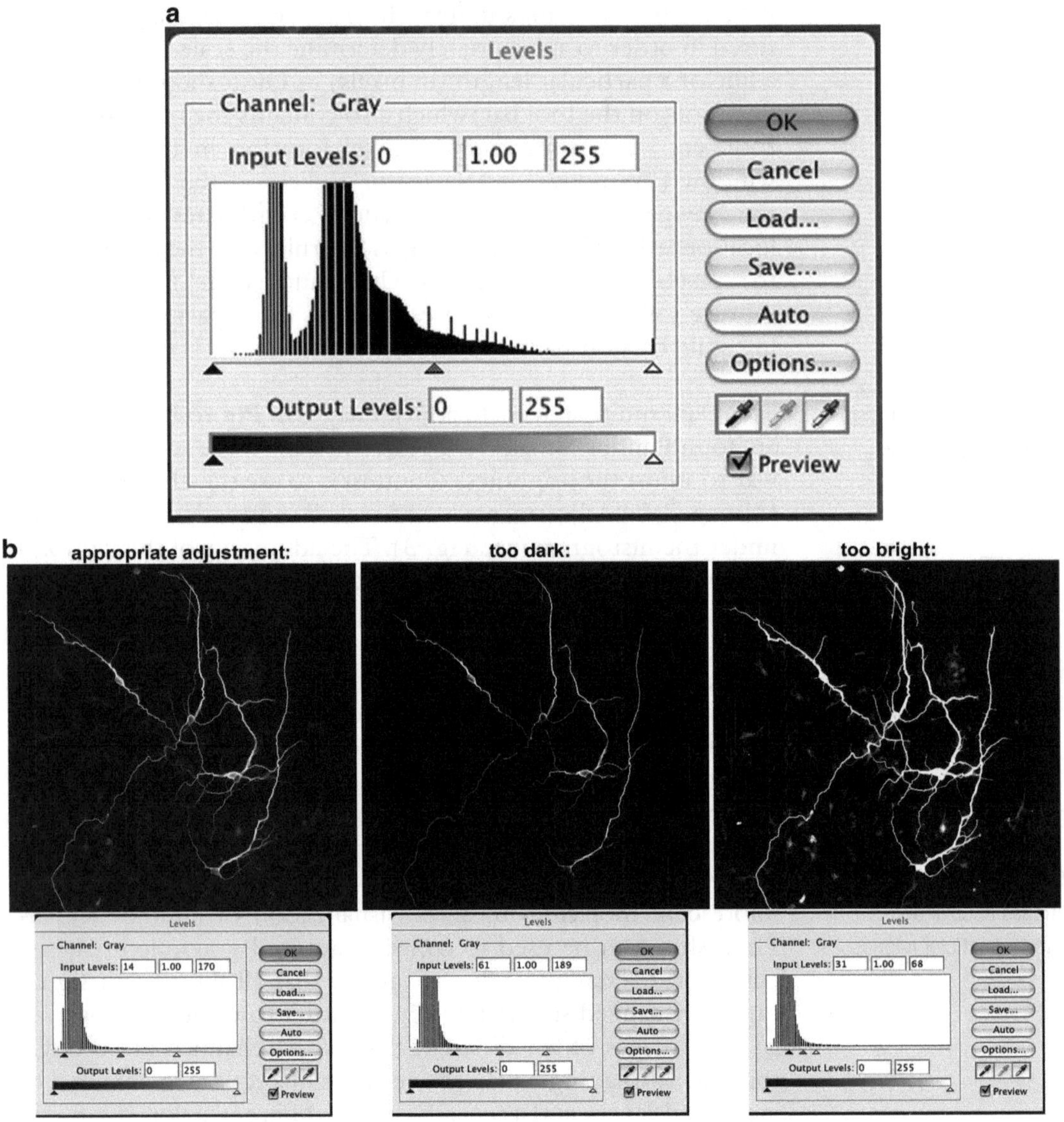

Fig. 3. Adjusting the brightness/contrast of an image. (**a**) By using the histogram to adjust brightness/contrast levels, you can more accurately adjust the intensity of the image without altering the data. (**b**) Examples for images that have been appropriately and inappropriately adjusted.

(change the number in the box on the right from 255 to 0), for a Green image, make the Red and Blue output levels 0, and for a Blue image, make the Red and Green output levels 0. These steps can be repeated for different layers within the same document to create layers that are of different colors. To overlay differently colored layers, position one colored layer directly above the other colored layer (in the "Layers" window) and then change the button under the "Layers" tab from "Normal" to "Screen". You should now see both layers overlaid. Be sure, however, to keep the original, unchanged B&W files (raw data) of all images for future reference or alternative image production.

3.4.3. NIH Image (ImageJ)

NIH Image (or ImageJ) is a free program available for download that can be used to quantify a wide variety of parameters in an image. In addition to the basic features of ImageJ, there are Macros that others have created (or you can write yourself) that expand the functionality of the program. For details and downloads see: http://rsb.info.nih.gov/nih-image/Default.html.

Measurements in ImageJ

In order to do measurements in ImageJ, you must first know the scale of your picture in real dimensions (see "Scale Bar" section above). An easy way to convert this information to a scale in ImageJ is to draw a line across the entire width of your image (use the straight line tool on the toolbar). Then, go to "Analyze" → "Set Scale" and set the known distance to the numerical value and unit of length for your image width (for example, the width of the image described above would be 580 μm). Keep the Pixel Aspect Ratio as 1 and use "um" for "μm". If you are analyzing multiple images that were taken under the same conditions and thus have the same scale, you can check "Global" in the "Set Scale" window and the scale will be automatically applied to all the images. After setting the scale, the length of any line drawn and measured will be given in the desired units. To measure an element in your image, you can draw a line (straight, segmented, or freehand) and then click on "Analyze" → "Measure". "Analyze" → "Set measurements" allows you to decide what parameters will be measured. Note that you can also choose other types of shapes (other than a line) and measure parameters such as area.

3.4.4. Image Pro

The following section will describe using Media Cybernetics' Image-Pro plus (www.mediacy.com) to photograph snapshots of a field of interest using a digital camera.

1. Turn on digital camera.
2. Open Image Pro.
3. Under Acquire select "Video/Digital Capture."
4. The following window will allow you to preview and snap pictures directly from the camera.
5. Click Start Preview.
6. Adjust the exposure time to brighten image without over-saturating digital feed (most digital camera drivers have a configuration setting to provide live saturation warnings).
7. Snap the image when you are satisfied with the previewed image.
8. Save image (a ".tiff" file format is recommended for preserving image detail).

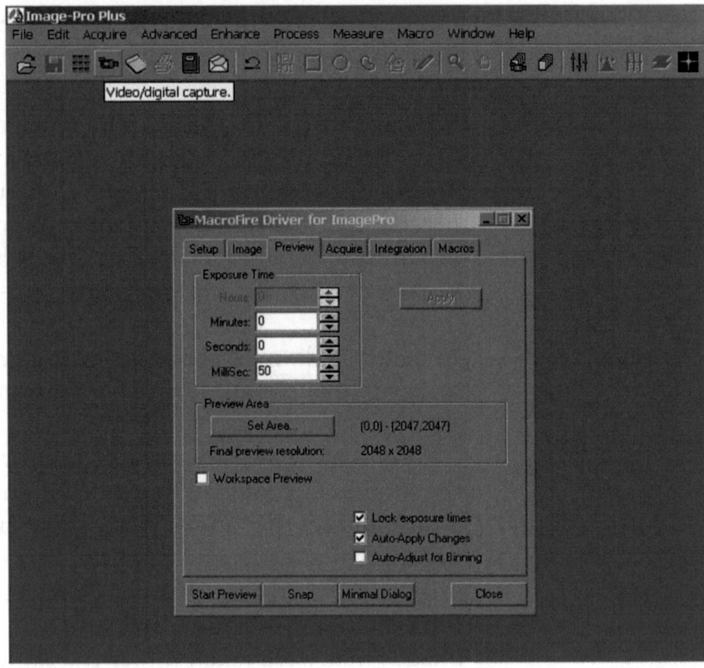

Fig. 4. Acquiring an image in Image Pro.

3.4.5. Using Image-Pro AFA Plug-In

The Image-Pro AFA Plug-in is a useful tool for organizing and managing multiple channels from a field of interest. Exposures may be optimized for each channel before imaging the field of interest as a set. After a set of images is obtained, the color composite tool may be used to pseudo-color and merge channels.

1. Open the "Advanced Acquisition window."
2. Click on preview (note: the exposure times for the preview are set for the first channel).
3. Adjust exposure times for each channel.
4. Check the boxes for the channels you want to photograph.
5. Click Acquire Set.
6. If the microscope used is fully automated it will automatically rotate the filter cubes and photograph the samples. If it is not, a prompt will ask you to manually turn the wheel between pictures.
7. Once the set has been acquired, you can save it using the "Set Manager."

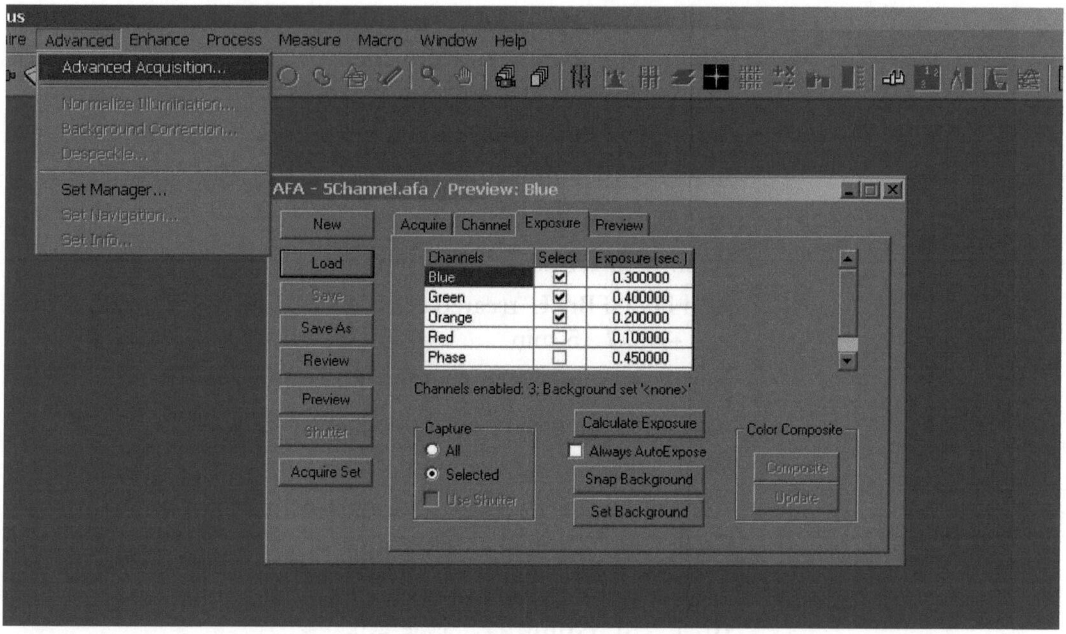

Fig. 5. The Image-Pro AFA Plug-in allows the user to optimize multiple channels from a field of interest.

4. Notes

1. General caution for antibody staining:

   Importantly, the cells should never be allowed to dry out, so you should not completely aspirate all the liquid from the well and you should always have the next solution at hand to add immediately after aspiration.

2. Fixative preparation and storage:

   Paraformaldehyde: With heat and basic pH, paraformaldehyde will depolymerize to a very pure form of formaldehyde. Solubilization of paraformaldehyde powder is often accomplished with heat and strong base but take care not to heat the solution above 55–60°C and add just enough base to depolymerize the paraformaldehyde (pH $\leq$ 10). If the solution goes over 65°C during preparation the formaldehyde degrades rapidly to formic acid and water. Therefore, do not use it as it will produce a strong autofluorescence in cells or tissues.

   37% *Formaldehyde*: Storage of 37% formaldehyde over several months also results in degradation to formic acid and water. Old formaldehyde stocks should be disposed of every 12–24 months in accordance with your institution's chemical policies.

10% *Buffered formalin*: 10% buffered formalin will pH drift due to degradation of formaldehyde to formic acid. Do not use if below pH 6.5 and rotate stocks regularly.

3. Summary of immunostaining procedure:

   Remove medium from cells, wash with DPBS +/+.

   Add Fixative, 10 min, RT.

   Wash with DPBS +/+, 2 × 5 min.

   Add HCl if BrdU-treated cells, 20–30 min, 37°C, wash DPBS +/+, 2 × 5 min

   Add Blocking Buffer, 15 min, RT, remove

   Add diluted primary antibodies, overnight, 4°C

   Wash DPBS +/+, 2 × 5 min.

   Add diluted secondary antibodies, 1 h, RT.

   Wash with DPBS +/+, 2 × 5 min.

   Add Hoechst (1:500 in DPBS +/+), 1–5 min, RT.

   Wash with DPBS +/+, 1 × 5 min.

   Mount and coverslip. Seal with nail polish.

   View on microscope.

4. Choosing the Right Antibodies:

   Most fluorescence microscopes have the ability to discern several unique fluorochromes using various optical filter arrangements. In designing a plan for co-staining for more than one antigen, it is important first to select primary antibodies of unique species or subtypes (i.e., Mouse IgG, Mouse IgM, Rabbit IgG, Goat IgG, Chicken IgG, Guinea Pig IgG, Rat IgG). If the primary antibodies for different antigens are from the same species and subtype, secondary antibodies will indiscriminately bind to both markers. For multiple antibody staining, care should be taken to use secondary antibodies that are highly specific for the class and species of primary antibody that needs to be detected. Some vendors provide secondary antibody reagents that are validated to have minimal cross reactivity to a wide spectrum of antibody classes and species (Jackson Immuno Research Laboratories is a reliable source). In addition, the fluorophores chosen for the secondary antibodies must match the particular filter sets present on your microscope to prevent optical overlap between the fluorophores (Table 1).

5. Antibody concentration:

   Most manufacturers provide recommendations for antibody concentrations for specific applications. When using an antibody for the first time, it's a good idea to try a range of concentrations around that recommended by the manufacturer.

For example, if the recommended concentration is 1:100, try a range from 1:10 to 1:1,000. If no recommended concentration is given, start with 1 μg/100 μL antibody dilution buffer.

6. Antibody aspiration and washing:

   To save time, set up your aspirator accordingly: Attach a Pasteur pipet to the end of your aspirator. Now, place a disposable P100 or smaller pipette tip on the end of the attached Pasteur pipette and replace only the P100 tips for each aspiration. A used aspirator tip greatly increases the likelihood of cross-contaminating adjacent wells.

7. Background staining:

   A sample may have a high level of background fluorescence or fluorescent debris. Here are some possible remedies for resolving this common problem and further discussion of a few specific causes of background staining particularly useful for tissue staining.

   (a) Spin the antibodies to remove precipitates before adding the antibody to the sample (see steps 3 and 4 of Day 2 above).

   (b) Use fresh antibodies. Over time antibodies will degrade and increase the amount of background and nonspecific staining. To avoid multiple freeze–thaw cycles, aliquot the antibody upon receipt into smaller working volumes.

   (c) Primary antibody and/or secondary antibody concentrations are too high.

   (d) Increase DPBS rinsing time or number of washes.

   (e) Incorrect blocking serum or insufficient blocking time. One can also try blocking with IgG-free BSA rather than animal serum (use 5% w/v in DPBS for blocking buffer and 1% w/v in DPBS for antibody dilution buffer).

   (f) Cell cultures were stressed during growth. Refine growth conditions.

   (g) Attempt to use a different antibody for the antigen (try to choose an antibody that recognizes a different epitope on the same molecule).

8. Species mismatch:

   Problem: Same-species antibodies yield high background. For example, when mouse primary antibodies are used on mouse tissues, detection with anti-mouse secondary antibodies will detect all mouse immunoglobulins that are native to the mouse tissue.

   Solution: Use species-mismatched primary antibodies or block the endogenous antibodies by *preincubating with an unconjugated secondary antibody*. When blocking, it is necessary to use Fab fragments and important to use a Fab preparation that matches the conjugated secondary antibody that will be used for detection. Vendors often sell unconjugated Fab preparations that match the detecting secondary antibody for this purpose.

Note: Why use Fab fragments for blocking endogenous Ig? Whole Ig is multivalent and a block with a multivalent antibody will leave many Fab ends unbound. Subsequent treatment with the primary antibody will simply bind these exposed ends and aggravate the background problems.

9. Fc Receptors in sample:

   Problem: Fc receptors expressed by cells nonspecifically bind primary and secondary antibodies. Particularly problematic for tissues that have been damaged and contain activated immune cells.

   Solution: Use Fab preparations for detection rather than whole antibodies or block using unconjugated Fc fractions that match both primary and secondary antibody preparations.

   Note: When using Fab fragments for detection, the secondary antibody must be one that recognizes a Fab fragment. Typically, the secondary antibody used will recognize light-chain rather than heavy chain and one must take care to determine the class of light chain present in the Fab fragment (i.e., either kappa or lambda light chain).

10. Generalized Background

    Problem: Very high overall background.

    Solution: Titrate antibodies (both primary and secondary) for optimum signal to noise. Primary or secondary antibody may recognize nonspecific antigens. To determine if the problem is with the primary or secondary antibody, prepare one sample that is treated with secondary antibody alone. If background is low, then problem is with primary antibody. If background is present in samples treated with secondary antibody alone, then the problem is with secondary antibody. In both cases, an alternate antibody should be tried (if available) or more aggressive means to improve specificity should be explored.

    Note: Secondary antibody background can be reduced if the vendor provides unconjugated preimmune serum from the same species (ideally collected from the same animal prior to immunization). This is used in the initial blocking step to bind all non-specific sites prior to the final detection using the conjugated secondary antibody preps.

11. Weak staining
    (a) Test the antibodies on known positive and negative controls.
    (b) Try another antibody to the same antigen.
    (c) Fixation and/or Permeabilzation – Follow the manufacturer's specific instructions for methods of fixation and permeabilization to use with the antibody in question. In addition, check the literature for papers that have used the antibody (and have nice images of immunostained cells) and follow the protocol verbatim (call or email the authors, if necessary, to get details). Most antibodies are sensitive to the type of fixation and or permeabilization used. In addition, the concentration of chemical used and the time of exposure can also be critical (it is possible to over-fix).
    (d) Increase the concentration of primary and/or secondary antibody.
    (e) Increase the time of the primary antibody incubation. If positively staining slides have faded over time, be certain the nail polish sealant on slides is intact and that the slides are being stored in a desiccated, cold environment.
12. Too much staining
    (a) Reduce primary antibody concentration.
    (b) Reduce primary or secondary antibody incubation period.
    (c) Attempt to use a different clone of antibody for the same antigen.
    (d) See notes on blocking in "Background staining" section above.

Acknowledgments

We gratefully acknowledge intellectual contributions from Lisa A. Flanagan, Richard Pepple, Boback Ziaeian, and Theo Palmer.

References

1. Li, F., Lu, S., Vida, L., Thomson, J.A., and Honig, G.R. (2001) Bone morphogenetic protein 4 induces efficient hematopoietic differentiation of rhesus monkey embryonic stem cells in vitro. *Blood* **98**, 335–342.
2. Pera, M.F., Filipczyk, A.A., Hawes, S.M., and Laslett, A.L. (2003) Isolation, characterization, and differentiation of human embryonic stem cells. *Methods Enzymol.* **365**, 429–446.
3. Vallier, L., Reynolds, D., & Pedersen, R.A. (2004) Nodal inhibits differentiation of human embryonic stem cells along the neuroectodermal default pathway. *Dev. Biol.* **275**, 403–421.

4. Fenderson, B.A., De Miguel, M.P., Pyle, A.D., & Donovan, P.J. (2006) Staining embryonic stem cells using monoclonal antibodies to stage-specific embryonic antigens. *Methods Mol. Biol.* **325**, 207–224.
5. Draper, J.S., Moore, H.D., Ruban, L.N., Gokhale, P.J., & Andrews, P.W. (2004) Culture and characterization of human embryonic stem cells. *Stem Cells Dev.* **13**, 325–336.
6. Hoffman, L.M. & Carpenter, M.K. (2005) Characterization and culture of human embryonic stem cells. *Nat. Biotechnol.* **23**, 699–708.
7. Loring, J.F. & Rao, M.S. (2006) Establishing standards for the characterization of human embryonic stem cell lines. *Stem Cells.* **24**, 145–150.
8. Ohnuki, M., Takahashi, K., & Yamanaka, S. (2009) Generation and characterization of human induced pluripotent stem cells. *Curr. Protoc. Stem Cell Biol.* **Chapter 4**, Unit 4A.2.

Chapter 16

Flow Cytometric Analysis of Human Pluripotent Stem Cells

Mirabelle S.H. Ho, Andrew Fryga, and Andrew L. Laslett

Abstract

Human pluripotent stem cells, human embryonic stem cells and induced pluripotent stem cells, represent an exciting new era in regenerative medicine and drug discovery. However, prior to their clinical translation, there is a need to gain an in-depth understanding of human pluripotent stem cell biology by characterizing these potentially heterogeneous populations of cells. Flow cytometry provides a rapid and efficient approach with which to isolate, purify, and study the functional properties of defined pluripotent stem cell types.

Key words: human embryonic stem cells, induced pluripotent stem cells, flow cytometry

1. Introduction

The discovery of human pluripotent stem cells (PSCs) has generated unprecedented global excitement. This excitement stems mainly from the potential of PSCs to undergo autonomous self-renewal while retaining the ability to differentiate down multiple cell lineages. These properties may eventually permit cell types derived from PSCs to be therapeutic options for debilitating conditions such as Parkinson's and Alzheimer's disease as well as platforms for drug discovery and developmental studies (1, 2).

The recent advent of induced pluripotent stem cells (iPSCs) promises to revolutionize the landscape of PSCs. iPSCs are created by the ectopic insertion of a cocktail of transcription factors associated with the governance of pluripotency (3). The capacity to create autologous iPSCs appears to circumvent the challenges of immune rejection and ethical dilemmas around the use of human embryos for the derivation of human embryonic stem cells (hESCs) (3, 4). Prior to iPSCs being used therapeutically, however, it will be important to establish the safety of these cells and their differentiated derivatives. Flow cytometry provides an

Philip H. Schwartz and Robin L. Wesselschmidt (eds.), *Human Pluripotent Stem Cells: Methods and Protocols*, Methods in Molecular Biology, vol. 767, DOI 10.1007/978-1-61779-201-4_16, © Springer Science+Business Media, LLC 2011

accurate, rapid, and efficient approach with which to identify distinct cell types at the single-cell level. An extension of this technology is the use of a fluorescence-activated cell sorter (FACS), which allows the isolation and enrichment of different cell types for subsequent studies. Flow cytometry operates on the principle of characterizing cells based on their size (forward scatter, FCS), cellular complexity (side scatter, SSC), and fluorescent properties.

Single-cell suspensions of PSCs can be labeled with one or more fluorophore-tagged antibodies at a time and multicolor analyses can be performed by taking advantage of the varied antibody classes and isotypes (e.g., IgG2a, IgM, etc.) or species of antibody derivation (e.g., mouse, goat, rabbit, etc.). Primary antibodies obtained from different isotypes or species are detected using isotype- and species-specific secondary antibodies that are conjugated to spectrally distinct fluorophores. When more than one primary antibody is used concurrently, proper single-antibody controls must be performed to ensure the absence of nonspecific cross-reactivity.

Currently, analyzers with multiple lasers (up to seven) are able to measure more than 20 fluorescent and two scatter parameters per cell at speeds of greater than 50,000 cells/s. Hence, in a short period of time, millions of data points can be generated. In addition, the accuracy of sorters, with purities of >99.5%, enable enrichment of very rare cell populations or elimination of contaminating subpopulations from a common population of cells. The ability to analyze multiple parameters in a high-throughput manner enables users to rapidly gain insight into the biology of the cells in question.

Antibodies reactive with cell surface-epitopes on undifferentiated PSCs are important tools with which to identify, isolate, characterize, and subsequently compare different cell lines as well as various culture conditions that contribute to the maintenance of pluripotency. Currently, there is no single marker that is entirely specific to PSCs (5). Researchers conventionally use a panel of antibodies to characterize pluripotent cells according to their surface epitope expression profile. These antibodies include extracellular surface antigens such as GCTM-2, CD9, Tra-1-60 as well as intracellular markers such as Nanog and OCT-4 (5). These epitopes are not exclusively expressed on PSCs, having been detected on a range of differentiated cell types (6, 7). Hence, it is noteworthy that these antibodies may only be used to detect PSCs prior to stem cell commitment and differentiation (5). We have demonstrated that hESC colonies can be subfractionated into separate populations (8, 9), and it is not unreasonable to assume that these subfractions possess distinct functional capabilities. Currently, the most stringent functional test for pluripotency involves injecting the cells in question into a blastocyst to generate chimeric offspring (10). While this test is feasible with mice, the same test cannot be applied to human cells due to appropriate ethical constraints. As a result, the gold standard test for pluripotency using human cells involves teratoma formation in which PSCs are injected into an immune-compromised strain of mice (11). Whilst this approach is able to qualitatively determine the ability of cells to give rise to cell types representing the three primitive germ layers of a developing embryo, it requires a period of between 8 and 12 weeks to achieve results.

In his recent review, Shinya Yamanaka highlighted the need for a rapid and sensitive test able to quantify pluripotency as well as to ascertain the safety of iPSCs (12). FACS followed by stem cell colony-forming assays may be able to form the basis of such a quick and sensitive test but requires the formal testing of whether stem cell colony-formation directly correlates to teratoma-forming ability. We have previously demonstrated that hESCs can be separated by FACS and subjected to stem cell colony-forming assays in vitro (9). The number of PSC colonies, where the majority of the cells in the colony co-stain for multiple PSC markers, generated within a 2-week period from each subfraction of the PSC culture provides a quantifiable measure of the stem cell colony-forming ability of each subfraction. Furthermore, the colony-forming assay allows one to compare existing and novel "stemness" biomarkers, individually or in combination. It can be used to accurately distinguish markers capable of identifying cells with the ability to form stem cell colonies. Below we present a detailed protocol applicable to either hESC or human iPSC cells for flow cytometric analysis, FACS separation and stem cell colony-forming assays post-FACS.

2. Materials

2.1. Cell Culture

1. ESC line: MEL1 (9).
2. iPSC lines: IMR90C2 and IPS-foreskin-CL1 (4).
3. 12-well (3.8 cm^2/well) tissue culture plates seeded with $1.2 \times 10^4/cm^2$ MEF cells.
4. 20% *KOSR–PSC medium*: prepared using Dulbecco's Modified Eagle's Medium DMEM/F12 (Invitrogen, Cat No. 11960-051) supplemented with 20% Knock-out Serum Replacer (KOSR, Invitrogen, Cat No. 10828), 2 mM L-glutamine, 0.1 mM MEM nonessential amino acids (NEAA), 0.1 mM β-mercaptoethanol, 1% penicillin/streptomycin (5,000 U/5,000 μg/mL) and basic FGF (FGF-2, R&D Systems, Minneapolis, MN. Cat No. 233-FB-025/CF). PSC cell culture medium is stored at 4°C for up to 14 days and supplemented daily with 10 ng/mL FGF-2 (for ESCs) or 100 ng/mL FGF-2 (for iPSCs).

5. 20% *FBS–PSC medium*: prepared using Dulbecco's Modified Eagle's Medium DMEM/F12 (Invitrogen, Cat No. 11960-051) supplemented with batch tested 20% fetal bovine serum (FBS), 2 mM L-glutamine, 0.1 mM nonessential amino acids, 0.1 mM β-mercaptoethanol, 1% penicillin/streptomycin (5,000 U/5,000 μg/mL) and basic FGF (FGF-2, R&D Systems, Minneapolis, MN. Cat No. 233-FB-025/CF). *FGF-2 is added only to FBS–PSC medium used for the washes and staining prior to colony-forming assays, not for routine culture.*
6. *Mouse embryonic fibroblast-conditioned medium (MEF-CM)*: supplement 20 mL of KOSR–PSC medium with 10 ng/mL FGF-2. Add to a 75-cm^2 tissue culture flask that has been seeded with 6×10^4 cells/cm^2 of inactivated MEFs. Conditioned medium (CM) is collected 24 h later. Repeat for up to 7 days. Filter sterilize CM and either use within 24 h or store at −20°C.

2.2. Flow Cytometry and FACS

1. TrypLExpress™ (Invitrogen, Cat No. 12604) for harvesting cells.
2. Dulbecco's Phosphate-buffered saline without calcium and magnesium (DPBS$^{-/-}$) (Invitrogen, Cat. No. 14190).
3. Primary antibodies: TG30 (1:1,000 dilution), GCTM-2 (1:100 dilution) provided by the protein facility at the Australian Stem Cell Centre, Victoria node (see Note 1).
4. Isotype control mouse IgG2a (BD Pharmingen, San Diego, CA. Cat No. 554126) and mouse IgM (BD Pharmingen, Cat No. 553472).
5. Secondary antibodies: Alexa Fluor® 488-conjugated goat anti-mouse IgG2a (αIgG2a-AF488) (Invitrogen, Cat. No. A21131) diluted 1:500, Alexa Fluor® 647-conjugated goat anti-mouse IgM (αIgM-AF647) (Invitrogen, Cat. No. A21238) diluted 1:1,000. Antibodies diluted in DPBS.
6. 40-μm filter mesh (Becton Dickinson, Cat No. 352235).
7. Propidium iodide (PI) (Sigma, Cat. No. P4864-10ML).
8. Rainbow calibration particles (Spherotech 8 peak Ultra Rainbow beads http://www.spherotech.com).

3. Methods

3.1. Multicolor Analyses Using Flow Cytometric Analysis and Cell Sorting

PSCs are cultured using standard techniques in 20% KOSR–PSC medium (see Chapter 8). A typical flow cytometric profile is shown by Fig. 1.

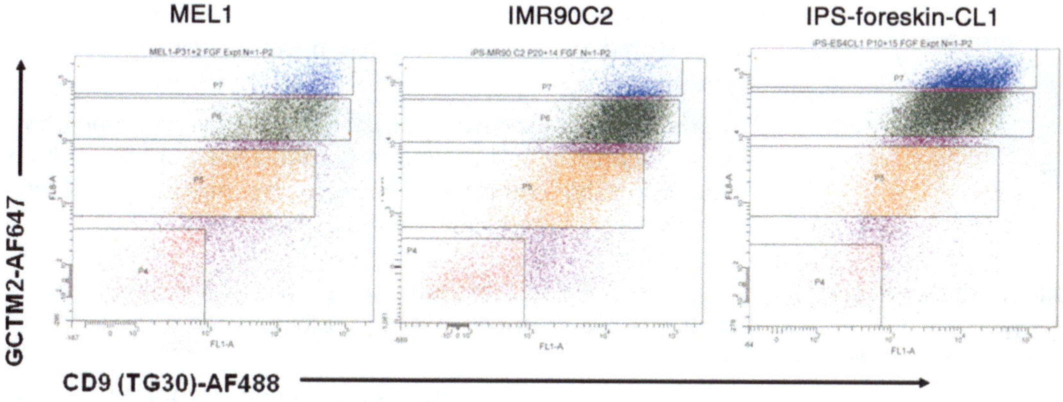

Fig. 1. Representative flowcytometric profile of ESCs and iPSCs. Both iPSC and ESC co-labeled with GCTM-2 and CD9 share a similar flowcytometric profile.

1. Carefully harvest PSCs using TrypLExpress™ and dissociate into a single-cell suspension by trituration.
   (a) Cells are washed twice with DPBS prior to addition of the TrypLExpress™ enzymatic buffer (0.067 mL/cm^2). Following a 5-min incubation at 37°C, pluripotent cells are gently agitated and lifted from the flask prior to trituration using a 1-mL pipette tip.
   (b) The cell suspension is then subjected to gentle centrifugation at 500 × *g* for 2 min. Supernatant is carefully aspirated and cells washed twice with 20% FBS–hESC medium in the same manner.
2. Gently resuspend the cell pellet in 2 mL of hESC medium supplemented with 20% FBS. The primary antibodies GCTM-2 and TG30 are added to a final dilution of 1:100 and 1:1,000, respectively, in this cell suspension.
3. The cell suspension is then placed horizontally in an ice-box and placed on rocking platform. Cells are incubated on ice and protected from light for 30 min–1 h.
4. Primary antibodies are washed off via gentle centrifugation at 500 × *g* for 2 min followed by two washes with 20% FBS–ESC medium.
5. After washing, the cell suspension is incubated in a similar fashion to step 2 with secondary antibodies αIgG2a-AF488 (1:500 dilution) and αIgM-AF647 (1:1,000 dilution) or similar fluorescently tagged secondary antibodies.
6. Secondary antibodies are washed off as described above in step 3.
7. Propidium iodide (PI) is added to the cell suspension just prior to analysis, at a final concentration of 1.0 μg/mL to discriminate dead cells.

8. Cell suspension is then filtered through a 40-μm filter mesh and stored on ice until analysis using an appropriate flow cytometer.
9. Gates for flow cytometric analysis are set based on negative isotype controls and the rainbow calibration particles, which enable reliable comparison between experiments. For problems with staining please see Notes 2–5.

3.2. Colony-Forming Assay

1. 24 h before FACS experiment, prepare 12-well (3.8 cm^2/well) tissue culture plates seeded with $1.2 \times 10^4/cm^2$ MEFs.
2. Day of FACS: Replace MEF medium with MEF-CM, supplemented with 10 ng/mL FGF-2.
3. Plate 30,000 FACS-sorted cells/well, in triplicate for each subfraction of cells. Immediately place into incubator, at 37°C in a 5% CO_2 humidified incubator and leave undisturbed for 48 h prior to the next medium exchange.
4. For the next 12 days, perform a routine daily medium exchange using MEF-CM supplemented with the appropriate concentration of FGF-2 (10 ng/mL FGF-2 for ESCs or 100 ng/mL FGF-2 for iPSCs.
5. To assess the colonies generated at day 14: colonies from each subfraction (see Fig. 2) are harvested and immunostained with TG30 and GCTM-2 antibodies as previously described in Subheading 3.1 and analyzed using flow cytometry. If colonies do not form please see Note 6 for troubleshooting. Alternatively, in situ immunocytochemistry analysis can provide a qualitative means with which to assess the stem cell colony-forming ability of the cells from each of the subfractions (see Chapter 15).

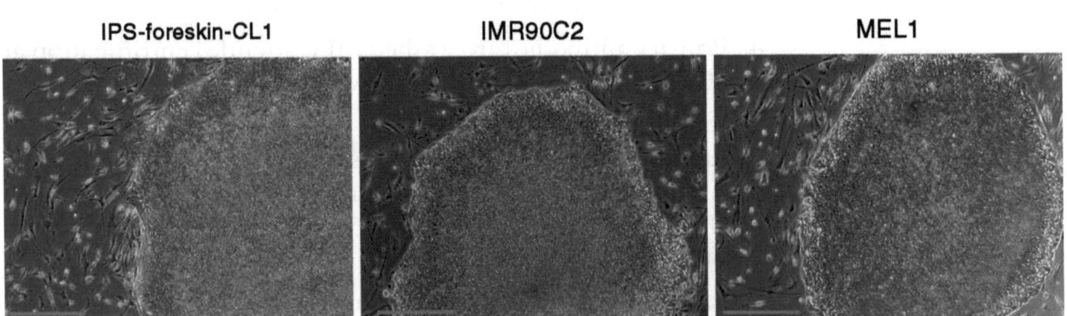

Fig. 2. Representative CD9-High/GCTM-2High hESC and iPSC colony morphology on day 14 after FACS. All colonies formed from either iPSC or hESC lines display a compact rounded morphology with defined edges that are typically observed in conventional hESC colonies. Scale bar = 1,000 μm.

4. Notes

1. TG30(CD9) antibody also available from Millipore # MAB4427, and TG343 an antibody which detects the same protein as GCTM-2 (and can be substituted for GCTM-2 in this assay) is also available from Millipore # MAB4346.
2. *Background staining*
   - A high level of background staining tends to be the result of nonspecific antibody binding. This is particularly relevant with intracellular staining (for example, POU5F1/OCT4). The following lists some approaches with which to counter the challenges of excessive background staining.
   - Use antibodies diluted freshly on the day of the experiment itself. Antibodies will degrade over a period of time resulting in an increase in the level of nonspecific binding to the secondary antibody and consequent background staining.
   - Optimize the concentrations of both primary and secondary antibodies.
   - Generally, where available, a monoclonal antibody will give more specific results than a polyclonal antibody for the same antigen of interest.
   - Include secondary antibody-only control samples to assess background staining.
   - Increase the concentration of FBS used in the wash buffer (i.e., from 20 to 30%) and/or increase dilution of antibody.
   - Test a range of clones of the antibody and from different suppliers for the same antigen of interest.
   - Filter antibodies pre-dilution to remove debris.
3. *Weak staining*
   - New antibodies should always be tested against known positive- and negative-cell controls to ascertain that the antibody in question is active and specific.
   - Perform a titration over a range of dilutions. An older batch of antibody known to be working well (i.e., specific) should be used as a positive control to determine optimal concentration of antibody required.
   - Increase primary antibody incubation time.
   - Use a brighter fluorophore (e.g., AF488 to replace FITC).
   - Test another antibody or batch of antibodies to the same antigen.
   - During immunostaining incubations, constantly agitate the cell suspension to prevent cell clumping which frequently occurs when a large volume of cells are involved.

Gentle agitation of cell suspension maximizes cell movement and consequently contact with antibodies.

- As hESCs are highly auto-fluorescent, it may be better to not set the photomultiplier tube (PMT) voltage too low for the fluorescence detector. It is acceptable for some of the cells to spill outside the first decade. This potentially facilitates the cytometer to achieve optimal resolution sensitivity. One would then expect to see an increase in the ratio of positive and negative cells compared to lower settings.

4. *Over staining*
   - Reduce the concentration of the primary or secondary antibody.
   - Optimize antibody incubation times.
   - Decrease PMT voltage for the fluorescence detector to place the brightest cells on scale.

5. *Controls and compensation*
   - The starting reference is conventionally the unstained control sample. The cells in this tube are subjected to the same staining procedures (i.e., washes and incubation conditions) without the addition of any primary or secondary antibodies. The unstained control permits the set up of the level of background fluorescence on the cytometer thus facilitating the detection (if any) of nonspecific staining in subsequent isotype control.
   - The isotype control permits the determination of the amount of nonspecific staining due to the class of antibody used.
   - When performing multicolor analysis, each fluorophore used must have a separate control tube. Compensation is applied to neighboring channels as each tube is run to eliminate or minimize any spillover.
   - When performing multicolor analysis, where possible, try to choose fluorophore with minimal spectral overlap.
   - Compensation: When the fluorophore FITC is used concurrently with PI, a FL1 vs. FL3 dot-plot should be used to monitor the amount of FITC spillover into the PI detector. An elevated level of FITC fluorescence may potentially bleed through the PI detector, making cells appear PI positive (i.e., dead). A small amount of FL3–%FL1 compensation will position the FITC-high events back into the viable gates.

6. *Absence of colonies in CFA analysis*
   - Always perform a cell titration (i.e., progressive doubling of cell number per well) prior to carrying out any further in vitro experiments.

- Immediately post-FACS place culture plates into a 37°C, 5% CO_2 incubator to equilibrate.
- Leave cells undisturbed in the incubator for at least 24–36 h prior to the next medium change. This allows the cells time to adhere to the MEF-coated plates.
- Whilst 20% KOSR supplemented hESC medium may be used for daily medium change, cell survival, and proliferation has been observed to benefit appreciably from the use of conditioned medium.

Acknowledgments

Methods described in this chapter were developed using funding from the Australian Stem Cell Centre (ASCC) and the NSW/VIC Government Stem Cell Research Grant Program 2008 to ALL. The authors would like to thank James Thomson, University of Wisconsin for the provision of the human iPSC lines, the Australian Stem Cell Centre's Core hESC Laboratories (Stem Core) for providing cell culture and support services as well as the Flow Core (a collaborative initiative between Monash University, the ASCC, and the Australian Regenerative Medicine Institute (ARMI)) for technical assistance and advice.

References

1. Maherali, N., and Hochedlinger, K. (2008) Induced pluripotency of mouse and human somatic cells, *Cold Spring Harb Symp Quant Biol* **73**: 157–162.
2. Menendez, P., Wang, L., and Bhatia, M. (2005) Genetic manipulation of human embryonic stem cells: a system to study early human development and potential therapeutic applications, *Curr Gene Ther* **5**: 375–385.
3. Takahashi, K., Tanabe, K., Ohnuki, M., Narita, M., Ichisaka, T., Tomoda, K., and Yamanaka, S. (2007) Induction of pluripotent stem cells from adult human fibroblasts by defined factors, *Cell* **131**: 861–872.
4. Yu, J., Vodyanik, M. A., Smuga-Otto, K., Antosiewicz-Bourget, J., Frane, J. L., Tian, S., Nie, J., Jonsdottir, G. A., Ruotti, V., Stewart, R., Slukvin, II, and Thomson, J. A. (2007) Induced pluripotent stem cell lines derived from human somatic cells, *Science* **318**: 1917–1920.
5. Laslett, A. L., Filipczyk, A. A., and Pera, M. F. (2003) Characterization and culture of human embryonic stem cells, *Trends Cardiovasc Med* **13**: 295–301.
6. Andrews, P. W., Goodfellow, P. N., Shevinsky, L. H., Bronson, D. L., and Knowles, B. B. (1982) Cell-surface antigens of a clonal human embryonal carcinoma cell line: morphological and antigenic differentiation in culture, *Int J Cancer* **29**: 523–531.
7. Pera, M. F., Filipczyk, A. A., Hawes, S. M., and Laslett, A. L. (2003) Isolation, characterization, and differentiation of human embryonic stem cells, *Methods Enzymol* **365**: 429–446.
8. Laslett, A. L., Grimmond, S., Gardiner, B., Stamp, L., Lin, A., Hawes, S. M., Wormald, S., Nikolic-Paterson, D., Haylock, D., and Pera, M. F. (2007) Transcriptional analysis of early lineage commitment in human embryonic stem cells, *BMC developmental biology* 7: 12.
9. Kolle, G., Ho, M., Zhou, Q., Chy, H. S., Krishnan, K., Cloonan, N., Bertoncello, I., Laslett, A. L., and Grimmond, S. M. (2009) Identification of Human Embryonic Stem Cell Surface Markers by Combined Membrane-Polysome Translation State Array Analysis and

Immunotranscriptional Profiling, *Stem Cells* **27**: 2446–2456.

10. Thomson, J. A., Kalishman, J., Golos, T. G., Durning, M., Harris, C. P., Becker, R. A., and Hearn, J. P. (1995) Isolation of a primate embryonic stem cell line, *Proc Natl Acad Sci USA* **92**: 7844–7848.
11. Thomson, J. A., Itskovitz-Eldor, J., Shapiro, S. S., Waknitz, M. A., Swiergiel, J. J., Marshall, V. S., and Jones, J. M. (1998) Embryonic stem cell lines derived from human blastocysts, *Science* **282**: 1145–1147.
12. Yamanaka, S. (2009) A fresh look at iPS cells, *Cell* **137**: 13–17.

Chapter 17

The Teratoma Assay: An In Vivo Assessment of Pluripotency

Robin L. Wesselschmidt

Abstract

A teratoma is a nonmalignant tumor comprised of a disorganized mixture of cells and small foci of tissue comprised of cells from all three of the embryonic germ-layers. By definition, a cell is pluripotent if it can differentiate into cells derived from all three of the embryonic germ-layers: ectoderm, mesoderm, and endoderm. In the teratoma assay, putative pluripotent stem cells (PSCs) are implanted into an immune-compromised mouse where they may proliferate and differentiate to form a teratoma. The PSCs grow at the implantation site supported by a complex mixture of factors from the local milieu, as well as circulating factors that are vital components of normal mammalian physiology. After a predetermined time of 6–12 weeks or when the tumor has reached sufficient size, it is removed and subjected to histopathological analysis. The teratoma may be further processed by immunocytochemistry and gene expression profiling. This chapter describes methods to generate teratomas through the implantation of putative PSC lines in the SCID mouse. Implantation at the following sites is described: (1) intramuscular, (2) subcutaneous, (3) under the testis capsule, and (4) under the kidney capsule.

Key words: embryonic stem cells, pluripotent stem cells, pluripotency assay, teratoma, testis capsule teratoma, kidney capsule teratoma

1. Introduction

While in vitro differentiation assays and in silico gene expression arrays are useful in assessing pluripotency, the gold standard remains the teratoma assay (1, 2). This in vivo assay provides a means to assess the developmental potential of human pluripotent stem cell (hPSC) lines at a level that cannot yet be achieved using in vitro and in silico assays. A pluripotent stem cell, by definition, is a cell that can differentiate into cells derived from all three embryonic germ-layers: ectoderm, mesoderm, and endoderm. When putative PSCs are transplanted in immune-compromised mice, they are exposed to a complex mixture of growth and extracellular matrix factors that cannot, so far, be fully replicated in a culture dish.

Philip H. Schwartz and Robin L. Wesselschmidt (eds.), *Human Pluripotent Stem Cells: Methods and Protocols*, Methods in Molecular Biology, vol. 767, DOI 10.1007/978-1-61779-201-4_17, © Springer Science+Business Media, LLC 2011

This mixture of factors promotes the growth and differentiation of the PSCs into teratomas; begin tumors that contain a complex mixture of cells and tissues derived from all three germ-layers (review (3)). The teratoma assay is part of the standard set of quality control and basic characterization assays used in hPSC laboratories. It is performed as part of routine culture evaluation, when new embryonic stem cell (ESC) or induced pluripotent stem cell (iPSC) lines are generated, and when PSCs are expanded and banked to make working stocks (4–6). Several engraftment sites have proven useful for teratoma production; however, the graft site, number of cells implanted, and the cell preparation has been shown to influence the type of somatic cells found in the teratoma, whether the teratoma is cystic or solid tumor, and the growth rate of the teratoma (7–11).

To date, the largest single study assessing the functional pluripotentiality of human ESCs via teratoma assay was performed by the International Stem Cell Initiative (ISCI) under which teratomas were generated and analyzed from 15 independent hESC lines (12). Investigators implanted cells from each of these hESC lines under the testis capsule of SCID mice and evaluated a total of 37 histological slides. Most of the hESC lines in the study produced teratomas. Ectodermal and mesodermal tissues predominated in the teratomas. Neural tissue was most often present as immature rosettes. Mesoderm included fibroblasts, capillaries, smooth muscle, striated muscle, cartilage, bone, and fat. Endodermal tissues included gland-like structures lined with columnar or cuboidal epithelium. Interestingly, three of the cell lines produced teratomas that contained foci of undifferentiated cells that had undergone malignant transformation into embryonal carcinomas, which when the cell lines were karyotyped were found to contain aneuploid cells. Excellent examples of histological sections of hESC-derived teratomas have been published (5).

This chapter describes the production of teratomas following implantation into four different sites in the SCID mouse: intramuscular injection in the lower flank, subcutaneous injection in the lower leg, implantation under the testis capsule, and implantation under the kidney capsule. Each of these sites is effective at generating teratomas, but with varying efficiency, and each requires a different level of surgical skill. In each case, PSCs are implanted into SCID mice, the mice are monitored for 6–12 weeks, and the tumor is harvested and analyzed for the appearance of cellular derivatives from all three germ-layers by a qualified clinical pathologist. Implantation under the testis or kidney capsule is major survival surgery and requires a high level of surgical expertise not required for intramuscular or subcutaneous injections. However, these surgical implantation methods have advantages in that they require fewer cells and use fewer animals than the injection procedures as they tend to be more efficient sites for engraftment. Recent reports, however, have suggested that resuspending the cells in Matrigel™ increases the teratoma efficiency after injection and may eliminate the need for surgery to achieve high rates of engraftment (10, 11).

Whichever site is chosen for implantation, PSC cultures should be undifferentiated and actively growing (see Chapter 8). Care should be taken to ensure the test sample is representative of the entire culture.

2. Materials

2.1. Injection: Intramuscular or Subcutaneous

1. 5 SCID-BEIGE mice/cell line or culture to be assayed (see Note 1).
2. Dulbecco's Modified Eagle's Medium/Nutrient Mixture F12 (DMEM/F12).
3. Sterile 1-cc syringe with 23 g, ½-in. needle.
4. Measuring calipers.

2.2. Implantation Under the Testis Capsule

1. 3–5 male-SCID-BEIGE mice/cell line to be tested.
2. Dulbecco's Modified Eagle's Medium/Nutrient Mixture F12 (DMEM/F12).
3. Glass capillary pipettes, pulled and fire-polished on one end for implanting cells under the capsule without rupturing the membrane.
4. Small Animal Clipper with #40 blade for shaving the mouse's abdomen prior to surgery.
5. 70% ethanol.
6. Betadine solution.
7. Sterile gauze pads.
8. Sterile surgical pack that includes: fine dissecting scissors, ser-refine clamp, blunt forceps, watchmaker's forceps #5, absorbable suture size 5-0 with an attached curved size 10 needle that is triangular and pointed, surgical stapler and 9-mm wound clips.
9. Anesthetic as directed by your veterinarian, institutional animal care and use committee, and biosafety committee (see Note 2).
10. Heating pad or slide warmer.
11. Clean cage for postoperative recovery.
12. Measuring calipers.

2.3. Implantation Under the Kidney Capsule

1. All items listed in Subheading 2.2 for surgical implantation under testis capsule.
2. 26-mm diameter Chalazion forceps (desmarres) optional (see Note 3).

2.4. Teratoma Tumor Harvest and Analysis

1. Sterile surgical pack containing dissecting scissors and forceps.
2. Sterile scalpel or razor blade.
3. Dulbecco's Phosphate-Buffered Saline (D-PBS).
4. Neutral buffered formalin 10%.
5. Liquid nitrogen for flash freezing tissue if RNA isolation is to be performed.

3. Methods

Regarding cell cultures: Ideally, a healthy hPSC culture in the log phase of growth is harvested (see Chapter 8). This can usually be achieved by harvesting the cells 1 or 2 days before they would routinely be subcultured.

Regarding the use of mice: Procedures that involve the use of live animals require institutional approval prior to initiating the experiment. Investigators are encouraged to work with their veterinarians and skilled animal care personnel who can provide them with surgical training and expert animal handling and care, utilizing best practices developed at their institution.

3.1. Teratoma Formation via Intramuscular or Subcutaneous Injection

Teratoma formation by injection of PSCs is the easiest method. No surgery is required and tumor growth can be monitored by visual observation and palpation. However, a large number of cells are required and only 25–50% of the mice develop tumors. Nevertheless, teratoma formation by injection has been used successfully to assess the pluripotency of hESCs and iPSCs (13–15). The efficiency of tumor formation may be improved by suspending the cells in extracellular matrix components such as Matrigel™ (10).

1. *Collect undifferentiated PSCs*: Harvest cells from 1 to 6 wells of a six-well dish. Lift the cells from the dish as you would when passaging using collagenase IV or dispase (see Chapter 8). Try not to carry along the feeder cells. Wash cells twice with DMEM/F12, by resuspending the cells in 5 ml of DMEM/F12 and spinning at 200 × *g*. After the first wash, count the cells while they are being spun down for the second time. Cells may not be single cells, estimate the number of cells/small clump. Resuspend the cells to a concentration of $1–2 \times 10^7$ cells/ml in DMEM/F12.

2. *Inject hPSCs*: Inject 50 μl of cell suspension either subcutaneously on the lower hind leg, near the ankle, or 50–100 μl into the thigh muscle using a sterile 1-cc syringe and a 23-g, ½-in. needle (see Note 4). Repeat the injection into each of five mice.
   (a) Inject: 0.5–1 × 10^6 cells/50 μl subcutaneous injection into the lower leg
   (b) Inject: 3–5 × 10^6/50 μl cells into the thigh muscle
3. *Observe the animals daily*. Watch for changes in appearance and behavior. Monitor the injection site for tumor growth, for about 6–12 weeks or until the predetermined experimental endpoint (see Note 5).
4. *Dissect the tumor*. When the tumor is palpable and about 5 mm in size, or the predetermined endpoint for tumor growth is met, euthanize the mouse and surgically remove the tumor.

3.2. Teratoma Formation via Implantation Under the Testis Capsule

Teratoma formation via implantation under the testis capsule has been used to assess pluripotency of hESC lines (16–18) and was the method chosen by the International Stem Cell Initiative (ISCI) to comparatively assess the pluripotency of 15 independent hESC lines (12). This engraftment site has advantages over subcutaneous and intramuscular injection sites and implantation under the kidney capsule; it does not require a large number of cells, the testis is not a vital organ, and the teratoma growth can be monitored by visual observation and palpation. Transplantation of cells under the testis capsule is a fairly straightforward operation, with the surgical setup similar to that of vasectomy.

Surgery requires preapproval, specialized training, and planning in order to ensure that the location of the surgery and method of anesthesia is in keeping with the institutional rules and regulations.

1. *Collect undifferentiated PSCs*: Using aseptic technique, manually dissect hPSC colonies into clumps of 200–400 cells. 10–15 clumps will be implanted/testis. Carefully collect the bits of colonies in a sterile 1.5-ml microfuge tube containing 1 ml of DMEM/F12. If using dispase or collagenase to dissociate the colonies, collect between 10,000 and 100,000 cells and wash the cells twice to remove the enzyme (see Note 6).
2. *Prepare an appropriate surgical location*. This is an approved location where survival surgery can be performed aseptically. Assemble sterile surgical instruments and supplies, as well as postoperative materials, such as clean cage and heating pad or slide warmer to aid in postoperative recovery.

3. *Anesthetize the mouse*: The choice of anesthesia should be determined by consulting with your local veterinarian and biosafety committee. Ideally the chosen anesthetic will keep the animal anesthetized for 20–30 min and have minimal negative effects on the animal as well as personnel (see Note 2).

   After administering the anesthetic, monitor the animal for slowed breathing and perform a reflex check by gently squeezing the rear paw and monitoring response. When the mouse is under anesthesia, it will not withdraw its paw and its breathing will be slow and shallow.
4. Place the mouse on its back on the prepared surgical surface.
5. Shave the lower abdomen.
6. Swab shaven area with 70% ethanol or Betadine solution.
7. Using aseptic technique and sterile instruments make a small incision (1–2 cm) in the lower abdomen at the height of the knees. First make an incision in the skin and then a slightly smaller incision in the wall of the abdomen.
8. Gently squeeze the scrotum to push the testis up into the abdomen.
9. Find the fat pad attached to the testis; using blunt forceps, gently pull the fat pad to remove the testis from the abdomen.
10. Using a small serrefine clamp immobilize the testis by clamping the fat pad to expose and stabilize in an accessible position for transplantation.
11. Under a dissecting microscope, carefully lift the testis capsule (membrane surrounding the testis) with a fine forceps and puncture it with the tip of a sterile 26-g needle. Then using the pulled and polished glass micropipette, inject about 25–30 μl of cell suspension under the testis capsule. Place the cells toward the back of the testis without puncturing it. This will help the cells remain inside the capsule when the testis is placed back into the abdomen.
12. Carefully release the serrefine clamp and gently push the testis back into the abdomen using a blunt forceps. The testis will descend into the scrotum on its own.
13. Suture the abdomen wall with 1–2 stitches of absorbable suture.
14. Close the skin with wound clips.
15. Place the animal into a prewarmed clean cage.
16. Observe closely until the mouse recovers from anesthesia and apply analgesics as necessary and advised by your veterinarian.

17. Place cage in the animal room.
18. Observe the animal carefully daily. Check for changes in appearance and behavior.
19. Remove wound clips, using the clip remover tool, as soon as the incision has healed, 1–2 weeks following surgery.
20. Monitor the testis for tumor growth. Depending on the number of cells implanted, the tumor is likely to be present at 6 weeks and can be grown for additional 6 weeks.
21. When the tumor is palpable, about 5 mm in size, or at the predetermined experimental endpoint, euthanize the mouse and remove the tumor for analysis.

3.3. Teratoma Formation via Implantation Under the Kidney Capsule

Implantation of adult and embryonic tissues under the kidney capsule has been used for many years to study tissue rejection and to obtain teratocarcinomas from early mouse embryos. Implantation under the kidney capsule has been reported to give the highest efficiency of tumor formation (10), but it requires a great deal of surgical skill since the kidney is a vital organ. This protocol was adapted from the one found in *Manipulating the Mouse Embryo*, Second Edition (19).

1. *Collect undifferentiated PSCs*: Using aseptic technique, manually dissect hPSC colonies into clumps of 200–400 cells. 10–15 clumps will be implanted under the capsule. Carefully collect the bits of colonies in a sterile 1.5-ml microfuge tube containing 1 ml of DMEM/F12. If using dispase or collagenase to dissociate the colonies, collect between 10,000 and 100,000 cells and wash the cells twice to remove the enzyme (see Note 6).
2. *Prepare an appropriate surgical location*. This is an approved location where survival surgery can be performed aseptically. Assemble sterile surgical instruments and supplies, as well as postoperative materials, such as clean cage and heating pad or slide warmer to aid in postoperative recovery.
3. *Anesthetize the mouse*: The choice of anesthesia should be determined by consulting with your local veterinarian and biosafety committee. Ideally the chosen anesthetic will keep the animal anesthetized for 30–45 min (see Note 2). After administering the anesthetic, monitor the animal for slowed breathing and perform a reflex check by gently squeezing the rear paw and monitoring response. When the mouse is under anesthesia, it will not withdraw its paw and its breathing will be slow and shallow.
4. Shave the abdomen of the mouse.
5. Swab with Betadine or 70% ethanol.

6. Working to the right side of the midline, make a 1–2 cm incision in the skin and a slightly smaller incision in the abdomen wall.
7. Find the fat pad that is connected to the kidney and using a blunt forceps, gently pull the kidney through the opening by the fat pad.
8. Immobilize the kidney using a Desmarres chalazion forceps. Allow the surface to dry for a few minutes.
9. Use a watchmaker's forceps to make a small hole in the capsule membrane.
10. Moisten the capsule with a small amount of sterile PBS and using moistened forceps make a pocket underneath the capsule.
11. Insert a capillary pipette containing the cells into the pocket and as far away from the tear as possible and deposit the cells into the capsule.
12. Release the kidney from the Demarres chalazion forceps and gently put it back into the body cavity using the blunt forceps.
13. Sew body wall with one or two stitches.
14. Close the skin with wound clips.
15. Place the animal in a clean, prewarmed cage for postoperative observation.
16. Observe the animal daily.
17. Remove wound clips after wound has healed, 7–10 days following surgery.
18. Euthanize the animal at the experimental endpoint, 6–12 weeks following implantation, and remove the tumor to be processed for analysis.

3.4. Teratoma Harvest and Analysis

Below is a brief description of three ways to analyze teratomas. Histopathology is the standard assay and should be carried out by a qualified pathologist. When histopathology, immunocytochemistry, and gene expression analysis are applied in combination to the analysis of the teratoma; however, the teratoma assay can also be used to enhance our understanding of development in addition to its utility as a key assay of pluripotency (2, 20).

(a) *Histopathology*: The tumor should be collected in D-PBS, washed three times, cut into pieces no more than 5-mm thick and fixed in 10% neutral buffered formalin. The fixed tumor is embedded in paraffin and slides are made. The tumor is sectioned (5–8 μm), fixed to slides, stained with hematoxylin and eosin, and evaluated by a pathologist (reviewed in (5)). TeratomEye, an automated assay system has been developed to identify the three representative tissue types – muscle, gut, and neural epithelia (21).

(b) *Immunocytochemistry*: Fixed tumor is embedded in paraffin and slides are made OR the tumor may be flash frozen in liquid nitrogen and then cryo-sectioned (2, 20). The slides are then processed for immunocytochemistry and labeled with lineage and cell-type-specific antibodies. Antibody staining is used to help identify early stage differentiated tissues that do not yet have identifiable morphology and can be useful in identifying the types of tissues arising from each germ layer, but it cannot take the place of careful histological analysis. Importantly, very few antibodies are specific for a single cell type.

(c) *Gene Expression Analysis*: Tissues are snap-frozen in liquid nitrogen and stored in sterile cryogenic vials until processed for RNA or DNA isolation. Standard protocols for RT-PCR or genome-wide microarrays are applied to the teratomas for gene expression analysis and compared to undifferentiated PSCs (6).

4. Notes

1. *Mice*: SCID-BEIGE (C.B-*Igh*-1b/GbmsTac-*Prkdcscid-Lystbg*N7)

   Severe combined immune deficient (SCID) mice are valuable xenotransplant models and have been used for many years to study immune rejection. While several strains of immunodeficient mice have been reported to support teratoma formation from hESCs including SCID, NOD-SCID, and nude mice, the SCID-BEIGE may be a superior recipient. The SCID-BEIGE is a double mutant created by breeding C.B-17 scid to the C57BL/6-bg strain. It carries both the scid mutation which causes the lack of B and T cells and the beige mutation which causes cytotoxic T cell and macrophage defects, as well as reduced natural killer cell activity.

2. *Anesthesia*: There are several choices for anesthesia. Some investigators prefer isoflurane and methoxyflurane, but these require specialized scavenging systems, since they are gaseous derivatives of ether. Others prefer Nembutal®, the brand name for injectable Phenobarbital sodium solution, which is under strict FDA control, yet others prefer zoletil-50/xylazine or ketamine-xylazine. All of these can be very effective when used appropriately. 2,2,2-tribromoethanol (Avertin) is another anesthetic, one that has been used for many years and if properly made and stored is a good and safe anesthetic. Recently, it has fallen out of favor because when it is not properly stored it can break down to form dibromoacetaldehyde and hydrobromic acid, both strong irritants and it has been

shown to effect the efficiency of transgenic mouse production (22). If using Avertin, protect it from heat and discard the unused solution after 2 weeks. With so many anesthetics available and new data regarding safety being made available, it is best to consult with the veterinarian at your institution to determine which anesthetic is believed to be the best one for this use given the particular laboratory setup, skill level, and regulatory issues at your institution.

3. The Chalazion forceps is a specialized surgical instrument that may facilitate the transplant of cells under the kidney capsule. It is used to hold the kidney in place while the cells are deposited in the capsule. It is designed with a solid bottom and open ring top, and has a thumb-screw mechanism to clamp the top and bottom. It was originally designed for ophthalmic procedures.
4. Keep the cells on ice until just before injection, when they are loaded into the syringe without a needle. This will limit the damage caused to the cells by the pressure generated when drawing them through a small needle.
5. *The endpoint for the experiment*: the experimental endpoint can be a predetermined length of time, 6 weeks, or when the teratoma has reached a certain size, 5 mm. Whatever endpoint is chosen, the health status of animals is the primary consideration. Since the animals are observed on a daily basis, it will be easy to determine if the animal is having a health issue and should be sacrificed before the predetermined endpoint. The number of cells transplanted has been shown to have an impact on how quickly the tumor forms. One wants to give the tumor enough time to develop into readily identifiable cells and tissue foci, hitting the right balance between the number of cells implanted and the time the tumor is allowed to develop is key to a good assay.
6. Implantation under the testis or kidney capsule do not require PSC colonies to be disaggregated to single cells. Small bits of colonies are ideal for these engraftment methods. Since the cells are collected in clumps, without dissociation to single cells, one may take a small aliquot of the cell-clump suspension and dissociate to single cells with trypsin, allowing one to determine the approximate number of cells transplanted.

Acknowledgments

Author wishes to thank many colleagues and friends, especially Philip Schwartz, Jeanne Loring, Martin Pera, and Melissa Carpenter, for many thoughtful discussions on stem cell biology.

References

1. Smith KP, Luong MX, Stein GS (2009). Pluripotency: Toward a gold standard for human ES and iPS cells. *J of Cellular Physiol.* **220**, 21–29.
2. Przyborski S. A (2005). Differentiation of human embryonic stem cells after transplantation in immune-deficient mice. *Stem Cells* **23**, 1242–50.
3. Andrews PW, Przyborski SA, Thomson JA (2001). Embryonal carcinoma cells as embryonic stem cells. *Stem Cell Biology*. Editors: Marshak DR, Gardner RL, Gottlieb D. New York: Cold Spring Harbor Laboratory Press. pp 231–265
4. Wesselschmidt RL (2007). Generation of human embryonic stem cell derived teratomas. *Human Stem Cell Manual: A Laboratory Guide*. First Edition. Editors: Loring JF, Wesselschmidt RL, and Schwartz PH. New York, London, San Diego: Elsevier. pp 162–170
5. Damjanov I and Ährlund-Richter L (2007). Characterization of human embryonic stem cell derived teratomas. In *Human Stem Cell Manual: A Laboratory Guide. Human Stem Cell Manual: A Laboratory Guide.* First Edition. Editors: Loring JF, Wesselschmidt RL, and Schwartz PH. New York, London, San Diego: Elsevier. pp 171–182
6. Gertow K, Przyborski S, Loring JF, Auerbach JM, Epifano O, Otonkoski T, Damjanov I, Ahrlund-Richter L (2007). Isolation of human embryonic stem cell derived teratomas for the assessment of pluripotency. *Curr Protoc Stem Cell Biol* **3**: Unit 1B.4 (1–29).
7. Cooke MJ, Stojkovic M, Przyborski SA (2006). Growth of teratomas derived from human pluripotent stem cells is influenced by the graft site. *Stem Cells and Dev* **15**:254–259.
8. Cao F, van der Bogt KEA, Sadrzadeh A, Xie X, Sheikh AY, Wang H, Connolly AJ, Robbins RC, Wu JC (2007). Spatial and temporal kinetics of teratoma formation from murine embryonic stem cell transplantation. *Stem Cells and Dev.* **16**: 883–892.
9. Shih CC, Forman SJ, Chu P, Slovak M (2007). Human embryonic stem cells are prone to generate primitive, undifferentiated tumors in engrafted fetal tissues in severe combined immunodeficient mice. *Stem Cells and Dev.* **16**: 893–902.
10. Prokhorova TA, Harkness LM, Frandsen U, Ditzel N, Schroder HD, Burns JS, Kassem M (2009). Teratoma formation by human embryonic stem cells is site dependent and enhanced by the presence of Matrigel. *Stem Cells and Dev.* **18**: 47–54
11. Lee AS, Tang C, Cao F, Xiaoyan X, van der Bogt K, Hwang A, Connolly AG, Robbins RC, Wu JC (2009). Effects of cell number on teratoma formation by human embryonic stem cells. *Cell Cycle* **8**: 2608–2612.
12. International Stem Cell Initiative (2007) Characterization of human embryonic stem cell lines by the International Stem Cell Initiative. *Nat Biotechol* **25**:803–816.
13. Thomson JA, Itskovitz-Eldor J, Shapiro SS, Waknitz MA, Swiergiel JJ, Marshall VS, Jones JM (1998). Embryonic stem cell lines derived from human blastocysts. *Science* **282**, 1145–47
14. Xu C, Inokuma MS, Denham J, Golds K, Kundu P, Gold JD, Carpenter MK (2001). Feeder-free growth of undifferentiated human embryonic stem cells. *Nat. Biotech.* **19**, 971–74.
15. Yu J, Vodyanik MA, Smuga-Otto K, Antosiewicz-Bourget J, Frane JL, Tian Shulan T, Nie J, Jonsdottier GA, Routti V, Stewart R, Slukvin II, Thomson JA (2007). Induced pluripotent stem cell lines derived from human somatic cells. *Science* **318**: 1917–1920.
16. Rubenoff BE, Pera MF, Fong C-Y, Trouson A, Bongso A (2000). Embryonic stem cell lines from human blastocysts: somatic differentiation in vitro. *Nat. Biotech.* **18**, 399–404.
17. Richards M, Fong CY, Tan S, Chan WK, Bongso A (2004). An efficient and safe xenofree cryopreservation method for the storage of human embryonic stem cells. *Stem Cells* **22**: 779–89.
18. Hovatta O, Mikkola M, Gertow K, Stromberg AM, Inzunza J, Hreinsson J, Rozell B, Blennow E, Andang M, Arhlund-Richter L (2003). A culture system using human foreskin fibroblasts as feeder cells allows production of human embryonic stem cells. *Hum Reprod* **18**: 1404–1409.
19. Solter D. (1994). Implanting tissues under the kidney capsule. *Manipulating the mouse embryo.* (2nd edition) Editors: Hogan, Beddington, Constantinti, Lacy. Cold Spring Harbor New York: Cold Spring Harbor Press.
20. Gertow K, Wolbank S, Rozell B, et al. (2004) Organized development from human embryonic stem cells after injection into immunodeficient mice. *Stem Cells Dev* **13**: 421–435.
21. Oh SKW, Chua P, Foon KL, Ng E, Chin A, Choo ABH, Srinivasan R (2009). Quantitative identification of teratoma tissues formed by human embryonic stem cells with TeratomEye. *Biotechnol Lett* **31**, 653–58.
22. Zeller W, Meier G, Bürki K, Panoussis B (1998). Adverse effects of tribromoethanol as used in the production of transgenic mice. Lab Anim. **132**: 407–13.

Chapter 18

Detection of Copy Number Variation Using SNP Genotyping

Gregory M. Cooper and Heather C. Mefford

Abstract

Genetic diversity among human genomes comes in many forms, including single nucleotide polymorphisms (SNPs) and small insertions and deletions on the order of one to several basepairs. More recently, large, >1 kb copy number changes have been identified as an important source of normal genomic variation as well as disease-causing variation. The ability to perform genome-wide discovery of large copy number variants (CNVs) has been facilitated by advances in two technologies – array comparative genomic hybridization and SNP genotyping platforms. Here, we discuss the general principles and strategies underlying CNV detection with SNP genotyping platforms, which are widely used and capable of providing both SNP and CNV genotyping information.

Key words: copy number variation, single nucleotide polymorphism genotyping, genomic variation, array comparative genomic hybridization

1. Introduction

Copy number variants (CNVs), defined as insertions, deletions, or duplications of sequence larger than 1 kb, are substantial contributors to human genomic diversity and are important factors in both normal (1) and disease (2, 3) variation. These include environmentally responsive traits like sensory perception (e.g., *opsins* and *olfactory receptors*), immune system function (e.g., Crohn's disease, psoriasis), severe early childhood diseases like developmental delay and autism, and neurological diseases like schizophrenia and epilepsy. Importantly, studies of CNV-trait associations have found evidence for the involvement of both common and rare CNVs in human disease. In the context of pluripotent stem cell development and analysis, the knowledge of CNVs in a given genome can be useful for several reasons. For example, CNVs can

Philip H. Schwartz and Robin L. Wesselschmidt (eds.), *Human Pluripotent Stem Cells: Methods and Protocols*, Methods in Molecular Biology, vol. 767, DOI 10.1007/978-1-61779-201-4_18, © Springer Science+Business Media, LLC 2011

affect expression of genes within (4) and near (5) the CNV, so expression data for genes affected by CNVs may be interpreted differently.

Owing to their size, heterogeneity, and sequence complexity, the accurate detection of CNVs in human populations is a technically challenging task. There are several methods that may be employed to detect CNVs. For targeted evaluation of one or a few genomic regions of interest, quantitative PCR (qPCR) (6–8) or multiplex ligation probe amplification (MLPA) (9) are commonly used. However, for more extensive, genome-wide analysis, there are two commonly used platforms: array comparative genomic hybridization (CGH) and single nucleotide polymorphism (SNP) genotyping arrays. Here we focus on the use of SNP arrays to detect copy number variation. The advantages of SNP-based CNV detection include.

1. Simultaneous ascertainment of SNP and CNV data (unavailable from CGH).
2. High-throughput sample processing, treatment, and quality control.
3. High-density of probes, with arrays ranging from hundreds of thousands to multiple millions.
4. Reasonable cost, typically ranging from tens to hundreds of dollars per sample, depending on probe density.

2. Materials

SNP microarray analyses typically require an input of 100 ng to 1 μg of genomic DNA, varying by the specific array/protocol employed and the manufacturer (see Note 1).

3. Methods

3.1. SNP Genotyping

Before discussing how to detect CNVs using SNP genotyping data, an understanding of the basic principles of SNP genotyping is required. SNPs are DNA sequence variants where a single nucleotide can differ among individuals (or chromosomes); most SNPs are di-allelic, meaning that there are two possible alleles (e.g., a "C" or a "T" at a given site). Microarray-based SNP genotyping platforms exploit fluorescence-based visualization of genomic DNA bound in an allele-specific manner to oligonucleotides fixed to a surface. While details vary substantially between different platforms, there are two critical pieces of information gathered for each

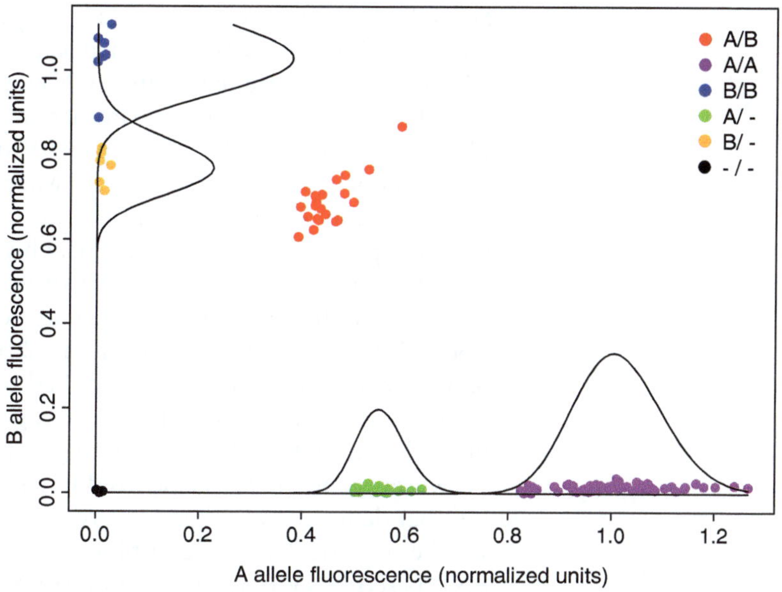

Fig. 1. Scatter plot of intensity information collected for a single SNP (rs10076425) assayed with an Illumina genome-wide SNP array on a collection of 126 (primarily HapMap[21]) samples. Each point corresponds to the intensity for a single sample; the *X*-axis indicates fluorescence intensity for the "A" allele while the *Y*-axis indicates intensity for the "B" allele. The three most populous clusters of samples correspond to the "AA" (*purple*), "AB" (*red*), and "BB" (*blue*) samples who are homozygous for the A-allele, heterozygous, or homozygous for the B-allele. Note the presence of three additional clusters corresponding to the hemizygous A- (*green*) individuals, hemizygous B- (*yellow*) individuals, or homozygous deletion carriers (*black*). The first two are heterozygous for a deletion allele while the latter have a copy number of zero at this location. Superimposed distributions (*black curves*) are estimated from the data and allow statistical separation of diploid from haploid samples. This figure is reproduced from Cooper et al. (12).

targeted genomic site: (1) the total fluorescence intensity for a site which reflects signal combined for both alleles in a given sample and (2) the allelic ratio providing the relative intensity measurements for the two alleles at each site (Fig. 1). For the vast majority of sites, individuals are diploid and will therefore be homozygous for one allele ("AA") or the other ("BB") or heterozygous with one copy of each ("AB"). Note that modern arrays also include many non-SNP (monomorphic sites) probes (e.g., (10)), which only provide total intensity data and are included so as to improve probe density in known or suspected CNV locations.

3.2. CNV Detection from SNP Data: Discovery vs. Genotyping

The task of CNV detection from SNP data can be broken down into two related but distinct challenges: CNV discovery, wherein variants are detected ab initio in a given genome without assumptions about their breakpoints, and CNV genotyping, wherein copy number status is assigned to a set of studied samples for given loci that are known (or suspected) to be copy number variable.

Several important consequences emerge from this distinction. First, CNV discovery is performed sample-by-sample and has the advantage of being able to detect CNVs anywhere in the genome, including for rare and de novo events unique to the given sample. However, owing to the large space of data being examined (if breakpoints are allowed to be anywhere, any pair of analyzed probes within a given chromosome is a candidate set of breakpoints), specificity must be extremely high to avoid an unacceptably large false discovery rate. Such stringent specificity is typically obtained at the expense of sensitivity to small (few probes) or noisy (intensities near detection threshold) sites. On the other hand, CNV genotyping can be applied to many samples simultaneously and can leverage the knowledge that a CNV exists at a given location to achieve both high specificity and sensitivity. However, genotyping is restricted to a priori defined sites, implying that de novo and other rare events outside of the targeted loci will be missed. Below, we outline the basic principles underlying SNP-based CNV detection, contrasting the discovery and genotyping challenges where appropriate.

3.3. CNV Genotyping

The conversion of intensity information into an estimate of copy number comprises several steps. First, the raw intensity information is normalized to account for systematic effects related to genotyping chemistry (e.g., differences in intensity between fluorophores), microscopy (e.g., location of a probe on a slide), and other factors (e.g., total intensity for a given slide). In addition, measurements are often obtained at multiple physical locations on an array corresponding to the same SNP (or genomic location for nonpolymorphic probes), and this information must be integrated to determine a single measurement for a given site. These steps are heavily sensitive to the specific platform used and typically handled by the manufacturer's software. After these normalization steps, intensity information for a given SNP is comparable across a set of samples, and for polymorphic sites, can be visualized as a two-dimensional scatter plot with intensity information for each of the two alleles (denoted from here forward as "A" and "B" for simplicity) plotted on a separate axis (Fig. 1).

Note that the position of a given sample in this two-dimensional space provides both total intensity (essentially distance to the origin) and allelic ratio (angle of the line joining the datum to the origin). The canonical SNP genotypes (assuming the probe is informative and analyzed in diploid samples) will generally stand out as three distinct clusters, with homozygous (AA or BB) individuals appearing on either of the two axes and heterozygous (AB) individuals appearing as a cluster toward the middle of this space. This clustering information forms the basis for SNP genotyping (10, 11) and can also be used to genotype CNVs via statistical evaluations of the relative locations (separation) and qualities (variance within a group) of the observed clusters (10, 12). Hemizygotes (A- or B-), for example, will appear as clusters of samples closer to the origin than homozygotes, because relative copy number, and therefore intensity, is reduced, and the allelic ratio indicates that the sample is homozygous (Fig. 1). "Null" individuals (i.e., samples that are homozygous for a deletion event) will typically appear near the origin reflecting the lack of any DNA binding for that sample at that location (Fig. 1). Samples bearing higher copy numbers may yield a diversity of cluster positions depending on total copy number and heterozygosity (Fig. 1). For example, individuals carrying a duplication of a given sequence actually have three copies of that sequence, and at heterozygous locations may be triply homozygous ("AAA" or "BBB") or exhibit a distortion in the allelic ratio ("AAB" or "ABB"). These latter cases can provide a powerful discriminatory signature to define duplication carriers, especially for CNV discovery (see below).

There are several critical caveats to consider. First, CNV genotyping using the above framework is frequency-dependent since it depends on the identification of clusters of individuals with the same copy number. Rare or individual CNV carriers in a sample series appear as outliers rather than in clusters, and alternative approaches are required to identify these events, although it is possible to genotype rare CNVs by exploiting this behavior and specifically looking for outlying samples (caution must be taken to contrast noisy samples from outliers that result from a change in copy number) (13). Second, the CNV genotyping process is sensitive to background intensities and probe-specific noise (e.g., cross-hybridization to other sites in the genome); clustering information from multiple SNPs is typically required to obtain robust copy number genotypes to avoid both probe-specific artifacts and reduce noise in genotype inference. It is also important that genotyping methods either have an automated method to identify "informative" probes or are only applied to predefined probe sets that are known to yield reliable copy number estimates; in most cases it is necessary to combine automated elimination of obviously bad probes with manual curation of potentially good probe sets. Finally, we note that copy number estimation is relative; absolute copy number is typically reliant on the assumption that a diploid copy number is predominant (often but not always true) (see Note 2). Even assignment of zero copies can be confounded by cross-hybridization, for example, and assignment of absolute copy number at high copies (>3) is particularly difficult as the ratios between copy number intensities become smaller (i.e., a change in copy number from 1 to 2 corresponds to a twofold change in intensity, while 5–6 is only a 20% change). In general, SNP-based CNV genotyping has not been shown to be accurate for higher copy ranges.

3.4. CNV Discovery

As noted above, clustering-based approaches to CNV detection can generally only apply to curated sets of probes at previously defined CNVs that are known or suspected to be common. Rare variant discovery is a distinct challenge that is usually accomplished by considering data from each individual separately, scanning across the genome to identify regions (sets of contiguous probes) that exhibit evidence for gain or loss of segmental DNA copy number. However, before such an analysis can be done, it is important that each probe be normalized so that intensity data are comparable between probes within a given sample (the normalization steps described above are taken to make intensities comparable across samples but within a given probe). Clustering of individuals is again applied here, except in this circumstance all individuals of a given SNP genotype are assumed to be the same copy number (ideally, common CNVs would be known and eliminated prior to such an analysis and for X-chromosome SNPs males and females would be treated separately). Subsequently, for each probe within each individual, a normalized intensity measure can be derived that describes the location of a given individual relative to the other samples. In Illumina genotyping, for example, the total intensity for a given probe in a given individual is reported as the "LogR Ratio," where a value of 0 indicates that a sample has a total intensity at the center of the cluster of individuals with the same SNP genotype (i.e., "AA" individuals), while positive and negative values indicate that a sample is above or below the mean intensity (14). Related to this is a normalized measure of the allelic ratio; in Illumina genotyping, this value is the "B-allele frequency" (BAF), so-called because it is inferred to be the fraction of the total intensity at a given probe that is derived from the "B" allele. For example, the BAF for "AA" individuals should be 0, for "AB" individuals should be 0.5, and for "BB" individuals should be 1, because in these samples 0, 50, or 100% of the total intensity comes from the B-allele, respectively. In practice, the center of the "AB" heterozygote cluster is used to define a BAF of 0.5, again because the absolute position of this cluster varies from probe to probe while it is assumed that probes are all capturing diploid locations.

After this probe-by-probe normalization is applied, the challenge lies in the identification of segments within a given genome that exhibit intensity and BAF information that is consistent with the presence of a CNV. Deletions, for example, should result in both negative LogR values and complete loss of heterozygosity (BAF of only 0 or 1; Fig. 2a). Duplications should manifest as positive LogR values and a skew in BAF at heterozygous sites to either 1/3 ("AAB") or 2/3 ("ABB") since the alleles are no longer in one-to-one proportion (Fig. 2b). Note that homozygous sites in a duplicated segment ("AAA" or "BBB") would still have BAF values of 0 or 1, and because of this polymorphic sites

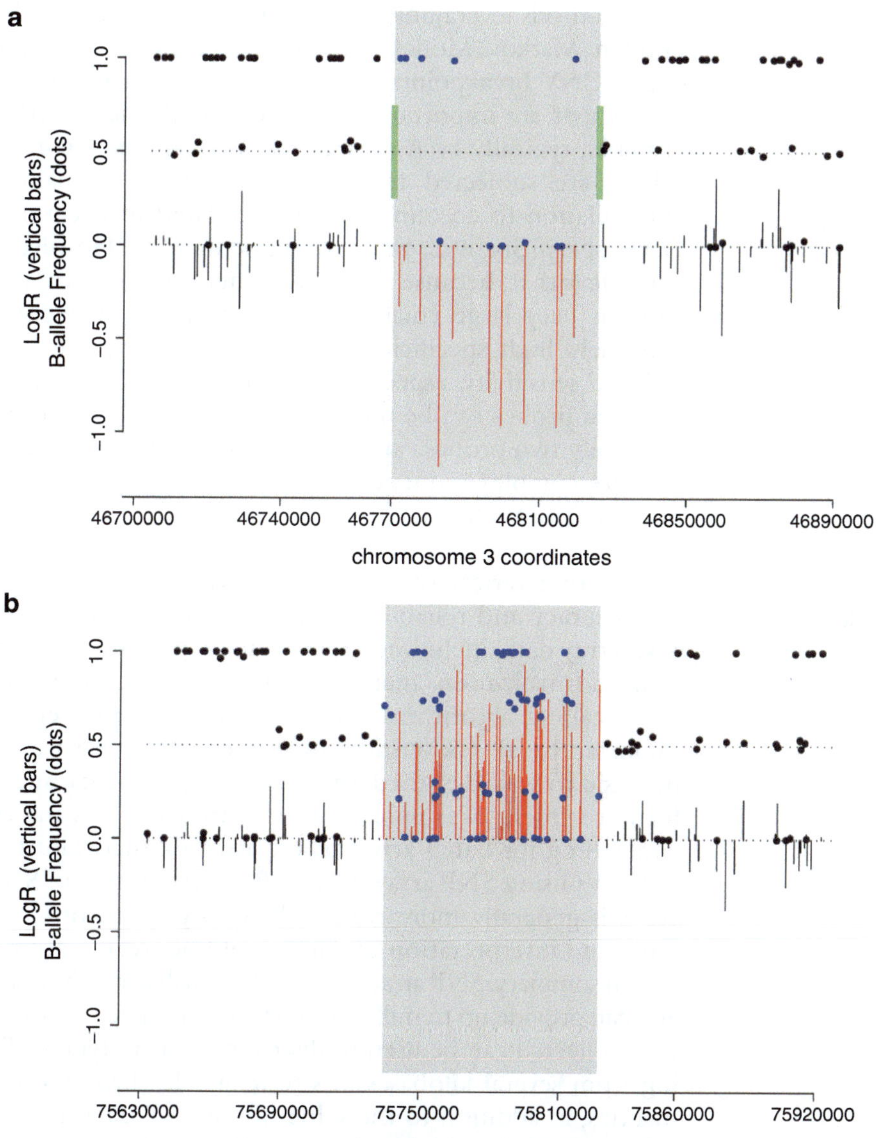

Fig. 2. CNVs "discovered" within Illumina genome-wide SNP array intensity data, adapted from Cooper et al. (12). (**a**) Example of a deletion event. Intensity data for all probes within the indicated genomic interval (*X*-axis) for a single sample are plotted. "LogR Ratio" and "B-allele Frequency" (14) are plotted as *vertical bars* and *filled dots*, respectively. The *gray box* indicates the deletion span inferred by computational segmentation of the SNP array data; probes internal to this box are colored *red* ("LogR Ratio") or *blue* ("B-allele Frequency"). *Green vertical bars* indicate the deletion borders defined by resequencing. Note that the LogR drops within the deletion and the B-allele Frequency values indicate a loss of heterozygosity. (**b**) Similar to (**a**), except an example of a duplication is highlighted. Note that "heterozygous" SNPs within the duplicated segment have a B-allele frequency of either ~1/3 or ~2/3, indicating that this individual carries three copies of this segment of the genome ("AAB" or "ABB").

are in general more informative than non-polymorphic sites; equivalently for deletions, loss of heterozygosity information is only useful at polymorphic sites. There are a variety of methods available to perform segmentation, with the most commonly used methods leveraging well-established statistical methods like Hidden Markov Models (for several examples of methods to detect CNV breakpoints see (12, 15–17)). There are numerous details that are important to consider in all these methods (for example, typically even normalized intensity data, e.g., "LogR" values, are subjected to further rounds of normalization or manipulation to account for effects related to allele frequency, probe specificity, etc., prior to segmentation). However, a general rule is that, because the breakpoints are unknown, the search space is very large (many potential pairs of breakpoints) and extremely high specificity is required. This typically results in reduced sensitivity, especially to small CNVs (i.e., variants spanning ten probes can be more robustly inferred than those spanning only two probes) and those (as for genotyping) embedded in more complex sequence (e.g., CNVs that change copy number state from 5 to 6).

3.5. Additional Considerations

There are a variety of critical contextual factors that influence the accuracy and reliability of CNV information inferred using SNP array data, including DNA quality, quantity and concentration; normalization methods; quality-control measures; and CNV calling algorithms used (see Notes 1–3). Perhaps most importantly, all the critical quality-control measures that are intrinsic to well-designed array-based experiments, such as uniform treatment of samples, randomization of cases and controls, controlling for batch artifacts, etc are also important to studies of CNVs using SNP array data; in fact, assignment of SNP genotypes is generally more robust than assignment of CNV genotypes and interpretation of data should be treated accordingly.

In summary, SNP arrays are widely available, relatively affordable and can provide up to millions of SNPs in a single experiment. The data collected can be used to discover and genotype CNVs ranging from several kilobases in size to whole chromosome abnormalities, in addition to the value of the SNP data. Furthermore, it is expected that this information will improve as maps of known CNVs become more comprehensive (10, 20), as density of SNP arrays increases, and as genotyping algorithms and data normalization approaches become more accurate (21).

4. Notes

1. DNA quality, quantity and concentration can affect fluorescence intensity levels and therefore inference of CNVs. Furthermore, whole-genome amplification steps can introduce systematic noise that may overwhelm legitimate CNV signal; this may be true even when SNP genotypes can be reliably inferred.

2. Data normalization must always be considered when interpreting CNV information, especially in the context of common CNVs. For example, if a deletion event is at high frequency in the population, then the assumption that most samples have a copy number of 2 does not hold and inappropriate assignment of absolute copy number may result.
3. Many algorithms that have been and continue to be developed for the discovery and/or genotyping of CNVs from SNP microarray data. It is important to note that some studies explicitly differentiate these tasks while others attempt to perform both simultaneously. Examples of algorithms include (but are not limited to) QuantiSNP (16), PennCNV (15), SCIMM (12), SCOUT (13), BirdSuite (10, 18), and others (19). Choice of any given algorithm and sets of parameters to apply is complex, but should take into consideration the platform used (e.g., Illumina vs Affymetrix), the goals of the study (e.g., common CNV genotyping vs. rare variant discovery), the density and spacing of probes, and the respective costs of false positive and false negative CNV assignments.

References

1. Perry GH, Dominy NJ, Claw KG, et al. (2007) Diet and the evolution of human amylase gene copy number variation. *Nat Genet* **39**:1256–60.
2. Mefford HC, Eichler EE. (2009) Duplication hotspots, rare genomic disorders, and common disease. *Curr Opin Genet Dev* **19**:196–204.
3. Fanciulli M, Petretto E, Aitman TJ. (2009) Gene copy number variation and common human disease. *Clin Genet* 77:201–13.
4. Henrichsen CN, Chaignat E, Reymond A. (2009) Copy number variants, diseases and gene expression. *Hum Mol Genet* **18**:R1-8.
5. Merla G, Howald C, Henrichsen CN, et al. (2006) Submicroscopic deletion in patients with Williams-Beuren syndrome influences expression levels of the nonhemizygous flanking genes. *Am J Hum Genet* **79**:332–41.
6. Porcher C, Malinge MC, Picat C, Grandchamp B. (1992) A simplified method for determination of specific DNA or RNA copy number using quantitative PCR and an automatic DNA sequencer. *Biotechniques* **13**:106–14.
7. Livak KJ, Schmittgen TD. (2001) Analysis of relative gene expression data using real-time quantitative PCR and the 2(–Delta Delta C(T)) Method. *Methods* **25**:402–8.
8. Wong A, Cortopassi G. Reproducible quantitative PCR of mitochondrial and nuclear DNA copy number using the LightCycler. (2002) *Methods Mol Biol* **197**:129–37.
9. Schouten JP, McElgunn CJ, Waaijer R, Zwijnenburg D, Diepvens F, Pals G. (2002) Relative quantification of 40 nucleic acid sequences by multiplex ligation-dependent probe amplification. *Nucleic Acids Res* **30**:e57.
10. McCarroll SA, Kuruvilla FG, Korn JM, et al. (2008) Integrated detection and population-genetic analysis of SNPs and copy number variation. *Nat Genet* **40**:1166–74.
11. Steemers FJ, Chang W, Lee G, Barker DL, Shen R, Gunderson KL. (2006) Whole-genome genotyping with the single-base extension assay. *Nat Methods* **3**:31–3.
12. Cooper GM, Zerr T, Kidd JM, Eichler EE, Nickerson DA. (2008) Systematic assessment of copy number variant detection via genome-wide SNP genotyping. *Nat Genet* **40**:1199–203.
13. Mefford HC, Cooper GM, Zerr T, et al. (2009) A method for rapid, targeted CNV genotyping identifies rare variants associated with neurocognitive disease. *Genome Res* **19**: 1579–85.
14. Peiffer DA, Le JM, Steemers FJ, et al. (2006) High-resolution genomic profiling of chromosomal aberrations using Infinium whole-genome genotyping. *Genome Res* **16**:1136–48.
15. Wang K, Li M, Hadley D, et al. (2007) PennCNV: an integrated hidden Markov model designed for high-resolution copy number variation detection in whole-genome SNP genotyping data. *Genome Res* **17**:1665–74.

16. Colella S, Yau C, Taylor JM, et al. (2007) QuantiSNP: an Objective Bayes Hidden-Markov Model to detect and accurately map copy number variation using SNP genotyping data. *Nucleic Acids Res* **35**:2013–25.

17. Shaikh TH, Gai X, Perin JC, et al. (2009) High-resolution mapping and analysis of copy number variations in the human genome: a data resource for clinical and research applications. *Genome Res* **19**:1682–90.

18. Korn JM, Kuruvilla FG, McCarroll SA, et al. (2008) Integrated genotype calling and association analysis of SNPs, common copy number polymorphisms and rare CNVs. *Nat Genet* **40**:1253–60.

19. Gamazon ER, Zhang W, Konkashbaev A, et al. (2009) SCAN: SNP and Copy number Annotation. *Bioinformatics* **26**:259–62.

20. Kidd JM, Cooper GM, Donahue WF, et al. (2008) Mapping and sequencing of structural variation from eight human genomes. *Nature* **453**:56–64.

21. International HapMap Consortium. (2005) A haplotype map of the human genome. *Nature* **437**:1299–320.

Chapter 19

Genome-Wide Epigenetic Analysis of Human Pluripotent Stem Cells by ChIP and ChIP-Seq

Michael J. Hitchler and Judd C. Rice

Abstract

Chromatin immunoprecipitation (ChIP) is used to evaluate the interaction of proteins and genomic DNA. In eukaryotic cells, the DNA is highly compacted with the evolutionarily conserved histone proteins (which together with DNA form the nucleosome) and other chromosomal-associated proteins to form the chromatin structure. Chromatin structure is dynamically regulated by several mechanisms including transcription factor binding and various posttranslational modifications of the histone proteins. The chromatin structure can be affected by environmental factors, such as those that induce differentiation or promote self-renewal in stem cells. Using very specific antibodies, one can evaluate the specific amino acids within the histones and each one of these modifications is associated with a distinct DNA-templated process, including transcription. Therefore, determining the location of transcription factors and histone modifications can yield important insights into the DNA-associated activities that are occurring at that particular region of the genome at that time. ChIP followed by high-throughput DNA sequencing (ChIP-Seq) provides a means to rapidly determine the precise genomic location of transcription factor binding sites and histone modifications on a genome-wide scale. Genome-wide mapping of histone modifications and chromatin-associated proteins have already begun to reveal the mechanisms responsible for regulating the pattern of gene expression in mouse embryonic stem cells. However, similar studies in human embryonic stem cells are currently lacking due to the difficulty in obtaining the large number of purified cells typically required for ChIP and ChIP-Seq experiments. Here, we describe a detailed method for determining the locations of specific histone modifications using only one million cells.

Key words: chromatin immunoprecipitation, ChIP, ChIP-Seq, histone methylation, human embryonic stem cells, chromatin, epigenome

1. Introduction

Eukaryotic cells package and organize their genetic information into chromatin, a complex structure consisting of DNA, the evolutionarily conserved histone proteins (which together with DNA

Philip H. Schwartz and Robin L. Wesselschmidt (eds.), *Human Pluripotent Stem Cells: Methods and Protocols*, Methods in Molecular Biology, vol. 767, DOI 10.1007/978-1-61779-201-4_19, © Springer Science+Business Media, LLC 2011

form the nucleosome), and other chromosomal-associated proteins. Chromatin structure is dynamically regulated by several different mechanisms including transcription factors and the various posttranslational modifications of the histone proteins. Specific histone modifying enzymes can covalently phosphorylate, acetylate, ubiquitylate, or methylate specific amino acids within the histones and each one of these modifications is associated with a distinct DNA-templated process, including transcription (1–3). Therefore, determining the location of transcription factors and histone modifications can yield important insights into the DNA-associated activities that are occurring at that particular region of the genome at that time. Increasing evidence indicates that concerted changes in transcription factor binding and histone modifications play a causal role in mammalian development by regulating gene expression patterns (4–6). Therefore, mapping their location will provide fundamental insights into the mechanisms influencing gene expression during discrete stages in development. Chromatin immunoprecipitation (ChIP) followed by high-throughput DNA sequencing (ChIP-Seq) provides a tool to accurately determine the location of histone modifications on a genome-wide scale. Genome-wide mapping of histone modifications and chromatin-associated proteins have already begun to reveal the mechanisms responsible for regulating the pattern of gene expression in mouse embryonic stem cells (7–11). However, similar studies in human pluripotent stem cells (hPSCs) are currently lacking due to the inability to obtain purified cells in sufficient number required for ChIP-Seq experiments. Here, we describe a detailed method for ChIP-Seq of histone modifications that requires only one million human PSCs. This method was developed using the HES3 (ES03), a human embryonic stem cell (hESC) line, but should be adaptable to the analysis of other PSC lines including induced pluripotent stem cell (iPSC) lines as well as their differentiated progeny. Since this method only requires a million cells for the generation of the ChIP-Seq libraries, we believe that it will be useful when isolating small bits of colonies or small cultures.

2. Materials

2.1. Preparation of Chromatin for ChIPs

1. 0.05% Trypsin with ethylenediaminetetraacetic acid (EDTA).
2. Human PSC growth medium, containing fetal bovine serum.
3. 37% Formaldehyde, molecular biology grade.
4. Solution of 1 M glycine.

5. Dulbecco's phosphate buffer saline, pH 7.4, no $CaCl_2$, no $MgCl_2$ (DPBS).
6. Nuclei isolation buffer: 10 mM HEPES, pH 7.4, 150 mM NaCl, 1.5 mM $MgCl_2$, 10 mM KCl, 0.5% NP-40, 0.5 mM dithiothreitol (with protease inhibitors, see Notes 1 and 2). Make fresh before use.
7. Nuclei lysis buffer: 50 mM Tris–HCl, pH 8.0, 10 mM EDTA, 1% sodium dodecyl sulfate (SDS) (with protease inhibitors, see Note 2). Store at room temperature.

2.2. Immuno-precipitation of Protein–DNA Complexes

1. ChIP grade antibody to the histone modification or chromatin-associated protein that is being assayed.
2. Whole IgG from rabbit serum (negative antibody control).
3. Protein A Dynabeads and magnetic stand (Invitrogen, see Note 3).
4. Immunoprecipitation dilution buffer (IP dilution buffer): 16.7 mM Tris–HCl, pH 7.5, 150 mM NaCl, 3.3 mM EDTA, 1% Triton X-100, 0.1% SDS, 0.5% sodium deoxycholate (Na-Doc) (with protease inhibitors added as in Note 2). Store at 4°C.
5. Diagenode Bioruptor or equivalent sonicator (see Note 4).

2.3. Washing and Elution of Bound Chromatin

1. Wash buffer A: 20 mM Tris–HCl, pH 8.1, 150 mM NaCl, 2 mM EDTA, 0.1% SDS, 1.0% Triton X-100. Store at 4°C.
2. Wash buffer B: 20 mM Tris–HCl, pH 8.1, 500 mM NaCl, 2 mM EDTA, 0.1% SDS, 1.0% Triton X-100. Store at 4°C.
3. Wash buffer C: 10 mM Tris–HCl, pH 8.0, 0.25 M LiCl, 1 mM EDTA, 1.0% Na-Doc, 1.0% NP-40. Store at 4°C.
4. TE buffer: 10 mM Tris–HCl, pH 7.5, 1 mM ETDA.
5. Elution buffer: 100 mM Tris–HCl, pH 8.0, 20 mM EDTA, 1% SDS.
6. Stock solution of 10 mg/ml RNAse A (Fermentas).
7. Stock solution of 5 M NaCl.

2.4. Purification of Immunoprecipitated DNA

1. Phenol/chloroform 1:1. Store at 4°C.
2. 10 mg/ml Glycogen (Fermentas).
3. 100% Ethanol, molecular biology grade.
4. Proteinase K.
5. 10× Proteinase K digestion buffer: 100 mM Tris–HCl pH 8.0, 1% SDS. Store at –20°C.

2.5. Analysis of Immunoprecipitated DNA

1. Quant-iT dsDNA Assay Kit, high sensitivity (see Note 5).
2. Fluorometric plate reader.
3. Primers designed to control genes or your regions of interest (see Note 6).
4. Real-time PCR machine.
5. 2× SYBR green PCR master mix.

2.6. Construction of Libraries for High-Throughput Sequencing (Illumina)

1. Mini-Elute PCR cleanup kit (Qiagen).
2. Enzymes for DNA end repair. We recommend using the End-it DNA End Repair Kit (Epicentre Biotechnologies).
3. Klenow (3′–5′ exo) Fragment (New England Biolabs) with NEB buffer #2.
4. Stock solution of 1 mM ATP.
5. 37°C Heat block or water bath.
6. Rapid T4 DNA ligase with supplied 2× Rapid Ligase buffer (Enzymatics).
7. Illumina sequencing adapters diluted 40× (Illumina).
8. High-fidelity DNA Polymerase such as *Pfx* DNA Polymerase (Invitrogen).
9. Stock solution of Illumina library amplification primers, 25 μM each (Illumina).
10. Conventional PCR machine.
11. Tris–Acetate–EDTA: Prepare 50× stock with 242 g Tris Base, 57.1 ml of Glacial Acetic acid, 100 ml of 0.5 M EDTA pH 8.0. Dilute 20 ml of 50× stock with 980 ml of deionized water before use.
12. Agarose, electrophoresis grade.
13. Agarose gel extraction kit such as the QIAquick Gel Extraction Kit (Qiagen).

3. Methods

Successful completion of ChIP assays is dependent upon three major factors: obtaining an optimal length of fixed chromatin, a highly specific ChIP grade antibody, and using washing conditions that maximize target recovery and minimize nonspecific background binding. In order to obtain consistent, reproducible results each of these three factors should be optimized with the cell line and antibody to be used for ChIP-seq prior to conducting experiments. This protocol was optimized using the HES3 (ES03) hESC line, but should be applicable to other PSC lines with the proper optimization for that particular cell line as described below.

Chromatin fragment size is determined by the percentage of formaldehyde used and the exposure time and optimized by altering each of these parameters and assessing the length of the fragment. Sonication conditions are determined empirically and vary, depending on the cell type, cell number, the percentage of formaldehyde used, and type of sonicator used as described in Note 4. Antibody specificity and background signal are assessed by including a pre-immune total IgG immunoprecipitation negative control. If the yield of DNA from the IgG control is low compared to the yield of DNA from the antibody of interest, then the experimental conditions used are optimal for your experiments. If they are not, a different ChIP antibody or different wash conditions may be necessary.

Overall success of these experiments is determined by quantitative real-time PCR (Q-PCR) analysis of ChIPed DNA. Usually, at least two regions are amplified, a positive control region where the histone modification is known to be present, and a negative control region that should have minimal or no signal. The IP sample should amplify significantly higher at the control region than the input and IgG control samples, but be similar to those at the negative control region. If these criteria are met, the Input, IP, and IgG control DNA can be used to create Illumina sequencing libraries. Prior to Illumina sequencing, the libraries should be tested by Q-PCR with the same amplified regions as above. The enrichment values will differ between these two experiments. However, if patterns of enrichment between libraries are similar to what is seen for the ChIP samples, the libraries are ready for sequencing.

3.1. Preparation of hPSC Chromatin for ChIPs

1. *Collection of human pluripotent stem cells (PSCs)*. Colonies of PSCs are processed to obtain single cell suspensions for chemical fixation (see below). Treat each dish with 2 ml of 0.25% trypsin/EDTA solution for ~3 min at 37°C, or until feeder cells become detached from the dish. Remove trypsin and feeders by suction being careful not to dislodge the PSC colonies still attached to the dish. Add 1 ml of fresh 0.25% trypsin/EDTA and incubate for ~3–5 min at 37°C to release PSCs. Quickly, but gently, resuspend cells in trypsin/EDTA, pipette the cells up and down several times to create a single cell suspension.
2. Inactivate trypsin by adding 4 ml of growth medium containing fetal bovine serum. Cells are then collected by centrifugation at 200 × *g* for 5 min at room temperature and the medium is removed by aspiration (see Note 7). Resuspend PSC pellet in 5 ml of growth medium and determine cell number using a hemocytometer or coulter counter.

3. *Formaldehyde fixation of PSCs*. Fix chromatin in PSCs by adding 135 μl of 37% formaldehyde (final concentration of 1%). Rotate mixture at room temperature for 10 min if ChIPing for histones, or 15 min if ChIPing for a transcription factor or other DNA-associated protein complex. Quench the formaldehyde fixation by adding 625 μl of 1 M glycine (final concentration of 0.125 M) and rotating cells for an additional 5 min at room temperature. Centrifuge the fixed PSCs for 5 min, 500 × *g* at 4°C and discard the supernatant.
4. *Extraction of PSC nuclei and chromatin*. Resuspend pelleted PSCs in ice-cold nuclei isolation buffer to a final concentration of 10^6 cells/ml and incubate on ice for 10 min. Pellet nuclei by centrifugation at 1,000 × *g* and 4°C for 5 min. Resuspend nuclei to a final concentration of 10^7 nuclei/ml in ice-cold nuclei lysis buffer. Separate fixed nuclei into 100 μl aliquots in prechilled 1.5-ml microcentrifuge tubes. Chromatin can be used immediately or stored at −80°C for up to 6 months.

3.2. Immunoprecipitation of Protein–DNA Complexes

1. Transfer 50 μl of Protein A Dynabeads slurry to a 1.5-ml microcentrifuge tube for each immunoprecipitation (IP) and IgG control IP. Add 1 ml of IP dilution buffer and invert tubes to mix.
2. Collect Protein A Dynabeads using the magnetic stand and remove liquid by aspiration.
3. Resuspend Protein A Dynabeads in 100 μl of IP dilution buffer and add up to 10 μg of ChIP grade antibody to the beads. For the IgG control IP, use an equivalent amount (up to 10 μg) of IgG rabbit serum. Form covalent antibody/Protein A complexes by rotating for 1 h at 4°C.
4. Collect Dynabeads and remove liquid by aspiration as above and resuspend them in 500 μl of ice-cold IP dilution buffer. Collect Dynabeads using the magnetic stand and remove IP dilution buffer by aspiration.
5. Add 500 μl of ice-cold IP dilution buffer containing 5 mg/ml of bovine serum albumin (BSA) and rotate for 1 h at 4°C to block nonspecific protein binding to the antibody Protein A/Dynabeads. Collect Dynabeads and remove liquid by aspiration.
6. Resuspend Dynabeads in 500 μl of ice-cold IP dilution buffer. Collect Dynabeads, remove liquid by suction. Resuspend the beads for ChIPs and IgG control in 100 μl of IP dilution buffer and keep on ice while preparing chromatin (see below).

3.3. Fragmentation of Chromatin for ChIPs

1. Thaw aliquots of chromatin (see step 4, Subheading 3.1), add 200 μl of ice-cold IP dilution buffer and mix gently by pipetting on ice.
2. Shear chromatin to the desired length by sonication. See Fig. 1 and Note 4.
3. Centrifuge sonicated chromatin at 10,000 × *g* and 4°C to remove insoluble material and cellular debris. Transfer the supernatant to a fresh prechilled 1.5-ml microcentrifuge tube. Save one of the sheared samples on ice to serve as "Input" for downstream processing and analysis.
4. Preclear fragmented chromatin to be used for IP reactions by adding 700 μl of fresh ice-cold IP dilution buffer and 50 μl of fresh (nonconjugated) Dynabeads; rotate for 1 h at 4°C.
5. Collect the conjugated antibody/Protein A Dynabeads using the magnetic stand and remove the IP dilution buffer by aspiration.
6. Collect Dynabeads used for preclearing by the magnetic stand and transfer the 1 ml of solution containing precleared chromatin to the tube with the conjugated antibody/Protein A Dynabeads.
7. Mix chromatin and conjugated antibody/Protein A Dynabeads by rotation overnight at 4°C.

3.4. Isolation of Immunoprecipitated Chromatin

1. In order to reduce the background for downstream analysis, it is essential to sequentially wash the immunoprecipitated material bound to the conjugated Dynabeads with buffers containing increasing salt concentrations. First, collect

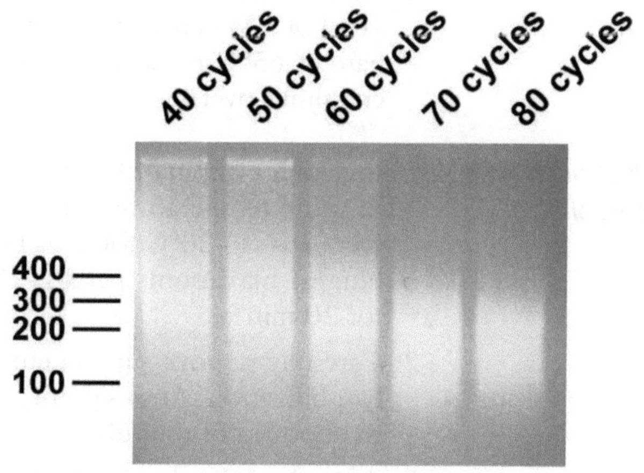

Fig. 1. Optimization of sonication conditions by Diagenode Bioruptor. Chromatin from 10^6 ES03 human embryonic stem cells sonicated in a Diagenode Bioruptor for the number of cycles indicated. Sonicated chromatin was treated with 20 μg of RNaseA for 30 min, the DNA was purified by phenol/chloroform extraction and quantified by PicoGreen. Smearing was visualized by fractionating 0.5 μg of DNA on a 2% TAE agarose gel.

Dynabeads and wash three times with 500 μl of Wash Buffer A; rotate 5 min at room temperature each time; use magnetic stand to collect beads and remove liquid by aspiration.

2. Wash three times with 500 μl of wash buffer B; rotate 5 min at room temperature each time; use magnetic stand to collect beads and remove liquid by aspiration.
3. Wash three times with 500 μl of wash buffer C; rotate 5 min at room temperature each time; use magnetic stand to collect beads and remove liquid by aspiration.
4. Wash once with 500 μl of TE buffer; rotate 5 min at room temperature, pellet beads gently by centrifugation at 500 × *g* for 1 min at room temperature (the TE buffer decreases the magnetic efficiency). Remove TE buffer by aspiration; Dynabeads are now ready for elution.
5. Elute the protein-associated chromatin by resuspending Dynabeads in 100 μl of elution buffer. Heat the suspension of Dynabeads to 65°C for 15 min. While eluting, vortex tubes periodically to keep the Dynabeads in suspension. After elution, pellet Dynabeads by centrifugation for 1 min at 500 × *g*. Carefully transfer supernatant to a new microcentrifuge tube to avoid contamination with Dynabeads. Repeat elution as described using a fresh 100 μl of elution buffer.
6. Combine the two elutions, and add 180 μl of TE buffer. For input samples, add 80 μl of TE for a final volume of 380 μl.
7. Digest any residual RNA in Input, ChIPed, and IgG control samples using 5 μl of 10 mg/ml RNase A, and incubating samples at 37°C for 10 min.
8. Reverse the protein–DNA crosslinks in all samples by adding 10 μl of 5 M NaCl (final concentration of 125 mM) and incubate at 65°C for at least 4 h. We recommend reversing the crosslinks overnight to insure the reaction is complete.

3.5. Purification of Immunoprecipitated DNA

1. Precipitate Input, ChIPed, and IgG control DNA with 1 ml of room temperature 100% ethanol, invert tubes to mix, and incubate at −20°C for 1 h. Pellet precipitated DNA by spinning at maximum speed (~15,000 × *g*) in a microcentrifuge for 20 min.
2. Carefully remove supernatant and dry the pellet by air or by lyophilization. At this point, a small pellet should be visible in the bottom of the tube.
3. After the pellet has been dried, resuspend Input and ChIPed DNA in 135 μl of room temperature TE buffer by pipetting up and down.

4. Remove proteins by digesting with 5 μl of 10 mg/ml Proteinase K in 15 μl of 10× Proteinase K buffer, mix by vortexing, and incubate for 2 h at 37°C.
5. Extract DNA with 380 μl of phenol/chloroform (1:1), mix thoroughly by shaking vigorously for 15 s. Separate the phases by centrifugation at maximum speed in a microcentrifuge for 10 min at room temperature.
6. Carefully transfer the upper aqueous phase by pipetting to a new 1.5-ml microcentrifuge tube and add 4 μl of 10 μg/ml glycogen as a carrier for subsequent precipitation.
7. Precipitate Input, ChIPed, and IgG control DNA with 1 ml of 100% ethanol, mix thoroughly and incubate at -20°C overnight.
8. Pellet ChIPed DNA by centrifugation at maximum speed in a microcentrifuge for 30 min at 4°C. At this point, a small clear pellet should be visible. Dry pellet as described in step 2 above, and resuspend it in 30 μl of deionized water. Samples can be stored indefinitely at –20°C.

3.6. Quantification of ChIPed DNA

1. The use of absorbance-based methods should be avoided since glycogen absorbs at 260 and 280 nm and, therefore, results in false quantification of the precipitated ChIPed DNA.
2. Make a working solution of PicoGreen quantification solution by diluting PicoGreen HS reagent 200-fold in dsDNA HS buffer. Enough reagent should be diluted so that all samples and points in the standard curve can be measured in duplicate. For example, for 20 assays dilute 20 μl of PicoGreen HS reagent in 4 ml of dsDNA HS buffer.
3. Prepare the low DNA concentration standards as described in Note 5.
4. Transfer 200 μl of the working solution into each well of an opaque 96-well-flat bottom plate.
5. Add 10 μl of each quantification standard to a separate well.
6. Quantify unknown samples by adding 2–4 μl of Input, or immunoprecipitated DNA to duplicate wells.
7. Mix all samples in the microplate by pipetting, and measure fluorescence in a fluorometric microplate reader.
8. Plot the fluorescence values vs. DNA concentration for standards and fit a line to the data points by linear regression. Using the slope and X-intercept of the line, determine the concentration of DNA in your unknown samples (Fig. 2). Recovery will depend on the histone modification or chromatin-associated protein being ChIPed. For IgG control

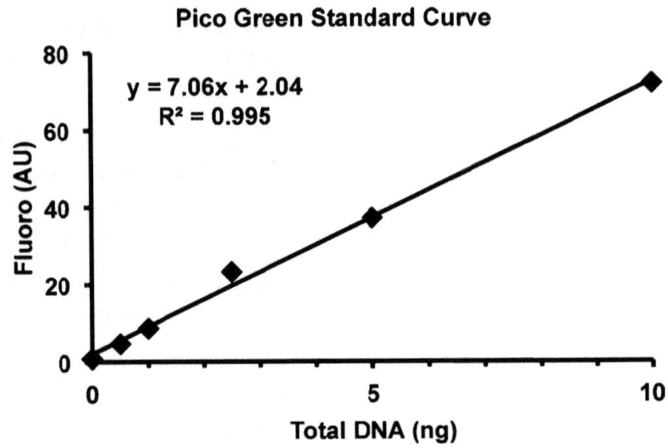

Fig. 2. PicoGreen quantification of DNA from chromatin immunoprecipitation experiments. Fluorescence is measured for a set of low concentration dsDNA standards (0, 0.5, 1, 2.5, 5, and 10 ng/μl), and plotted against the total amount of DNA in each well. A line is fitted to the fluorescence signal using linear regression analysis. Total DNA in each well for the Input and ChIPed samples can then be determined by inserting their fluorescence for *y*, and solving for *x*. Sample concentration is then computed by dividing the total amount of DNA in a well by the total volume of sample.

samples, the fluorescence measured should be approximately equal to background. Too much signal in IgG precipitations may indicate unacceptable background signal in immunoprecipitations. This problem can be solved by increasing wash time, or by decreasing the amount of Protein A Dynabeads used in immunoprecipitations.

3.7. Q-PCR Analysis of ChIPed DNA

1. To determine the success of the ChIP by Q-PCR, dilute 1 ng of Input and ChIPed DNA into 200 μl of ddH_2O (final concentration of 0.005 ng/μl). IgG samples may not be able to be diluted because their concentration is so low.
2. Prepare duplicate reactions for every sample with each primer set in the following manner: 10 μl of diluted DNA to 2.5 μl of oligo mix, and 12.5 μl of 2× SYBR green master mix. For information on the design and optimization of SYBR green quantitative PCR primer sets, see Note 6.
3. Load reactions onto the optical plate and run in the real-time PCR machine using the following program: One cycle of 95°C for 5 min, followed by 40 cycles of 95°C for 30 s and 60°C for 30 s.
4. Once the run is complete, use the real-time PCR software to select a threshold of signal that is within the linear range of all samples. The software will then report a cycle threshold (Ct) value for each sample. This value corresponds to the cycle at which the amplification of the sample reaches your selected threshold (see Fig. 3a).

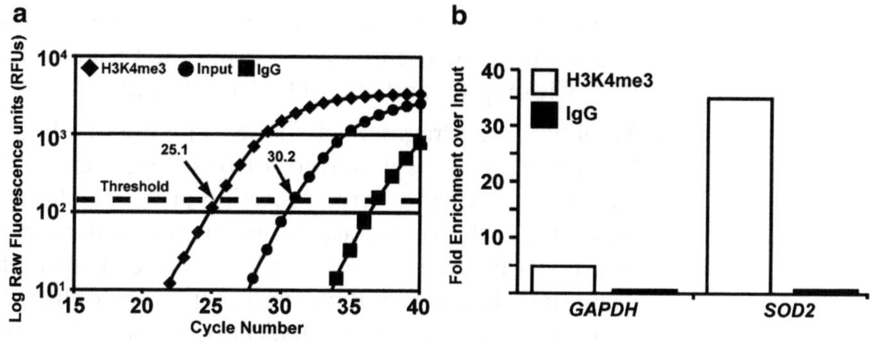

Fig. 3. Quantitative PCR analysis of H3K4me3 at a control gene in ES03 human embryonic stem cells. (**a**) Input (*filled circle*) and H3K4me3 immunoprecipitated (*filled diamond*) DNA from HES3 (ES03) were analyzed by SYBR Green real-time PCR with primers specific to the first exon of *SOD2*. A threshold within the linear range of amplification was selected (*dashed line*) and the cycle thresholds for each Input (30.2) and H3K4me3 (25.2) was determined. (**b**) Fold enrichment of H3K4me3 histones at *GAPDH* and *SOD2* in HES3 cells. Values were calculated using the following formula: $FE = 2^{(\text{Input Ct} - \text{IP Ct})}$.

5. Calculate the fold enrichment for each sample over input using the following formula: $2^{(\text{Ct Input} - \text{Ct IP})}$ (see Fig. 3b).
6. Enrichment values will differ depending on cell type, the histone modification/protein being immunoprecipitated, and amplicon location. During initial studies, we recommend to include a negative control amplicon where the histone modification of interest should not be present. This region should demonstrate no enrichment in Q-PCR experiments.

3.8. Library Construction for Illumina Sequencing

1. Transfer 5 ng of Input and ChIPed DNA to a fresh 1.5-ml microcentrifuge tube. Adjust the volume of the sample to 34 μl total using deionized water.
2. DNA fragments should then be end-repaired to create blunt ends using the End-It Kit (EpiCentre Biotechnology) following the manufacturer's protocol.
3. Stop the end-repair reaction by purifying DNA fragments with the Mini-Elute PCR Cleanup kit (Qiagen) and elute bound DNA using 32 μl of Qiagen EB buffer.
4. Create A-overhangs on the blunted Input and ChIPed DNA fragments using Klenow (3′–5′ exo) Fragment (New England Biolabs) in NEB Buffer 2 supplemented with 0.2 mM ATP for 30 min at 37°C. Cleanup and purify fragments using the Qiagen Mini-Elute PCR Cleanup kit as described above, except elute using 18.8 μl of EB.
5. Add Illumina sequencing adapters by mixing 40 pmol of Illumina Adapter Mix with the ChIPed DNA fragments and ligating them together using T4 DNA Ligase. We recommend using a rapid DNA ligase for this step and shortening

ligation time to 15 min at room temperature (Enzymatics). Again, purify ligated fragments using the Qiagen Mini-Elute PCR Cleanup kit and elute DNA in 39.2 μl.

6. Amplify ligated fragments by PCR. The use of a high-fidelity DNA polymerase is preferred for this step, such as *Pfx* polymerase (Invitrogen). Typical reaction protocols should be followed for each polymerase; however, the final concentration of primers in the reaction should be 0.5 μM. Amplify DNA using the following PCR program: One cycle of hot start activation (if required by your DNA polymerase with a 30 s, 98°C pre-denaturation). Proceed with 18 cycles of the following steps: 98°C for 10 s, 65°C for 30 s, 72°C for 30 s. Amplification is concluded with a final extension of 10 min at 72°C.

7. The amplification of all libraries is then visualized by fractionating 10 μl of the PCRs by agarose electrophoresis (1.5% TAE). The DNA smear should be clearly visible for each sample on the gel as shown in Fig. 4. The size range of the smear should be shifted by ~100 bp higher compared to the previous sonication optimization experiments (Fig. 1). The increased size is consistent with the addition of the Illumina sequencing adapters.

8. The remainder of each library PCR is then fractionated on a separate 1.5% agarose TAE gel to avoid cross-contamination. Lanes should be left empty between the DNA ladder and

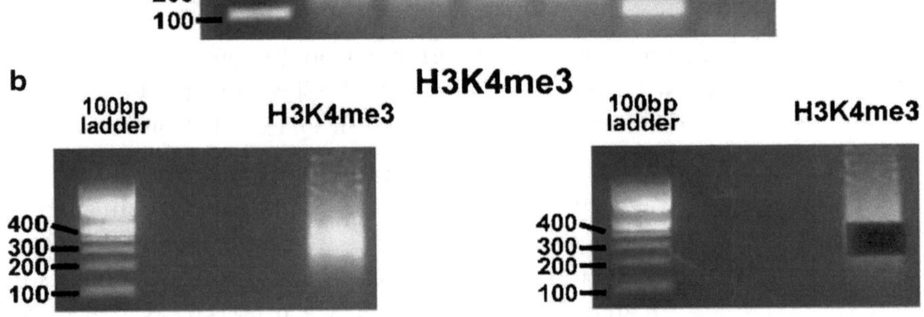

Fig. 4. Creation of ChIP-Seq libraries for Illumina sequencing. (**a**) PCR amplification of ChIP-Seq libraries, mock adapter ligation, and no template control. Notice the size distribution (smear) in all four libraries, and the presence of adapter dimer (~100 bp) in the no fragment Mock ligation control. (**b**) Each library is run on an independent 1.5% Agarose TAE gel for excision and gel purification to avoid cross-contamination. The gel piece that corresponds with fragments between 200 and 500 bp in size was then excised and gel purified. Care should be taken not to cut a gel piece that would capture the adapter dimer.

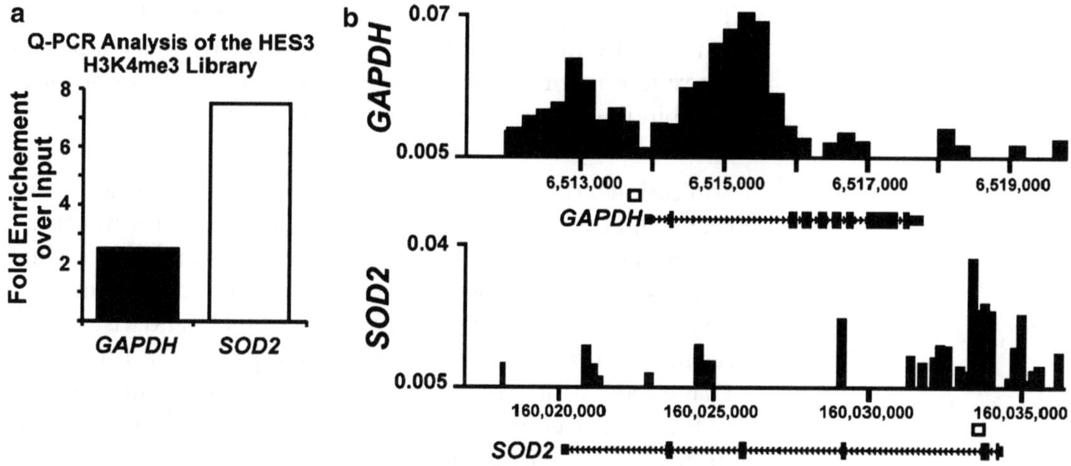

Fig. 5. Analysis of ChIP-seq libraries by Q-PCR and Illumina sequencing. (**a**) Fold amplification of *GAPDH* and *SOD2* in a H3K4me3 library over its matched Input library in HES3 (ES03). Values were calculated using the following formula: FE = $2^{(\text{Input Ct} - \text{IP Ct})}$. Note the decreased enrichment after library construction. (**b**) Close-up view of *GAPDH* and *SOD2* genes after Illumina sequencing of the H3K4me3 library. Gene structure was obtained from build 17 of the UCSC genome browser. *White boxes* indicate the location of the Q-PCR amplicons used in Figs. 3 and 5a.

each sample to avoid cross-contamination. The gel fragment corresponding to the desired range is then excised using a fresh razor blade for each sample to avoid contamination. For our experiments, we excise between 250 and 500 bp (see Fig. 3). This size is consistent with mono- and oligo-nucleosomes. DNA should then be isolated from gel fragments using the Qiaquick Gel extraction Kit (Qiagen).

9. Quantitate DNA using PicoGreen as described above (see Note 5).
10. Analyze library DNA using the same primers sets described in Subheading 3.5. The fold amplification of ChIP library over the input library will be less than the fold enrichment of the IP over input. However, trends in the relative amplification of separate amplicons should remain the same (see Fig. 5).

4. Notes

1. Cell lysis buffer should be prepared fresh daily. Alternatively, the buffer can be made with all components except NP-40, which can then be added just prior to use.
2. The addition of protease inhibitors is essential during experiments. Add a final concentration of 1 mg/ml aprotinin, 1 mg/ml leupeptin, 1 mg/ml pepstatin, and 1 mM PMSF to all buffers before use.

3. The use of magnetic beads is strongly recommended because of their ease of use and nonporous structure. This eliminates the requirement for preblocking and greatly reduces nonspecific binding during immunoprecipitations compared to agarose beads.
4. Sonication conditions must be empirically determined for each cell line, cell number, and sonicator. Conditions can be determined by sonicating fixed chromatin from your cell line using a constant power rating and various cycle numbers, reversing the crosslinks, purifying the DNA, and visualizing results by agarose electrophoresis. We prefer the use of the Diagenode Bioruptor because it is a closed system which reduces cross-contamination and sample loss during sonication, which is essential for working with small cell numbers.
5. PicoGreen is much more sensitive than A_{260} and is insensitive to protein contamination, ssDNA, glycogen, and nucleotides. Prepare dsDNA standards with the following concentrations of: 0, 0.05, 0.1, 0.25, 0.5, and 1.0 ng/μl. Set up a standard curve as instructed by the manufacturer's protocol. To quantitate samples, add between 2 and 4 μl of ChIPed DNA to each well. Determine DNA concentration as described in the manufacturer's protocol using a fluorometric microplate reader.
6. For quantitative real-time PCR analysis of ChIP experiments, we recommend the use of SYBR green-based assays. These assays are inexpensive because they utilize conventional oligos and are simple to optimize. Several computer-based algorithms can be used for designing SYBR green primers. Based on our experience, the following guidelines usually create successful SYBR green-based assays: an overall amplicon length between 55 and 70 bp with a melting temperature between 80 and 85°C; the individual oligos should have a T_m between 58 and 60°C; before primer sets can be used in quantitative studies, they should first be optimized following the procedures outlined in the *qPCR technical guide* from Sigma Life Sciences.
7. Human PSC lines should be cultured in medium and under conditions developed for that line. We used the hESC line ES03 for our experiments. These cells are routinely cultured in medium containing FBS on MEF feeder cells in organ-culture dishes. It is important to remove the feeder cells from the PSCs. Try to either differentially harvest the feeder cells and the PSCs using trypsin/EDTA solution or manually dissect the PSC colonies or bits of colonies from the culture and collect them in a tube, then dissociate them into single cells with trypsin/EDTA. After a single cell solution has been obtained, inactivate the trypsin by adding four volumes of medium containing FBS.

Acknowledgments

The authors wish to acknowledge the assistance of Martin Pera and Vicki Fox in the Eli and Edythe Broad Center for Regenerative Medicine and Stem Cell Research for their input on stem cell biology. We would also like to acknowledge the University of Southern California Epigenome Center for assistance in developing the library construction methods described in this chapter.

References

1. Rice JC, Allis CD. Histone methylation versus histone acetylation: new insights into epigenetic regulation. (2001) *Current opinion in cell biology.* **13**, 263–73.
2. Prigent C, Dimitrov S. (2003) Phosphorylation of serine 10 in histone H3, what for? *Journal of cell science.* **116**, 3677–85.
3. Gill G. (2004) SUMO and ubiquitin in the nucleus: different functions, similar mechanisms? *Genes & Development* **18**, 2046–59.
4. Surani MA, Hayashi K, Hajkova P. (2007) Genetic and epigenetic regulators of pluripotency. *Cell* **128**, 747–62.
5. Azuara V, Perry P, Sauer S, et al.(2006) Chromatin signatures of pluripotent cell lines. *Nat Cell Biol* **8**, 532–8.
6. Bibikova M, Laurent LC, Ren B, Loring JF, Fan JB.(2008) Unraveling epigenetic regulation in embryonic stem cells. *Cell Stem Cell* **2**,123–34.
7. Mikkelsen TS, Ku M, Jaffe DB, et al.(2007) Genome-wide maps of chromatin state in pluripotent and lineage-committed cells. *Nature* **448**, 553–60.
8. Mikkelsen TS, Hanna J, Zhang X, et al. (2008) Dissecting direct reprogramming through integrative genomic analysis. *Nature* **454**, 49–55.
9. Torres-Padilla ME, Parfitt DE, Kouzarides T, Zernicka-Goetz M. (2007) Histone arginine methylation regulates pluripotency in the early mouse embryo. *Nature* **445**, 214–8.
10. Barski A, Cuddapah S, Cui K, et al. (2007) High-resolution profiling of histone methylations in the human genome. *Cell* **129**, 823–37.
11. Orford K, Kharchenko P, Lai W, et al. (2008) Differential H3K4 methylation identifies developmentally poised hematopoietic genes. *Dev Cell* **14**, 798–809.

Chapter 20

Basic Approaches to Gene Expression Analysis of Stem Cells by Microarrays

Bernhard Schuldt, Qiong Lin, Franz-Josef Müller, and Jeanne Loring

Abstract

This chapter covers gene expression analysis by microarray to study and characterize stem cells. In a case-study scenario, we describe basic bioinformatic methodologies used to answer common questions in microarray experiments involving one or more stem cell populations. Service providers or departmental core labs usually carry out sample preparation, hybridization, and scanning of microarrays. Therefore, in this chapter, we focus on the state-of-the-art data analysis that avoids common pitfalls and introduces the reader to important controls that yield robust biologically relevant results. We describe evaluation of differentially expressed genes, clustering methods, gene-set enrichment analysis, and gene network discovery methods that can be used to formulate meaningful biological insights as well as suggest new wet lab experiments.

Key words: stem cells, microarray analysis, differential gene expression, gene set enrichment, protein–protein interaction network

1. Introduction

Pluripotent stem cells (PSCs) are one of the most widely studied mammalian cell types. Large, genome-wide datasets have been acquired in which the PSC transcriptome, genome, and epigenome have been extensively mapped with microarrays or whole-genome sequencing technologies. Genome-wide studies are attractive because they can be comprehensive and do not rely on investigator bias to drive the experimental design. It has become common practice to make microarray datasets publicly available via repositories such as NCBI GEO and ArrayExpress (1). These large datasets allow investigators to compare the activity of tens of

Philip H. Schwartz and Robin L. Wesselschmidt (eds.), *Human Pluripotent Stem Cells: Methods and Protocols*, Methods in Molecular Biology, vol. 767, DOI 10.1007/978-1-61779-201-4_20, © Springer Science+Business Media, LLC 2011

thousands of potential candidate genes across the samples and potentially identify relevant effects that would have been missed otherwise. However, for wet-bench biologists, "Is gene X expressed in PSCs?" and "Is gene Y more expressed in this cell type versus this other cell type?" are still the most frequently asked questions in the context of genome-wide datasets.

Microarray analysis faces various challenges such as measurement noise, uncontrolled biological factors, and low number of replicates. In addition, when testing a large number of hypotheses, as in microarray experiments, it is inevitable that spurious patterns are detected by pure chance. Despite these problems, there is a growing consensus on how these tools can be used (2).

State-of-the-art microarray analysis uses sophisticated normalization and background correction schemes to filter and correct for multiple hypothesis testing. It is not possible to evaluate all the methods independently. This means, for the practitioner, that one needs to evaluate the whole process for a specific question. As an example: the number of significantly differentially expressed genes that can be "found" depends on the preprocessing method, filtering, and data transformations (3). The often underappreciated power of the technology lies in choosing algorithms and in parameter tuning, which come with the responsibility to make sure that the bioinformatic pipeline used in a particular study is statistically sound and reproducible. A recent study (4) has showed that, even when the microarray data is publicly available, independent statisticians could not replicate a large fraction of statistics-based findings. For this reason, all the implicit "decisions" (hidden in parameter settings and cut-off points) need to be documented. The best way to deal with the problem is to use software tools that allow one to save the analysis parameters and generate script-based solutions that can be shared with colleagues.

With all of the aforementioned in mind, we, in this chapter, explain how to perform the following common bioinformatic procedures:

1. Loading data and collecting important information.
2. Class discovery: Hierarchical Clustering (HC) (5) and Principal Component Analysis (PCA) (6) to discover the class composition of the samples and to look for outliers and mislabeled samples.
3. Creating a list of differentially expressed genes using the *t*-test: Using this well-known classical statistical method will aide, as a filtering step to exclude noninformative genes from the following, often manual, analysis: The researcher looks for predicted gene "hits" and performs literature searches for "interesting" differentially expressed "candidates." We explain how a "hit" set can be constructed that offers a good compromise between scientific rigor and a manageable number of "hits."

4. Linking data structures to biological knowledge: Gene-set enrichment (7) and the MATISSE (8) algorithm is employed to relate the data to predefined gene sets and interaction networks. We use this step to algorithmically detect relevant gene sets and subnetworks that put the results of the microarray experiment in biological context.

Since we are providing an introduction to microarray data analysis, we do not cover statistical and biological validation in detail. However, it is important obtain independent biological validation using confirmatory bioinformatics on an independent data-set and prove of the resulting hypothesis in the wet lab. Microarray analysis is best understood by working with data.

2. Materials

2.1. Software

We assume that you have access to a computer with Internet browser, text and spreadsheet editors. The process described here can be performed with most bioinformatics software. However, in this chapter, we use MEV 4.4 (9) and Expander5 (10).

1. MEV, short for the TM4 MultiExperiment Viewer (http://www.tm4.org). We chose MEV for most of this tutorial since it is open source and available for all major operating systems. It is a stand-alone package for data analysis and visualization. A graphical user interface makes all options accessible.
2. We use Expander (EXpression Analyzer and DisplayER), http://acgt.cs.tau.ac.il/expander/, to access the MATISSE network analysis algorithm. Expander is provided free for academic use. It is possible to use it as an alternative to MEV for most of the basic analyses we describe. Its main strength lies in algorithms that link gene expression data to independently derived biological information in the large organism-specific databases provided by the developers.

2.2. Getting Data

Most bioinformatics departments and microarray cores have an established preprocessing pipeline by which they carry out quality control and data normalization. If this is not the case, you should familiarize yourself with the concepts of data extraction, background correction, transformation, and normalization of microarray data. The simplest approach to this is to follow a step-by-step tutorial for Bioconductor (11, 12) or use a predefined preprocessing workflow in GenePattern. For the Affymetrix platform, the raw data format is the "CEL file," which can be preprocessed by many software packages. The most widely used probe set summarization algorithms are MAS5 and RMA (13). For the Illumina platform, as used in this tutorial (see Note 1), we usually employ a very simple preprocessing procedure as suggested by Barnes (14) or follow the lumi procedure provided by the lumi package (12) in Bioconductor.

Regardless of which procedure one follows, it is most important to meticulously report all of the steps and all of the parameters of the preprocessing procedure, including the version of the applied software. For the methods applied in this chapter, a logarithmic or logarithmic-like transformation is used to "variance-stabilize" gene expression data (15). This has to be documented, since not all algorithms expect a logarithmic scale.

2.3. Tab-Delimited File Format

We suggest using a tab-delimited text file format as the basic format for all analysis steps. Microarray data is usually structured as a table in which each column represents a single array, and each row represents a single probe (or in the case of Affymetrix chips, a set of probes). The basic structure of these files is summarized in Table 1. In day-to-day practice, formatting and adjusting this table to the sometimes quirky file format requirements of diverse software packages takes a considerable amount of time and effort.

The file-format is called Tab Delimited Multiple Sample file (TDMS) in MEV and tab-delimited format in Expander (5).

We recommend that you always store the unique probe ID in the first column. Mapping of probes to genes can be done later and rather easily by most software based on the unique probe identifiers. Some software packages add additional descriptions and annotations to the file. If you encounter any file format problems, even within the same software, we recommend that you remove any additional information with the spreadsheet software and save a copy of the file as first troubleshooting step.

Table 1
Structure of tab-delimited data file

Header

Probes	Description	Array 1/sample 1	Array 2/sample 2	Array 3/sample 3
Probe 1	Gene A	Intensity value	Intensity value	Intensity value
Probe 2	Gene B	Intensity value	Intensity value	Intensity value
...	...	...	...	...

A header usually contains program-specific information, such as version, source of data, and number of rows and columns. Some formats use special characters or a fixed number of lines to identify the header. The next line contains the Column Identifiers. The following lines contain gene annotation and data. Sometimes more than one column or row are used for sample and probe annotation; this needs to be confirmed when using the same file with different software. Most bioinformatics software display a similar table when loading data

Example files in various formats can be downloaded at http://www.stemcellmatrix.org/ (16).

2.4. Gene Annotation

When connected to the Internet, MEV can automatically download annotation files for many popular chip platforms. Expander requires a conversion file to map probe IDs to NCBI Entrez gene IDs. This is a simple tab-delimited file containing probe IDs in the first column and Entrez gene IDs in the second.

If your array platform is not properly supported, you should provide annotation files as described in the MEV manual.

2.5. Gene Sets for GSEA

In addition to the gene expression data, we need a gene-set file for GSEA. These can be, for example, downloaded from the Molecular Signature DataBase (MolSigDB) (7) at the Broad Institute's Web site (http://www.broadinstitute.org/gsea/). In our day-to-day practice, we have found the C2 database, including more than 3,500 expertly curated gene sets, most useful. Most of these gene sets are attached to biological themes derived from peer-reviewed studies. If such a gene set is detected as being significantly up-regulated in our analysis, the background information contained in the associated manuscripts has often proven to be extremely helpful in discovering unexpected, but significant biological themes. Gene or probe annotation has to be provided by the user to translate the unique probe identifiers to ENTREZ or Gene Symbols.

3. Methods

Subheadings 3.1–3.5 are based on MEV 4.4. However, methods can be adapted to alternative software. Words set in *italics* refer to menus, buttons or options provided by the software and arrows "→" indicate submenus.

3.1. Loading the Data

1. Go to *Load Data→File*, *select expression data file*, and load the annotation from the file or choose to download annotation from the Internet by using the organism and platform selection menus (see Note 1). The loaded data is represented as a heat map. *Make sure that the sample annotation has not shifted.*
2. Go to *Display→Gene/Row Labels* and check the gene annotation. The EntrezGeneID or Gene Symbol is required for gene set enrichment analysis (GSEA).
3. If the raw data is not on a logarithmic scale you should perform a log2 transformation. When in doubt, check the documentation of your preprocessing pipeline to see if a logarithmic or generalized logarithmic transformation has been applied.

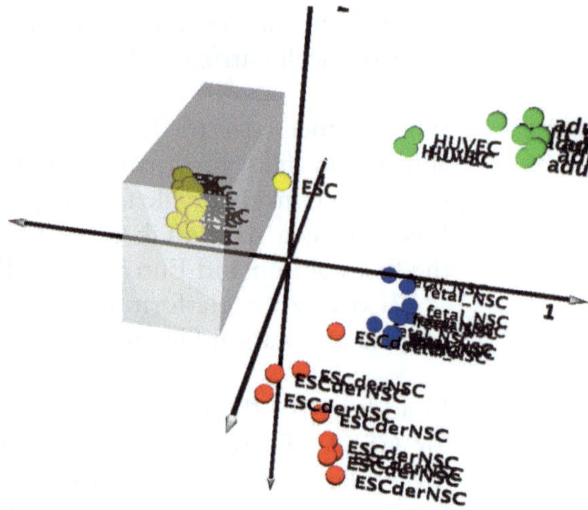

Fig. 1. Principal component analysis (PCA). This figure illustrates three-dimensional PCA. The *spheres* represent the microarray samples. The colors are determined according to the saved clusters from the hierarchical clustering and the results of clustering can be confirmed visually. The selection area (*in gray*) was used to create a new cluster and to remove outliers.

4. Go to *Display*→ *Set Color Scale* Limits, set the limits and the midpoint close to the values suggested by the software and press the *update limits* button. A small percentage of genes off scale is acceptable if it improves the display of values close to the mean.
5. Check the results in the *Original Data*/Expression Image folder on the left panel.
6. Normalization, that is centering the gene-expression to zero mean and dividing by standard deviation, helps to improve the display of the heat maps, but leads to worse results for many algorithms since some of the gene variance is lost. It is better to launch a new session with the data you would like to display. In that case go to Adjust Data→Gene/Row adjustment→Normalize gene row and set the color scale limits as described in step 4 before. You can store the images by choosing save image from the file menu (see Fig. 1).

3.2. Hierarchical Clustering

Hierarchical Clustering (5) is one of the most popular methods in gene expression analysis. Together with other methods, it can help to identify outliers and mislabeled experiments. In the context of stem cell research, it is often used to determine whether a given type of cell clusters with other stem cells, differentiated cells, or the feeder cells (see Note 2).

1. Go to *Clustering*→ *Hierarchical Clustering*. Check *Sample Tree* and choose the proper metric and linkage method. We prefer Pearson correlation and average linkage.

2. Check the results in the Analysis Results folder on the left panel. You will see the sample tree above the heat map.
3. Select clusters by right click in the subtree and choose *Store Cluster*. The clusters information will be available in the *Cluster Manager* section.

3.3. Principal Components Analysis

PCA is a data reduction technique that leads to an approximation of the data in lower dimensions. If the data contains 20,000 probes or probe sets for each sample, it allows interpret each sample as a point in a 20,000-dimension abstract space (see Note 3).

Since gene expression is, in most of the cases, highly correlated, for example many probes measure the same coregulated biological process; a large part of the information is redundant. It is therefore possible to reduce the 20,000 dimensions to a small set of often less than ten informative components, depending on experimental design. These components are sorted by the amount of variance in the data they explain. Tools for PCA usually report the percentage of variance each component explains and this helps to understand how many components are needed for further analysis. In exploratory analysis, the first three components are often sufficient, and we can use three-dimensional plots to learn more about the relation of different samples to each other.

1. Go to *Data Reduction → Principal Component Analysis.*
2. Select *Cluster Samples* and leave other parameters in default.
3. In the *Analysis Results* section, you can find *3D view* and *2D views.* To display the 3D view, right-click in *3D view* and *check Show spheres.* The spheres are colored according to stored clusters.
4. In addition, you can use *Selection area* to define new clusters based on the PCA result.
5. You can consult the *Eigenvalues* section for details on the mathematical results.

3.4. Differential Expression

We use the *t*-test to generate a list of genes that are differentially expressed between conditions (see Note 4).

1. From the *Statistics* menu, select *t-Test* and then select *Between subjects.*
2. Select two groups with the *Cluster Selection* or *Button Selection* menu.
3. We use *p*-values based on permutation and false-discovery control with an alpha of 0.01. *The proportion of the false-significant genes should not exceed 0.05* as starting values. As a result, MEV provides tables with significant genes.
4. It is important to keep in mind that we use the *t*-test as a filtering method to come up with a list of limited size. It is rather easy

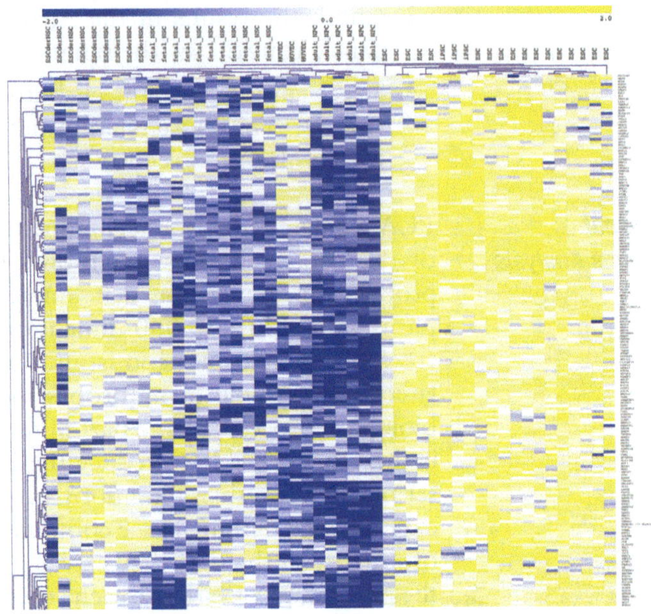

Fig. 2. Detail of a large heat map. A small set of significantly differentially expressed genes was selected with *t*-test as described in Subheading 3.4. We launched a new session with the selected genes. Gene/Sample hierarchal clustering was performed. The data was then normalized by Genes/Row. The corresponding color scheme was selected to obtain good visual display.

to come up with many hundreds of highly "significant" genes when different cell types are compared. If the main goal is to find a list with interesting genes, it is possible to repeat step 3 with new parameters (Fig. 2).

5. Alternatively, a *Volcano plot* (Fig. 3) is a useful tool to assess the number of genes that pass a combination of minimum fold-change and statistical-significance criteria. It can be used to define new subsets by right-clicking in the plot and choosing *use selection slider*. The selected genes can be used to launch a new session or to be stored as a gene cluster.
6. When an interesting list of genes has been identified, proceed as in 3.2 to generate a heat map with adapted colors.

3.5. Gene Set Enrichment Analysis

By using a predefined collection of gene sets instead of single genes, GSEA (7, 16, 17) is able to add more statistical power by "borrowing significance" from, for example, many members of a critical biological pathway (see Note 5). This basic concept has revolutionized microarray. GSEA provides a means to analyze novel data within the context of existing biological knowledge.

1. *Select Gene set enrichment analysis* from *Meta Analysis* menu.
2. Start by *Assign phenotype*. Typically, one would use one factor with two levels. Select the groups as was done with the *t*-test.

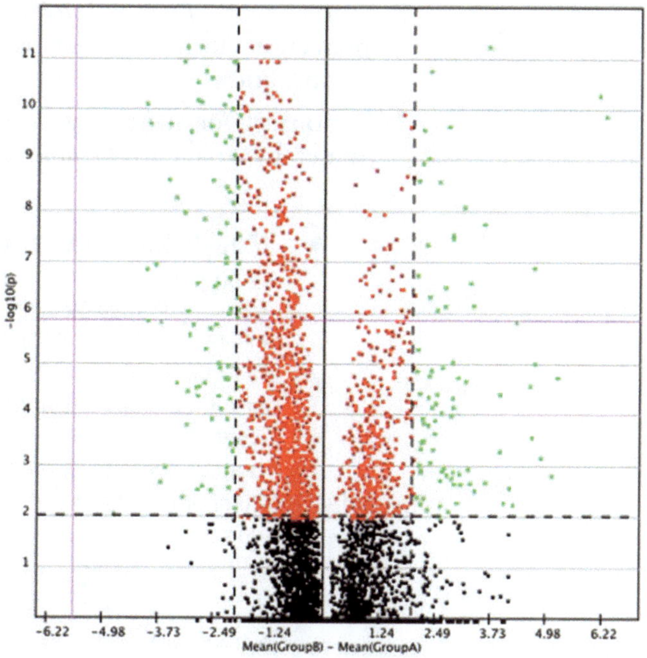

Fig. 3. Volcano plot. The volcano plot shows the fold change and p-value of each gene. Genes can be selected by selection sliders with defined criteria. In this case, genes with fold changes greater than 2 and –log(p) greater than 2 are selected and saved for further analysis. Since a base 10 logarithm is used, –log(p) scores of 2, 1.3, and 1 correspond to p values of 0.01, 0.05, and 0.1, respectively.

3. *Upload gene set* step: Browse the local hard disk to load the gene set files.
4. The enriched gene sets can be reviewed in *Table views→ Significant Gene set*. Heat maps for each set allow visual inspection of the behavior of the genes in a specific biological process or Pathway.

3.6. MATISSE

MATISSE is a module finding algorithm that looks for connected subnetworks that show high similarity. In addition to finding genes that show high similarity, the algorithm recruits nodes, called back nodes, from the network information to form connected components, this helps reveal relations between genes and gene products that might not be visible on the transcriptional level. A statistical testing procedure is used to obtain significant modules.

1. Load the expression data: *Open File→ New Session→ Expression Data→ Tabular Data files*. Select Organism (human in our case). If the ID of probes is not Entrez gene ID, you have to provide a file that maps probe ID to EntrezID.

2. Load network: *Data→Load network*. Use the .sif file ("simple interaction format") that is included in the Expander 5 distribution.
3. Optionally, for increasing computational speed: filter your dataset by going to *Preprocessing→Filter Probes→Variation*

a

Module Detection Info:

Algorithm: Matisse
Maximal module size input: 100
Minimal module size input: 4
Beta parameter input: 0.95
Front node source input: Filtered data
Overall Average Homogeneity: 0.691
Overall Average Separation: -0.037
Number of modules: 5
Number of independent nodes: 0

ID	Name	Size	Homogeneity
1	Module 1	41	0.619
2	Module 2	86	0.702
3	Module 3	26	0.667
4	Module 4	37	0.766
5	Module 5	11	0.624

b

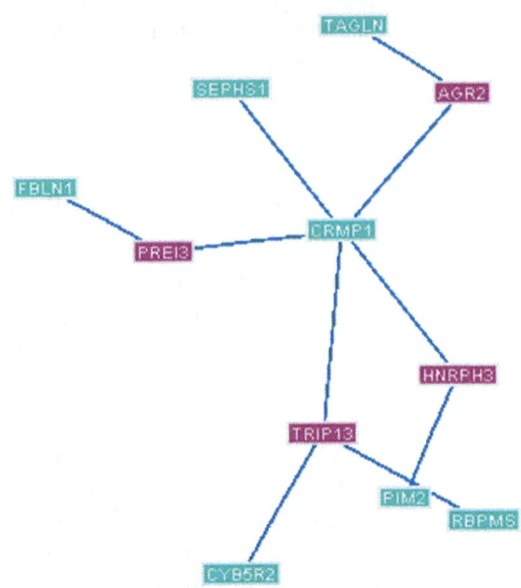

Fig. 4. Results of the MATISSE algorithm. (**a**) Matisse identified five Modules with sizes from 11 to 86 genes in the variance filtered subset of 1,000 genes. By selecting a module, Expander displays gene lists, heat maps, chromosomal positions, and network visualization for the selected module. (**b**) Interaction network for the smallest module. MATISSE can recruit additional nodes, called backnodes, to the module to form a connected network. These backnodes (AGR2, PREI3, TRIP13, HNRPH3) are colored in *pink*. Since they either were not in the dataset provided or were not found to be significant, they are possible targets for further research, perhaps they are the result of posttranscriptional interactions.

(see Note 6). Select, for example, 1,000 genes with the highest relative variation or select an interesting subset based on a *t*-test.

4. Then select *Grouping*→ *Network*→ *MATISSE*, select merge probes by gene IDs if the raw data is not gene-based. We choose module size between 4 and 100.
5. For each resulting module, the software provides the information about the gene list, expression matrix, a network of interactions, position on the chromosomes. This can be accessed on the right side panel. We show the results in Fig. 4.
6. The modules can be used for more sophisticated tests found in the *Group Analysis* Menu. Check Expander documentation for more information (see Notes 7 and 8).

4. Notes

1. Data file
   For this tutorial, we have selected samples from our study "Regulatory networks define phenotypic classes of human stem cell lines" (16). This data is available for download in different file formats on the Web site www.stemcellmatrix.org as described in the materials section.

   These samples represent adult human neural progenitors (HANSE cells), fetal neural stem cells (fNSCs) and human embryonic stem cells (hESCs) as well as induced pluripotent stem cells (iPSCs), hESC-derived NSCs, and human umbilical cord endothelial cells (HUVECs). The reader should be aware that such a rather coarse-grained selection of very different in vitro phenotypes will only allow rather general distinctions (e.g., that hESC and iPSC are very similar in the context of this analysis or that hESC-derived NSC are very different from HANSE cells or fNSCs). Such a small dataset will only allow the distinction of – figuratively speaking – apples from oranges, but may not be able to discern different kinds of apples from each other in the same analysis.
2. Clustering
   Compared to other clustering methods, Hierarchical Clustering is more sensitive to noise and might not lead to reproducible results (17, 18). Single linkage hierarchical clustering is often very sensitive to noise, e.g., technical variation from the hybridization process. If you do not get good results from such a simple analysis, additional algorithms such as k-means and Self-Organizing Maps might be worth trying. We had very good results by using NMF (19), which can be accessed

as part of the consensus clustering module in GenePattern (see also Note 6) and in future versions of MEV.

3. Principal Component Analysis
   PCA can be done for genes as well and it is sometimes useful to go beyond the first three components. Mathematically, PCA uses Singular Value Decomposition (SVD) on a transformed data matrix. It is possible to extract much more information using SVD and PCA than we did in this chapter. Check (20) for the application of SVD to relate gene expression to cell cycle that provides some informative plots.

4. Differential Expression
   The advantage of statistical tests over simple fold-change analysis is that, in addition to the difference of the means between the different groups, the variance is estimated to assess the significance of the results. For groups with an *N* less than five, the estimation of variance becomes unreliable; permutation-based tests become very grainy when the samples groups are too small. Introducing additional assumptions on the data structure can help to ameliorate the problem in some cases (see for example Chapter 7 in ref. 11), but in general it is better to use groups of sufficient size.

   Correction for multiple hypothesis testing (21) is extremely important for microarray experiments. A family-wise error control (FWER) can be alternatively used to the FDR approach. One common choice is Bonferroni correction with alpha of 0.05. The Bonferroni correction leads to smaller lists of differentially expressed genes and is sometimes required by reviewers.

   It can also be computed much faster since it is not limited to *p*-values coming from permutations.

5. Gene Set Enrichment Analysis
   The implementation of GSEA that we applied (17) is related to hypothesis testing and can have higher statistical power compared to *t*-test by looking for sets of genes that show a consistent behavior. On the original GSEA Web page (http://www.broadinstitute.org/gsea/), a stand-alone implementation of the original GSEA can be downloaded and be used as an alternative to the algorithm implemented in MEV.

6. MATISSE
   We have filtered our dataset for 1,000 high-variance genes to get quicker results. It is a good idea to increase the number of genes for an in-depth analysis. MATISSE is not deterministic; you might get slightly different results if you run the algorithm again and again. Test if increasing the number of runs leads to more stable results for your dataset.

   If you are interested in other uses of the MATISSE algorithm, for example the deMATISSE variant for using differential expression to find interesting subgraphs, you should use the original MATISSE software (8).

7. Alternative Software

   We suggest evaluating alternative tools such as GenePattern and Bioconductor.

   GenePattern (22) is a server-based interface that allows the user to access specific microarray analysis modules from R and other software via a graphical interface in their browser. You can find a Web-based version of GenePattern on http://genepattern.broadinstute.org/gp. We suggest the .gct file format for data exchange.

   Bioconductor (23) is an extension to the statistical programming language R. It provides a comprehensive and up-to-date collection of packages for processing, analyzing, and visualizing gene expression data. For many biologists there is a steep learning-curve associated with learning R, but there are good step-by-step guides available (for example (11)) that allow one to perform many tasks without a being an expert R user.

8. Validation

   This chapter is an introduction to exploratory data analysis for microarray data. The workflow presented here applies primarily to exploratory data analysis. The focus is on finding relations and gene "hits" in the data and methods to determine how to make these findings reproducible. Most researchers will not find the time for extensive formal training in bioinformatics and statistics. Many stem cell microarray experiments are relatively easy to perform, and the consequences of suboptimal design and analyses are less profound when compared to other fields, such as clinical trials for drug safety. The findings will usually be followed up with a new set of wet-lab experiments, and the results are presented to the research community for review. If your work relies on published results and the microarray data is available, for example on NCBI GEO, it is a good idea to first try to replicate the results with your own bioinformatics tools. We have found that discussing results and methods with colleagues and seeking the help of experts is the best way to learn microarray analysis skills.

Acknowledgments

Bernhard Schuldts work has been supported by Bayer Technology Services GmbH and the Deutsche Forschungsgemeinschaft through grant GSC 111. Qiong Lin acknowledges support by the DFG Priority Program 1356. Franz-Josef Müller is supported by an Else-Kröner Fresenius Memorial Fellowship.

References

1. Barrett T, Troup DB, Wilhite SE, Ledoux P, Rudnev D, Evangelista C, u. a. (2009) NCBI GEO: archive for high-throughput functional genomic data. *Nucl Acids Res.* **1**;37(suppl_1):D885–890.
2. Allison DB, Cui X, Page GP, Sabripour M. (2006) Microarray data analysis: from disarray to consolidation and consensus. *Nat. Rev. Genet.* 7:55–65.
3. Hackstadt AJ, Hess AM. (2009) Filtering for increased power for microarray data analysis. *BMC Bioinformatics.* **10**:11.
4. Ioannidis JPA, Allison DB, Ball CA, Coulibaly I, Cui X, Culhane AC, u. a. (2009) Repeatability of published microarray gene expression analyses. *Nat. Genet.* **41**:149–155.
5. Eisen MB, Spellman PT, Brown PO, Botstein D. (1998) Cluster analysis and display of genome-wide expression patterns. *Proc. Natl. Acad. Sci. USA.* **895**(25):14863–14868.
6. Pearson K, Pearson, K. On Lines and Planes of Closest Fit to Systems of Points in Space. (1901) *Philosophical Magazine.* **2**:559–572.
7. Subramanian A, Tamayo P, Mootha VK, Mukherjee S, Ebert BL, Gillette MA, u. a. (2005) Gene set enrichment analysis: A knowledge-based approach for interpreting genome-wide expression profiles. *Proceedings of the National Academy of Sciences of the United States of America.* **102**:15545–15550.
8. Ulitsky I, Shamir R. (2007) Identification of functional modules using network topology and high-throughput data. *BMC Syst Biol.* **1**:8.
9. Saeed AI, Sharov V, White J, Li J, Liang W, Bhagabati N, u. a. (2003) TM4: a free, open-source system for microarray data management and analysis. *BioTechniques.* **34**:374–378.
10. Shamir R, Maron-Katz A, Tanay A, Linhart C, Steinfeld I, Sharan R, u. a. (2005) EXPANDER - an integrative program suite for microarray data analysis. *BMC Bioinformatics.* **6**:232.
11. Hahne F, Huber W, Gentleman R. (2008) Bioconductor Case Studies. 1. Aufl. Springer, Berlin.
12. Du P, Kibbe WA, Lin SM. (2008) lumi: a pipeline for processing Illumina microarray. *Bioinformatics.* **24**:1547–1548.
13. Irizarry RA, Bolstad BM, Collin F, Cope LM, Hobbs B, Speed TP. (2003) Summaries of Affymetrix GeneChip probe level data. *Nucleic Acids Res.* **31**:e15.
14. Barnes M, Freudenberg J, Thompson S, Aronow B, Pavlidis P. (2005) Experimental comparison and cross-validation of the Affymetrix and Illumina gene expression analysis platforms. *Nucleic Acids Res.* **33**:5914–5923.
15. Bolstad B, Irizarry R, Astrand M, Speed T. (2003) A comparison of normalization methods for high density oligonucleotide array data based on variance and bias. *Bioinformatics.* **19**:185–193.
16. Müller F, Laurent LC, Kostka D, Ulitsky I, Williams R, Lu C, u. a. (2008) Regulatory networks define phenotypic classes of human stem cell lines. *Nature.* **455**:401–405.
17. Jiang Z, Gentleman R. (2007) Extensions to gene set enrichment. *Bioinformatics.* **23**:306–313.
18. Garge NR, Page GP, Sprague AP, Gorman BS, Allison DB. (2005) Reproducible clusters from microarray research: whither? *BMC Bioinformatics.* **15**;6 Suppl 2:S10.
19. Brunet JP, Tamayo P, Golub TR, Mesirov JP. (2004) Metagenes and molecular pattern discovery using matrix factorization. *Proc Nat Acad Sci USA.* **101**:4164.
20. Alter O, Brown PO, Botstein D. (2000) Singular value decomposition for genome-wide expression data processing and modeling. *Proc Nat Acad Sci USA.* **97**:10101–10106.
21. Benjamini Y, Hochberg Y. (1995) Controlling the false discovery rate: a practical and powerful approach to multiple testing. *Journal of the Royal Statistical Society. Series B (Methodological).* 289–300.
22. Reich M, Liefeld T, Gould J, Lerner J, Tamayo P, Mesirov JP. (2006) GenePattern 2.0. *Nat. Genet.* **38**:500–501.
23. Gentleman RC, Carey VJ, Bates DM, Bolstad B, Dettling M, Dudoit S, u. a. (2004) Bioconductor: open software development for computational biology and bioinformatics. *Genome Biol.* **5**:R80.

Chapter 21

Human Pluripotent Stem Cells: The Development of High-Content Screening Strategies

Sean P. Sherman, Jackelyn A. Alva, Kaushali Thakore-Shah, and April D. Pyle

Abstract

High-content screening (HCS) permits simultaneous observation and analysis of multiple cellular variables including cell morphology, survival, and differentiation in live cells at the single-cell level, at the level of the culture well, and across the entire culture. By combining high-throughput technologies such as robotics, chemical libraries, and automated high-resolution microscopy, scientists are able to evaluate a much broader array of experimental conditions than can be studied using conventional cell biological techniques that study fewer parameters at any one time. Thus, HCS assays provide a means to vastly improve our basic understanding of stem cell biology. We have developed a HCS assay that allows the study of the effects of hundreds of small molecules in parallel. The protocol described in this chapter was developed to assess the effects of small molecules on the survival, proliferation, and expression of pluripotent markers following single-cell dissociation of human embryonic stem cells, but can be applied to the study of other types of stem cells including induced pluripotent stem cells. A detailed protocol for the setup of HCS assays and the parameters used to identify chemical modifiers of survival in human pluripotent stem cells, as well as secondary assays used to validate the small-molecule "hits" obtained during the high-content screen, are described.

Key words: human embryonic stem cells, pluripotent stem cells, high-content screening, chemical genomics, cell fate, survival, differentiation, small molecules, HCS

1. Introduction

A key requirement for the use of human pluripotent stem cells (hPSCs) in therapeutic applications is the ability to direct their differentiation toward specific cellular fates. Developing differentiation protocols that reliably and reproducibly generate the desired differentiated cell type is a major goal of many research programs. We have developed a high-content screening (HCS)

Philip H. Schwartz and Robin L. Wesselschmidt (eds.), *Human Pluripotent Stem Cells: Methods and Protocols*, Methods in Molecular Biology, vol. 767, DOI 10.1007/978-1-61779-201-4_21, © Springer Science+Business Media, LLC 2011

method that allows one to examine the effects of many small molecules in parallel. Bringing together high-throughput technologies – robotics, large chemical libraries, and automated high-content screening – enables the evaluation of a much broader array of experimental conditions than can be reasonably studied using conventional cell biology techniques. The protocol described in this chapter was developed to assess the effects of small molecules on the survival, proliferation, and expression of pluripotent markers following single-cell dissociation of human embryonic stem cells (hESCs) (1). This is a key step in many experiments that use hPSCs, such as directed differentiation or gene targeting, and a major limitation to wide-spread use and development of these techniques in hPSCs. To improve the survival of individual hPSCs, we developed a HCS assay to identify small molecules that regulate survival and self-renewal in hESCs. The HCS assay described in this chapter can be used to identify small molecules capable of improving hPSC survival and self-renewal and can be adapted to study other aspects of PSC fate, including differentiation or reprogramming of somatic cells to PSCs. Since dissociated hPSCs exhibit poor survival, screening assays carried out as described here will have a low background because most of the cells will die and can be used to identify small molecules that result in higher cell survival. After the high-throughput primary screen is completed, secondary assays are performed using flow cytometry and conventional cell culture to confirm the results from the HCS screen. We describe in detail protocols that have worked well in our hands.

2. Materials

2.1. Human Pluripotent Stem Cell Culture Medium and Reagents

1. *Cells*: Three 6-well plates of hPSCs that are healthy and actively growing on feeder cells, expressing pluripotency markers (not differentiated) and ready for subculture.
2. *hPSC medium*: Dulbecco's Modified Eagle's Medium: Nutrient Mixture F-12 (DMEM/F-12) supplemented with 20% KnockOut Serum Replacement (KOSR, Invitrogen), 0.1 mM Nonessential Amino Acids (NEAA), 1 mM L-Glutamine + 0.1 mM 2-Mercaptoethanol [add 7 μl (14.3 M) to 5 ml L-Glutamine (200 mM)], and 4 ng/ml basic Fibroblast Growth Factor (bFGF, Invitrogen). Store in the dark at 4°C for up to 2 weeks.
3. *MEF medium*: Dulbecco's Modified Eagle's Medium (DMEM) supplemented with 10% fetal bovine serum (FBS), 0.1 mM NEAA, and 1 mM L-Glutamine. Store in the dark at 4°C for up to 2 weeks.
4. Collagenase Type IV (Invitrogen) as a 1 mg/ml solution in DMEM/F-12. Filter-sterilize before use.

5. Gelatin: 0.1% (w/v) gelatin from porcine skin, type A, in H_2O, stored at room temperature. Sterilize by autoclave prior to use.
6. DPBS: Dulbecco's phosphate-buffered saline 1×. Store at room temperature.

2.2. High-Content Screening Assay Reagents

1. Trypsin: 0.05% trypsin/EDTA solution.
2. Trypsin inhibitor: Soybean trypsin inhibitor, dissolved in DMEM/F-12 to a concentration of 1 mg/ml. Make a fresh solution each use and filter-sterilize before use.
3. Mesh filters: 40-μm nylon cell strainers (BD Biosciences, Franklin Lakes, NJ).
4. 384-well plates: Greiner micro-clear bottom 384 well plates (Cat. No. 781096, Greiner Bio-one, Germany) or Matrical clear bottom 384-well MatriPlate (Cat. No. MGB101-1-1, Matrical Bioscience, Spokane, WA).
5. High-throughput workstation: SAGIAN Core system (Beckman Coulter, Indianapolis, IN) including an ORCA rail-mounted robotic arm (Beckman Coulter), a Biomek FX Laboratory Automation Workstation (Beckman Coulter), and a 384-pin tool for liquid transfer (V&P Scientific, San Diego, CA). The Core system is integrated and operated using SAMI software (Beckman Coulter).
6. Screening libraries: such as the Prestwick Chemical Library (Prestwick Chemical, France), BIOMOL Bioactive Lipid Library (Enzo Life Sciences International, Plymouth Meeting, PA), or BIOMOL Kinase Inhibitor Library (Enzo) (see Note 1).

2.3. Antibodies and Reagents for High-Content Analysis of hPSC (see Note 2)

1. DPBS: Store at room temperature.
2. Paraformaldehyde: Prepare a 4% (w/v) solution fresh for each experiment by diluting 16% (w/v) paraformaldehyde (Electron Microscopy Sciences, Hatfield, PA) in PBS. Use at room temperature.
3. Permeabilization solution: 0.1% (v/v) Triton X-100 in DPBS.
4. Blocking solution: 10% (v/v) normal goat serum in DPBS.
5. Antibody dilution buffer: 1% (v/v) normal goat serum in DPBS.
6. Primary antibody: mouse anti-human Oct4 (Cat. No. sc-5279, Santa Cruz Biotechnology, Santa Cruz, CA) (see Note 3).
7. Secondary antibody: goat anti-mouse IgG conjugated to FITC (Pierce, Rockford, IL).
8. Nuclear stain: 0.05% (v/v) Hoechst 33342 in PBS.

2.4. High-Content Data Acquisition and Analysis Components Required (see Note 4)

1. Automated plate reader: Image-Xpress^{micro} (Molecular Devices, Sunnyvale, CA).
2. Image analysis software: MetaXpress (Molecular Devices).
3. Data analysis software: AcuityXpress (Molecular Devices).

2.5. Hit Confirmation by Flow Cytometry

1. Collagenase, Type IV, as a 1 mg/ml solution in DMEM/F-12. Filter-sterilize before use.
2. Trypsin: 0.05% trypsin-EDTA solution.
3. Trypsin inhibitor: Soybean trypsin inhibitor, dissolved in DMEM/F-12 to a concentration of 1 mg/ml. Make a fresh solution each time and filter-sterilize before use.
4. DPBS: Store at room temperature.
5. Flow Cytometry Buffer: 1% bovine serum albumin (BSA) in DPBS.
6. Primary antibody: mouse anti-SSEA-4 (Developmental Studies Hybridoma Bank, Iowa City, IA).
7. Secondary antibody: goat anti-mouse IgG, phycoerythrin conjugated (Cat. No. sc-3798, Santa Cruz Biotechnology).
8. Annexin V: FITC Apoptosis Detection Kit (Cat. No. 556547, BD Biosciences).
9. FITC BrdU Flow Kit (Cat. No. 559619, BD Biosciences).
10. Paraformaldehyde: Prepare a 4% (w/v) solution fresh for each experiment by diluting 16% (w/v) paraformaldehyde (Electron Microscopy Sciences) in PBS. Use at room temperature.

3. Methods

3.1. Pluripotent Stem Cell Preparation

Prior planning is required to obtain enough cells for a high-content screen. However, no special culture conditions are required to prepare cells for the HCS if the screen is to be performed on feeders. The cultures should be in log-phase of growth, healthy, and expressing all the markers of pluripotent stem cells prior to initiating the high-content screen. We typically use two to three 6-well plates of 80–90% confluent hESCs on MEFs, which yield six to seven 384-well screening plates. It is important to prepare replicates of each screening plate, as this minimizes false positives. In general, it is also a good idea to build the assay around at least two independent stem cell lines.

3.2. High-Content Screening Assay: Setting Up the Assay Plates

1. Coat 384-well screening plates with gelatin and plate MEFs, 40 μl/well, at a concentration of 7.5×10^4 cells/ml of MEF medium. Incubate overnight at 37°C (see Note 5).
2. Prepare 384-well plates for hESCs by removing MEF medium, washing with DPBS, and adding 40 μl hESC medium per well.
3. Remove medium from the two 6-well plates of hESC and wash once with DPBS.
4. Add 1 ml of trypsin/EDTA to each well of hESCs and incubate for 5 min at 37°C (see Note 6).
5. Briefly and gently pipette trypsin/EDTA up and down to help rinse cells off of the bottom of the plate.
6. Add a volume of 1 mg/ml trypsin inhibitor equal to the volume of trypsin added in step 4.
7. Gently pass the dissociated cells through a sterile 40-μm mesh filter into a sterile centrifuge tube and spin for 5 min at $200 \times g$.
8. Remove the supernatant and suspend hESCs in 5–10 ml hESC medium for counting. Add enough hESC medium to dilute hESCs to a final concentration of 5×10^5 cells/ml.
9. Add 10 μl (~5,000 cells) of hESCs to each well of the 384-well plate (total volume 50 μl/well).
10. Screening compounds are then added to the freshly prepared screening assay plates at a concentration of 10 μM using an automated robotics system.
11. Incubate hESCs with the compounds for 4 days at 37°C and assay as described below at the end of the 4-day incubation period (see Note 7).

3.3. High-Content Screening Assay: Immunostaining for Pluripotency Factors

1. Remove medium from 384-well plates and wash cells once with 50 μl/well DPBS.
2. Fix cells by adding 30 μl/well 4% paraformaldehyde and incubate for 30 min at room temperature.
3. Remove paraformaldehyde from cells. Permeabilize cells by adding 30 μl/well 0.1% Triton X-100 for 5 min at room temperature.
4. Remove permeabilization solution and wash cells with 50 μl/well DPBS.
5. Block nonspecific binding by adding 30 μl/well 10% goat serum in DPBS. Incubate for at least 1 h at room temperature.
6. Remove blocking solution. Add 30 μl/well primary antibody (Oct4) diluted 1:100 in DPBS containing 1% goat serum. Incubate for 2 h at room temperature.
7. Remove primary antibody and wash the cells one time with 50 μl/well DPBS.

8. Add 30 μl/well secondary antibody (anti-mouse IgG-FITC) diluted 1:500 in DPBS. Incubate for 2 h at room temperature in a drawer or other location away from light.
9. Remove secondary antibody and wash cells with 50 μl/well DPBS.
10. Add 50 μl/well Hoescht 33342 (Invitrogen) diluted 1:2,000 in DPBS. Wrap the plates in foil to protect from light and store at 4°C. Allow plates to warm to room temperature before reading to prevent condensation. For optimal results, the plates should be read within 1–2 days.

3.4. High-Content Assay: Data Acquisition and Analysis

3.4.1. Data Acquisition

1. We use MetaXpress software for data acquisition. Initial program setup should be performed with the assistance of hardware technicians or screening center administrators. Parameters that need to be determined for your particular HCS include the following: (1) plate type, (2) objectives to use, and (3) wavelengths to be acquired (see Note 8). In MetaXpress these settings are accessed by selecting "Load Settings" from the "Plate acquisition and control" window, under the *Screening* menu.
2. Clean plates before inserting by wiping the bottom of the plate with ethanol and drying.
3. Remove the lid from the 384-well plate and insert into the ImageXpress. The loading stage is accessed by pressing "Stage load and eject" within the MetaXpress software.
4. Select a well with the brightest expected signal to calibrate exposure times. Move the camera to this well by selecting it from the "Well to visit" tab in the "Plate acquisition setup" window.
5. From the "Plate acquisition setup" window select the W1 tab, which will be labeled with the first fluorophore to be acquired, Hoechst in the assay described here (see Note 9).
6. On the "Plate acquisition and control" window, press "Autofocus." This will focus on detectable cells and an image will appear showing a current snapshot. Focus can be tuned using the adjustment buttons in the "Plate acquisition and control" window.
7. In the W1 tab of the "Plate acquisition setup" window press the "Auto-expose" button. This will attempt to determine an optimal exposure time, and an image will appear showing a snapshot with your new focus and exposure settings. Exposure time can also be adjusted manually in this window.
8. Repeat steps 5–7 with wavelength W2 (FITC in this assay) to establish focusing and exposure time for the second fluorophore. If your screen uses more wavelengths, they should each be set up in this fashion.

9. In the "Plate acquisition setup" window return to the "well to visit" tab. Select all the wells you wish to acquire images from. Wells can be selected or deselected individually or by row or column. To collect multiple images from each well, check the "multiple sites" box and specify which (or all) regions of the well to acquire.
10. In the "Plate acquisition and setup" window, assign a unique name to your experiment and press "save settings" to save the focusing and exposure settings set.
11. To acquire data, press the "Acquire" button on the "Plate acquisition and control" window. For experiments spanning multiple plates you can swap plates and acquire using the same settings.
12. After acquiring images from each well, the number of total cells and pluripotent stem cells can be counted using the multiwavelength cell-scoring module in MetaXpress. The specific parameters will need to be adjusted for individual assays; however, the following settings were used in our published study: Total cells were counted by indentifying cells in the Hoechst channel with a minimum width of 6 μm (5 pixels) and a maximum width of 29 μm (22 pixels). Pluripotent cells were counted by identifying previously scored cells that were stained positively for Oct4 (FITC channel) with a minimum width of 10 μm (8 pixels) and a maximum width of 23 μm (18 pixels).

3.4.2. Data Analysis

1. Calculate the Z-factor score for each experiment (see Note 10). We calculated Z-factor scores for our screen using previously published methods (2, 3). On average, we obtained a Z-factor score of 0.5, which represents a high-quality screening assay. It is important that replicates of screening plates be built into the analysis, as this will also minimize false positives. In general, it is also a good idea to build in at least two stem cell lines. In our experiments we repeated and obtained similar Z-factor score and hits in both H9 and HSF1 hESCs.
2. The MetaXpress image data can then be opened and analyzed using an Acuity Express analysis package or similar HCS analysis software.
3. In order to compare data from multiple screening plates and/or assays, the data must be normalized. This is accomplished by a series of calculations known as centering and scaling. For each plate to be normalized, the mean and standard deviation of fluorescence intensity is calculated over all wells. The resulting mean is then subtracted from each well, resulting in a

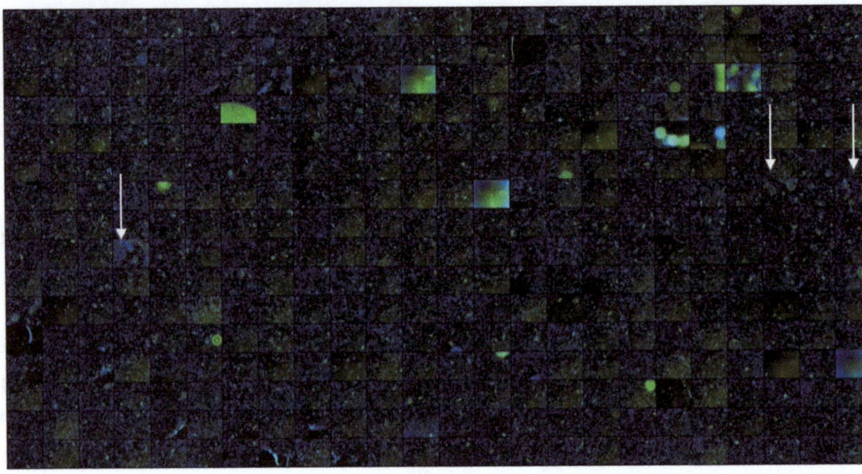

Fig. 1. Representative image of a 384-well hESC screening plate. In this HCS assay, MEFs and hESCs were plated as described in the Methods. *Arrows* represent wells with increased OCT4-positive hESCs that represent potential small-molecule "hits" that increase stem cell numbers in this HCS. Wells with *bright green* are typically a result of small molecules that promote cell death by the end of the 4-day assay (often resulting in autofluorescence). The positive "hits" (ex. *Arrows*) would need to be verified in secondary assays as described in section.

plate with an overall mean of zero. The value in each well is then divided by the standard deviation of the plate, resulting in scaled data that can be compared across multiple plates and/or experimental runs. This calculation is summarized by the following equation:

$$S_n = (x_n - \overline{x}) / \sigma$$

where x_n = fluorescence intensity in well n, $\overline{x}$ = mean fluorescence intensity of all wells on the plate, σ = standard deviation of all wells on the plate, Sn = centered and scaled fluorescence value in well n.

4. Compounds/wells that resulted in OCT4-positive hESCs that were at least 3 standard deviations away from the average were considered hits and are examined in secondary assays as described below (Fig. 1).
5. Once the data is scaled, the next approach is to develop heat maps to determine potential hits (see Note 11 and Fig. 2). Heat maps can be generated using the Euclidean clustering function in the Acuity software.
6. Clustering of compounds into groups can also be performed in the Acuity software program.
7. Small-molecule compounds that were considered hits were also examined for more chemical information using PubChem and SciFinder Scholar (see Note 12).

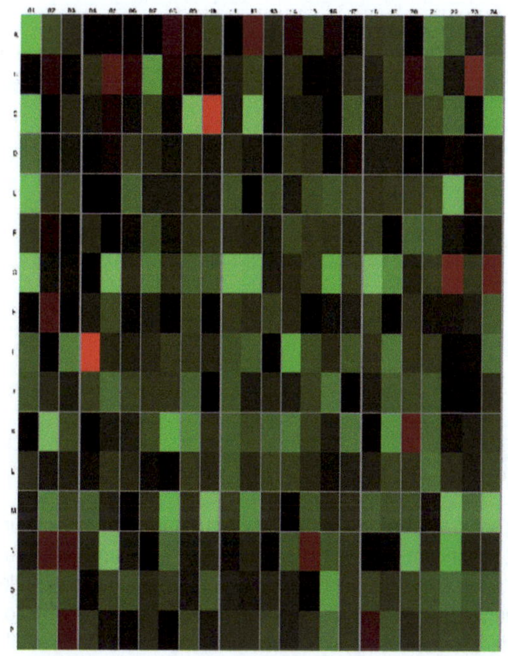

Fig. 2. An example Heat Map analysis of HSF1 hESCs from a 384-well biomol screening plate. It is important to perform heat map analysis of individual plates for both quality control and data analysis purposes. This heat map shows one 384-well plate from HSF1 hESCs treated with the Biomol library. Targets with the potential to increase hESC numbers are shown in *red*.

8. Another important resource that enables visualizing potential chemical networks that your compound may fit is the use of MetaDrug software analysis (see Fig. 3).

3.5. Hit Screens: Secondary Assays for Evaluating Potential Small-Molecule "Hits"

Once a hit is identified, it is imperative to employ additional techniques to verify the effect observed during the HCS assay. In this case, we assayed small molecules and measured *Oct4* expression. While we can correlate the expression level of *Oct4* to hESC survival and/or self-renewal, the change in the number of *Oct-4*-positive cells or colonies within the screen may be a result of changes in cell proliferation, cell death, or self-renewal.

3.5.1. Hit Assay: Flow Cytometry

Flow cytometry is commonly used to examine cells at the single-cell level, including protein expression, and DNA content. Because it can be difficult to assess the state of pluripotent cells based solely on colony morphology, we use flow cytometry to quantify the effects of screening hits on PSCs. By culturing PSCs under normal conditions in 6-well plates and adding the small molecules identified during the HCS screen as "hits", we have been able to validate hits that actually improve the number of PSCs in culture (using the surface markers SSEA-4 or Tra-1-81) and improve survival (decrease in surface-accessible Annexin V as a marker of apoptosis).

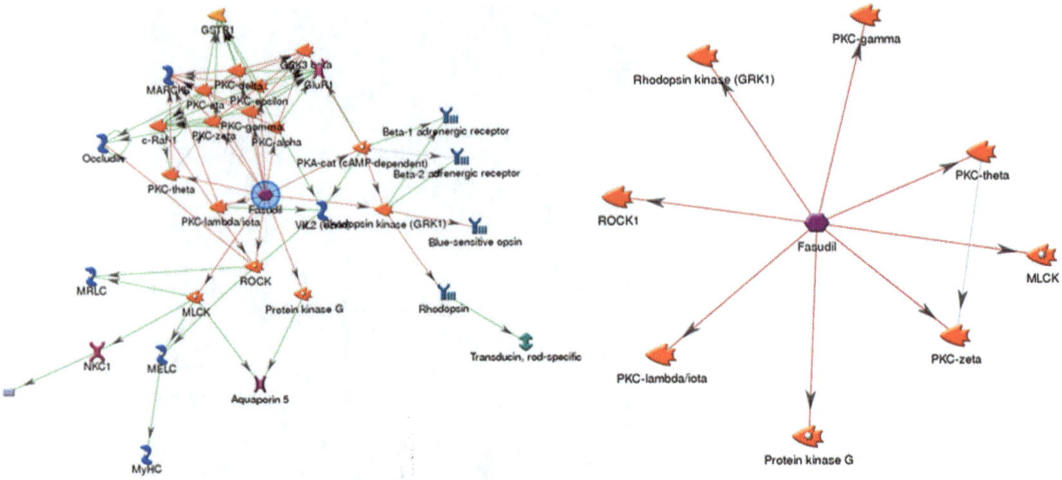

Fig. 3. MetaDrug software analysis of small-molecule targets of one of our HCS screening hits, Fasudil. In our HCS assays, one of the most significant hits was Fasudil (HA-1077), a small molecule inhibitor of Rho-kinase (ROCK) (1). We used MetaDrug/Gene Go to create a network around this novel small-molecule target that improves hESC survival. To the *left* is a simplified network with FASUDIL showing potential direct interactions or the mechanism of Fasudil inhibition. The image to the *right* is a more complex subnetwork created around Fasudil and its targets. This provides additional targets that can be validated using secondary assays described in the next section.

3.5.1.1. Flow Cytometry to Assess Proliferation

1. Collect hESCs with collagenase and plate with hESC medium in 6-well plates.
2. When hESC colonies are of medium size, add BrdU (10 μm; see FITC BrdU Flow Kit manual) for 3 h at 37°C.
3. Rinse cells with DPBS, dissociate with trypsin, and label with anti-SSEA-4 to identify hESCs.

3.5.1.2. Hit Assay: Flow Cytometry to Assess Cell Survival

1. Dissociate hESCs with trypsin and plate at a density of 100,000 cells per well of a 6-well plate.
2. After 4 days, stain hESCs with anti-SSEA-4 and Annexin V.

3.5.2. Hit Assay: Self-Renewal Over Time in Culture

1. Dissociate hESCs with trypsin and plate at a density of 500,000 cells per well of a 6-well plate.
2. After 5 days, dissociate the hESCs and stain to determine the number of SSEA-4-positive cells per well.
3. Repeat step 1 (for at least 5) passages and track the number of SSEA4-positive cells over time.

3.5.3. Hit Assay: Pathway Validation Using RNA Inhibition

For those hits that have a known target gene, RNA inhibition (RNAi) can be used to silence the gene and validate the screening results. For transient silencing, we have used siRNA nucleofection, shown to yield ~30–60% transfection efficiency (4). For stable silencing, we have used lentivirus-mediated shRNA silencing (see Note 13). Use of RNAi in validation helps eliminate any influence from the physical screening process (such as feeder-cell effect) and can be used to determine a direct effect on hESCs. In addition, many small molecules and compounds used in high-content screens can affect multiple targets. RNAi also enables more specific inhibition of a putative target.

4. Notes

1. There are an ever-increasing number of chemical libraries available. For the purpose of our studies, we have chosen libraries that comprise compounds with known biological targets, making them useful for assay validation. Other compound libraries are available, including the libraries available at the UCLA Molecular Screening Shared Resource: http://www.mssr.ucla.edu/lib.html .
2. The use of reporter lines provides an alternative strategy to immunostaining, described here to visualize cells in high-content screens. For example, a PSC line that expresses OCT4-GFP could be used in the assay described to identify compounds that promote PSC cell growth and proliferation, while limiting differentiation. In the case of OCT4-GFP expression is high in undifferentiated PSCs and is downregulated as the cells differentiate. Fluorescent reporter proteins can be designed to express in a cell-type specific manner for lineage tracking experiments.
3. We have found that the pluripotency factor Oct4 provides bright labeling of hESCs in screening assays. Other potential markers of PSCs include alkaline phosphatase, NANOG, and Sox2.
4. In addition to the Beckman Coulter and Molecular Devices screening equipment we used to develop this protocol, other high-content screening equipment is available from various vendors, including but not limited to Thermo Fisher, BD Biosciences, Caliper Life Sciences, Perkin Elmer, and Qiagen.
5. Inactivated MEFs can be purchased or primary MEFs can be inactivated by irradiation or treatment with Mitomycin C. We batch test one vial of a MEF preparation on normal hESC culture plates to determine plating efficiency and ensure that they support the culture of hESC prior to use in HCS assays.
6. In order to more accurately dispense an equal number of hESCs into each well of the screening plates, and to study the effects of screening compounds on individual hESCs, hESC colonies are dissociated using trypsin rather than collagenase.
7. We achieved the same results whether we changed the medium daily or allowed the cells to remain in the same medium for

the entire 4 days of culture with the compound. Also, we had lower contamination levels when the medium was not changed during the 4-day incubation period. The sporadic contamination we observed when medium was changed daily is believed to result from using a shared sample-handling robot in a non-sterile environment to transfer screening compounds.

8. Basic screening parameters, such as plates, objectives, filters, and light sources to be used, should be set up according to the manufacturers' specifications and after empirical determination by the user, and should remain constant once established. Establishment of a screening platform will need to be set up by Molecular Devices and/or a qualified screening center. See the NIH Web site below for more information on possible screening centers in the event that you do not have access to an appropriate screening center.
   (a) https://mli.nih.gov/mli/mlpcn/mlpcn-probe-production-centers/
   (b) http://nihroadmap.nih.gov/molecularlibraries/
9. Image collection is very sensitive and can easily be disrupted by any type of contaminant such as oil, dirt, or fingerprints on the bottoms of screening plates. The bottoms of screening plates should be cleaned with 70% ethanol prior to image acquisition or if any of the above are observed. The background levels of staining should be determined and appropriate dilutions used to avoid nonspecific staining or background. Be careful not to focus your exposure settings on background (debris or contamination) that may be in a screening well. This will result in acquisition of data that is not coming from hESC staining. Longer exposures could also result in increased acquisition time. Keep in mind that increased exposure time can also damage your sample and result in photo bleaching. Therefore, if exposure settings require longer than a minute, it may be necessary to repeat the experiment. Importantly, as you will be measuring signal intensity in the analysis, it is important not to overexpose your sample as this will result in false positives and inaccurate results.
10. It is important to calculate the Z-factor score for each experiment. This will determine whether your screening setup and experiment is accurate enough to find hits in the screen. Importantly, it will tell you how repeatable your screens are from day-to-day. This is crucial because PSC properties can change over time. It is important to determine the baseline survival in your PSC line and culture conditions. The Z-factor score helps in the development of consistent screening protocols and analysis.
11. It is important to analyze each plate via heat map to ensure that there is no "edge effect", which is likely to result in

off-target effects. This is shown as a total row of green or red. It is unlikely that all small molecules in a row would result in a hit. Instead, it is more likely that there was a problem in the setup or there was contamination. An edge effect could also be caused by improper calibration of the plate reader. This experiment would need to be repeated to verify that these small molecules are real hits. Remember that heat maps provide only a quick glance at potential hits. Any increase or decrease found in using this analysis must be verified by going back to the original image and making sure that this particular well did not have contamination or aberrant changes in fluorescence intensity that would create artifacts.

12. So far, we have only screened biological libraries with known targets. Additional screens will be performed with small molecules whose targets are not directly known.

13. In the course of our RNAi experiments, we found that culturing PSCs on Matrigel™ rather than MEFs enhanced transduction efficiency, as the MEFs tended to be transduced rather than the PSCs. In addition, we found that passing cells through a 100-μm filter during passaging resulted in smaller-sized colonies that were more effectively transduced.

Acknowledgments

S.P.S is supported by a CIRM training grant TG2-01169, A.D.P. is supported by an NIH R21 (1R21NS064855-01) from the Roadmap Initiative, CRCC, and JCCC Seed Grant. J.A. is supported by a UC Regent's President's Postdoctoral Fellowship. We would also like to acknowledge the UCLA MSSR, in particular Dr. Robert Damoiseaux for assistance with the HCS assay development. Thanks are due to Dr. Carrie Miceli and Dr. Stan Nelson and labs for use of the MetaDrug/GeneGo software. We would also like to thank Dr. Donald Kohn and Dr. Amander Clark for consultation on establishment of efficient lentiviral protocols for hESCs in our lab.

References

1. Damoiseaux R, Sherman SP, Alva JA, Peterson C, Pyle AD (2009). Integrated chemical genomics reveals modifiers of survival in human embryonic stem cells. *Stem Cells* **27**, 533–542.
2. Malo N, Hanley JA, Cerquozzi S, Pelletier J, Nadon R (2006). Statistical practice in high-throughput screening data analysis. *Nat Biotechnol* **24**, 167–175.
3. Zhang JH, Chung TD, Oldenburg KR (1999). A high-throughput homogeneous assay for reverse transcriptase using generic reagents and time-resolved fluorescence detection. *J Biomol Screen* **4**, 67.
4. Hohenstein KA, Pyle AD, Chern JY, Lock LF, Donovan PJ (2008). Nucleofection mediates high-efficiency stable gene knockdown and transgene expression in human embryonic stem cells. *Stem Cells* **26**: 1436–1443.

Chapter 22

Quantitative Proteome and Phosphoproteome Analysis of Human Pluripotent Stem Cells

Javier Muñoz and Albert J.R. Heck

Abstract

Understanding the signaling pathways governing pluripotency and self-renewal is a prerequisite for better controlling stem cell differentiation to specific fates. Reversible protein phosphorylation is one of the most important posttranslational modifications regulating signaling pathways in biological processes. Global analysis of dynamic changes in protein phosphorylation is, therefore, key to understanding signaling at the system level. Here, we describe a generic mass spectrometry (MS)-based phosphoproteomics strategy applied to monitor phosphorylation dynamics after bone morphogenetic protein 4 (BMP4)-induced differentiation of human embryonic stem cells (hESCs). Our method combines the use of strong cation exchange (SCX) and titanium dioxide (TiO_2) for phosphopeptide enrichment, high-resolution MS for peptide and protein identification, and stable isotope labeling by amino acids in cell culture (SILAC) for quantification. This approach allows us to identify thousands of phosphorylation sites and profile their relative abundance during differentiation. This systems-biology-based approach provides new insights into how human pluripotent stem cells exit the pluripotent state.

Key words: human embryonic stem cells, human pluripotent stem cells, mass spectrometry, quantitative proteomics, SILAC, phosphorylation, titanium dioxide (TiO_2)

1. Introduction

The pluripotency and capability for self-renewal of human pluripotent stem cells (hPSCs) provide, in essence, a unique source for regenerative medicine. However, all potential clinical applications of pluripotent cells are limited because we do not know how to control their developmental fate. The molecular mechanisms that drive differentiation into specific lineages remain poorly understood. All PSCs seem to share a similar core transcriptional regulatory network, involving OCT4 (1), SOX2 (2, 3), and NANOG (4),

Philip H. Schwartz and Robin L. Wesselschmidt (eds.), *Human Pluripotent Stem Cells: Methods and Protocols*, Methods in Molecular Biology, vol. 767, DOI 10.1007/978-1-61779-201-4_22,

which promotes the expression of PSC-related genes and represses lineage-specification genes. However, differences exist between cells of different origin. For instance, mouse embryonic stem cells (mESCs) and human embryonic stem cells (hESCs) differ significantly in their growth requirements: bone morphogenetic protein 4 (BMP4) and leukemia inhibitory factor (LIF) sustain self-renewal in mESC (5), whereas hESCs require basic fibroblast growth factor (bFGF) and transforming growth factor β (TGF-β)/Activin A signaling (6, 7). Exactly how differentiating cells downregulates the transcription factors that control self-renewal is still largely unclear. Transcriptional regulation of NANOG by TGF-β/Activin A and BMP-responsive SMADs has recently been demonstrated. In undifferentiated hESCs, SMAD2/3 dominates through TGF-β signaling, whereas SMAD1/5/8 becomes activated upon BMP-induced differentiation. These SMADs bind the proximal promoter of NANOG with opposing effects. SMAD2/3 promotes, but SMAD1/5/8 inhibits, NANOG expression (8).

Although numerous approaches based on molecular genetics or cellular biology have provided key insights for individual genes or proteins, recent advances in high-throughput technologies have enabled the study of a greater variety of biological processes in an unbiased manner, revealing some unanticipated findings. Several studies have been reported wherein hESCs were extensively characterized at the transcriptome level (9, 10), describing the complete set of genes that are expressed by pluripotent cells. Moreover, recent epigenetic studies have depicted the global chromatin state of hESCs and how this pattern becomes remodeled during differentiation (11, 12). However, transcriptomics and epigenetics represent only two facets in our understanding of the full biological process and are complemented by proteomic approaches (13), which can provide critical information about the protein content in terms of localization, activation state, and abundance.

Reversible protein phosphorylation is an essential regulatory mechanism involved in countless cellular processes. Mass spectrometry (MS)-based proteomics is currently one of the most powerful technologies for dissecting stimulus-dependent dynamics of phosphorylation events in living cells (14, 15). Nevertheless, the low abundance of phosphoproteins in most samples (i.e., whole-cell lysate) and the often low stoichiometry of the modification (sometimes less than 1%) make the study of protein phosphorylation one of the most challenging fields in proteomics. Enrichment of phosphorylated species is a prerequisite to tackle the enormous dynamic range of this posttranslational modification, and many strategies have proven successful, as extensively reviewed by Macek et al. (16). In this chapter, we describe the use of a double-enrichment strategy that takes advantage of the negative nature of the phosphoryl group. We combine a prefractionation step using low-pH strong cation exchange (SCX) (17) with titanium dioxide (TiO_2) chromatography (18). The initial fractionation takes advantage of the fact that at low pH most amino acids are neutral and phosphoryl groups are negative while TiO_2 has a preference for chelation to the phosphate moiety. The SCX/TiO_2 approach, when coupled with high-resolution mass spectrometers, such as LTQ-Orbitrap, allows the identification of thousands of phosphorylation sites (18). However, charting of phosphorylation sites is just a start, as protein phosphorylation is a highly dynamic process that often is only present very briefly (for instance during kinase activation). So, performing differential quantitative analyses is essential. At the present time, proteomic quantitative strategies normally involve the use of stable isotopes to generate "light" and "heavy" samples that are biochemically identical but can be distinguished at the MS level. Subsequently, relative protein abundances can be calculated from comparison of their peak intensities (peptide ion abundances). Currently, stable isotope-labeling by amino acids in cell culture (SILAC) is one of the most common methods for quantitative proteomics (19). SILAC labels cellular proteomes through normal metabolic processes, incorporating nonradioactive stable isotope-containing amino acids in newly synthesized proteins.

Although SILAC has some limitations (20), recently it has been shown to be adaptable to hESC cells, enabling accurate and sensitive quantitative proteomics experiments. Here, we present a detailed description of the methodology used by our group to monitor phosphorylation changes that take place during early hESC differentiation (21). In this method, by using SILAC, metabolically "heavy" labeled hESC are compared with unlabeled "light" differentiated cells at different time points following BMP-4 addition (Fig. 1). The combination of an SCX/TiO_2 enrichment approach with high-resolution MS resulted in, using stringent criteria, the identification and quantification of more than 5,000 proteins and ~3,000 unique phospho-sites during BMP4-induced differentiation. The global study of phosphorylation dynamics revealed several potential downstream effectors that are activated or inactivated upon BMP4-dependent differentiation, providing new insights into the molecular mechanisms involved during this process. The methods described here can be also used to study the signaling pathways governing other aspects of hESC biology such as self-renewal (7) or somatic reprogramming into induced pluripotent stem cells (iPSCs) (22).

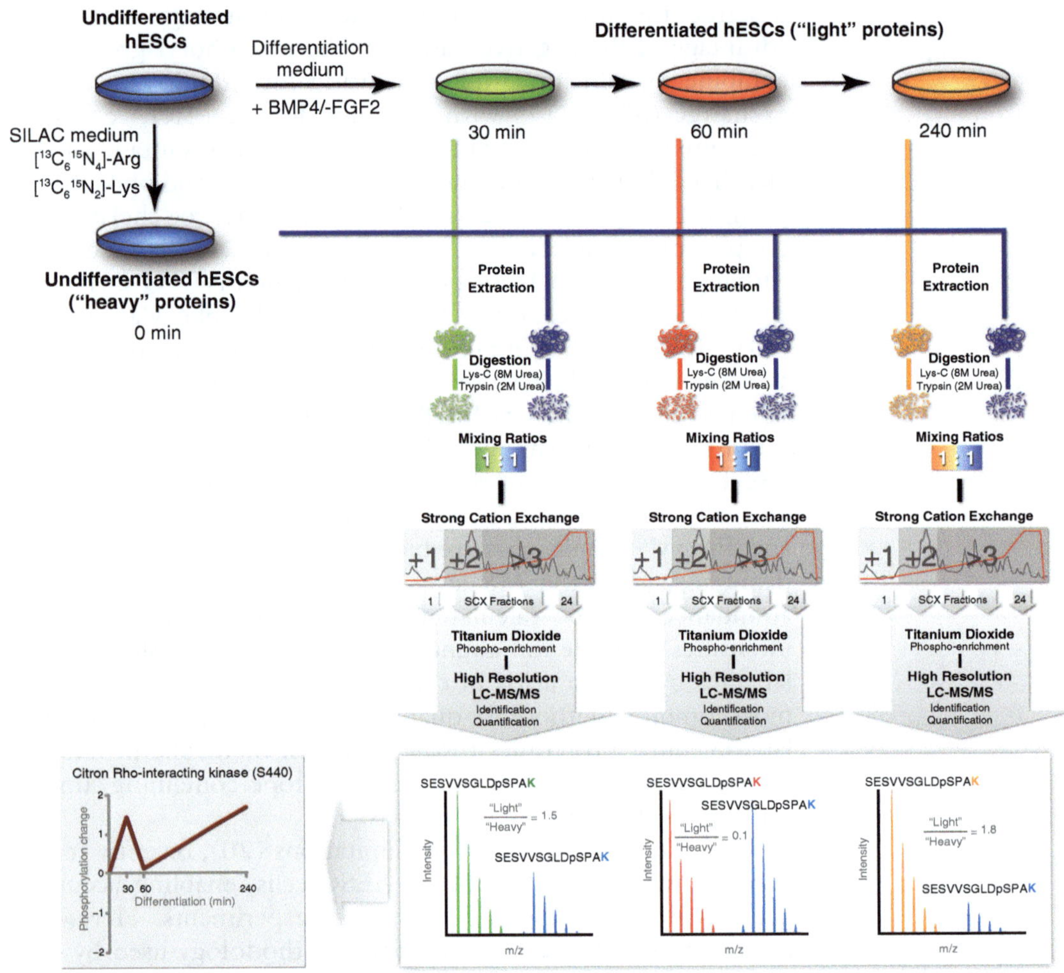

Fig. 1. Overview of methods used to monitor global phosphorylation dynamics during hESC differentiation. The analysis of the proteome and phosphoproteome upon BMP4-induced differentiation enables the study of molecular downstream effectors regulating pluripotency and self-renewal processes. The strategy combines a double fractionation technique by combining strong cation exchange and titanium dioxide to enrich in phosphorylated peptides and high-resolution MS to detect phosphorylation changes. Quantitative data is obtained through metabolic labeling of hESC using the SILAC method. The approach allowed us to monitor changes in 5,222 proteins and 3,201 phospho-sites after BMP4 addition, revealing multiple and novel phosphorylation networks spanning different signaling pathways, kinases, and transcription factors, such as SOX2 (21).

2. Materials

2.1. SILAC Labeling and Differentiation of hESC

1. *hESC line* HUES-7 (23), p20–25, cultured under feeder-free conditions on Growth Factor Reduced Matrigel (BD Biosciences)-coated flasks in MEF-conditioned medium (24).
2. *MEFs* (passage #5), mitotically inactivated with Mitomycin C and seeded at 6×10^4 cells/cm^2 in 75-cm^2 tissue culture flasks.

3. *Differentiation medium*: DMEM:F12 GlutaMAX with N2 and B27 supplement, 1% nonessential amino acids, 100 μM β-mercaptoethanol, Penicillin–Streptomycin, supplemented with 50 ng/mL BMP4 (R&D Systems).
4. *SILAC medium*: DMEM:F12 GlutaMAX without arginine and lysine (custom made, Invitrogen), supplemented with 147.5 mg/L [$^{13}C_6$,$^{15}N_4$]-arginine and 91.25 mg/L [$^{13}C_6$,$^{15}N_2$] Lysine, 15% KO Serum Replacement (Invitrogen), 100 μM β-mercaptoethanol, 4 ng/mL basic fibroblast growth factor (bFGF; PeproTech), 1% nonessential amino acids, and penicillin–streptomycin.

2.2. In-Solution Digestion

1. Lysis buffer: 7 M urea, 2 M thiourea in a solution of 25 mM ammonium bicarbonate, pH 8.2. One tablet of protease inhibitors (Complete mini, Roche) and one tablet of phosphatase inhibitors (PhosStop, Roche) per 10 mL of buffer. Alternatively, phosphatase tablets can be replaced by freshly prepared 10 mM potassium phosphate, 5 mM sodium fluoride, and 5 mM sodium orthovanadate (common phosphatase inhibitors). This buffer should be prepared fresh and protease and phosphatase inhibitors added shortly before use.
2. Reducing agent: prepare a fresh stock solution of 200 mM dithiothreitol (DTT) in 25 mM ammonium bicarbonate, pH 8.2.
3. Alkylating agent: prepare a fresh stock solution of 200 mM iodoacetamide (IAA) in 25 mM ammonium bicarbonate, pH 8.2.
4. Lys-C (Waco Chemicals): prepare and aliquot the enzyme according to the instructions from the manufacturer at a concentration of 2 μg/μL. The enzyme can be stored at –80°C for several months.
5. Modified sequencing-grade trypsin (Roche): prepare and aliquot the enzyme according to the instructions from the manufacturer at a concentration of 2 μg/μL. The enzyme can be stored at –80°C for several months.
6. Bradford, Lowry, or BCA protein assay.

2.3. Peptide Desalting

1. SepPak buffer A: 0.1 M acetic acid.
2. SepPak buffer B: 80% acetonitrile, 0.1 M acetic acid.
3. Reverse-phase C_{18} SepPak SPE cartridges (Waters). 1-cc cartridges are recommended for 1 mg of starting material.

2.4. Strong Cation Exchange

1. SCX buffer A (20% acetonitrile, 0.05% formic acid, pH 3.0).
2. SCX buffer B (500 mM KCl in 20% acetonitrile, 0.05% formic acid, pH 3.0).
3. Strong cation exchange is performed using two in-line coupled Zorbax BioSCX-Series II columns (0.8 mm ID × 50 mm L,

3.5 μm) (Agilent Technologies), a FAMOS autosampler (LC-packings) and a Shimadzu LC-9A binary pump and a SPD-6A UV-detector (Shimadzu).

4. After injection, the first 5 min are run isocratic at 100% SCX buffer A followed by a linear gradient of 1% min^{-1} of SCX buffer B.

2.5. 2D-(TiO_2/RP)-HPLC and Mass Spectrometry

1. Bi-phasic trap column consists of three separate precolumns as follows (Fig. 2a):
   - 30 mm L × 100 μm ID Aqua C_{18} (Phenomenex).
   - 5 mm L × 100 μm ID Titanium Dioxide (TiO_2) (GL Sciences Inc.).
   - 30 mm L × 100 μm ID Aqua C_{18} (Phenomenex).
2. Analytical column: 200 mm L × 50 μm ID ReproSil-Pur C_{18}-AQ, 3 μm 120 Å.
3. HPLC solvent A: 0.1 M acetic acid and 0.46 M formic acid.

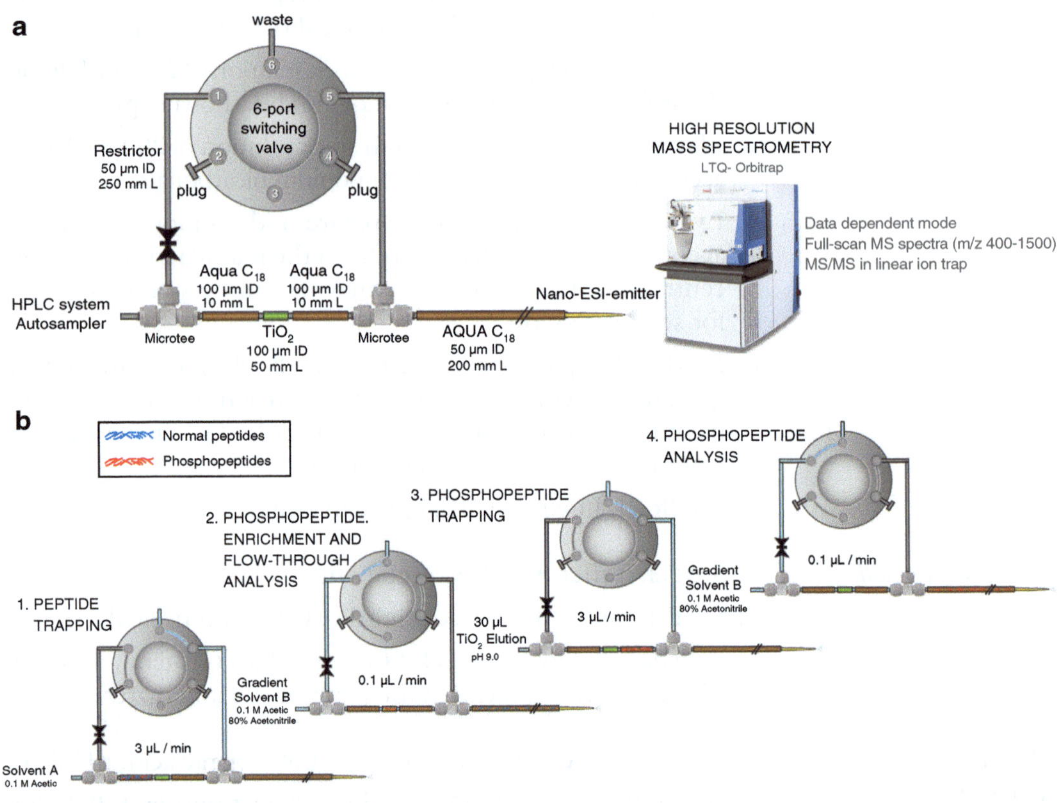

Fig. 2. Schematic presentation of the 2D-TiO_2/RP-nanoflow-HPLC and mass spectrometry technique used to enrich and analyze phosphopeptides (18). (**a**) Design of the vented column system with the triple precolumn scheme. Phosphopeptides are selectively captured by the TiO_2 material and subsequently analyzed by high-resolution mass spectrometry. (**b**) Sequential configuration of the 6-port valve system during analysis.

4. HPLC solvent B: 80% acetonitrile containing 0.1 M acetic acid and 0.46 M formic acid.
5. TiO_2 elution buffer: 250 mM ammonium bicarbonate, pH 9.0 (adjusted with ammonia), containing 10 mM sodium phosphate, 5 mM sodium orthovanadate, and 1 mM potassium fluoride.
6. LTQ Orbitrap mass spectrometer (Thermo Fisher) operating in positive ionization, data dependent mode, automatically switching between MS and MS/MS. Full scan MS spectra (from m/z 400 to 1,500) are acquired in the Orbitrap with a resolution of 60,000 at m/z 400 after accumulation to target value of 500,000. The three most intense ions at a threshold above 5,000 are selected for collision-induced fragmentation in the linear ion trap at normalized collision energy of 35% after accumulation to a target value of 10,000.

2.6. Computational Analysis

1. Xcalibur software (Thermo Fisher). XDK component must be installed.
2. MSQuant open source software (http://msquant.sourceforge.net/). It requires the .NET framework (Microsoft).
3. Mascot v2.2 search engine (Matrix Science).

3. Methods

The present method was optimized for HUES-7 hESC (21), but should be adaptable to other PSC lines with minor modifications. Since the aim of the study was to understand the initial response after differentiation had been triggered by BMP4, early time points (30, 60, and 240 min) after BMP4 addition were taken. The success and comprehensiveness of the method will largely depend on the amount of PSC starting material. The methodology described here uses 1 mg of protein lysate, which should be sufficient to identify and quantify approximately 3,000–4,000 unique phosphopeptides. Preparing all samples that need to be analyzed on the same day under identical conditions will reduce experimental variability and enhance accuracy of quantitation (25).

3.1. SILAC Labeling and Differentiation of hESC

1. HUES-7 hESCs are cultured under feeder-free conditions on Matrigel. A detailed explanation on how to transfer hESC to feeder-free culture can be found in (Chapter 10) and Braam et al. (26) At this point, hESCs are split in two, one population will be "heavy" labeled with SILAC medium and the other will be differentiated upon addition of BMP4 (Fig. 1).

2. Undifferentiated "heavy" labeled hESCs: SILAC medium is harvested daily for 7 consecutive days (first batch discarded) and resupplemented with 4 ng/ml bFGF after each collection. Growing the hESCs for 1 week (~5 population doublings) in the heavy MEF-conditioned medium results in complete labeling of the cells with the heavy stable isotopes (an aliquot should be saved to check labeling efficiency by MS) (Fig. 1).
3. Differentiated "light" unlabeled hESCs: rapid differentiation of hESCs is induced by BMP4 addition in the absence of bFGF (27). Conditioned medium is then replaced by differentiation medium and the cells are harvested at various times thereafter: 30, 60, and 240 min (Fig. 1).

3.2. In-Solution Digestion

1. Harvest cells by centrifugation at 2,500 × *g* for 10 min at 4°C. Wash the cell pellets with PBS and centrifuge again. Approximately, 10^7 unlabeled differentiated cells (~1 mg of protein) at the three time points (30, 60, and 240 min) and 3×10^7 labeled undifferentiated cells are necessary for this study.
2. Cell pellets are lysed by addition of 500 μL of lysis buffer and vortexed for 10 min to ensure complete protein solubilization.
3. Sonicate the samples three times for 30 s at 4°C (80% amplitude and 0.8 cycles) to enhance cell disruption and fragment DNA chains.
4. Spin down cell debris by centrifugation at 2,500 × *g* for 10 min at 4°C. Collect and transfer supernatants to new tubes.
5. Protein concentration is determined by Bradford, Lowry, or BCA assays. Pipette the volume necessary to obtain 1 mg of protein, since the following steps are optimized for such an amount of starting material.
6. Reduction: cysteine bonds are reduced with DTT. Add 2 μL of the 200 mM DTT stock solution per 100 μL of protein lysate (final concentration 4 mM) and incubate at 56°C for 25 min. Do not use temperatures higher than 60°C as urea can decompose into isocyanic acid leading to artifactual carbamoylation of free amino groups (protein N-termini and side chains of lysine residues) (28).
7. Alkylation: sulfhydril groups are then blocked with IAA. Add 4 μL of the 200 mM IAA stock solution per 100 μL of protein lysate (final concentration 8 mM) and incubate at room temperature for 30 min in the dark.
8. First digestion is with endoproteinase Lys-C: it is known that Lys-C retains activity upon the addition of up to 8 M urea; therefore, this enzyme is used to partially digest the sample

when proteins are unfolded at high concentrations of chaotropic agents. Add 5 μg of Lys-C to the protein solution (final ratio of 1:200 enzyme:substrate) and incubate the mixture for 4 h at room temperature.

9. Second digestion is with trypsin (Lys-C and Arg-C specificity): the activity of this enzyme is compromised when urea concentrations higher than 2 M are used. Samples are diluted fourfold with 25 mM ammonium bicarbonate, pH 8.2, to lower the urea concentration to 2 M and mixed with 20 μg of modified sequencing grade trypsin for further digestion overnight at 37°C (final ratio of 1:50 enzyme:substrate).
10. Quench the reaction by acidification with TFA to a final concentration of 0.2% (pH must be <3, add more TFA if necessary).

3.3. Peptide Desalting

1. The resulting peptides must be desalted and concentrated prior SCX fractionation. C_{18} SepPak solid-phase extraction cartridges are used for this purpose.
2. Condition the cartridges with 6 mL of 100% acetonitrile followed by 6 mL of SepPak buffer B.
3. Equilibrate the cartridges with 6 mL of SepPak buffer A to remove residual organic component from the system.
4. Load peptides at low flow rate (1 mL/min).
5. Wash the cartridges with 6 mL of SepPak buffer A. Repeat this step twice.
6. Elute the peptides from the columns with 3 mL of SepPak buffer B at low flow rate (0.5 mL/min).
7. Eluates are then dried completely in a Speedvac centrifuge. Resuspend the samples in 100 μL of 10% formic acid.

3.4. Determining Mixing Ratios

1. In our experience, current methods to calculate protein concentration (i.e., Bradford, Lowry, or BCA) are not sufficiently precise to accurately mix proteins in a 1:1 ratio ("light":"heavy"). We, therefore, recommend using MS for this purpose.
2. Mix 1 μL of the 30 min "light" differentiated hESC sample with 1 μL of the 0 min "heavy" undifferentiated hESC sample and analyze the mixture with normal reverse-phase LC settings in the LTQ-Orbitrap or any other appropriate mass spectrometer. Process the data as described below and quantify the most abundant proteins present in the sample, which will give a precise estimation for the proper mixing ratios. Repeat the procedure for the other two SILAC mixtures (0 min versus 60 min and 0 min versus 240 min).

3. Mix the three SILAC mixtures according to the MS-determined mixing ratios. At this point the samples are ready for SCX fractionation.

3.5. Strong Cation Exchange

1. Equilibrate the SCX system by running a blank (often solvent A).
2. Inject sample and collect the first 24 SCX fractions (1 min each, i.e., 50 μL elution volume) from the fractionation. These will contain the majority of the peptide material.
3. Dry down fractions completely (lyophilizing is preferred) and resuspend samples in 10% formic acid. At this point, the samples are ready for 2D (TiO_2/RP)-HPLC and mass spectrometry.

3.6. 2D (TiO_2/RP)-HPLC and Mass Spectrometry

1. Flow-through analysis: Inject samples onto the TiO_2/RP based nano-LC system. Peptides will be trapped at 3 μL/min in 100% HPLC solvent A on the first C_{18} trap column. The subsequent linear acetonitrile gradient (HPLC solvent B) will displace the peptides from the first C_{18} trap and those peptides with no TiO_2 affinity (mainly nonphosphorylated peptides) will be separated in the analytical column at a flow rate of 100 nL/min in a 70-min gradient from 0 to 40% of HPLC solvent B, whereas phosphopeptides will be retained in the TiO_2 precolumn (see Fig. 2b). The low flow rate during the gradient improves binding of phosphopeptides (18).
2. Elution analysis: phosphorylated peptides are subsequently eluted from the TiO_2 resin onto the second C_{18} trap with the injection of 30 μL of TiO_2 elution buffer followed by an injection of 20 μL of 5% formic acid (Fig. 2b). Phosphopeptides are then chromatographically separated using a 70-min gradient (0–40% of HPLC solvent B) at 100 nL/min.

3.7. Computational Analysis

1. Peakpicking: both flow-through and elution ".raw" files from all SCX fractions are converted to .mgf (Mascot Generic Format) files using DTASuperCharge which is an application used to convert Thermo .raw files to Mascot search input files in a format suitable for use with MSQuant. For MS/MS noise reduction, the Smartpicking algorithm is used with the following parameters: Max. Search level = 5 and Segment size [Th] = 200.
2. Database Search: spectra contained within the .mgf are then submitted to an in-house licensed Mascot 2.2 search engine against the Swiss-Prot fasta database with taxonomy filter for *Homo sapiens.* Carbamidomethyl cysteine is set as fixed modification; protein N-acetylation, N-terminal pyroglutamate, oxidized methionine, and phosphorylation of serine, threonine, and tyrosine are set as variable modification. The "SILAC R10 – K8" quantitation mode is used. Precursor tolerance is initially set at 15 ppm, whereas fragment ion tolerance is set at 0.6 Da.

3. False discovery rates (FDR) are calculated for each fraction by repeating the search against a decoy database consisting of the same proteins with reversed sequences (29). Calculate the proper Mascot Score threshold to achieve an FDR of 1% at the peptide level.

4. Quantitative analysis. Save all Mascot results in peptide summary view as an .html file by using Microsoft Internet Explorer 7 (or earlier version) and associate them with their corresponding .raw files in MSQuant software. The quantitation is based on the comparison of peak areas (extracted ion chromatograms) of both "light" and "heavy" peptides (see Fig. 3a). Manual validation of the quantitation is highly advised, especially for those peptides with low signal-to-noise ratios or when unrelated isotope clusters overlap with the SILAC pair being quantified (this is particularly frequent when complex samples like whole lysates are analyzed).

5. The presence of multiple phosphorylatable residues within a peptide sequence makes phosphorylation site localization difficult (30). The PTMscoring algorithm (31) included in MSQuant software is, therefore, used to calculate the probability of correct assignment for all identified phosphopeptides (Fig. 3b).

a

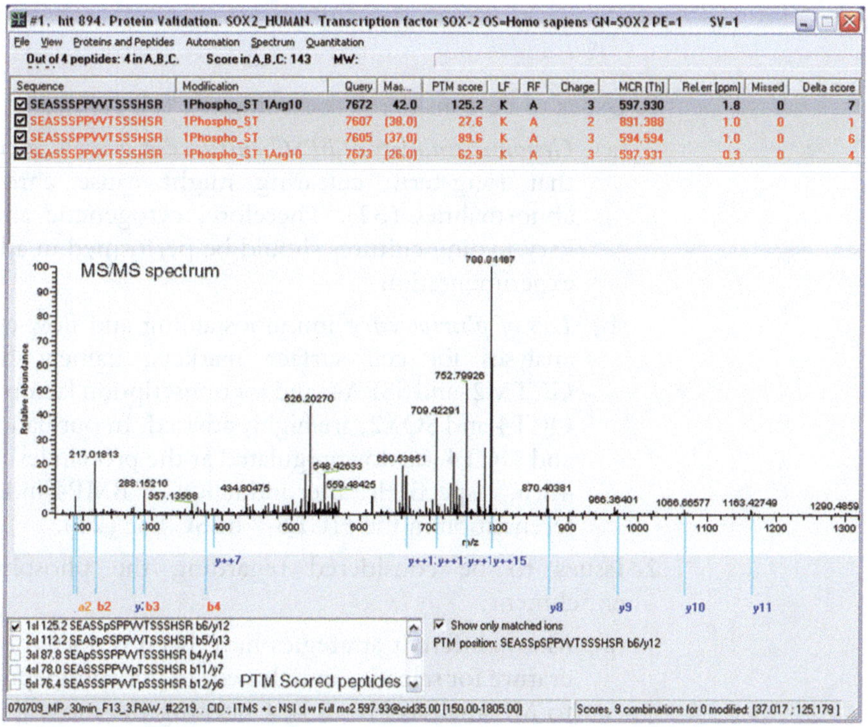

Fig. 3. Identification and quantification of phosphorylation sites in the SOX2 transcription factor using MSQuant software. (**a**) View of the Protein Validation interface. Display of phosphopeptides identified in SOX2. Fragmentation spectra are processed for site assignment revealing three phosphorylation sites in S249, S250, and S251 (21). (**b**) View of the Protein Quantitation interface. By comparing peak intensities of the "light" and "heavy" phosphopeptides, relative quantification is achieved.

b

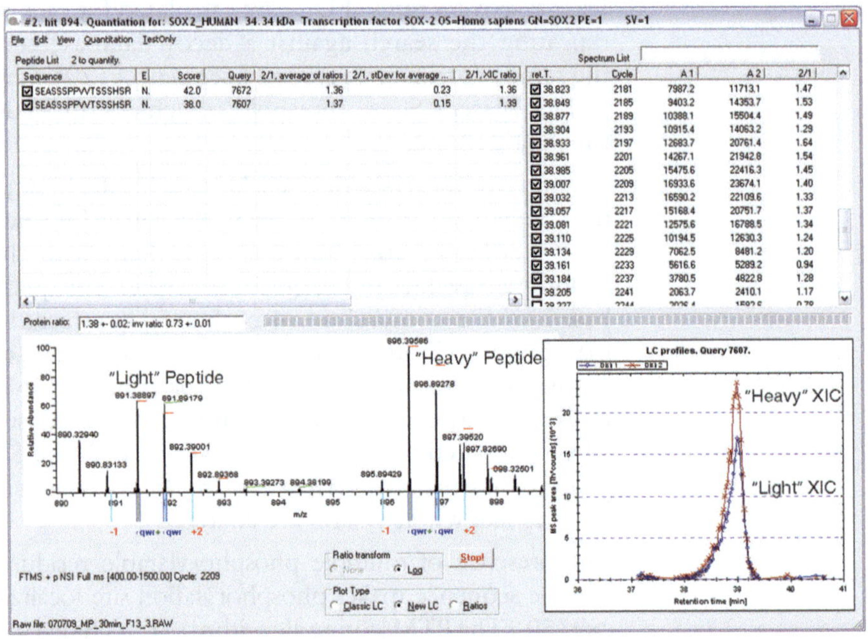

Fig. 3. (continued)

4. Notes

1. Issues to be considered regarding the cell culture of hPSC:
   - (a) *Genomic stability of hPSC cultures*: it is well documented that long-term culturing might cause chromosome abnormalities (32). Therefore, cytogenetic analysis by karyotyping cultures should be performed in advance of experimentation.
   - (b) *Loss of pluripotency*: immunostaining and flow-cytometry analyses for cell surface markers, namely, Tra-1-60, GCTM2, and SSEA4, and for transcription factors, namely, OCT4 and SOX2, are highly advised. In our hands, SOX2 and OCT4 are downregulated at the protein level, 24 and 48 h, respectively, after induction of BMP4-induced differentiation in the HUES-7 hESC line (21).
2. Issues to be considered regarding the phosphopeptide enrichment:
   - (a) Many different strategies have been reported in the literature for selective enrichment of phosphopeptides prior to MS analysis (16, 33). Choosing the most appropriate depends on the goal of the study and type of samples to be analyzed. Here, we present a strategy that combines offline SCX fractionation with online TiO_2 chromatogra-

phy, which has proved highly robust, automatable, and sensitive (18), factors especially important for large-scale quantitative studies. Alternatives to this approach include the use of offline TiO_2 in combination with 2,5-dihydroxy benzoic acid (DHB) to reduce binding of acidic peptides (31), offline immobilized metal affinity chromatography (IMAC) (34) that makes use of the affinity of negatively charged phosphopeptides toward certain trivalent metal ions (i.e., Fe^{3+}, Ga^{3+}, Zr^{3+}) or antibody-based enrichment when studying tyrosine signaling (35, 36).

3. Issues to be considered regarding the MS analysis:
   (a) Quantitative studies benefit from the latest generation of high-resolution mass spectrometers since peptide isotope clusters are easily resolved, improving the issues related to overlapping protein peaks that hamper quantification (37). Therefore, the use of low-resolution MS will result not only in a drop in the peptide identification rate but also in the precision of the quantification.
   (b) Despite the use of high-resolution mass spectrometers used for this sort of quantitative approach, time-dependent errors in the system may affect the mass accuracy of these instruments (38) and compromise peptide identification. Therefore, checking statistical mass accuracy (defined as the mass accuracy estimated from a statistical distribution of mass errors) (39) is recommended.
   (c) Although 70-min gradients with 200 mm × 50 μm reverse phase columns were used for this study, improvements in the LC settings toward longer gradients and longer analytical columns will enhance the sensitivity and dynamic range of the analysis. Ultra-high-pressure RPLC hyphen ated to an LTQ-Orbitrap Velos reveals a linear relation between peak capacity and number of identified peptides (45). Current settings in our group (i.e., 3-h gradients in 400 mm × 50 μm columns) have increased peptide identifications by 40–50%.
4. Issues to be considered regarding the quantitative analysis:
   (a) Arginine and lysine are probably the most common "heavy" amino acids used in SILAC experiments. However, the accuracy of quantification can be compromised by the metabolic conversion of arginine to proline in eukaryotes, with different kinetics in different cell lines. This may result in the generation of multiple satellite peaks for proline-containing peptides in the "heavy" state. A partial solution is the use of [$^{15}N_4$]-arginine in the "light" condition as well (20). Therefore, both conditions will be equally affected by arginine-to-proline

artifacts. Other alternatives are to empirically reduce the arginine concentration to minimize the conversion to proline (40), which may, however, compromise the viability of the cells, or the mathematical corrections for all the proline-containing peptides (14).

(b) Despite the fact that SILAC is probably one of the most sensitive, accurate, and robust approaches in MS-based quantitative proteomic studies, chemical labeling strategies can also be employed when metabolic labeling is not feasible. Recently introduced stable isotope dimethylation (41, 42) labels the N-terminus and side-chain amines of peptides from digested samples and has been used in phosphoproteomic studies (36). iTRAQ (isobaric tag for relative and absolute quantitation) represents an alternative where up to eight samples can be mixed and analyzed in a single MS experiment. The quantification in this case is performed at the MS/MS level by examining the relative intensities of the corresponding reporter ions.

(c) Although MSQuant was the quantitation platform chosen in this study, other software could be used as well (43). The new MaxQuant suite (44) is specially designed for SILAC experiments and dramatically improves automation, but it is limited to Thermo Fisher instruments (LTQ-Orbitrap and LTQ-FT-ICR).

Acknowledgments

Parts of the work described here were supported by the Bsik programs "Dutch Platform for Tissue Engineering," "Stem Cells in Development and Disease," the Netherlands Proteomic Center, and the FP6 EU Program Heart Development and Heart Repair. We also acknowledge the contributions from the Mummery and Heck groups, especially Dennis van Hoof, Stefan Braam, Martijn Pinkse, Shabaz Mohammed, and Jeroen Krijgsveld.

References

1. Niwa H., Miyazaki J., Smith A. G. (2000) Quantitative expression of Oct-3/4 defines differentiation, dedifferentiation or self-renewal of ES cells. *Nat Genet* **24**, 372–6.
2. Avilion A. A., Nicolis S. K., Pevny L. H., Perez L., Vivian N., Lovell-Badge R. (2003) Multipotent cell lineages in early mouse development depend on SOX2 function. *Genes Dev* **17**, 126–40.
3. Yuan H., Corbi N., Basilico C., Dailey L. (1995) Developmental-specific activity of the FGF-4 enhancer requires the synergistic action of Sox2 and Oct-3. *Genes Dev* **9**, 2635–45.
4. Mitsui K., Tokuzawa Y., Itoh H., Segawa K., Murakami M., Takahashi K., et al. (2003) The homeoprotein Nanog is required for maintenance of pluripotency in mouse epiblast and ES cells. *Cell* **113**, 631–42.

5. Ying Q. L., Nichols J., Chambers I., Smith A. (2003) BMP induction of Id proteins suppresses differentiation and sustains embryonic stem cell self-renewal in collaboration with STAT3. *Cell* **115**, 281–92.
6. Vallier L., Alexander M., Pedersen R. A. (2005) Activin/Nodal and FGF pathways cooperate to maintain pluripotency of human embryonic stem cells. *J Cell Sci* **118**, 4495–509.
7. Xu R. H., Peck R. M., Li D. S., Feng X., Ludwig T., Thomson J. A. (2005) Basic FGF and suppression of BMP signaling sustain undifferentiated proliferation of human ES cells. *Nat Methods* **2**, 185–90.
8. Xu R. H., Sampsell-Barron T. L., Gu F., Root S., Peck R. M., Pan G., et al. (2008) NANOG is a direct target of TGFbeta/activin-mediated SMAD signaling in human ESCs. *Cell Stem Cell* **3**, 196–206.
9. Hirst M., Delaney A., Rogers S. A., Schnerch A., Persaud D. R., O'Connor M. D., et al. (2007) LongSAGE profiling of nine human embryonic stem cell lines. *Genome Biol* **8**, R113.
10. Skottman H., Mikkola M., Lundin K., Olsson C., Stromberg A. M., Tuuri T., et al. (2005) Gene expression signatures of seven individual human embryonic stem cell lines. *Stem Cells* **23**, 1343–56.
11. Pan G., Tian S., Nie J., Yang C., Ruotti V., Wei H., et al. (2007) Whole-genome analysis of histone H3 lysine 4 and lysine 27 methylation in human embryonic stem cells. *Cell Stem Cell* **1**, 299–312.
12. Zhao X. D., Han X., Chew J. L., Liu J., Chiu K. P., Choo A., et al. (2007) Whole-genome mapping of histone H3 Lys4 and 27 trimethylations reveals distinct genomic compartments in human embryonic stem cells. *Cell Stem Cell* **1**, 286–98.
13. Cox J., Mann M. (2007) Is proteomics the new genomics? *Cell* **130**, 395–8.
14. Gruhler A., Olsen J. V., Mohammed S., Mortensen P., Faergeman N. J., Mann M., et al. (2005) Quantitative phosphoproteomics applied to the yeast pheromone signaling pathway. *Mol Cell Proteomics* **4**, 310–27.
15. Kratchmarova I., Blagoev B., Haack-Sorensen M., Kassem M., Mann M. (2005) Mechanism of divergent growth factor effects in mesenchymal stem cell differentiation. *Science* **308**, 1472–7.
16. Macek B., Mann M., Olsen J. V. (2009) Global and site-specific quantitative phosphoproteomics: principles and applications. *Annu Rev Pharmacol Toxicol* **49**, 199–221.
17. Beausoleil S. A., Jedrychowski M., Schwartz D., Elias J. E., Villen J., Li J., et al. (2004) Large-scale characterization of HeLa cell nuclear phosphoproteins. *Proc Natl Acad Sci USA* **101**, 12130–5.
18. Pinkse M. W., Mohammed S., Gouw J. W., van Breukelen B., Vos H. R., Heck A. J. (2008) Highly robust, automated, and sensitive online TiO2-based phosphoproteomics applied to study endogenous phosphorylation in Drosophila melanogaster. *J Proteome Res* **7**, 687–97.
19. Ong S. E., Blagoev B., Kratchmarova I., Kristensen D. B., Steen H., Pandey A., et al. (2002) Stable isotope labeling by amino acids in cell culture, SILAC, as a simple and accurate approach to expression proteomics. *Mol Cell Proteomics* **1**, 376–86.
20. Van Hoof D., Pinkse M. W., Oostwaard D. W., Mummery C. L., Heck A. J., Krijgsveld J. (2007) An experimental correction for arginine-to-proline conversion artifacts in SILAC-based quantitative proteomics. *Nat Methods* **4**, 677–8.
21. Van Hoof D., Munoz J., Braam S. R., Pinkse M. W., Linding R., Heck A. J., et al. (2009) Phosphorylation dynamics during early differentiation of human embryonic stem cells. *Cell Stem Cell* **5**, 214–26.
22. Takahashi K., Tanabe K., Ohnuki M., Narita M., Ichisaka T., Tomoda K., et al. (2007) Induction of pluripotent stem cells from adult human fibroblasts by defined factors. *Cell* **131**, 861–72.
23. Cowan C. A., Klimanskaya I., McMahon J., Atienza J., Witmyer J., Zucker J. P., et al. (2004) Derivation of embryonic stem-cell lines from human blastocysts. *N Engl J Med* **350**, 1353–6.
24. Braam S. R., Denning C., van den Brink S., Kats P., Hochstenbach R., Passier R., et al. (2008) Improved genetic manipulation of human embryonic stem cells. *Nat Methods* **5**, 389–92.
25. Zhang G., Fenyo D., Neubert T. A. (2009) Evaluation of the variation in sample preparation for comparative proteomics using stable isotope labeling by amino acids in cell culture. *J Proteome Res* **8**, 1285–92.
26. Braam S. R., Denning C., Matsa E., Young L. E., Passier R., Mummery C. L. (2008) Feeder-free culture of human embryonic stem cells in conditioned medium for efficient genetic modification. *Nat Protoc* **3**, 1435–43.
27. Pera M. F., Andrade J., Houssami S., Reubinoff B., Trounson A., Stanley E. G., et al. (2004) Regulation of human embryonic stem cell dif-

ferentiation by BMP-2 and its antagonist noggin. *J Cell Sci* **117**, 1269–80.

28. Righetti P. G. (2006) Real and imaginary artefacts in proteome analysis via two-dimensional maps. *J Chromatogr B Analyt Technol Biomed Life Sci* **841**, 14–22.
29. Elias J. E., Gygi S. P. (2007) Target-decoy search strategy for increased confidence in large-scale protein identifications by mass spectrometry. *Nat Methods* **4**, 207–14.
30. Beausoleil S. A., Villen J., Gerber S. A., Rush J., Gygi S. P. (2006) A probability-based approach for high-throughput protein phosphorylation analysis and site localization. *Nat Biotechnol* **24**, 1285–92.
31. Olsen J. V., Blagoev B., Gnad F., Macek B., Kumar C., Mortensen P., et al. (2006) Global, in vivo, and site-specific phosphorylation dynamics in signaling networks. *Cell* **127**, 635–48.
32. Maitra A., Arking D. E., Shivapurkar N., Ikeda M., Stastny V., Kassauei K., et al. (2005) Genomic alterations in cultured human embryonic stem cells. *Nat Genet* **37**, 1099–103.
33. Lemeer S., Heck A. J. (2009) The phosphoproteomics data explosion. *Curr Opin Chem Biol* **13**, 414–20.
34. Villen J., Beausoleil S. A., Gerber S. A., Gygi S. P. (2007) Large-scale phosphorylation analysis of mouse liver. *Proc Natl Acad Sci USA* **104**, 1488–93.
35. Rikova K., Guo A., Zeng Q., Possemato A., Yu J., Haack H., et al. (2007) Global survey of phosphotyrosine signaling identifies oncogenic kinases in lung cancer. *Cell* **131**, 1190–203.
36. Boersema P. J., Foong L. Y., Ding V. M., Lemeer S., van Breukelen B., Philp R., et al. (2009) In depth qualitative and quantitative profiling of tyrosine phosphorylation using a combination of phosphopeptide immunoaffinity purification and stable isotope dimethyl labeling. *Mol Cell Proteomics*.
37. Ong S. E., Kratchmarova I., Mann M. (2003) Properties of 13 C-substituted arginine in stable isotope labeling by amino acids in cell culture (SILAC). *J Proteome Res* **2**, 173–81.
38. Petyuk V. A., Jaitly N., Moore R. J., Ding J., Metz T. O., Tang K., et al. (2008) Elimination of systematic mass measurement errors in liquid chromatography-mass spectrometry based proteomics using regression models and a priori partial knowledge of the sample content. *Anal Chem* **80**, 693–706.
39. Zubarev R., Mann M. (2007) On the proper use of mass accuracy in proteomics. *Mol Cell Proteomics* **6**, 377–81.
40. Bendall S. C., Hughes C., Stewart M. H., Doble B., Bhatia M., Lajoie G. A. (2008) Prevention of amino acid conversion in SILAC experiments with embryonic stem cells. *Mol Cell Proteomics* **7**, 1587–97.
41. Boersema P. J., Aye T. T., van Veen T. A., Heck A. J., Mohammed S. (2008) Triplex protein quantification based on stable isotope labeling by peptide dimethylation applied to cell and tissue lysates. *Proteomics* **8**, 4624–32.
42. Hsu J. L., Huang S. Y., Chow N. H., Chen S. H. (2003) Stable-isotope dimethyl labeling for quantitative proteomics. *Anal Chem* **75**, 6843–52.
43. Lau K. W., Jones A. R., Swainston N., Siepen J. A., Hubbard S. J. (2007) Capture and analysis of quantitative proteomic data. *Proteomics* **7**, 2787–99.
44. Cox J., Mann M. (2008) MaxQuant enables high peptide identification rates, individualized p.p.b.-range mass accuracies and proteome-wide protein quantification. *Nat Biotechnol* **26**, 1367–72.
45. Köcher T., Swart R., Mechtler K. (2011) *Anal Chem.* 83(7), 2699–704. Epub 2011 Mar 9. PMID: 21388192 [PubMed – in process]

Part V

Genetic Manipulation (Gene Modulation)

Chapter 23

Lentivirus-Mediated Modification of Pluripotent Stem Cells

Ruchi Bajpai

Abstract

Relatively safe, HIV-1-based lentiviral vectors have served as an efficient means of transducing human embryonic stem cells (hESCs). Here we describe the variety of lentiviral vector systems available with the basic strategy for designing viral vectors and methods for generating viruses for efficiently infecting and selecting transduced hESCs.

Key words: human embryonic stem cells, lentivirus, infection, inducible, TRIP, TET-ON, hPFK-GFP, H2B-GFP, glonal

1. Introduction

The therapeutic potential of human embryonic stem cells (hESCs) has been the driving force behind understanding their ability to self-renew and for developing methods for lineage-specific differentiation. These attempts have also provided a unique opportunity to begin understanding the mechanisms regulating early human development that have remained a mystery up until now, but hold the key to defining the rules of differentiation. Even the most modest screening assay designed to identify differentiation factors and develop conditions that drive lineage-specific differentiation would benefit enormously from transgenic reporter lines. While a comprehensive analysis of the genes functioning in maintaining pluripotency and inducing differentiation require complementary gain- and loss-of-function experiments, they can be facilitated by genetic modification of undifferentiated hESCs. HIV-1-derived lentiviral vectors have served as a robust and efficient means of introducing transgenes (including both reporter genes such as GFP and cDNAs of interest) or short-interfering RNA constructs into stem cells.

Philip H. Schwartz and Robin L. Wesselschmidt (eds.), *Human Pluripotent Stem Cells: Methods and Protocols*, Methods in Molecular Biology, vol. 767, DOI 10.1007/978-1-61779-201-4_23, © Springer Science+Business Media, LLC 2011

Upon infection, these replication deficient, self-inactivating viruses can integrate into the host genome and are stably inherited, allowing for FACS- or antibiotic-based selection of transduced cells and the establishment of transgenic lines (Fig. 1). Described herein are the components of the "safe" lentiviral vector, the elements of vector design, the method of virus production and efficient infection of hESCs, with specific examples, to derive stable stem cell lines.

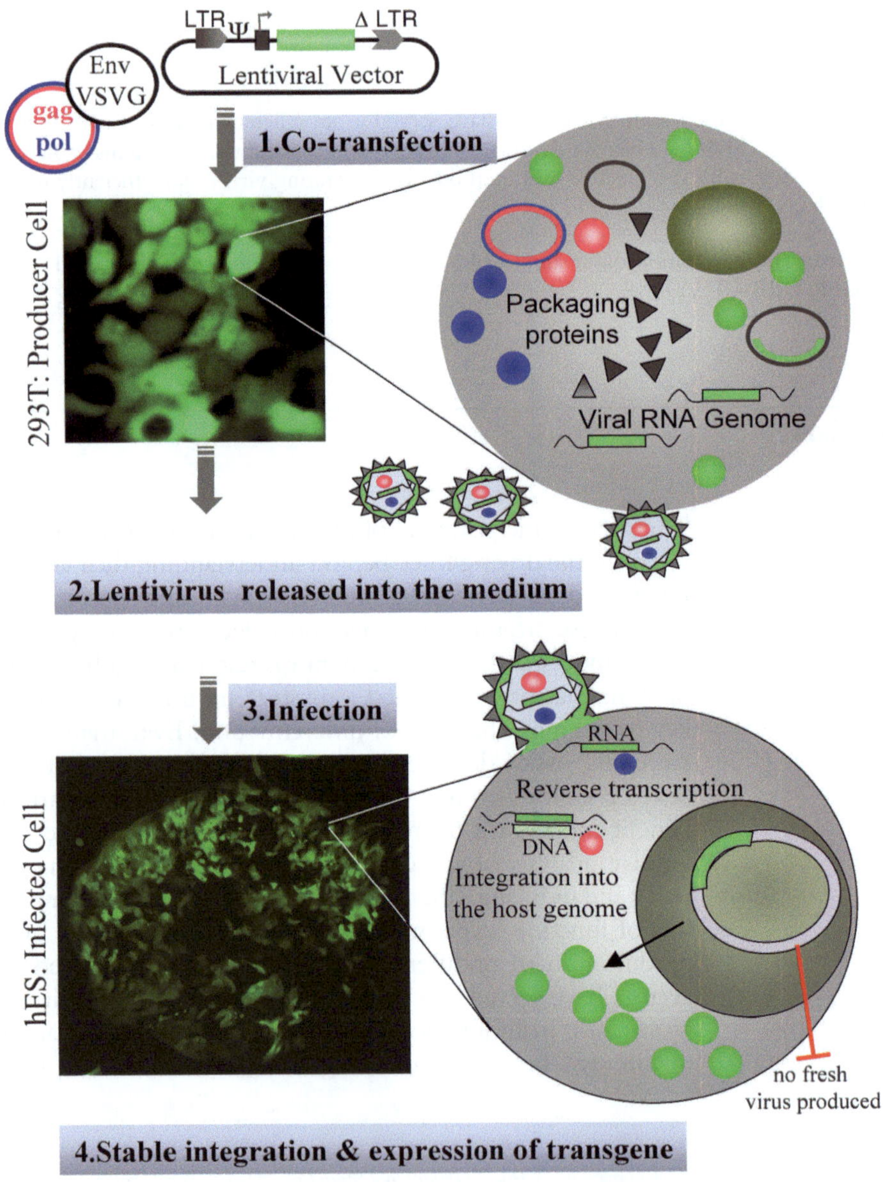

Fig. 1. Schematic representation of the process of lentivirus production in 293T cells (*top*) and transgene expression in infected human ESCs (*bottom*).

1.1. HIV-Based Lentiviral Vectors

The ability of a lentivirus to deliver its RNA genome into the infected cell where it is converted into double-stranded DNA facilitating its integration into the host genome is the basis for using lentiviruses as a shuttle vector for gene delivery (Fig. 1) (1, 2). An added advantage of using lentiviral vectors is repeated or simultaneous delivery of more than one transgene, because infection of a cell with one virion does not inhibit subsequent infections. In spite of being HIV-1-based, the commonly available vectors are relatively safe to use.

1.2. Safety Features

The following modifications have been implemented into the vector backbone and helper plasmids to minimize the risk of producing potentially hazardous replication-competent recombinants (RCR): (1) Deletion of HIV-I envelope genes, which are required for cell type-specific infection; (2) Elimination of the accessory genes required for virulence of the "unmodified" wild-type virus; and (3) Separation of the "transfer" vector (carrying only the promoters and genes of interest) from the "packaging" proteins (encoded by genes essential for the production, encapsulation, and subsequent integration of the viral genome). The packaging genes are split between two, three, or four plasmids (Table 1) and do not get included in the mature virion which, when assembled, includes their protein products. In addition, deletion of the promoter within the 3′ long-term repeat (LTR) that gets copied at the 5′ end, makes it a self-inactivating virus. In the later generation packaging systems, the HIV-1 promoter in the 5′LTR, which drives the initial round of virus production, is replaced with a non-HIV promoter.

Thus, a combination of elimination or separation of the *cis*-acting elements from the *trans*-acting proteins has significantly improved vector safety and virtually eliminated the concern for the formation of pathogenic, replication-competent virus during vector production or target cell infection.

1.3. Elements of Vector Design

Lentiviruses can infect both actively dividing and nondividing (cycling, quiescent, or terminally differentiated) cells at high efficiency and stably integrate into the genome without incurring cellular toxicity (3–6). The flanking long-terminal repeats (LTR) and the *psi* sequence are essential for packaging and infection of the nearly 10 kb viral genome. With most of the HIV genes deleted, this simple transfer vector now has a cloning capacity of approximately 9 kb and can be conveniently used for generating single- and dual-promoter constructs (see Note 1). Additional short elements such as central PPT and WPRE have been included to improve vector integration and transgene expression. Strategically-designed vectors with optimal promoters, regulatory elements, genes of interest, shRNAs, and selectable markers can be used to modulate gene expression in pluripotent stem cells

Table 1
Components of lentiviral packaging systems

HIV transfer vector	Packaging plasmids	Envelope	Remarks
First generation			
Transgene with *cis* regulatory elements	All HIV genes except *env*	VSVG	
Second generation			
Transgene with *cis* regulatory elements	*env* and accessory genes deleted	VSVG	Replication incompetent (RI)
Third generation			
Deletion of enhancer in U3 at the 3′LTR, cPPT and WPRE added	essential genes split into 1–5 plasmids[a] *env* and accessory genes deleted	VSVG	RI, self inactivating improved transgene expression reduced risk of RCR
5′ promoter replaced Deletion of enhancer in U3 at the 3′LTR cPPT and WPRE added	*tat* mutated[b] *env* etc. deleted, essential genes split into 2–5 plasmids	VSVG	minimized risk of RCR RI, self inactivating improved transgene expression

Transfer vectors are shuttle plasmids that can be amplified in bacteria and contain the lentiviral genome, marked by long-term repeats (LTRs) on either end, that is packaged in the mature virion and transmitted to the infected host cell but cannot replicate (e.g., pLV, Sin18PRRL, pCSCG, pTRIP-II, pLKO, pSICO, etc.).

Packaging Vectors include all the *trans*-acting proteins required for the production of the viral RNA genome, its encapsulation to form an infective particle and subsequent reverse-transcription and integration into the host genome. These genes are not part of the viral genome and are not transmitted.

VSVG: a plasmid encoding G envelope protein of vesicular somatitis virus-like pMD2.G or pCMV-VSVG;

cPPT: central polypurine tract improves infection and transduction of viruses; WPRE: woodchuck hepatitis viral PRE, improves transgene expression; RCR: replication competent retrovirus, requires TAT protein for transactivation of the viral LTR and synthesis of env as well as accessory genes for virus production.

[a]Packaging plasmids: pCMVΔR8.74, psPAX2 (single construct with *gag, pol, rev, tat*); p-gag-pol, pRSV-rev (*tat included*); p-gag, p-reverse transcriptase and integrase, pRSV-rev (*tat included*) etc.

[b]Packaging vectors with chimeric 5′LTR: pMDLg/pRRE, pRSV-rev (tat mutated, cannot be used to package tat-dependent vectors).

(PSCs) and provide a straightforward method for detecting the resulting changes in self-renewal as well as differentiation. Table 2 shows a variety of cassettes that have been successfully used in hESCs and can be used as a "mix-n-match" scheme to generate a wide range of vectors.

1.4. Selectable Markers

Reporter proteins are used to identify and select the infected cells and to visualize and enrich a specific subpopulation of differentiating cells when paired with a lineage-specific promoter. These include: fluorescent proteins such as eGFP and its variants and turbo RFP, which allow one to easily track the promoter activity in infected cells. They also function as readouts of coexpressed proteins or small-interfering RNAs (see Note 2). Cell surface proteins, such as the extracellular domain of mouse CD8 and CD2 proteins, can serve well for antibody-based enrichment of the

Table 2
Lentiviral vector components: "mix-n-match"

Promoter	Selectable Markers			Altering gene dosage	
constitutive or lineage restricted	Fluorescent reporter proteins	resistance genes for antibiotic selection	Cell surface proteins	Over-expression	knock-down
hPGK EF1α UbC	eGFP mCherry turbo RFP etc	Puromycin Neomycin Blasticidin Hygromycin	mCD2 mCD8	gene of interest i.e cDNA or rtTA3	shRNA mir-shRNA miRNA
OCT4 REX Nestin T(Brach.) Sox17 TetO					

pTRIPZ vector : TetO-turbo RFP-mir-shRNA- UbC-rtTA3-ires-Puromycin R

Table 3
Antibiotic selection of hESCs

Antibiotic resistance gene	Antibiotic used in medium with knockout serum replacement	Concentration range for selecting H9 hESCs (μg/mL)
Neomycin resistant genes	G418/neomycin	100–300
Puromycin *N*-acetyl transferase	Puromycin	0.5–3
Blasticidin resistant genes (i.e., BSD, a deaminase)	Blasticidin	2–5
Hygromycin phosphotransferase	Hygromycin	20–40
N/A (Tet response element)	Doxycyclin	0.2–1

expressing cells. Antibiotic-resistance genes, which have been traditionally used in mammalian cell culture experiments, work well in PSC experiments. The range of antibiotic sensitivity for H9 hESCs is listed in Table 3 (see Note 3).

1.5. Constitutive Promoters

These stably-integrated vectors can maintain sustained expression of the transgene during prolonged proliferation and subsequent differentiation. Some commonly used promoters are listed below.

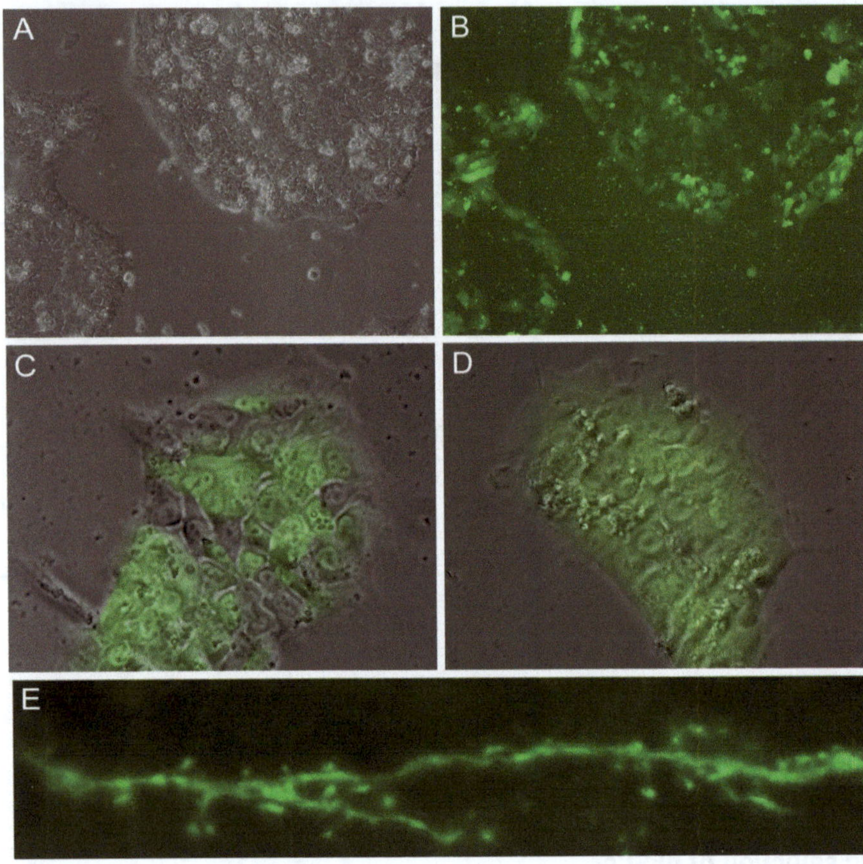

Fig. 2. Stable expression of GFP driven by hPGK promoter in hESCs and their derivatives. (**a**, **b**) Phase and fluorescence images of the same field of hESCs, 3 days post-infection. *Notice that greater than 50% of the cells are infected using this high virus-to-cell ratio.* Excess viral particles are evident as green fluorescent dots where there are no cells. (**c**) Heterogeneous colonies of infected and uninfected cells obtained after the first passage postinfection. (**d**) Several colonies uniformly expressing GFP can be manually dissected after the second passage postinfection. *Accutase is the preferred dissociation agent for passaging hESCs as small clusters (3–10 cells) that easily result in clonal colonies.* (**e**) A GFP expressing differentiating human neuron detected in an adult mouse brain 4 months after transplantation of hESC-derived neural precursors in the ventricles of neonatal mice.

The hPGK (human phosphoglycerate kinase) promoter allows for sustained high level transgene expression in undifferentiated stem cells, progenitors, and their terminally differentiated progeny. It is an ideal choice for fluorescently tracking transplanted cells in vivo (Fig. 2) (7). The EF1α (elongation factor 1 alpha) and CAAG promoters are constitutive promoters similar to the hPGK promoter with easily detectable, but comparatively lower, levels of transgene expression. The UbC (Ubiquitin C) promoter functions as a weak promoter in hESCs. eGFP expression driven by UbC promoter is often undetectable; however, sufficient antibiotic resistance is transcribed to allow for drug selection of infected cells. The weak activity of this promoter makes it ideal for expressing limited amounts of transactivator proteins, thereby reducing consequences of nonspecific interactions. The CMV (cytomegalovirus) promoter shows strong initial expression of transgenes; however, it is eventually silenced in most PSC clones.

1.6. Regulated Promoters

Lineage-specific regulation of transgene expression, whether through a selectable marker or a gene of interest, can be achieved in pluripotent cells through the use of endogenous tissue-specific enhancers and promoters (see Note 4). A few characterized promoters that have been shown to mark different lineages are Rex1 and OCT4 (undifferentiated cells), Nestin and Musashi (neural precursors), Brachyury (early mesendodermal precursors), alpha Myosin heavy chain and MLC2V (cardiac precursors), and Sox17 and PDX1 (endodermal cells) (8–14) (see Note 4).

Temporally-regulated conditional promoters respond to exogenously added inducers such as doxycycline (i.e., TRIPZ vectors) or *Cre*-recombinase (i.e., pSICO vectors). The doxycycline-inducible promoter has the added advantage of transient activation of gene expression (see Note 5). A library of human shRNAs in pTRIPZ is available from Thermo Fisher/Open Biosystems, where RFP can be replaced with a cDNA of interest to generate an inducible overexpression construct.

1.7. Some Examples

Described herein is a method for preparing third generation, replication-deficient, self-inactivating viral particles with extremely low probability of RCR formation, using a single HIV packaging construct as well as the transfer vector and VSVG pseudotyping plasmids. Examples of transfer plasmids used are SIN18-PGK-GFP and SIN18-PGK-H2B-mCherry, which constitutively express fluorescent proteins with a distinct subcellular localization pattern. SIN18-OCT4-GFP drives the expression of green fluorescence in pluripotent cells that is rapidly downregulated upon differentiation and serves as a good readout of the undifferentiated state (15). pTRIPZ is a dual promoter vector with a constitutive promoter (UbC) driving the expression of the reverse tet activator(rtTA3) and the puromycin resistance gene. In the presence of doxycycline, rtTA3 binds to the tet-regulated promoter (tetO), resulting in the expression of red fluorescence protein (turbo RFP) and the nontargeting shRNA present in its 3′ UTR (16).

2. Materials

2.1. Basic Plasmids for Lentivirus Generation

Transfer vector and packaging plasmids of transfection quality will be required. Commercial DNA purification kits (Qiagen, Marligen etc.) or cesium chloride purification can be utilized to prepare high-quality DNA (see Note 6).

1. Transfer vector (for example, pTRIPZ).
2. Packaging plasmid (for example, pCMVΔR8.74).
3. Envelope plasmid [for example, pMD2.G (VSVG)].

2.2. HEK293T Culture

1. HEK293T cells are commonly used for virus production. Their quality is crucial for obtaining high viral titers. Both the parental cell line and a high viral yield variant called 293FT are readily available from ATCC (see Note 7).
2. 293T cells are cultured in high-glucose DMEM, supplemented with 10% FBS, pyruvate, glutamine, and antibiotics (see Note 8).
3. 0.25% Trypsin–EDTA for passaging cells.

2.3. Lentivirus Production

1. *2× HeBS transfection buffer*: Dissolve 8.2 g sodium chloride, 5.8 g HEPES free acid and 0.15 g disodium hydrogen phosphate heptahydrate in a final volume of 500 mL with water. pH is adjusted to 7.05 with 5N NaOH. Filter sterilize. Aliquot and freeze (see Note 9).
2. 2.5 M calcium chloride solution made in water, filter sterilized and aliquoted. It can be frozen or stored at 4°C without a significant effect on transfection.
3. 5-mL sterile polystyrene tubes (BD 352003).
4. Vortex with medium high setting.
5. 2-mL pipettes.
6. Serum-free virus harvesting medium: Ultraculture medium (Bio-Whittaker # 12–725 F) supplemented with additional glutamine to a final concentration of 4 mM and antibiotics (see Note 10).
7. 0.22 μm, low-protein-binding syringe filters (Nalgene 190–2520) or steriflip filters (Millipore # SCGP00525).

2.4. Lentivirus Concentration

1. Ultra-clear centrifuge tubes (Beckman 344058).
2. Swinging bucket rotor (Beckmann SW28 or SW32).

2.5. Infection and Culture of Human Embryonic Stem Cells

1. Polybrene (hexadimethrine bromide, Sigma H9268).
2. PSC culture medium of choice (refer to other chapters in this volume).
3. Pluripotent stem cells (for example, H9 hESCs).
4. Accutase (Millipore SCR005).

2.6. Selection and Amplification of Transduced Cells

1. Upright fluorescence microscope.
2. Puromycin dihydrochloride (Sigma P8833) or other antibiotics of choice (see Table 3).
3. Doxycycline (see Table 3).

3. Methods (see Notes 11–16)

Lentiviral vectors designed for a variety of downstream applications are an optimal means of genetically modifying pluripotent cells. VSVG-pseudotyped viral stocks are prepared by transient cotransfection of 293T cells using a three-plasmid system. Infectious lentiviral particles harvested at 48 and 72 h posttransfection can be directly used for establishing clonal lines of infected pluripotent cells or concentrated prior to infection to increase efficiency, especially in the case of low yield, large constructs (see Fig. 1). Aliquots of concentrated virus can be frozen at −80°C.

The modified cells, which are labeled with fluorescent proteins, can be enriched manually or using a flow cytometry-based approach. Incorporating antibiotic-resistance genes within the vector allows for convenient drug selection. This protocol can be successfully scaled-up or -down as long as the ratios of the individual components are maintained.

3.1. Seeding Cells for Transfection

1. Trypsinize rapidly proliferating HEK293T cells after rinsing with DPBS.
2. Inactivate trypsin with an equal volume of HEK293T culture medium and spin down the cells at 200 ×*g*.
3. Plate sufficient cells in 10-cm tissue culture dishes in culture medium to achieve nearly 80–90% confluence at the time of transfection.
4. Cells can be transfected with calcium phosphate at the time of initial plating or up to 16 h post-plating (see Note 17).

3.2. Lentivirus Production

DAY 1: Transfection of HEK293T cells (2)

1. Add 10 μg transfer vector, 6.6 μg pCMVΔR8.74, and 3.3 μg pMD.G-VSVG (i.e., ratio of 3:2:1) in a 5-mL round bottom tube and make up to 450 μL with sterile water. Mix well.
2. Add 50 μL of 2.5 M calcium chloride solution to the DNA mixture in a dropwise fashion.
3. With a 2-mL pipette, add 500 μL of 2× HeBS buffer, dropwise with constant vortexing.
4. With the same pipette, gently transfer the transfection mixture evenly across the plate of 293T cells.
5. Incubate the cells in a 37°C CO_2 incubator for 16–24 h.

DAY 2: Medium replacement

Replace the transfection medium with serum-free virus harvesting medium and continue incubation at 37°C in a CO_2 incubator.

DAY 3, 4: Virus collection

1. Collect viral supernatant from the transfected plates every 24 h and replace with fresh medium. Each cotransfected cell releases approximately ten viral particles per day and virus yields range from 10^5 to 10^7 per mL.
2. Clear the viral supernatant by low speed centrifugation (300 × *g* in a swinging bucket rotor).
3. Filter the supernatant through a 0.22-μm syringe filter to remove cellular debris and viral aggregates (see Note 18). The supernatant is now ready for infection or can be stored for a couple of days at 4°C prior to concentration.
4. Virus can be harvested for up to 4 days, although maximum virus production is between 28 and 48 h posttransfection.

3.3. Lentivirus Concentration (Optional)

1. Pre-chill the rotor and the swinging buckets.
2. Label ultra-clear centrifuge tubes and sterilize with 70% ethanol. Allow *all* the ethanol to evaporate.
3. Add up to 34 mL of pooled lentiviral supernatant (harvested and filtered over several days or from multiple plates that were transfected with the same construct) in each tube. Six viral preps can be concentrated at once. All tubes must be filled and balanced.
4. Centrifuge at 30,000× *g* for 2–2.5 h at 4°C with the brake off.
5. Carefully transport the tubes back to the hood and discard the supernatant directly into a container with bleach in one quick motion.
6. Resuspend the viral pellet (may be translucent or invisible) in the remaining medium (200–400 μL) or additional 0.1% BSA by constant shaking for 2–12 h at 4°C (see Note 19).
7. Transfer the resuspended virus to a microcentrifuge tube and centrifuge at 15,000× *g* for 5 min to remove insoluble particles.
8. Make 20–50 μL aliquots in clearly-labeled microcentrifuge tubes. The concentrated virus is ready for infection or can be frozen at −80°C for later use (see Note 20).
9. Calculate the titer of the concentrated virus by transducing test cells with serially-diluted viral preps and directly estimating the infected cells by fluorescence monitoring, antibiotic selection, or RT–PCR-based methods.

3.4. Infection and Culture of Human Embryonic Stem Cells

1. Equal volumes of unconcentrated virus and culture medium can be added to freshly passaged cells in culture dishes with 5–10 μg/mL polybrene. For increased infection efficiency, infect cells in suspension with concentrated virus (see Note 21).

2. Collect PSCs as small clumps using Accutase or alternate methods (17).
3. Transfer the clusters to a 5-mL polystyrene tube in ~400 μL medium. Add polybrene to a final concentration of 5–10 μg/mL.
4. Add concentrated lentivirus at a multiplicity of infection (MOI) of 10–100 making sure not to exceed twice the starting amount. Loosely cap the tube and incubate in a 37°C incubator for 1–2 h.
5. Transfer the entire contents of the infection tube to an appropriate stem cell culture dish (coated with gelatin/Matrigel™ or feeder cells) with additional medium sufficient to just cover the cells.
6. Replace with fresh medium after 24 h.

3.5. Selection and Amplification of Transduced Cells

Since only a subset of the hESCs are modified, they can be selected by a variety of methods, which are described below with examples.

Manual enrichment of hESC modified with viruses bearing fluorescent markers (Fig. 3a). This works well to obtain clonal populations at low MOI.

1. Identify groups of labeled cells using a fluorescence-equipped dissecting/inverted microscope.
2. Dissect the region with a 20-μL disposable tip fixed at the end of a 1-mL syringe.
3. Transfer the dissected bit of colony to a fresh culture dish with prewarmed medium (see Note 22).

FACS sorting of hESCs expressing endogenous fluorescent proteins (Fig. 3b, c) (12). This method can be adapted to sort cells labeled by antibodies recognizing cell surface proteins.

1. Prepare 96-well plates by coating with Matrigel™ and seeding MEFs and a high density of unlabelled hESCs.
2. Culture for 2 days.
3. Prior to sorting the transduced PSCs, gamma irradiate the prepared plate at 3,000 rads so that the parental hESCs can serve as metabolically active, but replication deficient, support cells. Rinse with DPBS and replace fresh hESC medium.
4. Make a single-cell suspension of transduced hESCs using Accutase.
5. Sort the cells directly into 100% knockout serum replacement supplement.
6. Serially dilute the cells with prewarmed culture medium and transfer to the prepared plate for amplification (see Note 23).

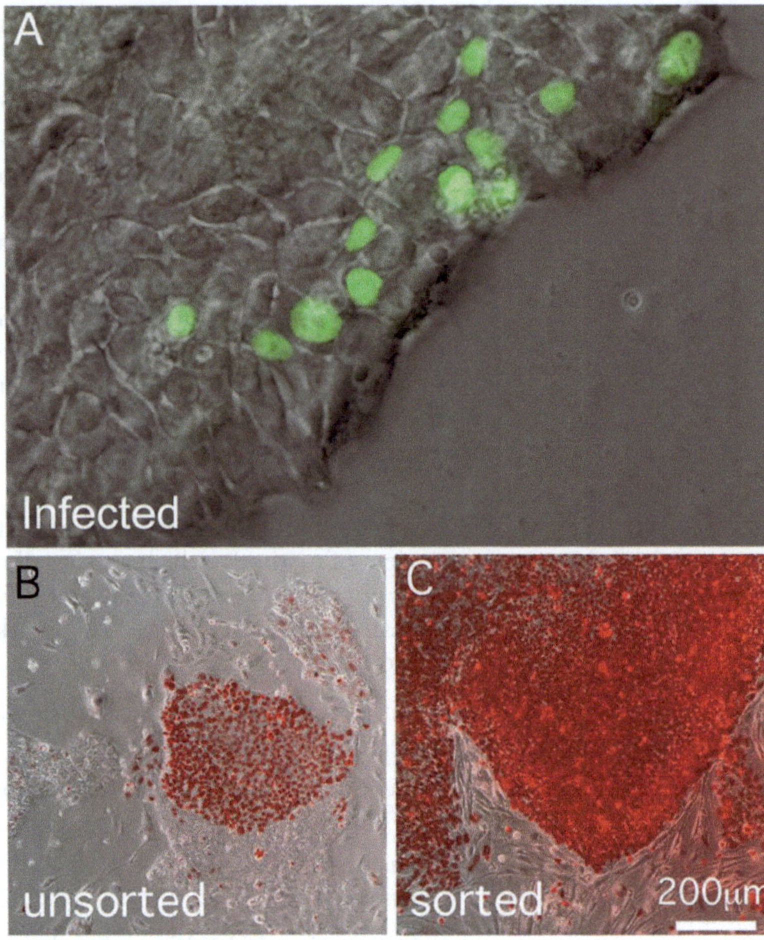

Fig. 3. Clonal populations of infected hESCs. (**a**) A group of 12 hESCs expressing nuclear GFP, imaged 7 days after infection with a lentivirus containing hPGK promoter-driven, histone 2B-GFP, fusion protein. Virus used at an MOI of ten resulted in less than 20% colonies containing clusters of 5–16 fluorescent cells each, likely representing a clonal population. (**b**) A similar amplified, unsorted culture of H2B-mcherry expressing hESCs. (**c**) A clonal population of H2B-mCherry expressing hESCs obtained by sorting single fluorescent cells into wells coated with Matrigel™, MEFs, and irradiated hESCs.

Antibiotic Selection of hESCs expressing antibiotic-resistance genes (see Table 3 and Note 24) (16).

1. Determine the minimal concentration of antibiotic sufficient to kill unmodified cells by titrating antibiotic at various concentrations on non-transduced cells.
2. Begin antibiotic selection of engineered cells 4–5 days after infection, to allow survival of clusters of infected cells as opposed to isolated single cells (Fig. 4). Alternatively, incorporate the appropriate dose of the antibiotic into the culture medium at the time of subculture/passaging. Only transduced cells will adhere and form colonies.

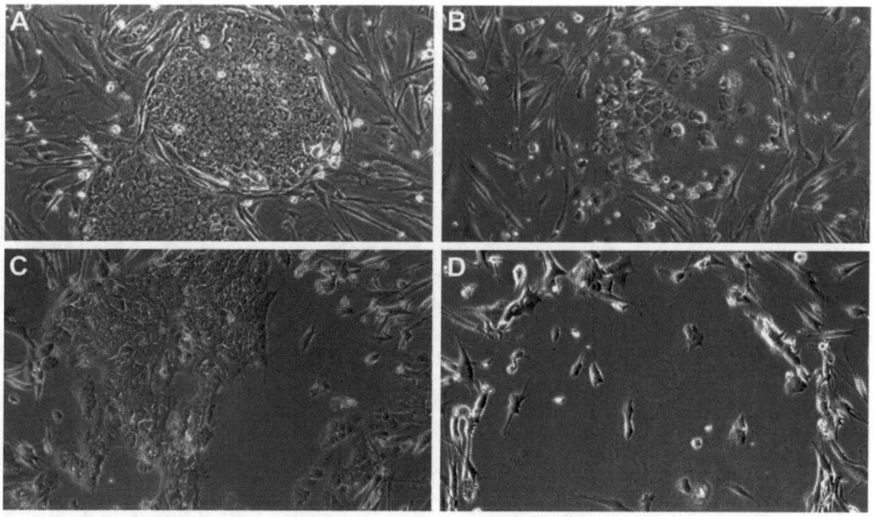

Fig. 4. Antibiotic selection of pTRIPZ-infected hESCs. (**a**, **b**) Phase-contrast images of infected cells cultured in the absence or presence, respectively, of 2 μg/mL puromycin for 36 h. (**c**) High-magnification image of cells under puromycin selection for 6 days. (**d**) Uninfected hESCs do not survive at the same concentration of puromycin. Cells imaged after 6 days in antibiotic-supplemented medium.

3. The use of promoters of genes expressed only in undifferentiated cells such as Oct4 or Rex1 can serve the dual purposes of selecting transduced and undifferentiated hESCs.
4. A caveat with using dual promoters is that they are often differentially silenced in unpredictable patterns (Fig. 5). It is, therefore, essential to screen for clones with validated expression patterns.

4. Notes

1. Increasing the vector length significantly decreases the viral packaging efficiency.
2. IRES often functions poorly in our hands. Whenever possible we tag the protein of interest with a fluorescent protein, while the shRNA of interest is incorporated in the 3′UTR of the fluorescent protein. An alternate method is to generate a fusion protein with 2A peptide sequence inserted in between that gets cleaved, separating the two.
3. It is advisable to estimate the dose response or kill curve for each cell line with respect to the antibiotic to be used for selection. This enables using minimal concentration of antibiotic during selection and maintenance of clones.

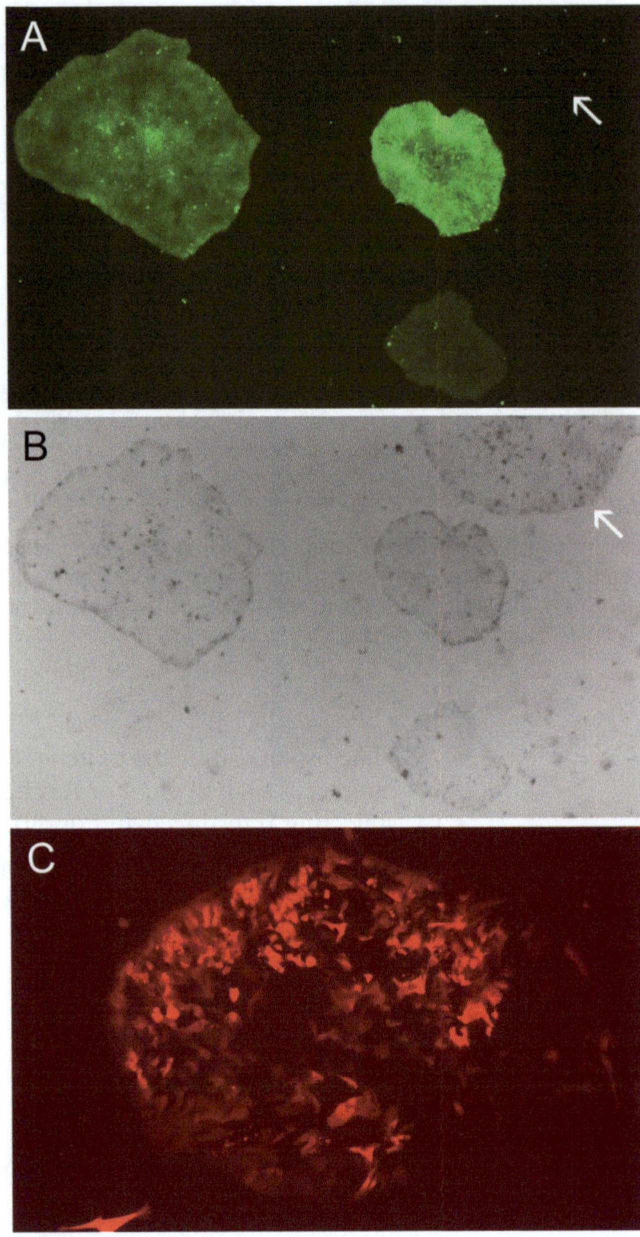

Fig. 5. Transgene expression varies in antibiotic-selected clones. (**a**, **b**) Fluorescence and brightfield images of hESCs selected for neomycin resistance for three passages after infection with SIN18-hPGK-GFP-Rex-NeoR virus. Greater than 80% of the cells express GFP, although the expression varied considerably between colonies and a few mixed-expression colonies were also detected. *White arrow* points to a neomycin-resistant colony with undetectable GFP expression. (**c**) Fluorescence image of pTRIPZ-infected hESCs, selected for puromycin resistance and amplified. In the first, as well as subsequent passages after selection, only 40–60% of the puromycin-resistant cells express red fluorescent protein in response to doxycyclin addition. It is possible to sort or manually enrich doxycyclin-regulatable clones.

4. It is important to establish promoter–reporter fidelity by testing the expression of the endogenous protein with immunostaining.
5. We highly recommend pTRIPZ vector for doxycycline inducible expression. H9 cells infected with TRIPZ have visible RFP expression within 8–24 h of doxycycline addition and can revert back to the original state of undetectable RFP in approximately 3 days after doxycycline withdrawal.
6. The transfer vectors, in spite of being multicopy bacterial plasmids, often have low yields due to their large size. We routinely do maxi-preps with 200 mL cultures grown in terrific broth.
7. Virus compatible HEK293Ts essentially are clonal lines screened for virus production.
8. Avoid overcrowding the HEK293T cultures. Passage 1:10 every 3–4 days. There is often decrease in transfection and virus production from extensively passaged cells. It is then advisable to go back to earlier frozen passages.
9. Transfection efficiency varies significantly between batches of HeBS. We empirically test several batches ranging from pH 7.0 to pH 7.3.
10. Virus can be harvested in 293T culture medium, human stem cell culture medium, or serum-free medium. We prefer serum-free medium to avoid serum-induced differentiation in stem cells.
11. Extra care and precaution are advisable while working with lentiviruses, especially if the vector includes potentially hazardous transgenes such as oncogenes.
12. Always wear personal protective equipment and avoid using sharps.
13. Extra precaution must be taken if using syringe filters to avoid spills.
14. Decontaminate all work surfaces, culture medium, stocks, and other infectious material with 0.1% sodium hypochlorite (10% Clorox bleach) prior to disposal.
15. In the case of exposure to skin due to spill or splash, wash thoroughly with soap and water for at least 15 min and immediately report to safety officer.
16. Follow your institution's safety guidelines.
17. Transfection can be done with Lipofectamine™ 2000 or other lipid-based reagents. If using these, follow manufacturer's recommendations for media specifications, plating time, and cell densities.

18. It is advisable to filter the supernatant using 0.22-μm filters. It may result in marginal viral losses, but virtually eliminates viral aggregates that cause nonuniform infection or serve as nucleation points for larger insoluble viral aggregates that are formed during the concentration process.
19. We place them in a bucket of ice on a shaker in the cold room overnight.
20. Virus can be concentrated up to a 100-fold by this method. Alternate methods using PEG precipitation often result in 10- to 15-fold concentration.
21. This method considerably reduces the excessive loss of virus to the easily infectable feeder cells and increases the effective concentration of the virus thereby increasing the infection efficiency for the same MOI.
22. When used with a warming stage, large numbers of labeled cells can be collected in relatively short periods of time without compromising the quality of the cells. However, it may be difficult to get a pure population of labeled cells.
23. After sorting, the majority of the cells either die or differentiate.
24. Use of antibiotic markers makes it necessary to use feeder cells harboring the same resistance genes.

Acknowledgements

RB thanks Maria Barcova and Mark Mercola for personal communications regarding antibiotic selection and generously sharing lentiviral constructs and Satyajit Mitra and Joanna Wysocka for their constant encouragement and support.

References

1. Naldini, L., Blomer, U., Gage, F. H., Trono, D., and Verma, I. M. (1996) *Proc Natl Acad Sci USA* ***93***, 11382–8.
2. Zufferey, R., Nagy, D., Mandel, R. J., Naldini, L., and Trono, D. (1997) *Nat Biotechnol* ***15***, 871–5.
3. Pfeifer, A., Ikawa, M., Dayn, Y., and Verma, I. M. (2002) *Proc Natl Acad Sci USA* ***99***, 2140–5.
4. Gropp, M., Itsykson, P., Singer, O., Ben-Hur, T., Reinhartz, E., Galun, E., and Reubinoff, B. E. (2003) *Mol Ther* ***7***, 281–7.
5. Ma, Y., Ramezani, A., Lewis, R., Hawley, R. G., and Thomson, J. A. (2003) *Stem Cells* ***21***, 111–7.
6. Zaehres, H., Lensch, M. W., Daheron, L., Stewart, S. A., Itskovitz-Eldor, J., and Daley, G. Q. (2005) *Stem Cells* ***23***, 299–305.
7. Bajpai, R., Coppola, G., Kaul, M., Talantova, M., Cimadamore, F., Nilbratt, M., Geschwind, D. H., Lipton, S. A., and Terskikh, A. V. (2009) *Cell Death Differ* ***16***, 807–25.
8. Ben-Shushan, E., Thompson, J. R., Gudas, L. J., and Bergman, Y. (1998) *Mol Cell Biol* ***18***, 1866–78.
9. Yang, H. M., Do, H. J., Oh, J. H., Kim, J. H., Choi, S. Y., Cha, K. Y., and Chung, H. M. (2005) *J Cell Biochem* ***96***, 821–30.
10. Yamaguchi, M., Saito, H., Suzuki, M., and Mori, K. (2000) *Neuroreport* ***11***, 1991–6.

11. Keyoung, H. M., Roy, N. S., Benraiss, A., Louissaint, A., Jr., Suzuki, A., Hashimoto, M., Rashbaum, W. K., Okano, H., and Goldman, S. A. (2001) *Nat Biotechnol* *19*, 843–50.
12. Kita-Matsuo, H., Barcova, M., Prigozhina, N., Salomonis, N., Wei, K., Jacot, J. G., Nelson, B., Spiering, S., Haverslag, R., Kim, C., Talantova, M., Bajpai, R., Calzolari, D., Terskikh, A., McCulloch, A. D., Price, J. H., Conklin, B. R., Chen, H. S., and Mercola, M. (2009) *PLoS One* ***4***, e5046.
13. Subramaniam, A., Jones, W. K., Gulick, J., Wert, S., Neumann, J., and Robbins, J. (1991) *J Biol Chem* ***266***, 24613–20.
14. Melloul, D., Marshak, S., and Cerasi, E. (2002) *Diabetes* 51 *Suppl 3*, S320–5.
15. Katkov, II, Kim, M. S., Bajpai, R., Altman, Y. S., Mercola, M., Loring, J. F., Terskikh, A. V., Snyder, E. Y., and Levine, F. (2006) *Cryobiology* ***53***, 194–205.
16. Bajpai, R., Chen, D. A., Rada-Iglesias, A., Zhang, J., Xiong, Y., Helms, J., Chang, C. P., Zhao, Y., Swigut, T., and Wysocka, J. *Nature* ***463***, 958–62.
17. Bajpai, R., Lesperance, J., Kim, M., and Terskikh, A. V. (2008) *Mol Reprod Dev* ***75***, 818–27.

Chapter 24

Nucleofection of Human Embryonic Stem Cells

Helen Fong, K.A. Hohenstein Elliott, Leslie F. Lock, and Peter J. Donovan

Abstract

The ability to realize the full potential of human pluripotent stem cells (hPSCs) as tools for understanding human development and advancing the field of regenerative medicine is dependent on efficient methods to genetically manipulate these cells. There are several methods for introducing foreign DNA into cells such as electroporation, lipid-based transfection technology, and viral transduction. We describe here a method to transfect human embryonic stem cells (hESCs) using nucleofection technology. This unique method uses the Nucleofector II Device that combines the use of a cell type-specific Nucleofector Solution and preprogrammed electrical parameters to efficiently deliver DNA into the cell nucleus. The use of this technology allows high-efficiency transfer of nucleic acids into hESCs enabling both transient and stable manipulation of gene expression in these cells.

Key words: human embryonic stem cells, pluripotent stem cells, nucleofection, transfection, transgene expression, RNA interference, neurotrophins

1. Introduction

Human pluripotent stem cells (hPSCs) are an invaluable resource for the study of mammalian development and a promising source of cells for regenerative therapy, predictive toxicology, disease modeling, and drug screening. Their potential lies in their ability to self-renew and to differentiate into cell types derived from each of the three primary germ layers (1–3). Understanding how hPSCs maintain these unique characteristics is important for achieving the full potential of these cells. Thus, genetic manipulation of hPSCs is likely to be a powerful method in determining how these characteristics are regulated. High-efficiency transfection of short-interfering RNA (siRNA), expression vectors, and targeting constructs would allow in-depth analysis of gene regulation (4).

Philip H. Schwartz and Robin L. Wesselschmidt (eds.), *Human Pluripotent Stem Cells: Methods and Protocols*, Methods in Molecular Biology, vol. 767, DOI 10.1007/978-1-61779-201-4_24, © Springer Science+Business Media, LLC 2011

Transfection of human embryonic stem cells (hESCs) is difficult without compromising the pluripotent state and is generally inefficient (5, 6). In addition, most transfection methods require dissociation of cells into a single cell format; survival of hESCs is often poor in these conditions. The addition of neurotrophin-3 (NT-3), neurotrophin-4 (NT-4), and brain-derived neurotrophic factor (BDNF) has shown to be effective in increasing clonal cell survival (7). Other factors and inhibitors, such as the Rho-associated kinase (ROCK) inhibitor, may also contribute to clonal cell survival but have not been well-tested in the specific transfection method described in this chapter (8). Nevertheless, by supplementing hESC medium with NT-3, NT-4, and BDNF, we show here an improved transfection technique for individual hESCs using the Amaxa Nucleofector II Device. This nucleofection technique efficiently delivers DNA to the cell nucleus by using a cell type-specific Nucleofector Solution in combination with preprogrammed electrical parameters. This method of transfection allows the development of true clonal cell lines from a population of nucleofected single cells while preserving their pluripotent state.

2. Materials

2.1. Human Embryonic Stem Cell Culture

1. *Human embryonic stem cell medium (hESC medium)*: Dulbecco's Modified Eagle's Medium: Nutrient Mixture F-12 (DMEM/F-12) supplemented with 20% Knockout serum replacement, 1× GlutaMAX-I, 1× MEM nonessential amino acids (all from Invitrogen), and 0.1 mM 2-mercaptoethanol (Fisher Scientific, Waltham, MA).
2. Basic fibroblast growth factor (bFGF) (Invitrogen) is dissolved at 20 μg/mL in 0.1% bovine serum albumin fraction V (Fisher Scientific) in Dulbecco's phosphate-buffered saline (DPBS) with Ca/Mg (Invitrogen) and aliquots are stored at −20°C according to manufacturer's instructions. bFGF is used at a concentration of 4 ng/mL in hESC medium.
3. Brain-derived neurotrophic factor (BDNF), neurotrophin-3 (NT-3), and neurotrophin-4 (NT-4) (Peprotech, Rocky Hill, NJ) are each dissolved at 100 ng/mL in sterile, distilled water and stored at −20°C. They are used at a concentration of 10 ng/mL in hESC medium.
4. Solution of 0.05% trypsin–ethylenediamine tetraacetic acid (EDTA) (1×) (Invitrogen) stored at −20°C according to manufacturer's instructions.
5. Soybean Trypsin inhibitor (Invitrogen) is dissolved at 1 mg/mL in DMEM-F12, incubated in a 37°C water bath until

completely dissolved, and then filter-sterilized with a 0.22-µm filtration unit. Trypsin inhibitor should be made fresh before use.

6. 40-µm cell strainer (BD Biosciences, San Jose, CA).

2.2. Nucleofection Preparation

1. RPMI medium 1640 (Invitrogen).
2. 0.1% Gelatin solution: gelatin from porcine skin, type A (Sigma-Aldrich, St. Louis, MO) is dissolved in MilliQ water, autoclaved, and stored at room temperature.
3. *Mouse embryonic fibroblast medium* (*MEF medium*): DMEM supplemented with 10% fetal bovine serum (FBS), GlutaMAX-I, and MEM nonessential amino acids (all from Invitrogen).
4. Primary mouse embryonic fibroblasts, strain CF1, mitomycin c-treated, passage 3 (MEF) are grown on gelatin-coated, 6-well, tissue culture-treated plates.
5. Amaxa Mouse ES Cell Nucleofector Kit (Lonza, Basel, Switzerland): Mouse ES Cell Nucleofector Solution, 0.5 mL Supplement 1, certified cuvettes, and certified pipettes. Combine 0.5 mL of Supplement 1 with the Mouse ES Cell Nucleofector Solution and store at 4°C.
6. pmaxGFP vector, 0.5 µg/mL (Lonza).

3. Methods

Due to the fragile nature of hESCs, it is important to complete each nucleofection reaction within 20 min or less. To minimize the procedure time, it is recommended that all materials required during the nucleofection process be readily available. All cells should also always be kept in a 37°C, 5% CO_2 incubator unless otherwise noted. In addition, the nucleofection procedure, including preparation of reagents and hESCs, should be performed in a cell culture hood to maintain sterile conditions.

The ultimate purpose of any transfection method is to introduce DNA of interest into the cells and to determine if the DNA has been successfully delivered. This particular technique delivers the DNA directly to the cell nucleus. Therefore, it is recommended that cells also be transfected with the provided pmaxGFP vector in a 5:1 ratio of DNA to pmaxGFP vector to visually identify positively transfected cells.

3.1. Preparing Reagents for hESC Nucleofection

1. Nucleofected hESCs are grown on a monolayer of MEFs (Fig. 1a). One day prior to the experiment, plate the MEFs on tissue culture treated, 6-well plates precoated with 0.1% gelatin solution, and incubate in a 37°C incubator.

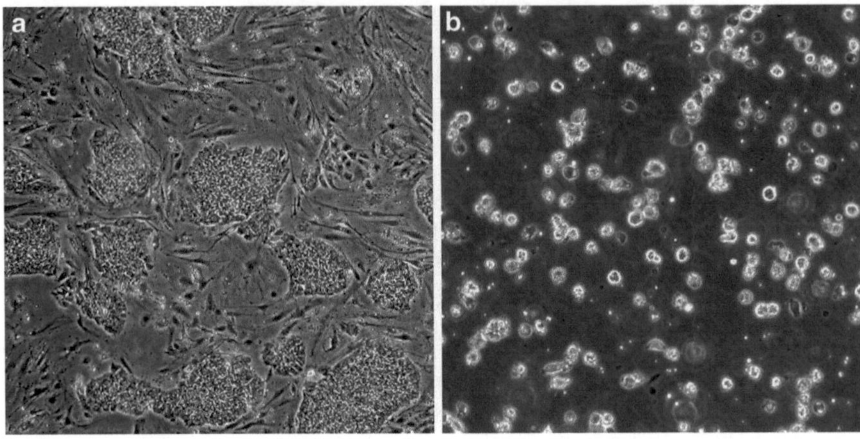

Fig. 1. Preparations of hESCs for nucleofection. (**a**) hESCs are grown to 60–70% confluency on a layer of MEFs as shown in a brightfield image at 4×. (**b**) hESCs are trypsinized to single cells prior to nucleofection.

2. On the following day, count the hESCs. Approximately 2×10^6 cells are required per reaction (Fig. 1). To determine the cell number, aspirate the medium from 1 well of a 6-well plate, rinse one time with DPBS, dispense 1 mL of 0.05% trypsin–EDTA, and incubate at 37°C for 5 min (see Note 1). Inactivate the trypsin–EDTA by dispensing 1 mL of 1 μg/mL of trypsin inhibitor. Using a 1,000-μL pipettor gently dissociate into single cells and filter the suspension through a 40-μm cell strainer to exclude cell clumps. An example of the trypsinized, single cells are shown in Fig. 1b. Count the cells using a hemacytometer and discard the cells, since this well is used to approximate the number of cells/well only (see Note 2).
3. Place Nucleofector II Device into the cell culture hood and turn on the device.
4. Aliquot 500 μL of RPMI medium into 1.5-mL microcentrifuge tubes. Keep these tubes warm by storing in the 37°C incubator prior to beginning the experiment. Warm the Mouse ES Cell Nucleofector Solution with Supplement 1 to room temperature (see Note 3). Approximately 100 μL of the Solution will be used per reaction.
5. Prepare the MEF plates, which were plated on the previous day, for nucleofection by washing one time with DPBS. Then add 2 mL/well of hESC medium supplemented with 4 ng/mL of bFGF and 10 ng/mL each of NT-3, NT-4, and BDNF and incubate at 37°C.

3.2. Preparing hESCs for Nucleofection

1. After counting the hESCs, trypsinize the correct number of wells required for the experiment and prepare as described in Subheading 3.1. Once the cells have been filtered, pellet the cells by centrifugation at $200\times g$ for 5 min at room temperature.

2. After centrifugation, visually inspect the tube to ensure the cells have been properly pelleted at the bottom of the tube. Remove the supernatant by aspiration leaving only the pellet at the bottom. Resuspend the pellet with prewarmed Mouse ES Cell Solution with Supplement 1 by gently pipetting up and down 2–3 times with a 1,000-μL pipettor. Use 100 μL of the Solution per reaction (see Note 4). Then incubate the tube containing the resuspended cells in a 37°C incubator for 5 min.
3. While the cells are incubating, prepare the DNA of interest to be nucleofected. It is recommended that 2–4 μg of DNA per nucleofection reaction be used (see Note 5). The amount of DNA, however, may be optimized depending on the conditions (see Note 6).

3.3. Nucleofection of hESCs

1. Set the program on the Nucleofector II to A-23 (see Note 7). Remove nucleofection cuvettes and pipettes from packaging and keep handy in the cell culture hood. Take the microcentrifuge tubes containing prewarmed RPMI medium from the incubator and place them in the cell culture hood.
2. After the hESCs have incubated in the Mouse ES Cell Solution for 5 min, remove from incubator, add the desired amount of DNA or RNA (see Note 8), and mix by gently pipetting up and down 2–3 times with a 1,000-μL pipettor. Pipette 100 μL of the mixture into the cuvette and cap it. Gently tap the cuvette to eliminate any bubbles and to ensure the mixture covers the bottom of the cuvette. Place the cuvette into the cuvette holder in the Nucleofector II Device and press the "X" button to start the program.
3. Once the "OK" appears on the screen of the Nucleofector II, the nucleofection program is complete. Quickly remove the cuvette from the cuvette holder. Using the plastic nucleofection pipette, transfer 500 μL of the prewarmed RPMI medium from its microcentrifuge tube to the cuvette. Gently transfer the entire mixture from the cuvette back into the microcentrifuge tube and incubate at 37°C for 5 min.
4. After 5 min, gently transfer the nucleofected cells into one well of the 6-well MEF plate containing hESC medium supplemented with bFGF, NT-3, NT-4, and BDNF. Incubate at 37°C.
5. Continue to culture the cells 24–48 h after nucleofection, replacing the medium daily with fresh hESC medium supplemented with bFGF, NT-3, NT-4, and BDNF. There will be some cell death during the 24 h following the procedure. Figure 2 shows cultures of hESCs over a period of 4 days following nucleofection.

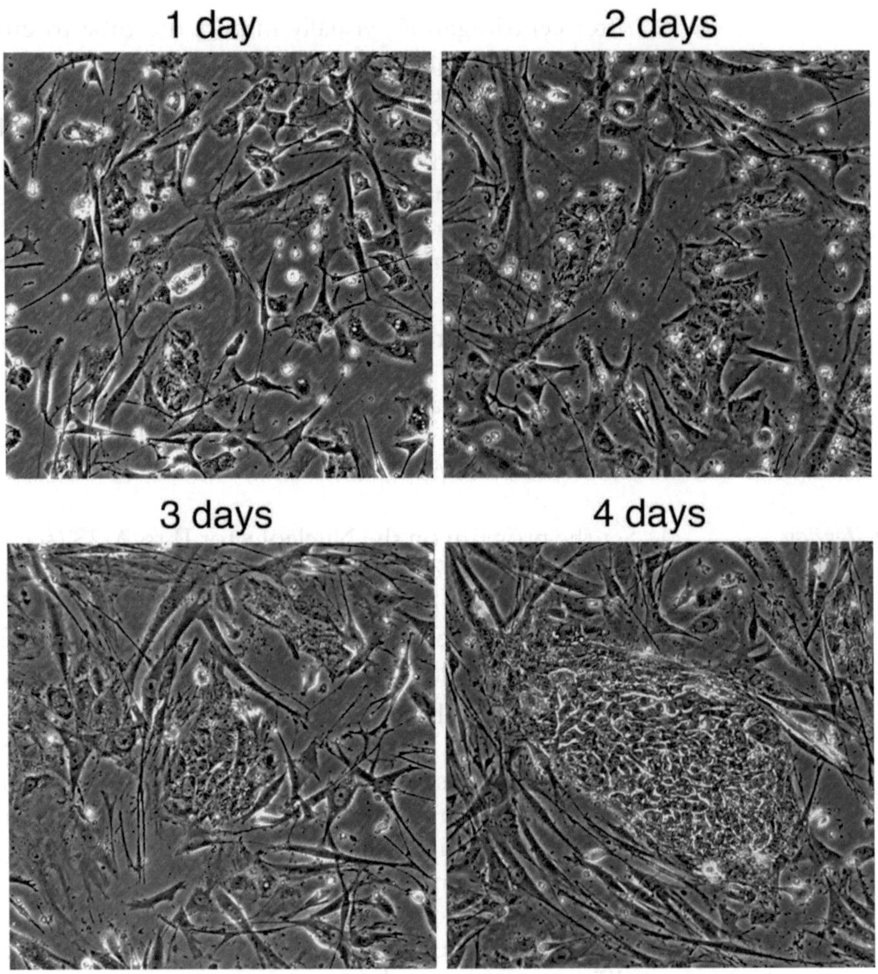

Fig. 2. Nucleofected hESCs in culture. Phase-contrast images of hESC at 10× are shown over a period of 4 days. Single cells seen after 1 day of nucleofection, which multiply over time to form colonies of hESCs.

6. If nucleofected cells contain a selectable marker, selection can begin 96 h after transfection to obtain stably transfected hESC colonies (see Note 9). An example of hESCs nucleofected with the pmaxGFP vector is shown in Fig. 3 (see Note 10).

4. Notes

1. If the hESCs have been plated at equal density in each well of the 6-well plate, then trypsinizing and counting a single well can serve as a reasonable representation of the number of cells present in each well.
2. hESCs should not be kept in trypsin for an extended period of time during cell counting as they may become unhealthy

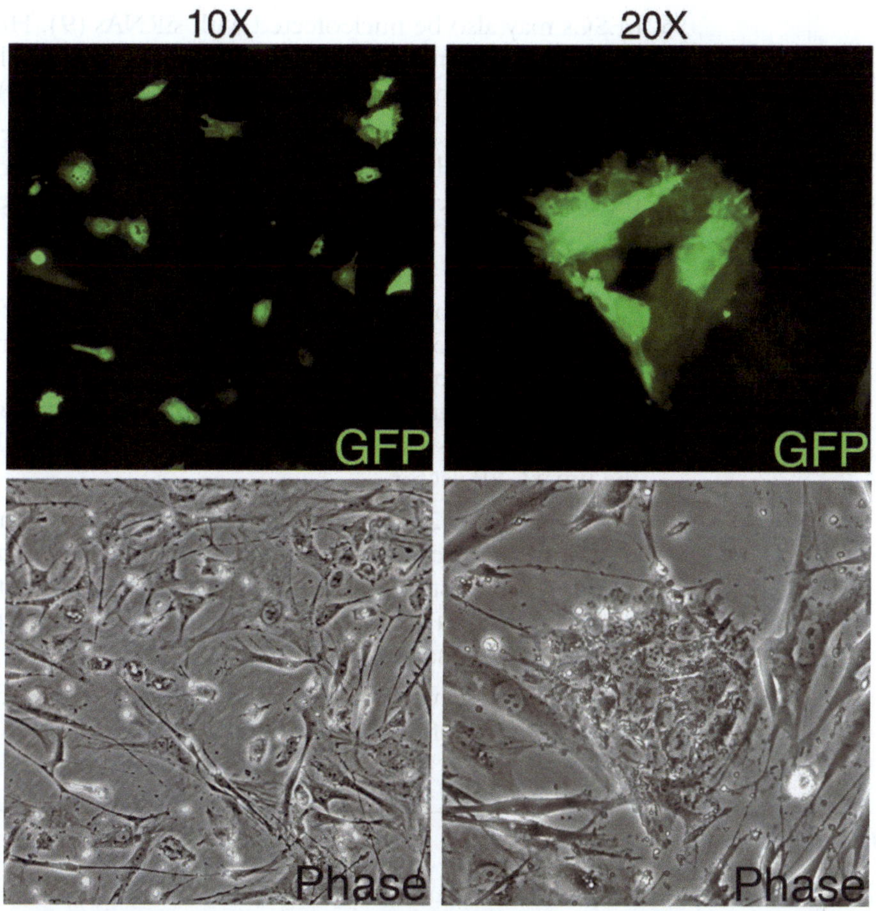

Fig. 3. hESCs nucleofected with pmaxGFP vector. Single GFP-positive hESCs (*top left*) and corresponding phase-contrast image (*bottom left*) are shown at 10×. A GFP-positive colony (*top right*) and corresponding phase-contrast image (*bottom right*) are shown at 20×.

for nucleofection. Thus, we usually discard the cells after counting and start with a fresh set of cells when beginning the nucleofection experiment to ensure maximum nucleofection efficiency and cell survival.

3. Human Stem Cell Nucleofector Kits from Lonza are also available for use with the Nucleofector II Device but we have not tested their efficiencies. Therefore, it is important to use the Mouse ES Cell Nucleofector Kit for this protocol.

4. The hESC mixture will be slightly cloudy at this point. It is important to be particularly gentle when pipetting the mixture as the cells are especially fragile. Rough or excessive pipetting will result in increased cell death.

5. It is recommended that the volume of DNA added be kept under 4 μL. We have found that increasing this volume results in undesired differentiation of nucleofected hESCs.

6. hESCs may also be nucleofected with siRNAs (9). However, the concentration must be optimized depending on the conditions, due to the possibility of off-target effects (10).
7. Other programs are also available on the Nucleofector II, but we have found that the A-23 program in combination with the Mouse ES Cell Nucleofector Solution is the most efficient for introducing DNA into hESCs (4).
8. We recommend including two controls in addition to the experimental samples. A sample of nucleofected cells containing no DNA should be included as the negative control to ensure that the procedure itself is not modifying the cells in any way. As a visual indicator and as a measure of transfection efficiency, we also recommend nucleofecting the cells with the pmaxGFP vector.
9. G418 sulfate at 25 μg/mL can be used 96 h after transfection, if hESCs were nucleofected with a construct containing a neomycin resistant gene. G418 concentration can then be increased to 50 μg/mL after 1 week and then to 100 μg/mL after 10 days posttransfection.
10. If the pmaxGFP vector has been used, the GFP should be visible within 4 h of nucleofection.

Acknowledgements

We would like to thank Dr. Raymond Ching-Bong Wong and Marek Mandau for helpful comments on this manuscript. This work was supported by funding from the National Institutes of Health (HD149488; HD47675), the California Institute for Regenerative Medicine (RC1-00110), and by institutional funds from the University of California, Irvine to P.J.D.; by a CIRM Predoctoral Training Fellowship (T1-00008) and by the Cellular and Molecular Biosciences Graduate Program at the University of California, Irvine to H.F.; by the Human Genetics and Molecular Biology Training Program (NIH 2T32GM07814) at Johns Hopkins University School of Medicine to K.A.H.E.; and by institutional funds from the University of California, Irvine to L.F.L.

References

1. Donovan, P. J., Gearhart, J. (2001) The end of the beginning for pluripotent stem cells *Nature* **414**, 92–7.
2. Thomson, J. A., Itskovitz-Eldor, J., Shapiro, S. S., et al. (1998) Embryonic stem cell lines derived from human blastocysts *Science* **282**, 1145–7.
3. Trounson, A., Pera, M. (2001) Human embryonic stem cells *Fertil Steril* **76**, 660–1.
4. Hohenstein, K. A., Pyle, A. D., Chern, J. Y., Lock, L. F., Donovan, P. J. (2008) Nucleofection mediates high-efficiency stable gene knockdown and transgene expression in human embryonic stem cells *Stem Cells* **26**, 1436–43.

5. Lakshmipathy, U., Pelacho, B., Sudo, K., et al. (2004) Efficient transfection of embryonic and adult stem cells *Stem Cells* **22**, 531–43.
6. Kobayashi, N., Rivas-Carrillo, J. D., Soto-Gutierrez, A., et al. (2005) Gene delivery to embryonic stem cells *Birth Defects Res C Embryo Today* **75**, 10–8.
7. Pyle, A. D., Lock, L. F., Donovan, P. J. (2006) Neurotrophins mediate human embryonic stem cell survival *Nat Biotechnol* **24**, 344–50.
8. Watanabe K., Ueno M., Kamiya D., et al. (2007) A ROCK inhibitor permits survival of dissociated human embryonic stem cells *Nat Biotechnol* **25**, 681–6.
9. Fong, H., Hohenstein, K. A., Donovan, P. J. (2008) Regulation of self-renewal and pluripotency by Sox2 in human embryonic stem cells *Stem Cells* **26**,1931–8.
10. Tschuch C, Schulz A, Pscherer A, et al. (2008) Off-target effects of siRNA specific for GFP *BMC Mol Biol* **9**, 60.

Chapter 25

Nonviral Gene Delivery in Neural Progenitors Derived from Human Pluripotent Stem Cells

Sujoy K. Dhara, Anirban Majumder, Mahesh C. Dodla, and Steven L. Stice

Abstract

Human pluripotent stem cells (hPSCs) have been used to derive self-renewing neural progenitor (NP) cell lines. Here we describe methods to genetically modify these cells. Detailed methods for transfection and nucleofection in PSC-derived NP cells are presented. We have shown that nucleofection results in higher yield of GFP+ NP cells as compared with transfection. However, nucleofection leads to higher cell death than transfection. Application of these methods allows for the development of novel tools to study human development and cellular differentiation. Genetically modified NPs have direct application in neural imaging, tracking neural cells, and for drug delivery to target organs using neural progenitor cells as carriers.

Key words: neural progenitors, human pluripotent stem cells, genetic manipulation, transfection, nucleofection

1. Introduction

Human embryonic stem cells (hESCs) have three remarkable features: (a) they are pluripotent, and therefore, have been used to generate cells and tissues of different lineages; (b) they are highly prolific, and therefore, can supply unlimited numbers of cells that are required for research and therapy; and at the same time (c) they retain normal genetic architecture, unlike many highly prolific transformed cell lines (1–3). Overexpression of a specific set of transcription factors in somatic cells has been shown to reprogram the somatic cell and generate induced pluripotent stem cells (iPSCs) (4–6). Both hESCs and hiPSCs are considered invaluable resources for studying early human embryonic development and differentiation. These human pluripotent stem cells

Philip H. Schwartz and Robin L. Wesselschmidt (eds.), *Human Pluripotent Stem Cells: Methods and Protocols*, Methods in Molecular Biology, vol. 767, DOI 10.1007/978-1-61779-201-4_25,

(PSCs) can form tissue types belonging to all three embryonic germ layers (ectoderm, mesoderm, and endoderm) (1, 2, 4–6) as well as extraembryonic ectoderm (7).

Neural cells are formed from ectoderm (reviewed in ref. 8) and significant effort has been made to differentiate PSCs directly into the three fully differentiated neural cell types: neurons, astrocytes, and oligodendrocytes (9–11). In addition, stem cell-like populations committed to the neural lineage, known as neural progenitors (NPs), have been derived from PSCs. These highly prolific PSC-derived NPs express markers such as NESTIN, MUSASHI-1, SOX2, PSA-NCAM, and CD133, which are characteristic of NP cells. PSC-derived NPs can be maintained in culture over several propagations in a serum-free medium without compromising their potential to generate all three types of terminally differentiated neural cells (12–15). This self-renewing ability of PSC-derived NPs avoids the difficulties, time, and expense of maintaining hPSCs in continuous culture as well as the time and expense required for the differentiation of NPs from hPSCs each time one wants to study NP biology (12). NPs offer the opportunity to study cellular differentiation into region-specific neural cell types (16), neural circuitry, neurodegenerative diseases, and they can be used in drug and toxicology screening programs (17), as well as delivery vehicles for drugs. However, to fully explore these possibilities, one needs optimized tools that are suited to enhancing these applications.

Genetic manipulation of NP cells is required to fully utilize these specialized cells for understanding the molecular mechanisms that drive their proliferation and differentiation. Modifying the genome of these cells allows one to specifically alter the expression of endogenous genes as well as express exogenous genes, including fluorescent reporter proteins that allow for the tracking of the modified cells in transplant models. These important technologies are also part of a strategy that allows the study of gene function in the context of the genetic background of the NP cell line. Though gene delivery techniques have been developed for many cell types including hPSCs (18) and adult stem cells (19–23) (see the review in ref. 24), until very recently, these methods were not optimized for NP cells derived from hPSCs (25).

DNA delivery systems can be broadly classified into viral and nonviral methods (26). Further, the nonviral (nonbiological) gene delivery methods are of two types: transfection and electroporation. Transfection is the method of DNA delivery into cells using various chemicals. These chemicals include various nanoparticles (27), calcium phosphate (28), various hydrophobic polymers (such as oligosaccharides or cyclodextrin), cationic peptides (29), cationic lipids (30), cationic polymers including poly

L-lysine, chitosans (31), or dendrimers (32). During the transfection process, temporary holes are formed in the cell membrane allowing DNA to pass through. Among the cationic polymers used for transfection, linear poly-ethylenimine (PEI) molecules are available with a high cationic-charge density potential and protonatable amino nitrogen atoms. The cationic charge makes PEI an excellent DNA condensing and gene-delivery agent, whereas the protonatable amino nitrogen makes the polymeric association an effective "proton sponge" escaping lysosomal degradation of the packaged DNA (33). ExGen500 (Fermentas Life Sciences), a commercially available PEI molecule, has been used for transfection in hESCs (18) and, recently, for hESC-derived NP cells (25).

Electroporation, or electropermeabilization, is the other type of nonviral gene delivery system. In electroporation, the cell membrane is permeabilized under an electrical field. Cells are suspended in an electroporation solution, placed into a cuvette and a brief, but high-voltage electrical charge is applied to the cuvette, causing temporary holes to form in the cell membrane that allow DNA to enter the cells (34). Recently, a proprietary modification of electroporation and transfection methods known as nucleofection (Amaxa and Lonza) has been shown to be successful in transfecting many cell types including hESCs (35–37) and PSC-derived NPs (25). Though gene delivery using viral systems (known as transduction) is often described as the most efficient method for hPSCs, the viral system requires elaborate measures to ensure bio-safety in laboratory practices. Hence, in many laboratories, nonviral gene delivery by transfection or nucleofection is preferred. Moreover, we have found both transfection and nucleofection to be useful in experiments where a small number of cells are available. Here we describe our methods to genetically modify hESC-derived NP cells by transfection and nucleofection.

2. Materials

2.1. Neural Progenitor Cells and Cell Culture Reagents

1. *Human neural progenitors (NP) cells*: STEMEZ hNP™ (Cat# hNP17009, Aruna Biomedical, Athens, GA). Here onwards they are referred to as NP cells. Other sources of neural progenitor cells may work equally well. See (Chapter 30) for methods to differentiate NPs from PSCs.
2. *Human neural progenitor (NP) propagation medium*: Protect all components from light. Store all thawed components and formulated medium at 4°C, and avoid wide variation in

temperature. For 50 mL medium, mix the following components and pass through a 0.22 μm sterile filter:

(a) 0.5 mL (1%) penicillin–streptomycin (5,000 U/5,000 μg stock).

(b) 1.0 mL (1×) ArunA ANS™ (Cat# hNP7011.2, ArunA Biomedical, Athens, GA) (see Note 1).

(c) 0.5 mL (2 mM) L-glutamine (200 mM stock).

(d) 20 μL (20 ng/mL) FGF2 (Cat# F-0291, Sigma-Aldrich).

(e) 50 μL (10 ng/mL) LIF (Cat# LIF1010, Millipore).

(f) Makeup to 50 mL with ArunA AB2™ (Cat# hNP7011.3, ArunA Biomedical) (see Note 3).

3. Trypan blue solution (0.4% solution), for live and dead cell counting.
4. Propidium iodide (PI, Cat#556463, BD Pharmingen) for counting dead cells in flow cytometry.
5. Poly-L-ornithine hydrobromide (Cat# P3655, Sigma-Aldrich): Prepare a 10 mg/mL stock in sterile tissue culture grade water and store at −20°C in aliquots to avoid repeated freeze-thaw.
6. Laminin from Engelbreth-Holm-Swarm murine sarcoma basement membrane (aqueous solution, 1 mg/mL stock (Cat# L2020, Sigma-Aldrich)). Store at −20°C and thaw on ice.
7. Tissue culture 35 × 10 mm dish.
8. Cell scraper.
9. Dulbecco's Phosphate-Buffered Saline ($DPBS^{-/-}$) without calcium and magnesium.
10. Dulbecco's Phosphate-Buffered Saline ($DPBS^{+/+}$) with calcium and magnesium.

2.2. Reagents and Equipment for Transfection and Nucleofection

Store all reagents at 4°C. For long-term storage, plasmid DNA vectors should be stored at −20°C. Once the Nucleofector® Supplement is added to the Nucleofector® Solution, it is stable for 3 months at 4°C.

(a) Plasmid maxGFP® Vector (provided with Amaxa Nucleofector kits as control) in Tris EDTA buffer (see Note 4).

(b) ExGen500 (Cat# R0511, Fermentas Life Sciences).

(c) Sodium chloride (NaCl, ≥99.5% pure cell culture tested) 150 mM in sterile tissue culture-grade water (Cat# SH30529.02, ThermoFisher).

(d) Nucleofector® Rat Neural Stem Cells Kit (Cat# VPG-1005) (Lonza) (see Note 5).

(e) The Nucleofector® Device (Cat#AAD-1001, Lonza).

3. Methods

3.1. Preparation of Poly-L-Ornithine–Laminin-Coated Dishes

Poly-L-ornithine–laminin-coated dishes provide an appropriate extracellular matrix for NP cells to adhere and propagate. Prepare dishes (see Note 6) coated with poly-L-ornithine and laminin as described below:

1. Thaw poly-L-ornithine and laminin on ice and dilute them in sterile tissue culture-grade water.
2. Dispense 2 mL of poly-L-ornithine (20 μg/mL) per 35 mm dish and incubate at 37°C for an hour.
3. Aspirate poly-L-ornithine and rinse the dish once with 1 mL of water.
4. Dispense 2 mL of laminin solution (5 μg/mL) per 35 mm dish and incubate at 37°C for 1 h. Store dishes at 4°C until needed. Use within 3 weeks.

3.2. Thawing of Human Neural Progenitors

1. Bring NP propagation medium to 37°C or to room temperature.
2. Warm poly-L-ornithine–laminin-coated dishes in 37°C incubator for 15–30 min before plating the cells.
3. Remove a vial of NP cells from cryogenic storage.
4. Add 0.5–1.0 mL of freshly prepared, warmed NP medium directly to the cryogenic vial and allow the frozen cells melt at room temperature in the tissue culture hood.
5. While waiting for the cells to thaw, add 4 mL of NP medium to a 15 mL centrifuge tube.
6. As soon as cells start to thaw, gently transfer them to the 15 mL centrifuge tube using a 5 mL pipette.
7. Centrifuge at 200 × *g* for 4 min at room temperature.
8. Remove the supernatant and add 2 mL of fresh NP medium.
9. Mix the cells gently, using 5 mL pipette, and transfer to a prepared 35 mm dish.
10. Place the cells in a 37°C, 5% CO_2 incubator.
11. Replace with fresh NP proliferation medium every 2 days.

3.3. Subculture of Human Neural Progenitors

1. Once the (STEMEZ hNP™) NPs become about 90% confluent, aspirate the medium.
2. Add 2 mL of pre-warmed propagation medium. Use a cell scraper to detach the cells from the dish and triturate them manually using a 5 mL pipette (see Note 7).
3. Count live cells using hemacytometer and Trypan blue staining to identify dead cells. Add 10 μL of cell suspension and 10 μL

Trypan blue solution in a microfuge tube, mix well by using 20 μL pipette. Transfer 10 μL of cell mixture to Neubauer's chamber and count unstained and blue stained cells in each of 0.1×0.1 mm blocks in the chamber.

4. Count the live cells. The concentration of live cells is the (number of unstained cells $\times 2 \times 10^4$ cells/mL). Adjust the volume of medium to obtain a cell density of 1×10^6 cells/mL.
5. Add 1 mL (1×10^6 cells) of cell suspension to each of new 35 mm dish, and make up the volume in the dish to 2 mL with NP medium. When maintaining a high-density culture, NP cells can be propagated at 1:2 to 1:3 ratio. If required, the cells can be frozen at any passage and thawed later for further experiment (see Note 8).

3.4. Transfection Protocol of Human NP Cells Using ExGen500

The protocol described here is a modification of published protocols, one which was developed for hESCs (38) and one used to transfect NP cells using ExGen500 (25).

1. Grow NP cells to ~80% confluence on 35 mm poly-L-ornithine–laminin-coated dishes.
2. For each 35 mm dish, dilute 2 μg of DNA into 100 μL of sterile 150 mM sodium chloride solution. Vortex gently and then spin down the solution.
3. Add 10 μL ExGen500 to the DNA solution (not the reverse order) and vortex the solution immediately for 10 s, and then spin down briefly (see Note 9).
4. Incubate DNA–ExGen solution mixture at room temperature for 10 min (see Note 10).
5. While waiting on the incubation, replace the culture medium of the NP cells with 1 mL of fresh medium.
6. Uniformly disperse 100 μL of DNA/ExGen500 solution drop-by-drop into each dish. Gently swirl the dish to achieve an even distribution of DNA complexes.
7. Transfer to 37°C incubator and leave it overnight.
8. Remove culture medium after 24 h and wash three times with DPBS$^{+/+}$. Then add 2 mL of fresh medium.
9. Return plate to cell culture incubator. Transient transfection will reach its peak expression after approximately 48 h. For stable transfection, linearized plasmid should be used to avoid undesired fragmentation of transgene cassettes (see Note 11).

3.5. Nucleofection Protocol for Human NP Cells

All work should be done under aseptic conditions using sterile technique in a biosafety cabinet, except for the electroporation of the cells during nucleofection which can be performed on bench-top outside of the hood. Add the entire supplement to the Nucleofector® Solution before use. As with transfection, one may use a linearized plasmid to achieve higher levels of stable integration as described in Note 11.

1. Place the Nucleofector Device near the tissue culture hood, and set the program to A-33.
2. Harvest NP cells using the cell scraper and count live cells using Trypan blue (see Subheading 3.3, Steps 3 and 4) and wash with DPBS$^{+/+}$.
3. For each nucleofection, add 1.5 mL of propagation medium to a 35 mm poly-L-ornithine–laminin-coated dish and an additional 0.5 mL of medium to a sterile microfuge tube. Leave them in the tissue culture incubator to equilibrate.
4. Add the plasmid DNA (5 μg) to the nucleofection cuvette.
5. Resuspend one million NP cells in 100 μL of reconstituted Nucleofector Solution at room temperature and transfer to the cuvette containing the DNA. This will allow the DNA to be well dispersed in the cell suspension (see Note 12).
6. Dislodge bubbles by gently tapping the cuvette.
7. Cap the cuvette and insert it into the cuvette holder in the Nucleofector device and run the program by pressing the "X" button on the device.
8. Immediately after pulsing, take out the cuvette. Transfer 0.5 mL of pre-warmed medium from the microfuge tube to the cuvette using transfer pipette provided in the kit. Avoid repeated pipetting (see Note 13).
9. Gently transfer cell suspension back to microfuge tube and leave for 5–10 min in the incubator.
10. Label the tissue culture dish containing 1.5 mL of medium. Add the cell suspension from the microfuge tube to the dish drop-by-drop and place the dish in the incubator.

3.6. Assessment of Gene Delivery

In addition to the intended molecular changes introduced through transfection, one should also assess the impact of the gene delivery method by monitoring: (a) the efficiency of DNA delivery and (b) cell death. Efficiency of gene delivery measures how many cells received the construct and how many copies of transgene are present in each transgenic cell. Since gene delivery systems rupture the cell membrane, many of the cells die after the procedure. Thus, assessment of cell death after gene delivery should be considered a critical criterion for optimizing gene delivery techniques. Usually, more efficient methods are harsh and therefore, result in more cell death. The following is a procedure to assess the overall efficiency of a gene delivery system.

1. Forty-eight hours after transfection or nucleofection, examine the dishes, and take pictures of several random fields under bright field and (GFP) fluorescent microscopy (Fig. 1). The percent of GFP^{+} cells can be estimated from these pictures.

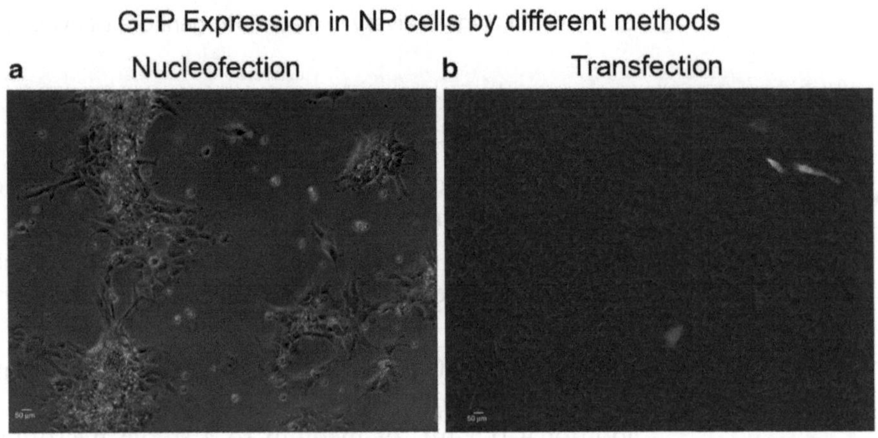

Fig. 1. Green fluorescent protein (GFP) expression in neural progenitors (NP) derived from human embryonic stem cells (hESCs). An EGFP-expressing DNA construct (pmaxGFP) was delivered into NP cells by two different methods: nucleofection and transfection with PEI (ExGen500). Fluorescent and bright field images were captured 48 h after the delivery of the DNA. Shown are the merged images of both bright and fluorescent fields. (**a**) Nucleofection was performed with "Rat Neural Stem Cell" nucleofection-specific reagent with electrical settings A33. (**b**) Transfection was done with ExGen500, adapting a protocol described previously for hESCs. Note that though nucleofection generates more GFP expressing cells, it also produces more dead cells than ExGen500-based transfection in culture.

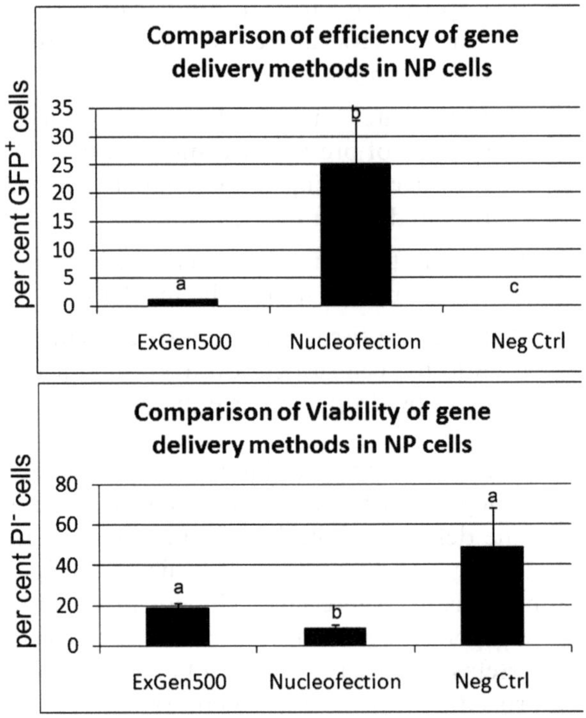

Fig. 2. Efficiency of gene delivery and viability of neural progenitor (NP) cells after genetic manipulation. Neural Progenitors were transfected with GFP-expressing DNA construct by nucleofection and ExGen500 transfection. Forty-eight hours posttransfection, flow-cytometric analysis was performed on harvested cells. To count dead cells, propidium iodide was added to cells prior to flow cytometry. Shown are bar diagrams of average percents (three replicates) of live cells (PI negative cells, *bottom panel*) and GFP^{+}/PI^{-} cells (*top panel*). Non-engineered NP cells were used as a negative control. The superscript on each bar indicates statistical differences for both methods with negative control between both treatments. (Reproduced from (25) with permission from Mary Ann Liebert, Inc.).

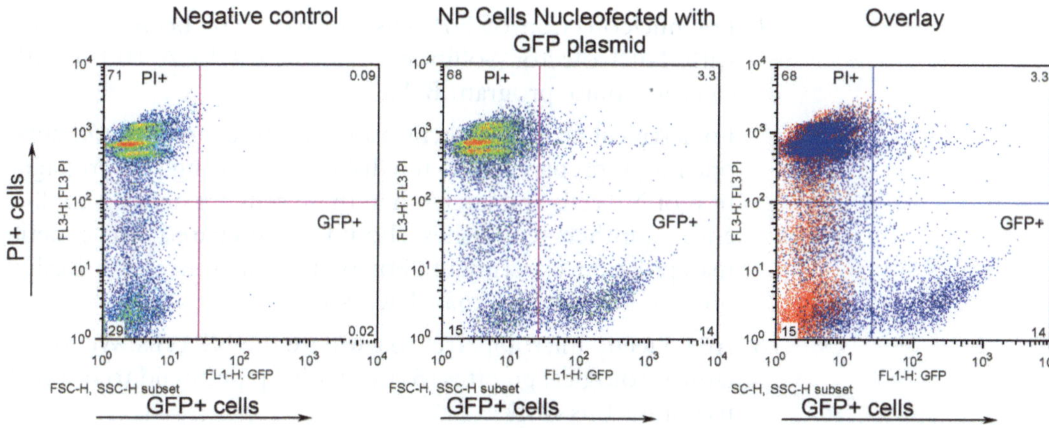

Fig. 3. Scatter plots of flow-cytometry data of nucleofected neural progenitors (NP) derived from human embryonic stem cells. Neural progenitor cells were nucleofected with plasmid maxGFP and flow-cytometry measurements were done with a FacsCalibur (BD Bioscience, San Jose, CA). Dead cells were counted by adding PI prior to running the samples in the cytometer. Data analysis was performed using FlowJo software (Tree Star, Ashland, OR). Shown are scatter plots for negative control cells (*left panel*), nucleofected cells (*middle panel*), and an overlay of both plots (*right panel*).

2. The percent of dead cells can be determined by flow-cytometry. Wash the cells with DPBS$^{-/-}$ and harvest them in DPBS$^{-/-}$ using cell scrapers (no enzymatic treatment required). Centrifuge at 250×*g* for 5 min at room temperature and store cells on ice before running flow-cytometry experiment.
3. Add PI (20 μg/mL) to the cell tubes just before performing flow cytometry to estimate percent of GFP$^+$ and PI$^+$ cells. GFP$^+$ cells will indicate efficiency and PI$^+$ cells will indicate dead cells caused by the gene delivery method (Figs. 2 and 3).

4. Notes

1. Alternative to ArunA ANS™: 1.0 mL B27 (50×) (Cat#17504-044, Invitrogen).
2. Alternative to ArunA AB2™: Neurobasal medium (Cat#21103–049, Invitrogen).
3. We have also used Plasmid DNA: pZsGreen1-N1 (Cat#632448, Clontech, Mountain View, CA) and found similar results with NP cells. For plasmid DNA isolation, we preferred using the endotoxin-free maxiprep kit for a better transfection rate. DNA quality: OD 260/280 ratio should be between 1.8 and 2.0.

4. For nucleofection of NP cells, we have also been successful with Nucleofector Solution Kit V (Cat# VCA-1003) with electrical pulse program B-16.
5. For a typical set of transfection and nucleofection experiments, one might prepare five to ten dishes for thawing and propagation of NP cells and later, when these cells are nearing confluence, another ten dishes should be prepared for genetic manipulation experiment. For better adherence, use freshly prepared dishes (less than 1 week old).
6. Avoid using micropipettes to triturate the NP cells. We found large serological pipettes (5 mL) with a pipette-aid to be most useful for this work.
7. NP cells can be expanded over many passages. Cytogenetic analysis shows that these cells have normal karyotype for at least 40 passages.
8. Further optimization of the ExGen500 to DNA ratio may increase transfection efficiency.
9. Longer incubation (up to 15–20 min) of DNA–ExGen solution does not seem to affect the transfection outcome. However, a minimum of 10 min incubation is required to allow the formation of DNA-PEI globules.
10. When making stable cell lines, linearize the plasmid. Plasmid pZsGreen1N1, containing a neomycin resistance gene, was digested with restriction enzyme ApaLI. Selection for resistance to neomycin gene was initiated 72 h post-DNA delivery with drug G418 (200 μg/mL) and continued for 2 weeks.
11. One can use up to 5×10^6 cells for each nucleofection experiment. Also, thorough mixing of the DNA with the cell suspension produces better nucleofection efficiency, but it should be done with minimal pipetting.
12. After applying the nucleofection pulse, white frothing can be seen on the top of the cell suspension inside the Nucleofector cuvette. This is due to the salt and other materials present in the solution. However, we did not find any adverse effect on the outcome of nucleofection efficiency due to frothing.
13. Repeated and harsh pipetting reduces cell viability.

Acknowledgements

The data used in this article were generated with research support from Department of Defense (N00014-08–1-0989) and National Institutes of Health (R41-NS053272-01). Dr. Steven L. Stice is the Chief Scientific Officer of ArunA Biomedical and holds financial interests in the company.

References

1. Thomson, J. A., Itskovitz-Eldor, J., Shapiro, S. S., Waknitz, M. A., Swiergiel, J. J., Marshall, V. S., and Jones, J. M. (1998) Embryonic stem cell lines derived from human blastocysts. *Science* **282**, 1145–7.
2. Mitalipova, M., Calhoun, J., Shin, S., Wininger, D., Schulz, T., Noggle, S., Venable, A., Lyons, I., Robins, A., and Stice, S. (2003) Human embryonic stem cell lines derived from discarded embryos. *Stem Cells* **21**, 521–6.
3. Mitalipova, M. M., Rao, R. R., Hoyer, D. M., Johnson, J. A., Meisner, L. F., Jones, K. L., Dalton, S., and Stice, S. L. (2005) Preserving the genetic integrity of human embryonic stem cells. *Nat Biotechnol* **23**, 19–20.
4. Takahashi, K., Tanabe, K., Ohnuki, M., Narita, M., Ichisaka, T., Tomoda, K., and Yamanaka, S. (2007) Induction of pluripotent stem cells from adult human fibroblasts by defined factors. *Cell* **131**, 861–72.
5. Yu, J., Vodyanik, M. A., Smuga-Otto, K., Antosiewicz-Bourget, J., Frane, J. L., Tian, S., Nie, J., Jonsdottir, G. A., Ruotti, V., Stewart, R., Slukvin, II, and Thomson, J. A. (2007) Induced pluripotent stem cell lines derived from human somatic cells. *Science* **318**, 1917–20.
6. Park, I. H., Zhao, R., West, J. A., Yabuuchi, A., Huo, H., Ince, T. A., Lerou, P. H., Lensch, M. W., and Daley, G. Q. (2008) Reprogramming of human somatic cells to pluripotency with defined factors. *Nature* **451**, 141–6.
7. Xu, R. H., Chen, X., Li, D. S., Li, R., Addicks, G. C., Glennon, C., Zwaka, T. P., and Thomson, J. A. (2002) BMP4 initiates human embryonic stem cell differentiation to trophoblast. *Nat Biotechnol* **20**, 1261–4.
8. Dhara, S. K., and Stice, S. L. (2008) Neural differentiation of human embryonic stem cells. *J Cell Biochem* **105**, 633–40.
9. Carpenter, M. K., Inokuma, M. S., Denham, J., Mujtaba, T., Chiu, C. P., and Rao, M. S. (2001) Enrichment of neurons and neural precursors from human embryonic stem cells. *Exp Neurol* **172**, 383–97.
10. Shin, S., Dalton, S., and Stice, S. L. (2005) Human motor neuron differentiation from human embryonic stem cells. *Stem Cells Dev* **14**, 266–9.
11. Keirstead, H. S., Nistor, G., Bernal, G., Totoiu, M., Cloutier, F., Sharp, K., and Steward, O. (2005) Human embryonic stem cell-derived oligodendrocyte progenitor cell transplants remyelinate and restore locomotion after spinal cord injury. *J Neurosci* **25**, 4694–705.
12. Shin, S., Mitalipova, M., Noggle, S., Tibbitts, D., Venable, A., Rao, R., and Stice, S. L. (2006) Long-term proliferation of human embryonic stem cell-derived neuroepithelial cells using defined adherent culture conditions. *Stem Cells* **24**, 125–38.
13. Dhara, S. K., Hasneen, K., Machacek, D. W., Boyd, N. L., Rao, R. R., and Stice, S. L. (2008) Human neural progenitor cells derived from embryonic stem cells in feeder-free cultures. *Differentiation* **76**, 454–64.
14. Hong, S., Kang, U. J., Isacson, O., and Kim, K. S. (2008) Neural precursors derived from human embryonic stem cells maintain long-term proliferation without losing the potential to differentiate into all three neural lineages, including dopaminergic neurons. *J Neurochem* **104**, 316–24.
15. Peh, G. S., Lang, R. J., Pera, M. F., and Hawes, S. M. (2009) CD133 expression by neural progenitors derived from human embryonic stem cells and its use for their prospective isolation. *Stem Cells Dev* **18**, 269–82.
16. Wilson, P. G., Cherry, J. J., Schwamberger, S., Adams, A. M., Zhou, J., Shin, S., and Stice, S. L. (2007) An SMA project report: neural cell-based assays derived from human embryonic stem cells. *Stem Cells Dev* **16**, 1027–41.
17. Hurst, J. H., Mumaw, J., Machacek, D. W., Sturkie, C., Callihan, P., Stice, S. L., and Hooks, S. B. (2008) Human neural progenitors express functional lysophospholipid receptors that regulate cell growth and morphology. *BMC Neurosci* **9**, 118.
18. Eiges, R., Schuldiner, M., Drukker, M., Yanuka, O., Itskovitz-Eldor, J., and Benvenisty, N. (2001) Establishment of human embryonic stem cell-transfected clones carrying a marker for undifferentiated cells. *Curr Biol* **11**, 514–8.
19. Tang, X., Cai, P. Q., Lin, Y. Q., Oudega, M., Blits, B., Xu, L., Yang, Y. K., and Zhou, T. H. (2006) Genetic engineering neural stem cell modified by lentivirus for repair of spinal cord injury in rats. *Chin Med Sci J* **21**, 120–4.
20. Honma, T., Honmou, O., Iihoshi, S., Harada, K., Houkin, K., Hamada, H., and Kocsis, J. D. (2006) Intravenous infusion of immortalized human mesenchymal stem cells protects against injury in a cerebral ischemia model in adult rat. *Exp Neurol* **199**, 56–66.
21. Kurozumi, K., Nakamura, K., Tamiya, T., Kawano, Y., Kobune, M., Hirai, S., Uchida, H., Sasaki, K., Ito, Y., Kato, K., Honmou, O., Houkin, K., Date, I., and Hamada, H. (2004) BDNF gene-modified mesenchymal stem cells

promote functional recovery and reduce infarct size in the rat middle cerebral artery occlusion model. *Mol Ther* **9**, 189–97.

22. Sun, M., Kong, L., Wang, X., Holmes, C., Gao, Q., Zhang, G. R., Pfeilschifter, J., Goldstein, D. S., and Geller, A. I. (2004) Coexpression of tyrosine hydroxylase, GTP cyclohydrolase I, aromatic amino acid decarboxylase, and vesicular monoamine transporter 2 from a helper virus-free herpes simplex virus type 1 vector supports high-level, long-term biochemical and behavioral correction of a rat model of Parkinson's disease. *Hum Gene Ther* **15**, 1177–96.
23. Tuszynski, M. H., Thal, L., Pay, M., Salmon, D. P., U, H. S., Bakay, R., Patel, P., Blesch, A., Vahlsing, H. L., Ho, G., Tong, G., Potkin, S. G., Fallon, J., Hansen, L., Mufson, E. J., Kordower, J. H., Gall, C., and Conner, J. (2005) A phase 1 clinical trial of nerve growth factor gene therapy for Alzheimer disease. *Nat Med* **11**, 551–5.
24. Phillips, M. I., and Tang, Y. L. (2008) Genetic modification of stem cells for transplantation. *Adv Drug Deliv Rev* **60**, 160–72.
25. Dhara, S. K., Gerwe, B. A., Majumder, A., Dodla, M. C., Boyd, N. L., Machacek, D. W., Hasneen, K., and Stice, S. L. (2009) Genetic manipulation of neural progenitors derived from human embryonic stem cells. *Tissue Eng Part A* **15**, 3621–34.
26. El-Aneed, A. (2004) An overview of current delivery systems in cancer gene therapy. *J Control Release* **94**, 1–14.
27. Link, N., Brunner, T. J., Dreesen, I. A., Stark, W. J., and Fussenegger, M. (2007) Inorganic nanoparticles for transfection of mammalian cells and removal of viruses from aqueous solutions. *Biotechnol Bioeng* **98**, 1083–93.
28. Graham, F. L., and van der Eb, A. J. (1973) A new technique for the assay of infectivity of human adenovirus 5 DNA. *Virology* **52**, 456–67.
29. Wyman, T. B., Nicol, F., Zelphati, O., Scaria, P. V., Plank, C., and Szoka, F. C., Jr. (1997) Design, synthesis, and characterization of a cationic peptide that binds to nucleic acids and permeabilizes bilayers. *Biochemistry* **36**, 3008–17.
30. Felgner, P. L., Gadek, T. R., Holm, M., Roman, R., Chan, H. W., Wenz, M., Northrop, J. P., Ringold, G. M., and Danielsen, M. (1987) Lipofection: a highly efficient, lipid-mediated DNA-transfection procedure. *Proc Natl Acad Sci USA* **84**, 7413–7.
31. Koping-Hoggard, M., Mel'nikova, Y. S., Varum, K. M., Lindman, B., and Artursson, P. (2003) Relationship between the physical shape and the efficiency of oligomeric chitosan as a gene delivery system in vitro and in vivo. *J Gene Med* **5**, 130–41.
32. Vincent, L., Varet, J., Pille, J. Y., Bompais, H., Opolon, P., Maksimenko, A., Malvy, C., Mirshahi, M., Lu, H., Vannier, J. P., Soria, C., and Li, H. (2003) Efficacy of dendrimer-mediated angiostatin and TIMP-2 gene delivery on inhibition of tumor growth and angiogenesis: in vitro and in vivo studies. *Int J Cancer* **105**, 419–29.
33. Godbey, W. T., Wu, K. K., and Mikos, A. G. (1999) Poly(ethylenimine) and its role in gene delivery. *J Control Release* **60**, 149–60.
34. Neumann, E., Schaefer-Ridder, M., Wang, Y., and Hofschneider, P. H. (1982) Gene transfer into mouse lyoma cells by electroporation in high electric fields. *EMBO J* **1**, 841–5.
35. Koch, P., Siemen, H., Biegler, A., Itskovitz-Eldor, J., and Brustle, O. (2006) Transduction of human embryonic stem cells by ecotropic retroviral vectors. *Nucleic Acids Res* **34**, e120.
36. Lakshmipathy, U., Pelacho, B., Sudo, K., Linehan, J. L., Coucouvanis, E., Kaufman, D. S., and Verfaillie, C. M. (2004) Efficient transfection of embryonic and adult stem cells. *Stem Cells* **22**, 531–43.
37. Siemen, H., Nix, M., Endl, E., Koch, P., Itskovitz-Eldor, J., and Brustle, O. (2005) Nucleofection of human embryonic stem cells. *Stem Cells Dev* **14**, 378–83.
38. Drukker, M., Dhara, S. K., and Benvenisty, N. (2005) Genetic engineering of human embryonic stem cells *in* "Human embryonic stem cells" (Odorico, J. S., Zhang, S.-C., and Pedersen, R. A., Eds.), pp. 215–30, Garland Science/BIOS Scientific Publishers, Oxford.

Chapter 26

Gene Targeting in Human Pluripotent Stem Cells

Ying Liu and Mahendra Rao

Abstract

Targeted homologous recombination (HR) is an essential tool in stem cell biology. It can be used to study gene function and is a highly developed technology in the mouse where precise genetic modifications are introduced into the genome via HR in mouse embryonic stem cells (mESCs). However, gene targeting has not been widely applied to the study of human pluripotent stem cells (hPSCs) due to its relatively low efficiency in human cell lines. To overcome this technical hurdle, we have developed and established a protocol that allows efficient gene targeting in hPSC lines. This chapter provides a detailed protocol for efficiently performing gene targeting in hPSCs by electroporation. The protocol describes methods for cell preparation, antibiotic selection, and excision of the selectable marker following gene targeting. While we can only target one allele at a time, HR covers a broad range of important applications including making knock-in reporter lines and knock-in lineage tracers, generating disease models that are caused by dominant mutants, repair of patient-derived induced PSCs that only involve a single allele mutation, and knocking out genes that are located on the X chromosome in male lines. When targeting to both alleles is needed, such as generation of a knockout cell line, the cells can be electroporated twice with targeting vectors designed to target each of the alleles. This protocol will find broad applications in generating lineage-specific reporter lines and point mutations in genetic repair in disease models using hPSCs.

Key words: pluripotency, genetic engineering, reporter cell line, homologous recombination, gene targeting, electroporation

1. Introduction

The ability to target a specific gene locus in the mouse and to alter it in a controllable fashion has fundamentally changed biomedical research and made mice the preeminent model for the study of gene function. However, unique signal transduction pathways in humans may not be revealed by using only mouse or other animal models. An approach that combines the advantages of established

Philip H. Schwartz and Robin L. Wesselschmidt (eds.), *Human Pluripotent Stem Cells: Methods and Protocols*, Methods in Molecular Biology, vol. 767, DOI 10.1007/978-1-61779-201-4_26, © Springer Science+Business Media, LLC 2011

mouse models with parallel experiments in human embryonic stem cells (hESCs) and/or induced pluripotent stem cells (hiPSCs) offers significant advantages over current methodologies. Homologous recombination (HR) has been achieved in several genes in hESCs, including Oct4, Hprt1, Mixl1, Fezl, Olig2, Isl1, Rosa, PIG-A (1–5). In iPSCs, PIG-A (5), LMNA (15), and Olig2 have been targeted (Liu and Rao, unpublished).

The targeting efficiency is determined by: the accessibility of the genetic loci in ESCs and iPSCs in the undifferentiated state, the origin of the DNA for targeting vectors, the design of the targeting vectors, as well as other unidentified factors. Reports have shown that both isogenic and nonisogenic DNA vectors can be used to successfully target genes in human cells [summarized in ref. 6) and hESCs (7, 8), and the use of isogenic DNA to build targeting vectors does not seem to significantly enhance targeting efficiency in human cells. This is an important observation as it suggests that there is not a requirement to generate targeting vectors from DNA isolated from specific hESC or hiPSC lines and allows increased flexibility in design and construction of targeting vectors using bacterial artificial chromosome (BAC)-based recombineering (9–13).

Another effective approach for manipulating the genome by gene targeting is to use zinc finger nucleases (ZFN). Researchers have utilized either integrase-defective lentiviral vectors (IDLVs) (14) or virus-free systems (5) to deliver ZFNs. ZFNs induce site-specific DNA double-stranded breaks (DSB) to trigger subsequent HR at desired genomic locations. ZFN-medicated gene targeting is a promising highly efficient protocol in ESCs and iPSCs, although two major concerns exist: ZFNs might cause chromosomal instability (14) and ZFN design is not easy.

Importantly, we can only target one allele at a time using the conventional HR protocol. This, however, already covers a broad range of important applications including making knock-in reporter lines and knock-in lineage tracers, generating disease models that are caused by dominant mutants, repair of patient-derived iPSCs that only involve a single allele mutation, and knocking out genes that are located on the X chromosome in male PSC lines. When targeting to both alleles is needed, such as generation of a knockout cell line, the cells have to be electroporated twice with targeting vectors designed to target each of the gene alleles individually. In these cases, ZFN-mediated targeting might have advantages: it is possible that ZFN-mediated HR may target both alleles at the same time, therefore circumventing the necessity of repeated electroporations. Recently, helper-dependent adenoviral vectors have also been shown to significantly increase the gene targeting efficiency (15).

One of the defining features of hESCs is that they are capable of very long-term growth in culture while maintaining a normal diploid karyotype. It is critical to maintain the cells in a diploid state even after stressful procedures such as electroporation. A simple question to ask is: how many rounds of electroporation can hESCs or hiPSCs endure without becoming aneuploid or picking up other chromosomal abnormalities? Current data shows that two rounds of electroporation do not cause any adverse effect on karyotype. However, repeated electroporation experiments do stress the cells and tend to cause chromosomal instability. Therefore, if more complex genetic manipulation is needed, close monitoring of the karyotype by G-banding and comparative genomic hybridization (CGH), as well as careful monitoring of the expression of pluripotency markers by quantitative-PCR and immunocytochemistry is necessary and highly recommended, to ensure appropriate interpretation of data with these gene-targeted clones.

In this chapter, we describe a protocol for performing HR experiments in hESCs and hiPSCs. This protocol has been successfully used to target a variety of human cell lines, including hESC lines BG01, BG01V, WA09 (H9), and iPSC lines derived from skin fibroblasts of normal individuals and patients with neurodegenerative diseases. First the cells are transfected by electroporation, then selected for antibiotic resistance; individual antibiotic resistant clones are picked, expanded, and homologous recombinants are identified by Southern blot analysis. Finally, included in this protocol is a method for removing the antibiotic resistance gene from HR clones.

2. Materials

2.1. Reagents and Supplies

Electroporation equipment and reagents

1. Targeting Vector.
2. Electroporator: Gene Pulser Xcell (Bio-Rad), or BTX ECM 630 (or BTX ECM 830).
3. Electroporation buffer: OptiPro SFM (Invitrogen, Cat# 12309-019) or Electroporation buffer (Millipore, Cat# ES-003-D).
4. FIAU (Moravek, Cat# M251).
5. Geneticin (G418).
6. 0.4-cm Electroporation cuvette (Bio-Rad, Cat#165-2088).
7. Modified dissecting microscope in biological safety cabinets (cell culture hood).

Cell culture media and reagents

1. Dulbecco's Modified Eagle's Medium (D-MEM) (high glucose, contains L-glutamine, and sodium pyruvate).
2. 0.25% Trypsin–EDTA solution.
3. Fetal bovine serum (ES cell qualified).

4. L-Glutamine (200 mM solution).
5. Nonessential amino acids 100× (NEAA, 10 mM solution).
6. Dulbecco's Phosphate-Buffered Saline (Ca and Mg free) (DPBS).
7. Dulbecco's Phosphate-Buffered Saline (with Ca and Mg) ($DPBS^{++}$).
8. Dimethyl sulfoxide (DMSO) (Sigma, Cat# D-2650).
9. β-Mercaptoethanol (2-ME).
10. D-MEM/F12 (1×) 1:1 with Glutamax (Invitrogen, Cat# 10565-018).
11. Knockout Serum replacement (KSR) (Invitrogen, Cat# 10828028).
12. Basic Fibroblast Growth Factor (bFGF) (Invitrogen Cat# PHG0026).
13. Collagenase Type IV (Invitrogen, Cat#17104-019).
14. Accutase Cell Dissociation Reagent (Invitrogen, Cat# A1110501).
15. Matrigel (BD Biosciences, Cat# 356230).
16. Geltrex (Invitrogen Cat# 12760-013).
17. MEF Conditioned Medium (see Chapter 31).
18. Antibiotic resistant mouse embryoic feeder cells (MEFs).

2.2. Media and Solutions

hESC Medium. DMEM/F12 (containing Glutamax), 20% KSR, 0.1 mM NEAA, 4 ng/mL bFGF, and 100 μM 2-ME. Mix reagents together in the top of a 2-μm, low-protein-binding filter unit and filter sterilize. Store at 4°C. Warm to 37°C prior to use. Discard any unused medium after 10 days.

MEF medium. DMEM, 10% FBS, 2 mM L-glutamine, 0.1 mM NEAA, and 100 μM 2-ME. Mix components in the top of a sterile 2-μm filter unit. Filter sterilize. Store at 4°C for up to 2 weeks, supplement with additional glutamine, if keeping longer than 1 month.

bFGF solution. Make 10 μg/mL of stock solution in DPBS with 2 mg/mL BSA. Store at −20°C in working aliquots. The shelf-life of the above stock solutions is 1 year at −20°C.

Collagenase solution. Prepare 1 mg/mL solution of collagenase in DMEM/F12. Filter sterilize and store at 4°C. This solution can be used for up to 2 weeks. Aliquot an appropriate amount each time it is to be used and warm the aliquot to 37°C prior to use.

3. Methods

This protocol describes a method to successfully electroporate human pluripotent stem cells. About 1-week prior to electroporation, PSCs are transferred from coculture with MEFs to feeder-free culture as described in Chapter 31. Two days prior to electroporation, drug-resistant MEFs are plated and prepared to accept the electroporated PSCs. Successful completion of this protocol requires careful upfront planning. The targeting vector, reagents, media, and cells must all be prepared and ready to go well in advance of the actual electroporation.

Two types of cells are carried in preparation of electroporation:

1. PSCs: under feeder-free conditions
2. MEFs: antibiotic resistant

Briefly, healthy, undifferentiated PSCs (Fig. 1) are moved from coculture with feeder cells to feeder-free culture in MEF-conditioned medium on Matrigel or Geltrex. The PSCs are allowed to proliferate under feeder-free conditions until the culture becomes 80% confluent, then they are harvested using Accutase to generate a single-cell suspension. The cells are electroporated with 30–50 µg of linearized targeting vector. After 21 days of selection with antibiotic, antibiotic-resistant clones are isolated. The clones are "picked" and placed into individual

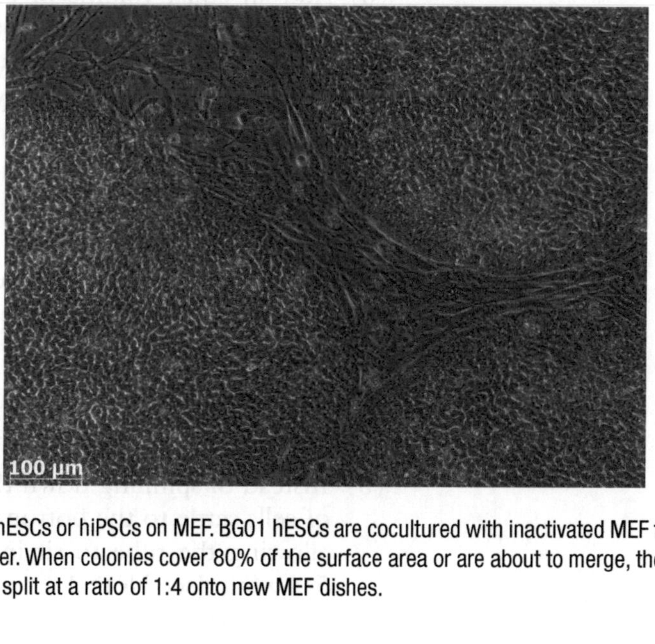

Fig. 1. hESCs or hiPSCs on MEF. BG01 hESCs are cocultured with inactivated MEF feeder cell layer. When colonies cover 80% of the surface area or are about to merge, the cells can be split at a ratio of 1:4 onto new MEF dishes.

wells of 24-well plates containing MEFs, where they are carefully monitored and passaged until enough cells are available to make a frozen stock vial and to isolate DNA for screening. Clones are screened for HR at the target locus. After HR clones are identified, the selectable marker is removed by transient expression of Cre recombinase and clones are banked for future use and characterization.

3.1. Targeting Vector

Construction of targeting vectors is a critical step in the successful generation of reporter and knockout cell lines by HR. However, vector design is dictated by the goals of individual research projects and is beyond the scope of this chapter. The reader is directed to the excellent references for guidance on generating targeting vectors (9–14).

3.2. Preparation of PSCs

PSCs are transferred from MEF coculture to feeder-free culture prior to electroporation.

One week prior to transferring PSCs: Prepare MEF-conditioned medium (MEF-CM) as described in Chapter 31 (this is roughly 2 weeks prior to electroporation).

One week prior to electroporation: Transfer PSCs onto Geltrex-coated dishes for feeder-free culture as described below.

1. Preparation of Geltrex-coated dishes:
   (a) Thaw a whole bottle of Geltrex at 4°C for 3 h or overnight to prevent polymerization (see Note 1).
   (b) Add an equal volume of cold DMEM/F12 to make a 100× stock solution and store desired aliquots at −20°C.
   (c) Before coating plates, thaw aliquots at 4°C. Add an appropriate volume of DMEM/F12 to make a 1× solution, and add to culture dishes to completely cover the surface area (e.g., 2–3 mL/60-mm dish), and incubate at room temperature for 1 h.
   (d) Coated dishes can be used immediately after coating or stored at 4°C for up to 2 weeks.
   (e) Avoid drying the dishes; remove coating solution immediately before use.
2. Passage of PSCs onto Geltrex-coated dishes:
   (a) Harvest cells using collagenase as described in Chapter 8.
   (b) Instead of spinning down the cell pellets, let the clumps of cells settle to the bottom of the tube. This is done by allowing the tube to sit in the tissue culture hood for 5–10 min (see Note 2).

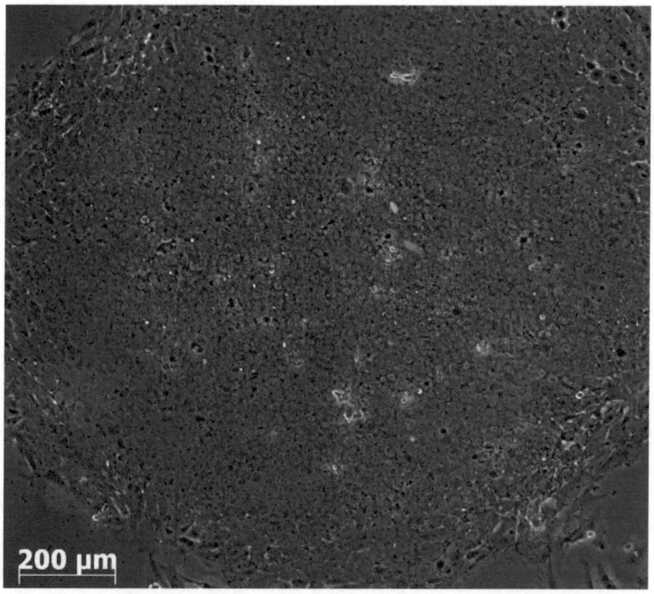

Fig. 2. hESCs or hiPSCs cultured in a feeder-free manner in MEF-conditioned medium. An hiPSC line grown on Geltrex substrate without MEFs in MEF-conditioned medium. When cells cover 80% of the surface area, they can be harvested by Accutase for electroporation.

(c) Remove supernatant, and aliquot cells at 1:4 ratio onto Geltrex-coated dishes in MEF-CM supplemented with fresh bFGF (4 ng/mL).

(d) Feed cells daily with MEF-CM freshly supplemented with bFGF.

(e) When cells reach ~80% confluency (Fig. 2), they can be used for electroporation.

3.3. Preparation of Antibiotic-Resistant MEFs

Two days before electroporation:

1. Prepare one 60-mm dish of antibiotic-resistant MEFs for each construct to be electroporated. Plate 6×10^5 cells/60-mm dish.
2. Place dishes in the incubator and allow MEFs to attach overnight.

One day before electroporation:

1. Replace MEF medium with 3 mL of fresh hESC medium.

3.4. Electroporation

1. Replace the hESC medium in the prepared antibiotic-resistant MEF dishes with 3 mL of MEF-CM (supplemented with fresh 4 ng/mL bFGF).
2. Place 30–50 μg of the linearized targeting vector DNA, which is at a concentration of 1 μg/μL, in a sterile 1.5-mL microcentrifuge tube.

3. Place 3 mL MEF-CM in a 15-mL conical tube and warm to 37°C (one tube/electroporation).
4. Harvest feeder-free PSCs. Two 60-mm dishes ($3–6 \times 10^6$ cells) for each electroporation.
   (a) Add 1 mL Accutase to each 60-mm dish of PSCs and incubate for 3–5 min until cells dislodge.
   (b) Triturate cells, using a 1-mL pipette, into single cells, add to 3 mL of DPBS and spin down pellet at $200 \times g$ for 5 min at room temperature.
5. Resuspend cells in 5 mL of DPBS and spin again at $200 \times g$ for 5 min at room temperature.
6. Remove DPBS and resuspend cell pellets in 800 μL OptiPro SFM (see Note 3).
7. Add cells to the 1.5-mL microcentrifuge tube containing the vector construct and mix by gently pipetting up and down three times with a 1,000-μL sterile pipette, and transfer the mixture to a sterile electroporation cuvette (0.4 cm gap).
8. Electroporate the cells once using the Bio-Rad Gene pulser II or Xcell system using the following conditions: 250 V, 250 μF (see Note 4).
9. Immediately transfer the electroporated cells to the prepared 15-mL conical tube containing MEF-CM. Gently tap the tube to mix. Incubate at room temperature for 5 min.
10. Transfer the electroporated cells to the 60-mm dish of prepared antibiotic-resistant MEFs. Distribute the cells evenly and return the dish to the incubator. Leave undisturbed overnight.
11. For the next 2 days, exchange the medium with 3 mL of MEF-CM that has been supplemented with 4 ng/mL bFGF.
12. *Begin Antibiotic Selection*: 72 h after electroporation, colonies are visible under microscope (Fig. 3). At this point, start positive selection.
13. Feed cells using fresh hESC medium containing appropriate antibiotics (e.g., G418, 25–50 μg/mL) daily. See Note 5 for approximate concentration of antibiotic for selection.
14. *Start negative selection using FIAU* (125 μM). Culture for 21 days under double (positive and negative) selection, in hESC medium. Exchange the medium daily.
15. *Isolate HR clones*: Count the number of clones on each plate and determine how many clones will be isolated. Based on the rate of occurrence for HR for different gene loci, the number of clones that needs to be isolated varies. At least 50 clones should be isolated. Based on the number of clones, prepare 24-well inactivated MEF plates 1–2 days before planned manual "picking" of the clones.

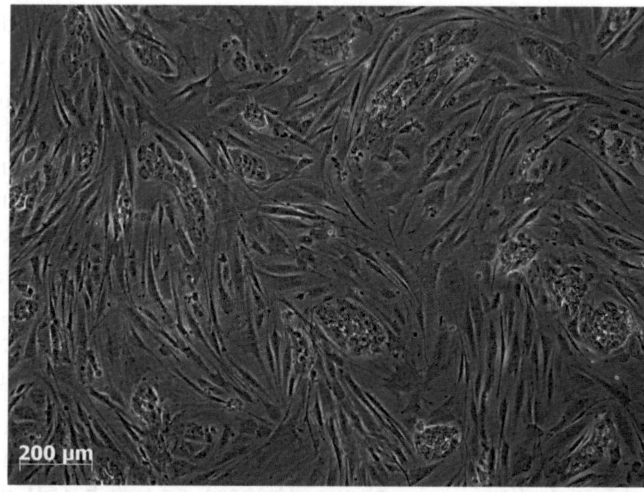

Fig. 3. hESCs or hiPSCs recover 72-h postelectroporation. Three days after electroporation, small colonies of BG01 cells started to emerge. At this point, positive selection begins and lasts for at least 21 days.

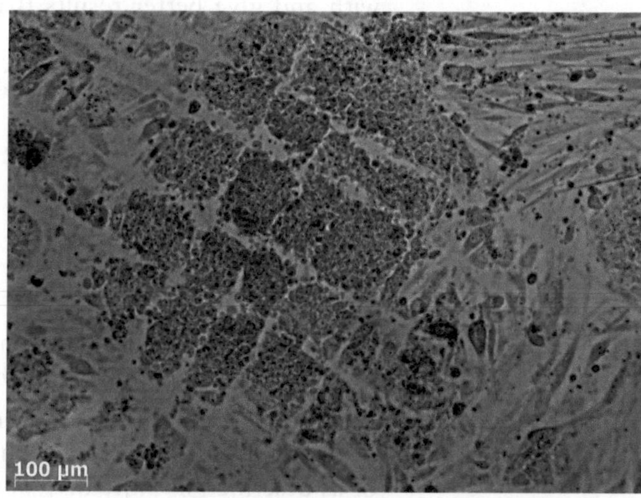

Fig. 4. Manual isolation of individual clones after selection. BG01 cells were electroporated and selected for 21 days, clones were ready to be picked and expanded individually. A cutting tool made from a 9-in. Pasteur pipette was made and colonies were cut into grid to be transferred onto a well of a 24-well plate, which had been seeded with inactivated MEFs.

16. When individual colonies are visible to the naked eye, but are not merging with neighboring colonies, they are ready to be isolated. Manually pick single clones using a glass tool made from a Pasteur pipette or a 25 3/8 gauge needle (see Note 6). Carefully grid and dislodge the colony using the tool (Fig. 4), and use a p200 pipette with a 200-μL tip to transfer the dissected clumps into 1 well of a 24-well plate. Make sure to

completely remove all dislodged pieces before dissecting the next clone to avoid cross-clonal contamination.

17. Continue to culture and expand individual clones under positive selection in hESC medium. Split cells every week at a 1:2 ratio. When cells reach approximately 5×10^6 in number, (~two 60-mm dishes), freeze and stock 1–2 vials per clone and continue culturing. This takes 2–3 weeks.
18. *Passage and bank cells.* For each clone, bank at least stock two batches, with two vials per batch, $2–3 \times 10^6$ cells/vial. Save cell pellets ($5–10 \times 10^6$ cells) to extract genomic DNA for Southern blot analysis to identify homologous recombinants.

3.5. Excision of the Selectable Marker by Transient Transfection with Cre Recombinase

1. Select one of the HR clones to thaw onto an inactivated MEF dish, then adapt to feeder-free culture as described above. Prepare one dish (60 mm) of the feeder-free PSC clone (80% confluency) for transfection.
2. One day before transfection, prepare 100-mm dishes of inactivated MEFs (100-mm dishes are more convenient to work with and give better results than vessels that have a smaller surface area).
3. Prepare 10 μg of supercoiled Cre plasmid DNA in a 1.5-mL microcentrifuge tube.
4. Incubate the feeder-free hESC clone culture with Accutase (see Note 7) for 3 min at room temperature. Gently triturate cells into single cells.
5. Spin down at $200 \times g$ for 5 min at room temperature.
6. Resuspend the cells in 800 μL 1× DPBS (without Ca, Mg). Record the cell number. Mix the cells and DNA, transfer the mixture to a 0.4-cm Gap electroporation cuvette.
7. Electroporate the cells once using Bio-Rad Gene pulser II or Xcell system using the following conditions: 250 V, 250 μF or as described in Note 3.
8. Based on the recorded cell number in step 6, calculate and plate 200 cells onto one 100-mm MEF dish with MEF-conditioned medium supplemented with high concentration bFGF (20 ng/mL). Put dishes back into a 37°C incubator (see Note 8).
9. Replace with fresh MEF-CM supplemented with high concentration bFGF (20 ng/mL) everyday for 3–5 days until colonies are visible under microscope.
10. Change medium with fresh hESC medium daily. Gradually reduce the concentration of bFGF from 20 to 4 ng/mL over a period of 2 weeks, when individual colonies are ready to be manually isolated (picked).

11. Manually isolate single clones as described for initial isolation of antibiotic resistant clones in Subheading 3.4.
12. Continue to culture and expand individual clones in hESC medium supplemented with basic FGF (4 ng/mL). Split cells every week at a 1:2 ratio. When cells reach about 5×10^6 in number (~two 60-mm dishes), prepare two stock vials for cryogenic storage and continue culturing. This takes 2–3 weeks.

3.6. Identification of Excision-Positive Clones by PCR

During the expansion of Cre'd clones, save cell pellets for DNA isolation and PCR to identify clones that can be shown to meet the following criteria: (a) the neo cassette has been excised from the target locus and (b) the Cre plasmid has not randomly integrated into the genome. Clones that no longer contain selectable markers and do not have spurious integration of the targeting vector or the cre plasmid are bona fide homologous recombinant clones in which one gene allele has been correctly targeted. These clones can be cultured and passaged in hESC medium to generate a master cell bank and subsequent working banks as described in Chapters 2 and 11.

4. Notes

1. When handling Geltrex, keep undiluted Geltrex on ice or in 4°C at all times.
2. When moving hESCs or hiPSCs to feeder-free culture, extra care is needed to avoid carrying over MEFs. Letting the clumps of cells settle in the conical tubes instead of spinning them down will largely eliminate MEFs which are single cells and settle slower than hESC or hiPSC clumps.
3. Other solutions can also be used as electroporation buffers including Millipore (Cat# ES-003-D), Dulbecco's Phosphate-Buffered Saline (Ca and Mg free), or Dulbecco's Phosphate-Buffered Saline (with Ca and Mg).
4. If the BTX ECM 630 electroporator is used, the parameters are 250 V, 250 μF; for the BTX ECM 830 electroporator, the parameters are 200 V, 10 ms, zap twice.
5. The concentration of the antibiotics used to select resistant clones is determined empirically. The effective concentration may vary among different hESC and iPSC lines as well. At the beginning of selection, a lower dose is always safer to not over-kill. The dosage can be increased gradually once the bulk culture of individual clones is established. Below is a rough dose range for different antibiotics for positive selec-

tion for readers' reference: Blasticidin 1–1.5 μg/mL, Hygromycin B 5–10 μg/mL, G418 25–200 μg/mL, and Zeocin 1–2.5 μg/mL.

6. When isolating individual clones, manual picking is necessary. We have been using 9-in. Pasteur pipettes to make a glass knife, which can grid and cut clones. Alternatively, a 25 3/8 gauge needle or a p20 pipette tip can be used. Avoid cross-clonal contamination. There should be about 50–70 colonies, maximum, per 60-mm dish.
7. Two enzymes are often used to harvest hESCs or hiPSCs to generate single cell suspension: Accutase or TrypLE, depending on the end user's preference. The homologous recombinant clones identified here do not have a positive selection cassette any more so they should not be cultured with antibiotic in the medium.
8. This step requires clonal-colony formation from single cells. While the efficiency is quite low, it is still practical. ROCK inhibitors (e.g., Y-27632) improve single cell survival of hESCs. However, how ROCK inhibitors work is not completely understood. Use this reagent only when necessary.

Acknowledgments

This work was supported by National Institute on Aging (NIA, NIH) Intramural Research Program, Intramural Research Training Awards, Robert Packard Center for ALS Research and Life Technologies Corporation.

References

1. Irion, S., H. Luche, P. Gadue, H. J. Fehling, M. Kennedy and G. Keller (2007) Identification and targeting of the ROSA26 locus in human embryonic stem cells. *Nat Biotechnol.* **25**, 1477–82.
2. Zwaka, T. P. and J. A. Thomson (2003) Homologous recombination in human embryonic stem cells. *Nat Biotechnol.* **21**, 319–21.
3. Ruby, K. M. and B. Zheng (2009) Gene targeting in a HUES line of human embryonic stem cells via electroporation. *Stem Cells.* **27**, 1496–506.
4. Bu, L., X. Jiang, S. Martin-Puig, L. Caron, S. Zhu, Y. Shao, et al. (2009) Human ISL1 heart progenitors generate diverse multipotent cardiovascular cell lineages. *Nature.* **460**, 113–7.
5. Zou, J., M. L. Maeder, P. Mali, S. M. Pruett-Miller, S. Thibodeau-Beganny, B. K. Chou, et al. (2009) Gene targeting of a disease-related gene in human induced pluripotent stem and embryonic stem cells. *Cell Stem Cell.* **5**, 97–110.
6. Sedivy, J. M., B. Vogelstein, H. L. Liber, E. A. Hendrickson and A. Rosmarin (1999) Gene Targeting in Human Cells Without Isogenic DNA. *Science.* **283**, 9.
7. Xue, H., S. Wu, S. T. Papadeas, S. Spusta, A. M. Swistowska, C. C. MacArthur, et al. (2009) A targeted neuroglial reporter line generated by homologous recombination in human embryonic stem cells. *Stem Cells.* **27**, 1836–46.
8. Davis, R. P., E. S. Ng, M. Costa, A. K. Mossman, K. Sourris, A. G. Elefanty, et al. (2008) Targeting a GFP reporter gene to the MIXL1 locus of human embryonic stem cells

identifies human primitive streak-like cells and enables isolation of primitive hematopoietic precursors. *Blood.* **111**, 1876–84.

9. Wu, S., G. Ying, Q. Wu and M. R. Capecchi (2008) A protocol for constructing gene targeting vectors: generating knockout mice for the cadherin family and beyond. *Nat Protoc.* **3**, 1056–76.
10. Datsenko, K. A. and B. L. Wanner (2000) One-step inactivation of chromosomal genes in Escherichia coli K-12 using PCR products. *Proc Natl Acad Sci USA.* **97**, 6640–5.
11. Lee, E. C., D. Yu, J. Martinez de Velasco, L. Tessarollo, D. A. Swing, D. L. Court, et al. (2001) A highly efficient Escherichia coli-based chromosome engineering system adapted for recombinogenic targeting and subcloning of BAC DNA. *Genomics.* **73**, 56–65.
12. Zhang, Y., F. Buchholz, J. P. Muyrers and A. F. Stewart (1998) A new logic for DNA engineering using recombination in Escherichia coli. *Nat Genet.* **20**, 123–8.
13. Zhang, Y., J. P. Muyrers, G. Testa and A. F. Stewart (2000) DNA cloning by homologous recombination in Escherichia coli. *Nat Biotechnol.* **18**, 1314–7.
14. Lombardo, A., P. Genovese, C. M. Beausejour, S. Colleoni, Y. L. Lee, K. A. Kim, et al. (2007) Gene editing in human stem cells using zinc finger nucleases and integrase-defective lentiviral vector delivery. *Nat Biotechnol.* **25**, 1298–306.
15. Liu, G. H., K. Suzuki, J. Qu, I. Sancho-Martinez, F. Yi, M. Li, et al. (2011) Targeted gene correction of laminopathy-associated LMNA mutations in patient-specific iPSCs. *Cell Stem Cell.* **8**, 688–94.

Chapter 27

Episomal Transgene Expression in Pluripotent Stem Cells

Michele M.P. Lufino, Anna R. Popplestone, Sally A. Cowley, Pauline A.H. Edser, William S. James, and Richard Wade-Martins*

Abstract

Herpes simplex type 1 (HSV-1) amplicon vectors possess a number of features that make them excellent vectors for the delivery of transgenes into stem cells. HSV-1 amplicon vectors are capable of efficiently transducing both dividing and nondividing cells and since the virus is quite large, 152 kb, it is of sufficient size to allow for incorporation of entire genomic DNA loci with native promoters. HSV-1 amplicon vectors can also be used to incorporate and deliver to cells a variety of sequences that allow extrachromosomal retention. These elements offer advantages over integrating vectors as they avoid transgene silencing and insertional mutagenesis. The construction of amplicon vectors carrying extrachromosomal retention elements, their packaging into HSV-1 viral particles, and the use of HSV-1 amplicons for stem cell transduction will be described.

Key words: HSV-1 amplicon, iBAC, Extrachromosomal vector, Stem cells, Gene expression vector

1. Introduction

Viral vectors represent an excellent tool for the genetic manipulation of stem cells as they are highly efficient at transgene delivery. Several viral delivery systems have been developed, such as vectors based on herpes simplex virus type 1 (HSV-1) which provide many advantages, including a high efficiency of transduction in a broad range of cell types. The helper-dependent HSV-1 vectors (amplicon vectors) that do not contain sequences coding for viral proteins offer lower immunogenicity and higher safety (1, 2). Amplicon vectors are characterized by a very large transgene capacity (up to ~152 kb) which allows the delivery of either small

*Molecular Neurodegeneration and Gene Therapy Research Group, Department of Physiology, Anatomy and Genetics University of Oxford, UK

Philip H. Schwartz and Robin L. Wesselschmidt (eds.), *Human Pluripotent Stem Cells: Methods and Protocols*, Methods in Molecular Biology, vol. 767, DOI 10.1007/978-1-61779-201-4_27, © Springer Science+Business Media, LLC 2011

cDNA-based expression cassettes or large genomic DNA loci, the latter capable of achieving physiologically regulated levels of transgene expression. Recently, a helper virus-free packaging system has been developed, allowing the production of amplicon stocks free from helper virus contamination (3).

HSV-1 amplicon vectors have been used to deliver several extrachromosomal expression systems, including vectors based on the Epstein-Barr virus (EBV) (4, 5), or the scaffold-matrix attachment region (*S/MAR*) sequence isolated from the human β-interferon gene (6, 7), and vectors based on sequences derived from human chromosomes (see Fig. 1a) (8). All of these vectors

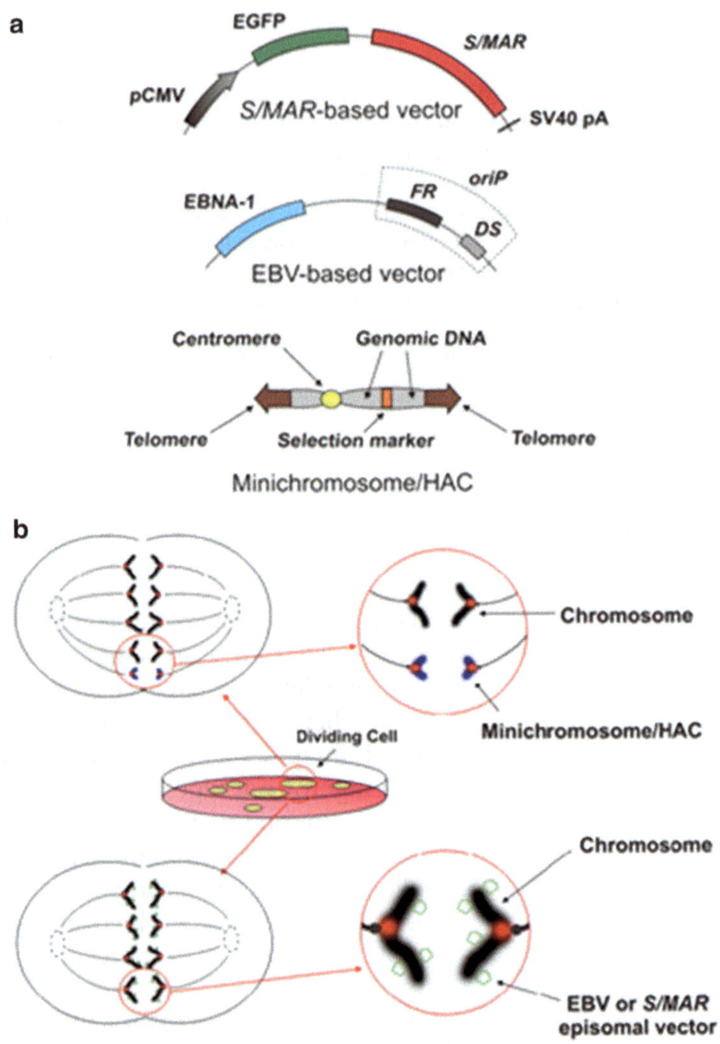

Fig. 1. Replication and retention mechanisms of episomal vectors. (**a**) Representation of three mammalian extrachromosomal systems and the elements responsible for their episomal replication and retention in cells. (**b**) Retention is mediated by interaction with spindle fibers (HACs) or by association with chromosomes (EBV and *S/MAR*-based vectors). Reproduced from ref. 16.

are capable of episomal replication and retention, which represent essential properties required to achieve long-term transgene expression from an extrachromosomal vector (16). By avoiding vector integration, episomal vectors overcome the problems of transgene silencing, host gene disruption and insertional mutagenesis which can lead to oncogenesis (9, 10). Episomal vectors represent an excellent approach for the genetic modification of stem cells because episomal replication eliminates genotoxicity and ensures vector retention in cells during stem cell division and expansion (see Fig. 1b).

In this chapter, we describe the construction of HSV-1-based amplicon vectors, including vectors carrying large genomic DNA loci, the preparation and purification of amplicon viral stocks, and the infection of stem cells in culture for transient and stable transgene delivery.

2. Materials

2.1. Amplicon Vectors: Design and Construction

1. Cre Recombinase and buffers (Novagen, cat. no. 69247).
2. Millipore MF-membrane filters 0.025 μm pore size – 25 mm (Millipore, Cat. no. VSWP02500).
3. SOC media (Invitrogen, Cat. no. 15544–034).
4. DH10B ElectroMAX (Invitrogen, Cat. no. 18290–015).
5. Electroporator Gene Pulser XCell (Biorad).
6. Cuvette 0.1 cm (Biorad, Cat. no. 1652089).
7. LB agar (Calbiochem, Cat. no. 12177). 400 mL of LB agar made up by dissolving 14.8 g of LB agar in deionized water and autoclaved.
8. Antibiotics: Ampicillin (Sigma, Cat. no. A9518) – 50 mg/mL solution made up in autoclaved MilliQ H_2O and filtered using Millipore Millex GP 0.22 μm (Cat. no. SLGP033RS); Kanamycin (Sigma, Cat. no. K4000) – 25 mg/mL solution made up in autoclaved MilliQ H_2O and filtered using Millipore Millex GP 0.22 μm; and Chloramphenicol (Sigma, Cat. no. C0378) – 15 mg/mL solution was made up using 70% ethanol.

2.2. BAC DNA Maxiprep

1. LB agar (Calbiochem, Cat. no. 12177) prepared as above.
2. Antibiotics: Ampicillin, Kanamycin, and Chloramphenicol, prepared as above.
3. LB Broth Miller (Novagen, Cat. no.: 71753–5), 1 L made up with 25 g LB broth with deionized water and autoclaved.
4. Qiagen Tip 500 Plasmid Maxiprep Kit (Cat. no. 12163). Buffers P1, P2, P3, QBT, QC, and QF are included in the kit.

Buffers P1 and P3 should be kept at 4°C and buffer QF should be heated at 55°C before use.

5. Beckman Rotors J10.5 and J17 for Beckman Avanti J-E centrifuge.
6. Kimwipes disposable wipers (Sigma, Cat. no. Z188956).
7. 250 mL Centrifuge bottles (Beckman, Cat. no. 356013).
8. Oakridge tubes (Beckman, Cat. no. 357003).
9. Isopropanol (VWR, Cat. no. 20842.323).
10. 70% Ethanol (Sigma, Cat. no. 32221) diluted using MilliQ H_2O.
11. Buffer Tris–EDTA (TE): 10 mM Tris–HCl (Sigma, Cat. no. T5941), 1 mM EDTA (Sigma, Cat. no. E5134). Sterilize by autoclaving. Store at 4°C or room temperature (RT).
12. Agarose (Sigma, Cat. no. A9539).
13. Tris Borate EDTA (Sigma, Cat. no. T4415): 10× stock diluted to 1× or 0.5× as required using deionized water.
14. PacI and buffers (NEB, Cat. no. R0547S).
15. Pulsed-Field Gel Electrophoresis (PFGE) (Biorad, CHEF-DR II).

2.3. Packaging of Amplicon Vectors into HSV-1 Virions

1. *Vero 2–2 cells growth medium*: DMEM [High Glucose (4.5 g/L), without l-glutamine] (PAA, Cat. no. E15-009), 10% fetal bovine serum (FBS) (Gibco, Cat. no. 10270), 1% penicillin–streptomycin (P/S) (Sigma, Cat. no. P4458), 1% l-glutamine (L-G) (Sigma, Cat. no. G7513), and 500 μg/mL G418 (Invitrogen, Cat. no. 10131–027).
2. *Vero 2–2 cells seeding medium*: DMEM, 10% FBS, 1% P/S, and 1% L-G.
3. OptiMEM (Invitrogen, Cat. no. 31985047).
4. Lipofectamine (Invitrogen, Cat. no. 18324012).
5. Plus Reagent (Invitrogen, Cat. no. 11514015).
6. *Packaging medium*: DMEM, 6% FBS, 1% P/S, 1% L-G, and 25 mM HEPES (Gibco, Cat. no. 15630).
7. Fisher Scientific Sonic Dismembrator Model 500, set to a 50% amplitude, with a tapered microtip, 2 mm tip diameter.
8. Hettich Rotanta 460R centrifuge.
9. SW32Ti swing-out rotor for use with Beckman ultracentrifuge L70.
10. Ultra Clear Beckman tubes (Beckman, Cat. no. 344061).
11. *Sucrose*: 30% and 60% solutions made up in PBS and filter sterilized (0.22 μm).
12. Hank's Balanced Salt solution (HBSS).

13. Phosphate-Buffered Saline (PBS) (Sigma, Cat. no. P4417). One tablet per 200 mL deionized water, autoclave before use.

2.4. Further Amplicon Purification Methods

1. Millipore Millex-HA 0.45 μm, syringe filter (Cat. no. SLHA033SS).
2. Sucrose solutions should be made up in PBS and filter-sterilized (0.22 μm).

2.5. Titration

1. *G16.9 cells growth medium*: DMEM, 10% FBS, 1% P/S, 1% L-G, and 200 μg/mL Hygromycin B (Invitrogen, Cat. no. 10687-010).
2. *G16.9 cells seeding medium*: DMEM, 10% FBS, 1% P/S, and 1% L-G.
3. *Titration medium*: DMEM, 2% FBS, 1% P/S, 1% L-G, and 25 mM HEPES.
4. *Fixative solution*: Dulbecco's phosphate-buffered saline (DPBS, Invitrogen, Cat. no. 14190169, without Calcium, Magnesium, or Phenol Red), 2% formaldehyde (Sigma, Cat. no. 533998-500mL), and 0.05% glutaraldehyde (Sigma, Cat. no. G6257-100mL).
5. *Staining solution*: DPBS, 5 mM potassium ferricyanide (Sigma, Cat. no. P8131), 5 mM potassium ferrocyanide (Sigma, Cat. no. P3289), and 2 mM $MgCl_2$ (Sigma Cat. no. M2670).
6. *X-gal*: 5-bromo-4-chloro-3-indolyl-β-d-galactopyranoside (Calbiochem, Cat. no. 203782). Resuspend in dimethylformamide (Sigma, Cat. no. D4551) to a final concentration of 100 mg/mL. Dilute 1:100 with staining solution just prior to use.

2.6. Infection of Stem Cells

1. 2% Gelatin (Sigma, Cat. no. G1393) diluted to 0.1% in DPBS.
2. Human/Mouse Dopaminergic Neuron Differentiation Kit (R&D Systems, Cat. no. SC001B).
3. Matrigel™, hESC-qualified (Becton Dickinson, Cat. no. 734–1440).
4. mTeSR™-1 medium (StemCell Technologies, Inc., Cat. no. 05851).
5. TrypLE™ Express (Invitrogen, Cat. no. 12604).
6. Antibodies:
   (a) Primary antibody goat anti-hOct4 (R&D systems Cat. no. Af1759).
   (b) Isotype control goat IgG (R&D systems Cat. no. AB 109-C).
   (c) Secondary antibody NL657-conjugated donkey anti-goat IgG (R&D systems Cat. no. NL002).

7. *FACS buffer*: DPBS, 0.01% sodium azide NaN_3, 1% FCS (Sigma, Cat. No. S2002), and 10 μg/mL human IgG (R&D systems, Cat. No. 1-001-A).
8. Paraformaldehyde (TAAB, Cat. No. P001).
9. Saponin (Sigma, Cat. No. S-7900).
10. ROCK inhibitor (Y-27632) (Merck/Cal Biochem, Cat. No. 688000).

2.7. Stable Clonal Cell Lines

1. Hygromycin B (Invitrogen, Cat. no. 10687-010).
2. 0.25% Trypsin–EDTA solution (Sigma, Cat. No. T3924).

3. Methods

3.1. Amplicon Vectors: Design and Construction

HSV-1 amplicons are gene expression plasmid vectors that contain the HSV-1 origin of DNA replication (*oriS*), DNA cleavage/packaging signals (*pac*) from HSV-1, and an expression reporter gene cassette used for titration, such as the enhanced green fluorescent protein (eGFP) under control of the immediate-early promoter pIE4/5. To achieve episomal retention in mammalian cells, additional sequences must be inserted into a basic amplicon vector. These include the *trans*-acting Epstein-Barr virus Nuclear Antigen-1 (EBNA-1) protein and the *cis*-acting *oriP* origin for the EBV-derived episomal system (4, 5); the *S/MAR* sequence under the transcriptional control of a cytomegalovirus promoter for pEPI-based vectors (6, 7); or α-satellite DNA for human artificial chromosomes (HACs) (8). A mammalian selection cassette can be inserted for the selection of long-term expressing, stably transformed clonal cell lines.

Construction of HSV-1 vectors carrying large genomic DNA loci is achieved through Cre-*loxP*-mediated recombination (hereafter referred to as retrofitting) between a plasmid carrying the amplicon elements and a bacterial artificial chromosome (BAC) carrying a genomic DNA locus to generate an infectious BAC vector, or iBAC (4, 11). To avoid the presence of two functional bacterial origins of replication in one plasmid after the retrofitting has taken place, the basic amplicon vector will carry a conditional replication origin (*R6Kγ*) which is active only in bacteria expressing the protein π (pi).

3.1.1. Retrofitting

What we describe as retrofitting is Cre-mediated site-specific recombination which joins together two vectors through *loxP* recognition by the Cre recombinase. When handling large BAC constructs it is very important to limit pipetting and to cut-off the ends of pipette tips to prevent shearing of the BAC DNA.

1. Dilute 10× Buffer for Cre reaction to 1×.
2. Dilute Cre enzyme 1:10 in 1× Buffer.
3. Prepare the Cre reaction mixture. (Always try two different BAC:amplicon vector ratios. The efficiency of successful retrofitting is dependent on the quality of the DNA preparation and can vary between preparations.)
   (a) Add 3 μL of 10× Buffer per reaction to a microfuge tube.
   (b) Add BAC DNA and amplicon vector DNA to ratios of 1 μg:50 ng and 1 μg:3 ng (BAC:amplicon) (see Note 1).
   (c) Add 1 μL of 1:10 diluted Cre enzyme per reaction.
   (d) Add H_2O to a final volume of 30 μL.
4. Incubate the reactions at 37°C for 1 h.
5. Stop the reaction by incubating the tubes at 70°C for 5 min.
6. Dialyze against double-distilled H_2O for 1.5–2 h on a 0.025 μm pore size 25 mm membrane filter at RT (see Note 2).
7. After dialysis recover the reaction (the volume will have reduced during the dialysis) and transfer to a fresh microfuge tube.
8. Electroporate 15 μL of reaction in 20 μL of DH10B bacteria in a 0.1 cm cuvette. Set electroporator at 1.8 kV, 25 μF, and 200 Ω. Then add 500 μL of SOC medium, mix by pipetting, and transfer reaction to a fresh tube to allow for bacterial recovery.
9. Shake at 37°C for 1 h at 225 rpm.
10. Plate the bacterial culture on 140 mm LB agar plates with antibiotics to select for the presence of both plasmids (see Note 3) and incubate at 37°C overnight until appearance of colonies.

3.2. DNA Maxiprep

Three vectors are used in the packaging protocol of HSV-1 amplicon vectors (see Subheading 3.3): pEBHICP27 (3), fHSVΔpacΔ27 0++ (3), and the amplicon vector, which can be a small 10–20 kb vector (5) or a high-capacity iBAC (4, 12). pEBHICP27 and small amplicon vectors can be prepared using the standard Qiagen maxiprep protocol. The following purification protocol is a modified Qiagen maxiprep protocol and is designed to achieve high yields of large vectors such as fHSVΔpacΔ27 0++ (178 kb) and iBAC vectors (~150 kb), although it can also be used for pEBHICP27 and small amplicon vectors. Use Qiagen Tip 500 Maxiprep kit.

1. Streak a small amount of bacterial stock on LB agar + antibiotics (see Note 3) and incubate at 37°C overnight.

2. In the morning, pick an individual colony, seed in a 1.5 mL LB + antibiotics starter culture and incubate for at least 8 h at 37°C. The medium should appear slightly cloudy at this point.
3. Tip the 1.5 mL starter culture into a 250 mL LB + antibiotics culture medium and incubate at 37°C overnight, shaking at 225 rpm.
4. The next day harvest the bacterial cells by centrifuging at 6,000 × *g* for 15 min at 4°C.
5. Aspirate the supernatant and resuspend the pellet in 15 mL of cold P1 solution + RNAse (see Note 4). To resuspend the pellet, allow the tubes to shake in a 37°C incubator at 225 rpm for 10 min.
6. Lyse the bacteria by adding 15 mL of P2 solution. Incubate for exactly 5 min, mix by swirling soon after adding P2 and then swirl again every 1–2 min (see Note 5).
7. After 5 min, add 15 mL of 4°C P3 solution and swirl to mix.
8. Put on ice for 20 min. Swirl lysate twice during the 20 min.
9. Centrifuge at 15,000 × *g* for 35 min at 4°C.
10. While waiting for centrifugation, prepare Tip-500 columns for DNA binding.
    (a) Equilibrate by adding 15 mL of QBT solution to the column and allow it to run through.
    (b) Fold a kimwipe tissue in half and insert in the column by pushing down with a finger. Cut excess tissue hanging over the edge of the column. The tissue will work as a fine filter to stop white lysate precipitate from entering the column (see Note 6).
11. Apply supernatant to the column and allow it to run through the tissue and column. After the material has passed through the column, squeeze the tissue into the column to recover the remaining material, taking care to prevent the white precipitate from falling into the column.
12. Wash the column twice with 30 mL of QC buffer.
13. Elute in Oakridge tubes with QF warmed up to 55°C as this will increase the DNA yield.
14. Add 10.5 mL of isopropanol to the eluted DNA, mix gently and centrifuge at 27,000 × *g* for 30 min at 4°C. Before centrifuging, mark the tube where the pellet is supposed to appear since often there is no visible pellet.
15. Carefully decant the supernatant (see Note 7), add 3.5 mL 70% ethanol without mixing and place the tubes back in rotor with the mark in the same position as for the previous spin.

16. Centrifuge again at 27,000 × *g* for 30 min at 4°C.
17. Carefully decant supernatant and let tubes air dry at RT for 5 min upside-down on a tissue and 10 min upright (see Note 8).
18. Add 250 μL of TE and flick the tube to loosen the pellet (see Note 9). Allow the DNA to dissolve in TE at 4°C overnight. Position the tube so that the TE covers the pellet.
19. The next day flick the tube again and spin at low speed (300 × *g* for 5 min) to collect the contents at the bottom of the tube. Transfer to a fresh tube and store the maxiprep at 4°C. Do not freeze as freeze/thaw cycles damage BAC DNA.
20. To check DNA quality:
    (a) Once quantified, digest ~400–500 ng of maxi with 3 U of enzyme (PacI for fHSVΔpacΔ27 0++) in a total volume of 15 μL and incubate at 37°C for 3–4 h.
    (b) Load the digest and run on PFGE with the following parameters: Volt/cm = 6, run time = 16 h, initial switch = 2 s, final switch = 16 s, temperature = 4°C, running buffer = 0.5× Tris Borate EDTA.
    (c) The next day, stain the gel with ethidium bromide. A band of ~178 kb should be visible for fHSVΔpacΔ27 0++ digestion.
    (d) On the same gel, also run ~400–500 ng of undigested DNA preparation to assess the quality of the supercoiled DNA. This should be visible as the highest band with an apparent size of ~250 kb (see Note 10).

3.3. Packaging of Amplicon Vectors into HSV-1 Virions

HSV-1 amplicons are produced using an improved helper virus-free system (3) utilizing a helper HSV-1 genome cloned into a BAC called fHSVΔpacΔ27 0++. The system has three features that prevent helper virus contamination of the amplicon preparation.

(a) fHSVΔpacΔ27 0++ lacks the essential HSV-1 packaging/cleavage sites (*pac*).

(b) In the unlikely event a recombination event places the *pac* signals within the helper BAC, the helper BAC can still not be packaged because it is oversized (178 kb) and is beyond the packaging capacity of HSV-1 virions. This has been achieved by adding extra copies of the ICP0 gene into the helper BAC.

(c) The HSV helper virus genome within the BAC plasmid is deleted for the essential viral gene ICP27, an additional safety feature. The ICP27 is supplied in *trans* from the additional plasmid, pEBHICP27, and from the Vero 2–2 cell line, a clonal cell line obtained from Vero cells and expressing ICP27.

This protocol, from transfection to amplicon harvesting, is performed over 5 days.

1. Grow Vero 2–2 cells in Vero 2–2 growth media until confluent.
2. Seed 1×10^6 Vero 2–2 cells per 6 cm dish in Vero 2–2 seeding media, and allow them to grow overnight.
3. The day after Vero 2–2 cells should be 80–90% confluent. Perform the cotransfection of the amplicon vector, the pEB-HICP27 plasmid, and the helper fHSVΔpacΔ27 0++ BAC as follows. This is for one 6 cm dish. For transfecting multiple dishes, increase the levels of each reagent to produce a bigger transfection mix. For each transfection, prepare two tubes, A and B. To tube A add, in order:
   (a) 250 μL OptiMEM per 6 cm dish.
   (b) 2 μg fHSVΔpacΔ27 0++.
   (c) 0.2 μg pEBHICP27.
   (d) 1.8 μg iBAC (see Note 11).
4. Mix tube A well and add 10 μL Plus reagent per transfection. Do not mix after adding Plus reagent as DNA may precipitate.
5. Leave tube A for 10 min.
6. To tube B add, in order:
   (a) 250 μL OptiMEM.
   (b) 23 μL Lipofectamine.
7. After the 10 min, add tube B to tube A drop-wise. Mix tube well.
8. Leave for complex formation for 30–40 min.
9. After the complex formation, add OptiMEM to reach a final volume of 1.5 mL per 6 cm dish.
10. Remove the medium from the Vero 2–2 cells, wash the cells twice with 2.5 mL OptiMEM, then add the transfection mix to the plates (1.5 mL/plate).
11. Leave for 4 h at 37°C. Swirl to mix the transfection every hour. Leaving longer than 4 h may cause toxicity to the Vero 2–2 cells.
12. Remove complex from the cells. Wash each dish three times with 2.5 mL OptiMEM, then add 3.5 mL of packaging media per 6 cm dish.
13. Let the packaging reaction proceed at 37°C (5% CO_2) for 60 h.
14. After 60 h, scrape the cells into their medium using a cell scraper. Place the scraped cells into a 50 mL tube, on ice, pooling from three to nine 6 cm dishes per 50 mL tube.

15. Freeze the tube in dry-ice + ethanol for 30 min, or at −80°C until you are ready to concentrate the amplicon. (Important: see Note 12.)
16. Thaw in ~30°C water. Break open the cells by using a sonicator set to a 50% amplitude for 15–30 s at 4°C: i.e. 20 s for 10.5 mL (three plates), 25 s for 21 mL (six plates), or 30 s for 31.5 mL (nine plates).
17. Spin down cellular debris by centrifuging at 1,200 × *g* for 10–15 min at 4°C in a bench-top centrifuge.
18. The amplicon is now ready for aliquoting and freezing, or further concentration (see Note 13).
19. For concentration, load on a 30% sucrose cushion into an ultracentrifuge rotor tube (5 mL for a SW32Ti rotor tube). Add the amplicon slowly over the 30% cushion. Balance the tubes and centrifuge for 3 h at 59,439 × *g* (22,000 rpm on SW32Ti rotor) at 4°C.
20. After centrifugation, carefully remove the supernatant and sucrose (rapid aspiration is best), the pellet will appear as a light brown-colored deposit at the bottom of the tube. Add 250 μL of either Hank's buffered salt solution (HBSS), PBS, or complete media (seeding media), flick the tube, and allow the amplicon to resuspend for 1 h or overnight at 4°C.
21. Resuspend the viral pellet by pipetting up and down several times, aliquot and store at −80°C. Prepare a separate 10 μL aliquot for stock titration.

3.4. Further Amplicon Purification Methods

During the amplicon extraction cell debris can co-purify with the amplicon viral particles and some cell types can show a slight toxic response to the debris. In this case, further purification of amplicons stocks can help reducing unwanted toxic effects.

3.4.1. Filtration

After step 16 of the packaging protocol (Subheading 3.3), the un concentrated amplicon lysate can be purified by filtration with 0.45 μm syringe filter and then aliquoted and frozen at −80°C or loaded on a sucrose cushion for further concentration (see Note 14).

3.4.2. Sucrose-Gradient Purification

The current protocol is an adapted version of the original HSV-1 sucrose-gradient purification protocol developed by W. Bowers et al. (13).

1. After step 17 of the packaging protocol (Subheading 3.3), transfer supernatant to a fresh tube and repeat centrifugation at 1,200 × *g* for 10–15 min at 4°C. This will greatly reduce the cellular debris in the amplicon viral prep.
2. Add 5 mL of PBS to a fresh ultracentrifuge tube.

3. Underlay the PBS with 5 mL of 30% sucrose, by slowly releasing it at the bottom of the tube. PBS and 30% sucrose should now be two separate layers, due to the higher density of 30% sucrose with respect to the PBS.
4. Underlay the 30% sucrose with 5 mL of 60% sucrose in the same way. Now three separate layers should be visible.
5. Add the amplicon slowly over the PBS. Balance the tubes and centrifuge for 1 h at 76,755 ×*g* (25,000 rpm on SW32Ti rotor) at 4°C.
6. Using a P1000 pipette or a glass pasteur, carefully extract the viral band which will appear as a gray interphase between 30 and 60% sucrose layers.
7. Put the extracted viral interphase in a 50 mL tube and slowly dilute with ~20 mL of cold PBS.
8. Load the PBS diluted gradient-purified amplicon prep onto 5 mL of 30% sucrose cushion as in step 19 of Subheading 3.3.
9. Balance the tubes and centrifuge for 1 h at 76,755 ×*g* at 4°C.
10. Aspirate the supernatant and resuspend the pellet as in steps 20 and 21 of Subheading 3.3. The viral stock is now ready for titration.

3.5. Titration

All amplicon vectors carry a reporter expression cassette specifically designed for titration of amplicon viral stocks. It is composed of a reporter gene such as green fluorescent protein (GFP) or β-galactosidase (β-gal) gene under the transcriptional control of the immediate-early HSV-1 promoter pIE4/5. The titration is performed on G16.9 cells, a Gli36-derived cell line that stably expresses the HSV-1 protein VP16, which enhances expression from the pIE4/5 promoter to give robust reporter gene expression.

1. Grow G16.9 cells in G16.9 growth medium until confluent.
2. Seed G16-9 cells in a 24-well plate in G16.9 seeding medium, at a density of 4×10^5 cells/well. Prepare four wells per amplicon preparation. It is important to have a confluent layer of cells the day after.
3. Defrost an amplicon aliquot and prepare stock dilutions. Starting amounts for titration of 5, 2, 0.5, and 0.1 μL of amplicon prep/well are suggested. All dilutions should be prepared in titration medium and made to add 250 μL of amplicon dilution per well.
4. Aspirate the seeding medium and add the amplicon dilution. Leave the infection for 24 h in a cell culture incubator (37°C, 5% CO_2).

5. Assay the infections by microscopy 24 h after infection (see Note 15). This assay is based on the assumption that every cell expressing the reporter gene has been infected by one amplicon particle only and so it represents one transducing unit (TU).
   (a) For GFP expression, use an inverted fluorescence microscope and pick an amplicon dilution with <100 green cells per field of view. Count several (e.g., 6) random fields of view, average the count, estimate how many fields of view there are in one well of a 24-well plate and estimate the titer of transducing particles per mL. On a Nikon Eclipse-TE inverted microscope using a 10× objective, there are 151 fields of view per well of a 24-well plate.
   (b) For β-gal expression, perform X-gal staining as follows.
      - Remove infection.
      - Wash once with PBS.
      - Fix cells by incubating with fixative solution for 10 min at RT.
      - Wash cells twice with PBS and add 300 μL of staining solution containing X-gal at 1 mg/mL.
      - Leave overnight at 37°C, then count the blue cells and estimate titer as in step 5a of Subheading 3.5.

3.6. Infection of Stem Cells

HSV-1 vectors can efficiently transduce stem cells. We have used amplicon vectors to transduce human and murine stem cells in both undifferentiated and terminally differentiated states. Several cell densities and multiplicities of infection (MOI – virus/cell ratio) can be used, depending on factors such as cell infectivity, amplicon stock titer, type of assay, etc. To increase viral transduction, plates with infected cells can be spun down soon after the addition of amplicon vector (see Note 16).

Here we describe the conditions used for the examples of stem cell transductions shown in Figs. 2 and 3.

3.6.1. Undifferentiated Human ES Cells

1. Seed ESCs on Matrigel™ in 24-well plates with an approximate density of 1×10^5 cells/well in mTeSR™-1 medium [with the addition of 10 μM ROCK inhibitor (Y-27632) during plating] (see Note 17).
2. Twenty-four hours later, confirm expression of stem cell markers by flow cytometry.
   (a) Detach cells from two wells by incubating cells with TrypLE™ Express for 5 min at 37°C.
   (b) Resuspend cells in FACS buffer containing 4% paraformaldehyde at a concentration of 1×10^5 cells/100 μL. Fix cells for 10 min.

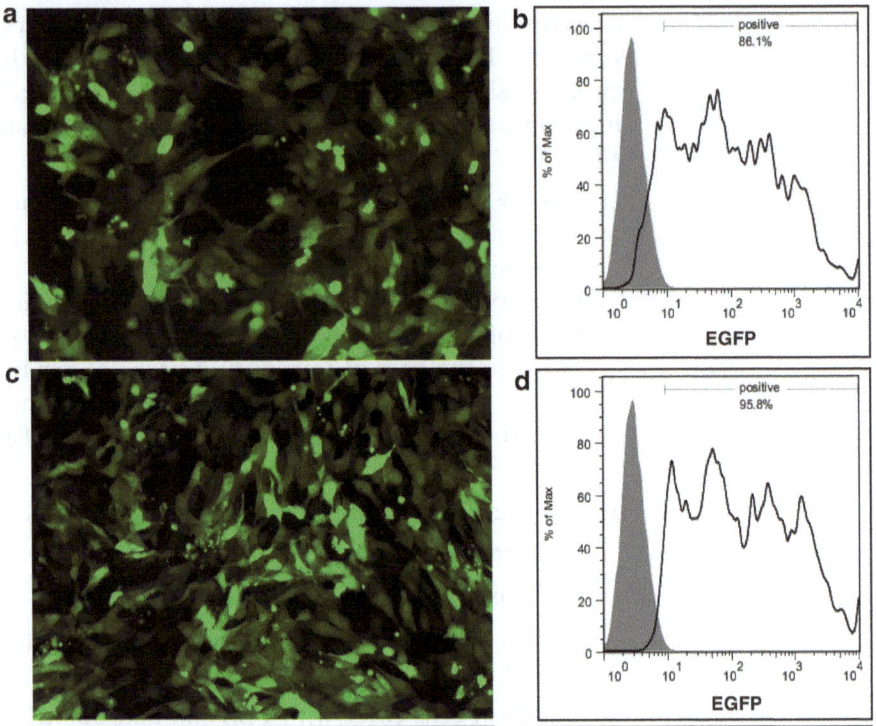

e

Amplicon preparation	% EGFP
Concentrated MOI 5	86.1
Concentrated MOI 10	95.8
Unconcentrated stock	60.8
Untransduced	0.5

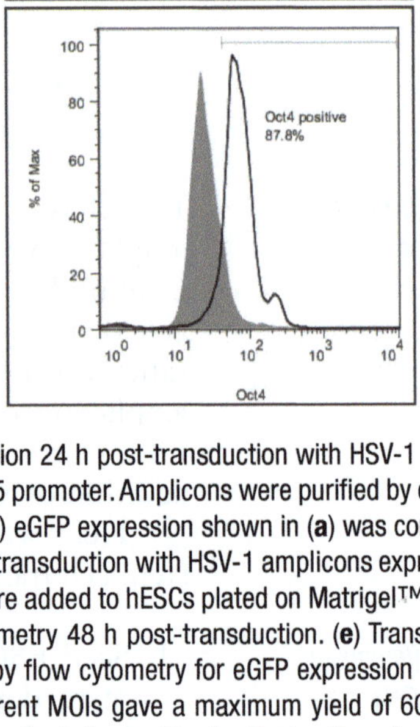

Fig. 2. Transduction of human embryonic stem cells. (**a**) eGFP expression 24 h post-transduction with HSV-1 amplicons expressing the eGFP reporter gene under the control of the HSV-1 IE4/5 promoter. Amplicons were purified by concentration and were added to hESCs plated on Matrigel™ at an MOI of 5. (**b**) eGFP expression shown in (**a**) was confirmed by flow cytometry 48 h post-transduction. (**c**) eGFP expression 24 h post-transduction with HSV-1 amplicons expressing the eGFP reporter gene. Amplicons were purified by concentration and were added to hESCs plated on Matrigel™ at an MOI of 10. (**d**) eGFP expression shown in (**c**) was confirmed by flow cytometry 48 h post-transduction. (**e**) Transduction of hESCs with amplicons prepared by different methods was analyzed by flow cytometry for eGFP expression 48 h post-infection. Transduction with unconcentrated amplicon stocks at different MOIs gave a maximum yield of 60.8% eGFP positive cells. (**f**) Non-transduced hESCs were confirmed by flow cytometry to be positive for Oct4 cell marker.

(c) Spin cells at 500 × *g* for 5 min at 4°C.

(d) Aspirate the supernatant and resuspend the cells in FACS buffer + 0.1% saponin. Leave for 30 min to permeabilize cells.

(e) Add 100 μL of the cell solution to the required number of wells of a V-bottomed microtiter 96-well plate.

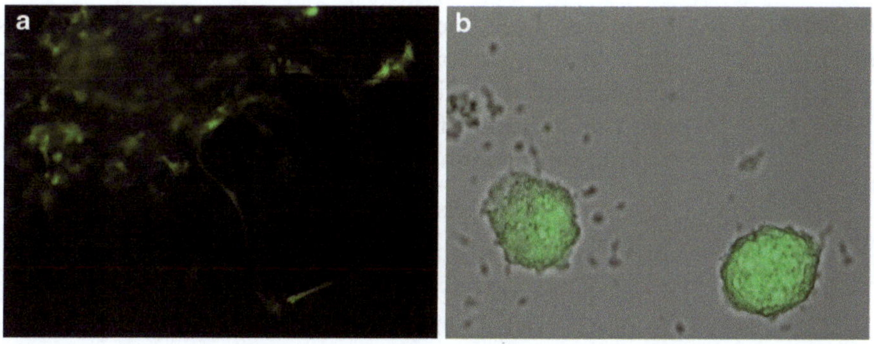

Fig. 3. Mouse and human embryonic stem cells: differentiation, transduction, and isolation of stable clonal cell lines. (**a**) Infection at an MOI of 20 of hESCs which have undergone 30 days of dopaminergic neuronal differentiation shows eGFP expression 48 h post-transduction. (**b**) Isolated mESC colonies after 2 weeks of hygromycin B selection.

(f) Spin cells at 500 × *g* for 5 min at 4°C.

(g) Aspirate supernatant and add 50 μL of antibody anti-Oct4 or isotype control antibody, diluted 1:10. Mix and incubate for 30 min at RT.

(h) Wash cells twice with FACS buffer then add 50 μL of secondary antibody, diluted 1:400 in FACS buffer.

(i) Incubate in the dark for 30 min at RT, wash twice with FACS buffer, transfer to FACS tubes, then analyse on a FACS machine, such as BD Calibur. The FACS graph in Fig. 2f shows that 87.8% of the analyzed cells were positive for the Oct4 marker.

3. Once expression of stem cell markers has been confirmed, trypsinize one of the wells and count cells to determine MOI.
4. Infect with an eGFP expressing amplicon vector at MOIs of 5 and 10. For best transduction results, use concentrated amplicon stocks.
5. Assess transduction efficiency after 24 h by reporter gene expression analysis (see Fig. 2a, c).
6. Trypsinize infected cells and perform quantitative analysis of eGFP-positive cells by flow cytometry (see Fig. 2b, d). Infection of hESCs at MOI 5 and 10 can yield a percentage of 86.1 and 95.8 of green cells, respectively, when using concentrated amplicon stocks (see Fig. 2e).

3.6.2. Human ESCs Differentiated to Dopaminergic Neurons

1. Differentiate human ESCs to dopaminergic neurons according to R&D Systems protocol found in the Human/Mouse Dopaminergic Neuron Differentiation Kit.
2. Infect differentiated cells in 24-well plates with an approximate MOI of 20.
3. Assess transduction efficiency after 48 h by reporter gene expression analysis (see Fig. 3a).

3.7. Stable, Clonal hESC Lines

Stable clonal cell lines carrying episomal amplicon vectors can be obtained by infection and selection for antibiotic resistance. We isolated stably transformed clonal cell lines from human ES cells, with the addition of 10 μM ROCK inhibitor (Y-27632) to the medium during cloning (14).

1. Seed 1×10^5 cells/well of 24-well plate.
2. Infect with an amplicon vector at an MOI of 2.
3. 48 h After infection, trypsinize the cells and do serial dilutions (1:10, 1:100) in three wells of a 6-well plate in medium containing hygromycin B (or a different antibiotic depending on the vector transduced) to a concentration of 125 μg/mL. (Alternatively hygromycin B can be added the day after seeding).
4. Two weeks after beginning the antibiotic selection, isolated colonies should be visible (see Fig. 3b). Pick reporter gene-positive individual colonies and expand into larger wells, in the presence of antibiotic selection.
5. Confirm the presence of the episomal vector in the selected clonal cell line.

4. Notes

1. A potential problem in this protocol is represented by the insertion of multiple copies of the amplicon vector (as a concatemer) into one copy of the BAC vector. This occurs as a result of the high efficiency of the Cre recombinase reaction. To decrease the occurrence of multiple insertions, it is suggested to try several BAC:amplicon ratios to achieve a low number of amplicon plasmid copies per BAC vector. One way to achieve this is to use decreasing amounts of amplicon plasmid DNA to establish the smallest DNA quantity able to give antibiotic resistant colonies. Even if multiple copies have been inserted, they can be easily identified by PFGE (see step 20 of Subheading 3.2).
2. This step reduces the buffer salts in the reaction which will increase the likelihood of a successful electroporation.
3. For DNA vector production, use final antibiotic concentrations of: ampicillin 50 μg/mL for pEBHICP27; kanamycin 25 μg/mL and chloramphenicol 15 μg/mL for fHSVΔpacΔ27 0++; ampicillin 50 μg/mL for amplicon vectors as pEHHG (4); chloramphenicol 15 μg/mL and ampicillin 50 μg/mL for BAC:amplicon retrofitted vectors (if the BAC has been retrofitted with an ampicillin resistant amplicon vector).

4. P1 solution is kept at 4°C after the addition of RNase A.
5. A key feature to avoid BAC DNA shearing during the lysis is to mix by swirling, instead of by inverting the tubes. Mix by swirling during the lysis but also after the addition of P3 solution.
6. Do not use gauze to filter bacterial lysate as this will let some debris through the column, which might block it and impede the solution from running through.
7. Rarely with fHSVΔpacΔ27 0++, but more often with small plasmids, the DNA pellet may detach at this point. In order not to lose the DNA pellet, the supernatant should be decanted in a 50 mL conical tube, which can then be visually checked for the presence of DNA precipitate. In case the pellet has detached and is visible in the tube, spin the falcon tube at 3,500 rpm (or max speed allowed by bench-top centrifuge) for 30 min at 4°C. After the spin, the supernatant can then be aspirated and the pellet resuspended in TE as in step 18 of Subheading 3.2. The same is valid for step 17.
8. Do not let the tubes dry for longer as this will result in a drier DNA pellet, more difficult to resuspend. If after 15 min the tube is not completely dry, remaining ethanol drops far from the pellet can be aspirated.
9. Do not vortex and always use cut off tips from this point onwards, to limit DNA shearing.
10. With constructs of this size, the supercoiled band runs higher than the linear or the open-circle forms.
11. This DNA quantity is advised for constructs of ~150 kb in size. For small amplicon vectors of ~10 kb in size, use 600 ng of DNA instead.
12. This step represents the only point where it is possible to stop and continue the protocol another day.
13. Large iBACs need further concentration to achieve average titers of $1–2 \times 10^7$. For 10–15 kb amplicons, vectors titers of 10^6 TU/mL can be obtained at this point, before concentration, but stocks can be further concentrated to obtain average titers of 10^8 TU/mL. The higher titers with smaller vectors are presumably due to more efficient transfection and packaging procedures.
14. Filtration can reduce the titer by two- to threefold.
15. In our experience of preparing BAC-based amplicons of 100–156 kb in size, titers of around $1–2 \times 10^7$ can routinely be obtained by concentrating the supernatant pooled from three 6 cm dishes and resuspending the pellet in ~250 μL of complete medium.

16. We have found that transduction efficiency can be increased by spinning down the plate with infected cells at 200 × *g* for 40 min at 20°C in a bench-top centrifuge.
17. Human ESCs need to be slowly adapted from coculture with MEFs to growth on Matrigel™ in mTeSR™-1 medium (15).

Acknowledgments

This work was supported by the Parkinson's UK Monument Trust Discovery Award; the Friedreich's Ataxia Research Alliance, Ataxia UK and the National Ataxia Foundation; the Medical Research Council and the Biotechnology and Biological Sciences Research Council. M.M.P.L. is an Ataxia UK Research Fellow, A.R.P. is a Medical Research Council student and S.A.C. is a Wellcome Trust Research Fellow.

References

1. Spaete, R. R., and Frenkel, N. (1982) The herpes simplex virus amplicon: a new eucaryotic defective-virus cloning-amplifying vector *Cell* **30**, 295–304.
2. Fraefel, C., Song, S., Lim, F., Lang, P., Yu, L., Wang, Y., Wild, P., and Geller, A. I. (1996) Helper virus-free transfer of herpes simplex virus type 1 plasmid vectors into neural cells *J Virol* **70**, 7190–7.
3. Saeki, Y., Fraefel, C., Ichikawa, T., Breakefield, X. O., and Chiocca, E. A. (2001) Improved helper virus-free packaging system for HSV amplicon vectors using an ICP27-deleted, oversized HSV-1 DNA in a bacterial artificial chromosome *Mol Ther* **3**, 591–601.
4. Wade-Martins, R., Smith, E. R., Tyminski, E., Chiocca, E. A., and Saeki, Y. (2001) An infectious transfer and expression system for genomic DNA loci in human and mouse cells *Nat Biotechnol* **19**, 1067–70.
5. Wang, S., and Vos, J. M. (1996) A hybrid herpesvirus infectious vector based on Epstein-Barr virus and herpes simplex virus type 1 for gene transfer into human cells in vitro and in vivo *J Virol* **70**, 8422–30.
6. Piechaczek, C., Fetzer, C., Baiker, A., Bode, J., and Lipps, H. J. (1999) A vector based on the SV40 origin of replication and chromosomal S/MARs replicates episomally in CHO cells *Nucleic Acids Res* **27**, 426–8.
7. Lufino, M. M., Manservigi, R., and Wade-Martins, R. (2007) An S/MAR-based infectious episomal genomic DNA expression vector provides long-term regulated functional complementation of LDLR deficiency *Nucleic Acids Res* **35**, e98.
8. Moralli, D., Simpson, K. M., Wade-Martins, R., and Monaco, Z. L. (2006) A novel human artificial chromosome gene expression system using herpes simplex virus type 1 vectors *EMBO Rep* **7**, 911–8.
9. Hacein-Bey-Abina, S., von Kalle, C., Schmidt, M., Le Deist, F., Wulffraat, N., McIntyre, E., Radford, I., Villeval, J. L., Fraser, C. C., Cavazzana-Calvo, M., and Fischer, A. (2003) A serious adverse event after successful gene therapy for X-linked severe combined immunodeficiency *N Engl J Med* **348**, 255–6.
10. Hacein-Bey-Abina, S., Von Kalle, C., Schmidt, M., McCormack, M. P., Wulffraat, N., Leboulch, P., Lim, A., Osborne, C. S., Pawliuk, R., Morillon, E., Sorensen, R., Forster, A., Fraser, P., Cohen, J. I., de Saint Basile, G., Alexander, I., Wintergerst, U., Frebourg, T., Aurias, A., Stoppa-Lyonnet, D., Romana, S., Radford-Weiss, I., Gross, F., Valensi, F., Delabesse, E., Macintyre, E., Sigaux, F., Soulier, J., Leiva, L. E., Wissler, M., Prinz, C., Rabbitts, T. H., Le Deist, F., Fischer, A., and Cavazzana-Calvo, M. (2003) LMO2-associated clonal T cell proliferation in two patients after gene therapy for SCID-X1 *Science* **302**, 415–9.
11. Hibbitt, O. C., Harbottle, R. P., Waddington, S. N., Bursill, C. A., Coutelle, C., Channon, K. M., and Wade-Martins, R. (2007) Delivery

and long-term expression of a 135 kb LDLR genomic DNA locus in vivo by hydrodynamic tail vein injection *J Gene Med* **9**, 488–97.

12. Wade-Martins, R., Saeki, Y., and Chiocca, E. A. (2003) Infectious delivery of a 135-kb LDLR genomic locus leads to regulated complementation of low-density lipoprotein receptor deficiency in human cells *Mol Ther* **7**, 604–12.
13. Bowers, W. J., and Federoff, H. J. (2006) Herpes simplex virus type 1-derived amplicon vectors, Gene Transfer: Delivery and Expression of DNA and RNA, A Laboratory Manual, pg. 227–254, Cold Spring Harbor Press, Cold Spring Harbor, NY.
14. Watanabe, K., Ueno, M., Kamiya, D., Nishiyama, A., Matsumura, M., Wataya, T., Takahashi, J. B., Nishikawa, S., Nishikawa, S., Muguruma, K., and Sasai, Y. (2007) A ROCK inhibitor permits survival of dissociated human embryonic stem cells *Nat Biotechnol* **25**, 681–6.
15. Ludwig, T. E., Bergendahl, V., Levenstein, M. E., Yu, J., Probasco, M. D., and Thomson, J. A. (2006) Feeder-independent culture of human embryonic stem cells *Nat Methods* **3**, 637–46.
16. Lufino, M. M., Edser, P. A., and Wade-Martins, R. (2008) Advances in high-capacity extrachromosomal vector technology: episomal maintenance, vector delivery, and transgene expression *Mol Ther* **16**, 1525–38.

Part VI

Differentiation

Chapter 28

The Generation of Embryoid Bodies from Feeder-Based or Feeder-Free Human Pluripotent Stem Cell Cultures

Alexander E. Stover and Philip H. Schwartz

Abstract

Embryoid body (EB) formation is a traditional method of inducing differentiation of pluripotent stem cells (PSCs). It is a routine *in vitro* test of pluripotency as well as the first stage in many differentiation protocols targeted toward the production of a specific lineage or cellular population, as in neural differentiation (see Chapters 29 and 30). The induction of differentiation via EB formation is fairly straightforward. However, depending on the specific PSC culture conditions – substrate, feeders, medium, and eventual cell type of interest – various methods are applied in order to most routinely obtain healthy EB cultures.

Key words: embryoid body, differentiation, aggrewell, ROCK inhibitor

1. Introduction

In this chapter, we present several protocols for EB formation. These are provided as general examples that have worked well in our hands and that can be used for both feeder-free and feeder-based culture systems. We also present methods for controlling the size of EBs through the use of Aggrewell dishes (1).

2. Materials

2.1. Reagents and Supplies

1. Matrigel™, reduced growth factor (BD Biosciences #354231).
2. Dulbecco's phosphate-buffered saline, without Mg_2 and Ca_2 (DPBS).
3. StemPro™ SFM kit (includes 50× supplement, DMEM-F12 with GlutaMax, and 25% BSA Solution, Invitrogen # A1000701).

Philip H. Schwartz and Robin L. Wesselschmidt (eds.), *Human Pluripotent Stem Cells: Methods and Protocols*, Methods in Molecular Biology, vol. 767, DOI 10.1007/978-1-61779-201-4_28, © Springer Science+Business Media, LLC 2011

4. 2-Mercaptoethanol (Invitrogen#21985-023).
5. Accutase (Millipore# SCR005).
6. Human bFGF/FGF2 (Stemgent #03-0002).
7. Y27632 ROCK Inhibitor (Stemgent # 04–0012). Prepare a 10 mM (1,000×) solution: dilute 2 mg of ROCK inhibitor in 643 μL of DMSO. Aliquot and store at −20°C.
8. Six-well vacuum gas plasma-treated tissue culture dishes (BD Falcon#353046).
9. Corning Costar Ultra-Low-Attachment six-well Plates (Corning# 3471).
10. 9-in. glass Pasteur pipettes (unplugged).
11. AggreWell™ 400 plates (Stem Cell Technologies #27845).
12. Centrifuge, swinging bucket style, high speed capable of 2,000 × *g*, with adapters to spin cell culture plates.

3. Methods

Although these protocols describe how to make EBs from cells that are grown in a six-well format, they can be scaled for other culture formats as well.

3.1. EB Formation via Cell Clusters

1. Exchange the medium in the well(s) to be used for differentiated with freshly prepared PSC medium.
2. Using a dissection microscope stationed within a laminar flow hood, mechanically dissociate healthy, undifferentiated colonies into medium-sized pieces using a needle, pipette tip, or fire-drawn glass Pasteur pipette as described in detail in Chapter 8 under mechanical subculturing (see Note 1).
3. After the desired colonies have been mechanically dissociated, collect in a tube and transfer, 1:1, to a low-attachment dish (see Note 2).
4. Place dish in the incubator.
5. The next day, EBs should be present. They are characterized by the pieces of colonies taking on a round appearance, with smooth borders. Irregularly shaped but smooth-bordered EBs may in fact be several EBs that have agglomerated.
6. Begin feeding on the second day and feed every other day thereafter.
   (a) To feed these suspension cultures, collect and pool the contents of the wells in a 15-mL conical tube.

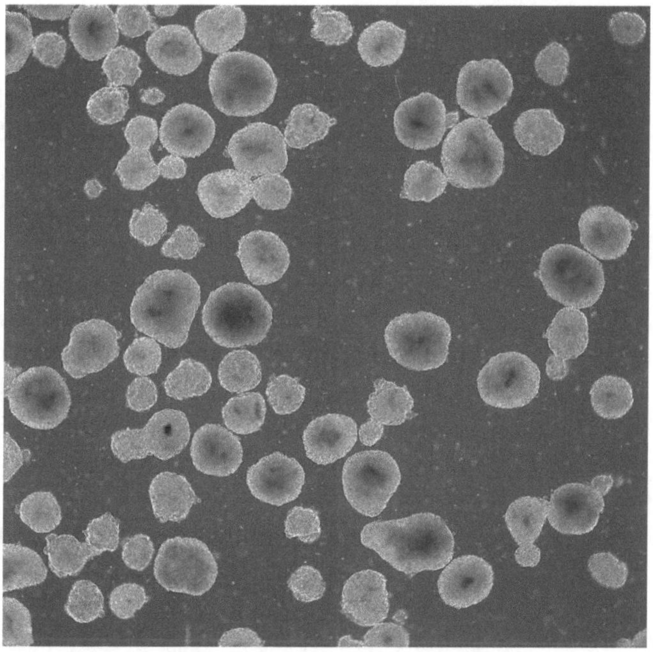

Fig. 1. Phase-contrast image with 4× objective showing EBs after 1 week.

(b) Allow the EBs to settle to the bottom, for about 5 min in the tissue culture hood at room temperature or in the incubator or water bath at 37°C.

(c) Carefully aspirate or pipette off the medium without disturbing the EBs, which have settled in the bottom of the tube.

(d) Add the same volume of fresh, warm medium to the tube and GENTLY resuspend the EBs – excessive trituration will break them apart.

(e) Dispense the EBs suspended in fresh medium into the ultra-low-attachment dish, and return to the incubator.

7. After 4–7 days, the EBs will be ready for plating onto slides for staining, or for further differentiation using a specific protocol (Fig. 1).

3.2. EB Formation via Dissociation to Single Cells

During EB formation from single cells, the cells are dissociated and resuspended in medium containing ROCK inhibitor to improve cell survival (2). The cells may be allowed to agglomerate and freely form EBs of varying size within an ultra-low-attachment dish or they may be seeded onto an Aggrewell™ dish, where their size can be more easily controlled.

Both variations of EB formation via single cells in suspension begin with the following:

1. Aspirate medium from the well(s) to be lifted, and gently rinse each well with 1 mL DPBS. Aspirate the DPBS.
2. Add 1 mL warm Accutase, place at 37°C for about 30 s, and then observe under a microscope to monitor cell dissociation (see Chapter 10).
3. When the cells appear to be phase-bright, or are noticeably lifting, immediately dilute the cells and Accutase using 9 mL of DPBS, and transfer the suspension to a 15 mL conical tube.
4. Rinse the well with 5 mL of DPBS to remove any remaining cells, pool with the contents of the 15 mL conical tube, and spin at 100*g* for 5 min.
5. At this point, both protocols call for the introduction of ROCK inhibitor (Y27632), which is a potent inhibitor of apoptosis (see note 3).

3.2.1. Free-Form EB Formation on Low-Attachment Surfaces

1. Resuspend the pellet in an appropriate volume of ROCK inhibitor-containing culture medium.
2. Add this suspension to a low-attachment dish. As a simple assay to determine EB-forming ability, plating the cells 1:1 from their original surface area usually produces satisfactory results. Increasing or decreasing the ratio will roughly alter the resulting sphere size, but not in a controlled, quantitative fashion.

3.2.2. Aggrewell™ Plating Method

Aggrewell™ plates make it easier to control the size of the EBs and reduce variability among EBs by controlling the initial starting size of each EB. The plates are similar to a standard 24-well dish, except that the center eight wells are textured to create hundreds of tiny, triangular microwells. When a cell suspension with a known concentration of cells is added to this plate, and the plate is then spun in an adapted centrifuge, each microwell collects a small, defined number of cells, which form an embryoid body. The EBs are removed after a period of 24–48 h for further culture. (The manufacturer's protocol and full technical data are available at http://www.stemcell.com.)

1. Resuspend the pellet in 2 mL of culture medium containing 10 μM ROCK inhibitor.
2. Count the cells using 10 μL of cell suspension, 10 μL of trypan blue, and a hemacytometer.
3. Refer to the chart in Table 1 (excerpted with permission from SCT) for detail on which seeding density should be used to obtain a particular EB size. Calculate the necessary volume of

Table 1
Table from stem cell technologies giving seeding densities for generating EBs of various sizes

	AggreWell™ 400 plate eight wells, each with approximately 1,200 microwells per well
Desired number of hESCs/hiPSCs per EB	Required number of hESCs/hiPSCs per well
50	6×10^4 cells
100	1.2×10^5 cells
200	2.4×10^5 cells
500	6×10^5 cells
1,000	1.2×10^6 cells
2,000	2.4×10^6 cells
3,000	3.6×10^6 cells
4,000	4.8×10^6 cells

media to obtain the desired density of cells, while taking into account that each well holds 1 mL of medium. Divide this volume in half, as half the volume will be used to remove air bubbles from the Aggrewell™ plate (see step 6 below). For example, to create eight wells of EBs that are approximately 500 cells each, would require 4.8×10^6 cells. Suspend 4.8×10^6 cells in 4 mL of medium, and proceed.

4. Fill each of the wells to be seeded with 500 μL of ROCK inhibitor-containing medium.
5. Place the Aggrewell™ plate (which does not yet contain cells) in a centrifuge with a swinging bucket rotor and the necessary adapters for spinning culture plates (Fig. 2). Spin at $2,000 \times g$ for 5 min. *This step is crucial*, as it removes bubbles that can become trapped in the microwells. The cells can be placed in an incubator to keep them warm during this time.
6. After the spin is complete, seed the cell suspension at the appropriate density.
7. Re-spin the plate at a lower speed of 150g for 5 min.
8. Place the plate in the incubator overnight.
9. The next day, the cells should have formed EBs (Fig. 3). If they do not appear to have done so, another 2–3 days may be necessary. They should not need to be fed during this time.
10. When EBs have formed, they can be removed from the Aggrewell™ microwells using gentle trituration and shaking of the plate and further cultured in a low-attachment plate (see Notes 4 and 5, Fig. 4).

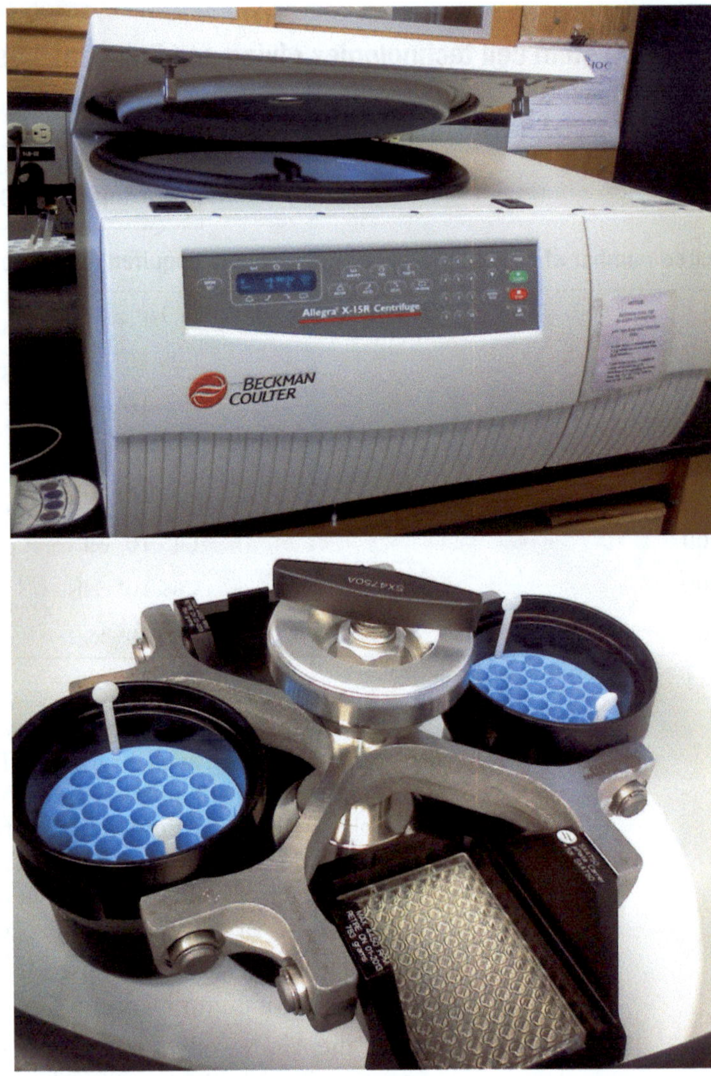

Fig. 2. Example of a low-speed centrifuge with adapter for spinning Aggrewell plates.

4. Notes

1. For EB formation, medium-large pieces tend to have the best survival. Extremely small colony chunks of only a few cells in size tend to have very low survival, and make very poor EBs.
2. If you have passaged a full six-well dish, transfer everything to a low-attachment six-well dish.
3. Without ROCK inhibitor, there is a strong likelihood that cells will die as a result of the shock of being put into suspension as single cells. We have noticed that some PSC cell lines that have been in culture for a long time are able to survive

Fig. 3. EBs forming in an Aggrewell plate.

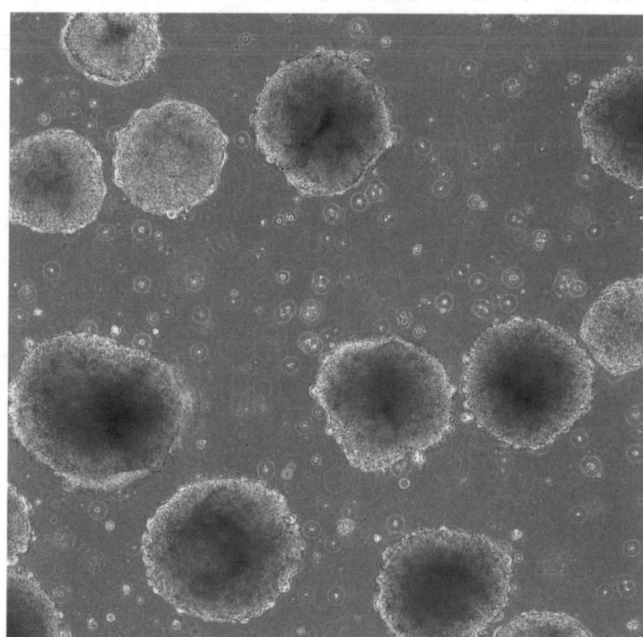

Fig. 4. EBs harvested from an Aggrewell plate (10× objective).

and form EBs without the use of ROCK inhibitor. However, this has only been at high seeding densities. For younger cell lines seeded at low densities, ROCK inhibitor is crucial.

4. We have found that the smaller the embryoid body size, the more difficult it is to remove them. Forceful trituration is effective but this often breaks the EBs apart. If small EBs are desired, they may not be easily or safely removed until they have grown to a larger size while still in the microwell. At this time, the effects of extended culture in the microwells are unknown.

5. Regardless of the method used to create them, EBs have a tendency to agglomerate once they are pooled into a low attachment dish. This adds a further complication to controlling sphere size. If the conjoined EBs become too large (greater than 1 mm across), the cells on the inside of the mass will have only restricted access to oxygen and nutrients, and viability will suffer. In some variations of differentiation protocols, limited conjoining of EBs is desired. For example, in our lab we have found that conjoined masses of 5–10 EBs tend to form rosettes well when seeded back onto a substrate. For other protocols, this may not be the case, however, and steps may be taken to prevent this. Other researchers have reported seeding EBs at an extremely low density, as well as routine gentle trituration to discourage aggregation.

Acknowledgments

This work has been funded by the National Institutes of Health (T15HL074286, R21MH087925, R01HD059967). The NCI Preclinical Repository supplied FGF-2. We also recognize Richard Pepple for proof-reading this chapter.

References

1. Ungrin M.D., Joshi C., Nica A., Bauwens C., and Zandstra P.W. (2008) Reproducible, Ultra-High-Throughput Formation of Multicellular Organization from Single Cell Suspension-Derived Human Embryonic Stem Cell Aggregates. *PLoS One* **3(2)**:e1565.

2. Watanabe K, Ueno M, Kamiya D, et al. (2007) A ROCK inhibitor permits survival of dissociated human embryonic stem cells. Nat Biotechnol;**25**:681–6.

Chapter 29

Derivation of Oligodendrocyte Progenitor Cells from Human Embryonic Stem Cells

Jason Sharp, Maya Hatch, Gabriel Nistor, and Hans Keirstead

Abstract

The directed differentiation of human pluripotent stem cells into specific, determined, and high-purity cell types can provide a means to study the cellular and molecular mechanisms of development and to generate cells for potential therapeutic applications. The ability to derive homogeneous cell populations obviates the need for transgene expression or cell sorting methods and can improve selection efficiency, lineage differentiation, cell viability, and clinical utility. Compared to undifferentiated pluripotent stem cells, high-purity cell phenotypes for clinical therapeutic strategies are expected to enhance engraftment, potentiate clinical efficacy, and decrease the risk of adverse effects such as dedifferentiation or teratoma formation. Clinical interest in the derivation of oligodendrocyte progenitor cells from pluripotent stem cells is based on research that demonstrates the effectiveness of progenitor cell transplants to improve outcomes after spinal cord injury. Here, we describe a protocol to generate oligodendroglial lineage-specific cells in high purity from human embryonic stem cells.

Key words: oligodendrocyte progenitor cells, human embryonic stem cells, high purity, directed differentiation, neurospheres

1. Introduction

Oligodendrocyte progenitor cells (OPCs) are the precursor cells of oligodendrocytes. OPCs are abundant in early development and although most mature to oligodendrocytes, a much smaller number remain resident in adults. Oligodendrocytes play a critical, supportive role in the central nervous system (CNS) via the insulation of axons with myelin sheaths that enable fast, saltatory conduction. This rapid conduction is important for such CNS functions as control of walking, perception of visual stimuli, and cognitive processes. When axons become demyelinated (i.e., lose their myelin sheaths) and/or oligodendrocytes are damaged or

Philip H. Schwartz and Robin L. Wesselschmidt (eds.), *Human Pluripotent Stem Cells: Methods and Protocols*,
Methods in Molecular Biology, vol. 767, DOI 10.1007/978-1-61779-201-4_29, © Springer Science+Business Media, LLC 2011

destroyed, as occurs due to trauma, disease, or toxins, axons cannot conduct signals properly and functional deficits can occur. This loss of function can be ameliorated through remyelination, a process that in large part involves the migration, division, and maturation of OPCs. However, evidence that OPCs express a number of neurotrophic factors and receptors for growth factors, neurotransmitters, and chemokines, indicates that OPCs, although progenitor cells, play a dynamic, integral role in the CNS.

Aside from their use as a research tool to understand such processes as CNS development and myelin formation, OPCs also have potential value as a therapeutic because they can restore CNS axonal conduction, can provide cellular support, and can remain resident in adult tissue. One approach to producing OPCs in a clinically viable manner is through differentiation of human pluripotent stem cells (PSCs), which include embryonic stem cells (ESCs) and induced pluripotent stem cells (iPSCs). The ability to control the differentiation of PSCs into populations of specific derivatives not only provides powerful methods with which to study cellular development, but enables generation of specific cells for cell replacement therapies. Based on concerns about therapeutic safety for use in humans, the derivation of high-purity end-stage cells, free of pluripotent stem cells, is essential (1). The protocol in this chapter addresses this issue by generating OPC cultures with greater than 90% purity. Although this protocol was developed for clinical application, the methods described here are intended for research purposes only.

Here, we describe an efficient way to produce OPCs from human PSCs; the specific example is derived from the hESC line, WA07 (WiCell). The protocol was first described by Nistor and colleagues (2) and stemmed from decades of research on development and the elucidation of detailed spatial and temporal patterns of morphogens (and serial induction) that occur in the primitive CNS. Hence, differentiation of PSCs into oligodendroglial progenitors is attained by using specific substrates and specialized media supplemented with specific morphogens at specific points in time.

2. Materials

2.1. Cell Culture Labware

1. Media filter flask (0.22 μm).
2. Teflon cell scrapers.
3. 75 cm^2 cell culture flasks.
4. Low adherent six-well plates (Corning).
5. Cell strainers (100 μm).
6. Lab-tek cell culture chamber slides (Permanox lab-Tek Chamber Slides, Nunc, # 177437).

2.2. Cell Culture Stock Solutions

1. Basic fibroblast growth factor (bFGF; human; Millipore) is dissolved in Dulbecco's phosphate-buffered saline (DPBS) with 0.5% (w/v) bovine serum albumin (BSA) at 10 μg/mL. Store in single-use, 20–100 μL, aliquots at –80°C.
2. Epidermal growth factor (EGF; human; Sigma) is dissolved in 0.1% (w/v) BSA at 20 μg/mL in KnockOut Dulbecco's Modified Eagle's Medium (KO-DMEM; Invitrogen). Store in single-use, 20–100 μL, aliquots at –80°C.
3. Insulin (bovine; Sigma) is dissolved in glacial acetic acid at 1 mg/μL and then diluted with water for embryo transfer (WET; Sigma) to 10 mg/mL. Store in single-use, 1 mL aliquots at –80°C.
4. Progesterone (Sigma) is dissolved in ethanol at 1 mg/mL and then diluted with Dulbecco's Modified Eagle's Medium (DMEM) to 63 μg/mL. Store in single-use 1 mL aliquots at –20°C.
5. Putrescine (Sigma) is dissolved at 10 mg/mL in DMEM. Stored in single-use, 1 mL aliquots at –20°C.
6. Retinoic acid – all *trans* – (Sigma) is dissolved at 20 mM in dimethyl sulfoxide (DMSO; Sigma). Stored in single-use, 20–100 μL aliquots at –80°C.
7. Sodium selenite (Sigma) is dissolved in WET at 50 μg/mL. Stored in single-use, 1 mL aliquots at –20°C.
8. Transferrin (human; Sigma) is dissolved at 50 mg/mL in DMEM. Stored in single-use, 1 mL aliquots at –80°C.
9. Triiodo-L-thryonine (T3; Sigma) is dissolved in 1 N NaOH at 100 mg/mL and then diluted in DMEM to 40 μg/mL. Store in single-use, 1 mL aliquots at –80°C.
10. Matrigel™, growth factor reduced (BD Biosciences, San Jose, CA) is diluted 1:1 (v:v) with cold (~4°C) KO-DMEM. Store in single-use, 1–2 mL, aliquots at –20°C (see Note 1).
11. Dulbecco's PBS (DPBS).

2.3. Cell Culture Media

1. Mouse Embryonic Feeder-Conditioned Medium (MEF-CM): maintain mouse embryonic fibroblasts (MEFs) with 50% (v/v) DMEM and 50% (v/v) KO-DMEM for 24 h then collect and filter the medium with a 0.22 μm filter flask. Store at –80°C. (see Note 2).
2. Glial restrictive medium (GRM): Add 974 mL (or 1 L) DMEM/F12 (with glutamine or Glutamax) (Invitrogen), 20 mL B27 supplement (Invitrogen), 1 mL insulin stock, 1 mL progesterone stock, 1 mL putrescine stock, 1 mL sodium selenite stock, 1 mL transferrin stock, and 1 mL T3 stock to a 0.22 μm filter flask and apply vacuum. GRM can be stored at 4°C.
3. Transition medium: Mix MEF-CM and GRM at a 1:1 ratio (v:v).

3. Methods

The OPC derivation protocol consists of four stages, with each stage defined by morphological features exhibited by the cultured cells. Observing these changes is a simple way to determine whether the protocol is working. However, morphological observations should always be confirmed by immunocytochemistry (see Subheading 3.12 below). In the first stage, undifferentiated hESCs (Fig. 1a) grow in colonies with a smooth looking surface. In the absence of an exogenous feeder layer, hESCs generate a subpopulation of migratory fibroblastic cells (extraembryonic endoderm derivatives). In the second stage, differentiation is started by culturing the cells on low-attachment substratum and treating with

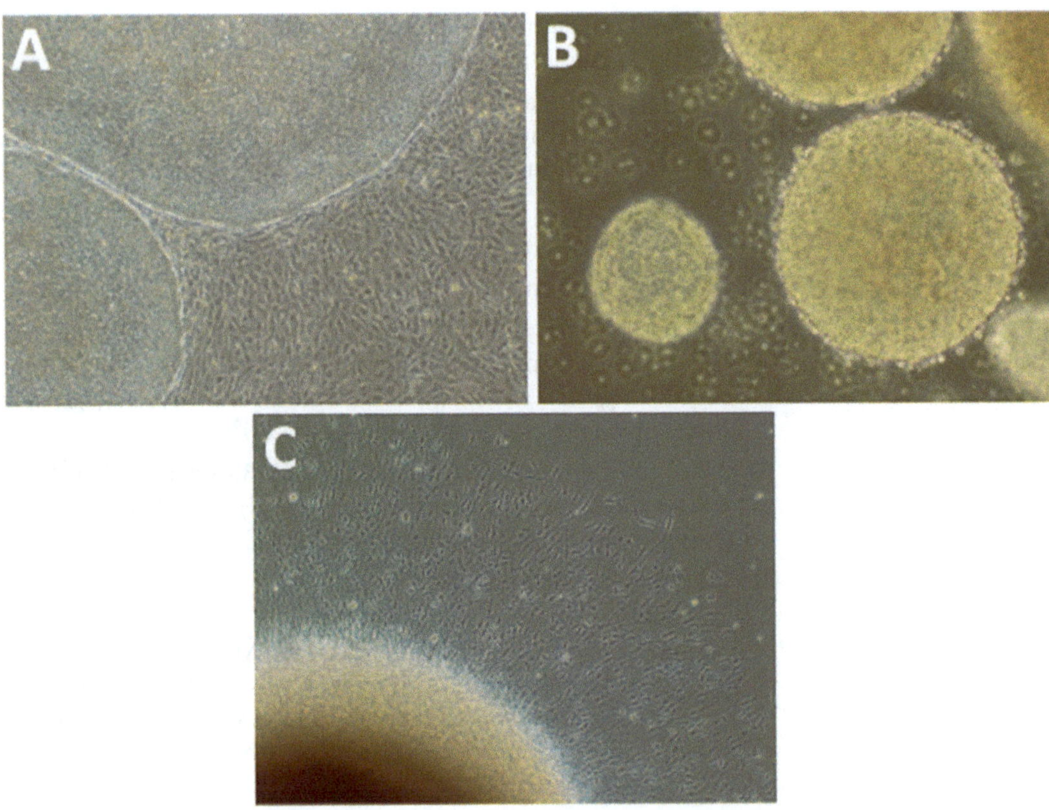

Fig. 1. Derivation of OPCs from hESCs. (**a**) The ideal morphology of hESC colonies for the start of the OPC differentiation protocol includes a well-defined, smooth, rounded border without protrusions. In addition, colonies have few, if any layers (notable *yellow* coloration), and are separated by stromal cells. (**b**) After growth on nonadherent substrate and exposure to RA, colonies take on a spherical morphology. *Yellow spheres* are composed of neuralized cells. Smaller aggregates, individual cells, and cell debris will be removed during successive feeds. (**c**) Plated spheres show extended processes and OPCs migrating from the spheres. OPCs continue to migrate from the spheres over time, so removal of the spheres is recommended to maintain desired cell concentration.

a restriction medium that induces neural differentiation. Small clusters of 20–100 cells, referred to here as "cellular aggregates," form after plating the partially dissociated hESC colonies in nonadherent culture plates. Isolated floating cells are usually not viable and are discarded at feeding with gravity or low-force centrifugation. In the third stage, yellow spheres/neural progenitors (Fig. 1b) form from the aggregates, acquire a spherical morphology, and are bright yellow. There will be small and large yellow spheres surrounded by cellular debris that includes nonselected, dark cell aggregates and floating cells. In the fourth stage, plating of yellow spheres/neural progenitors allows selection of viable cells, dissociation of the spheres, and further differentiation of neural progenitors into OPCs (Fig. 1c). A purification method is used by plating the yellow spheres on Matrigel™-coated flasks. Plating will eliminate dead or nonadherent cells and promote outward migration from yellow spheres. During this stage, migrating cells can exhibit either an epithelial or a bipolar morphology with short thick branches. The cultures will go through a differential panning process in which adherent cells (astrocytes, fibroblasts) attach to tissue culture plastic and the less adherent OPCs are collected and recultured. At the same time, the cell population is sampled for immunocytochemistry by plating them on laminin or Matrigel™-coated imaging slides or glass cover slips. Cells are tested for oligodendrocyte-lineage markers such as Olig1 and NG2. Some plated yellow spheres will extend long processes first, followed by migration of the OPCs along the radial branches over the next few days.

Variations from the given protocol can be performed with certain limitations. However, we recommend strict adherence to the described protocol, since timing and sequence of growth factor treatment is critical to success. The cultures are tolerant to changes in the concentration of supplements, however, changes can affect the outcome of oligodendrocyte vs. contaminant cell ratio and yield. However, some modification of this protocol has been successfully used by others. Izrael and colleagues (3) used a modification of this protocol that included the use of noggin to promote $O4^{+}$ oligodendrocytes in culture and increase myelin formation in vivo.

We recommend the use of morphological, immunocytochemical, and molecular characterization to control the quality of the PSC-derived OPCs. Quality control can prevent the continuation of aberrant cultures or experiments using low quality cells. In particular, we recommend that users examine a number of markers over a range of time-points and establish exclusion criteria. These markers and time-points can then be used to determine the quality of culture samples during subsequent differentiations. Table 1 gives an example an immunocytochemical profile of three cultures monitored at Days 10, 42, and 56. In addition to these markers, we recommend the use of Oct4, A2B5, Sox10 (molecular only), and Nkx2.2.

Table 1
Immunocytochemical characterization of cultured OPCs

% of cells expressing marker (day of differentiation)	Culture 1	Culture 2	Culture 3
Pax6 (Day 10)	98% ± 2%	97% ± 3%	96% ± 3%
Pax6 (Day 42)	>1%	>1%	>1%
Olig1 (Day 42)	83% ± 7%	84% ± 6%	88% ± 5%
NG2 (Day 42)	98% ± 2%	97% ± 3%	99% ± 1%
GalC (Day 56)	95% ± 4%	94% ± 6%	97% ± 2%
RIP (Day 56)	95% ± 2%	95% ± 5%	90% ± 6%
04 (Day 56)	85% ± 5%	82% ± 7%	80% ± 3%
GFAP (Day 56)	4% ± 4%	3% ± 3%	5% ± 3%
Tuj1 (Day 56)	8% ± 2%	7 ± 2%	6% ± 2%
BMP4 (Day 56)	0	0	0
SSEA4 (Day 56)	0	0	0

3.1. Coating Flasks with Matrigel™ Substrate

1. For a 2 mL aliquot of diluted Matrigel™, add 28 mL KO-DMEM for a total volume of 30 mL of 1/30 Matrigel™.
2. Add 5 mL of 1/30 Matrigel™ solution to each 75 cm² flask to be coated. Swirl the solution to coat the bottom of the flask and place at 4°C overnight.
3. At least 1 h before the flasks are to be used, remove the coating solution, wash once with KO-DMEM, and replace with 5–10 mL working medium. Place the flask in a CO_2 incubator for temperature and pH balance.

3.2. Thaw of PSC Cultures for OPC Differentiation

1. Prepare Matrigel-coated flask(s) prior to thaw of frozen PSCs aliquot(s) (see Note 3).
2. Warm 35 mL of MEF-CM in a 37°C water bath.
3. Thaw aliquot(s) of hESCs in 37°C water bath for 2 min and add to 9 ml of MEF-CM. Spin cells down at 200 × *g* for 4 min and resuspend in 5 mL of MEF-CM medium. Carefully break up the pellet by trituration. Do not break up the cell clusters.
4. Add the resuspended cells to the flask, and add 8 ng/mL of bFGF (add 0.8 μL of 10 μg/mL stock solution for each mL of medium). Place flask in incubator at 37°C with 5% CO_2.

3.3. Expansion and Passage of PSC Cultures for OPC Differentiation

1. Feed cells with 20–30 mL of freshly prepared MEF-CM containing 8 ng/mL bFGF every day.
2. Passage cells once per week at 1:4 to 1:6.
3. Before passaging, prepare plates, by coating with Matrigel™ as described above, 24 h in advance.
4. On the day of passage, prepare MEF-CM + 8 ng/mL bFGF.
5. Dissociate cells by adding 10 mL of 1 mg/mL collagenase IV to the flask and leave it in the incubator for 2–5 min.
6. Aspirate collagenase and wash cells with 10 mL DPBS. Aspirate DPBS and add 10 mL of new MEF-CM to the flask. Scrape the hESCs in the medium with a cell scraper.
7. Collect the cells and distribute according to desired splitting ratio into Matrigel™-coated flasks and add sufficient MEF-CM plus 8 ng/mL of bFGF to each.
8. Mark the passage number and return flasks to incubator.
9. Continue to feed the cells daily with fresh medium plus 8 ng/mL of bFGF.
10. For start of differentiation, grow culture(s) to 70–90% confluence (see Note 4).

3.4. Directed Differentiation of PSCs to OPCs: Transition, Day 1

1. Prepare 30 mL of transition medium for each 75 cm² flask and prewarm and pH balance the medium in the CO_2 incubator. Add 4 ng/mL bFGF (use 0.4 μL of 10 μg/mL bFGF stock for each mL of medium) just before use.
2. Treat PSCs with 10 mL per flask of 1 mg/mL collagenase IV for 2–5 min in incubator. Aspirate collagenase and wash cells with 10 mL of DPBS.
3. Aspirate DPBS, add 30 mL of prewarmed transition medium, and scrape cells with a scraper to dislodge. Collect cells in to 50 mL centrifuge tube and pipette cell aggregates (about three to five times) to slightly break up larger clumps.
4. Distribute cells evenly to six-well, low-attachment plates (5 mL cell suspension in each well). Incubate at 37°C, 5% CO_2.

3.5. Directed Differentiation of PSCs to OPCs: Transition, Day 2

1. Prepare 30 mL of transition medium + 4 ng/mL bFGF + 20 ng/mL EGF and 10 μM RA for each six-well plate (use EGF and RA stocks) (see Note 5).
2. Collect cells from each well of the six-well plate and combine in a 50-mL conical tube. Spin cells at 200 × *g* for 4 min.
3. Aspirate old medium and add 30 mL of new transition medium with supplements. Resuspend gently, without breaking up the cell clusters.
4. Distribute cells evenly in 5 mL to each well in the plate. Return cells to incubator.

3.6. Directed Differentiation of PSCs to OPCs: Aggregate Formation, Days 3–10

1. Feed the cultures every day.
2. Prepare 30–35 mL of GRM + 20 ng/mL EGF + 10 μM RA for each six-well plate of cells.
3. Remove debris from the culture by low-force centrifugation: collect cells in a 50-mL conical tube and centrifuge at 200 × *g* for 2 min.
4. Aspirate supernatant and add 30 mL of new GRM + EGF + RA medium. Do not dissociate clumps.
5. Redistribute 5 mL in each well of a six-well plate.

3.7. Directed Differentiation of PSCs to OPCs: Medium Aggregates, Days 11–15

1. Feed the cultures every other day (e.g., M-W-F).
2. Prepare 30 mL of GRM + 20 ng/mL EGF for each six-well plate of cells.
3. Collect the cells and perform the same procedures as described for days 3–10 with a low centrifugation force (200 × *g* for 1–2 min).

3.8. Directed Differentiation of PSCs to OPCs: Yellow Spheres, Days 16–27

1. Change the medium three times per week (e.g., M-W-F) and always return the cells to the incubator as soon as possible.
2. Prepare 30 mL of GRM + 20 ng/mL EGF for each six-well plate of cells.
3. Collect the cells in 50-mL tubes and allow them to settle by gravity without centrifugation for 5–10 min.
4. Aspirate the supernatant and add new medium on top of the cells.
5. Redistribute the clusters: agitate the tube and immediately collect 15 mL using a 25 mL pipette and distribute suspension quickly to the first three wells.
6. Agitate the tube again, collect the remaining 15 mL, and redistribute the volume in the other three wells.

3.9. Directed Differentiation of PSCs to OPCs: Plating Spheres, Day 28

1. Prepare Matrigel™ matrix-coated 75 cm^2 flask with 1:30 Matrigel™ in KO-DMEM 24 h in advance.
2. Prepare 30 mL of GRM + 20 ng/mL EGF, place the medium in the coated flask and establish proper temperature and pH for at least 1 h in the CO_2 incubator.
3. Collect cells from the six-well plate into a 50 mL centrifuge tube. Let the spheres settle to the bottom of tube for 5 min.
4. Aspirate old medium and add a small amount (~5 mL) of prewarmed GRM + EGF.
5. Place cells into the coated 75 cm^2 flask with the balanced medium.
6. Return to the incubator at 37°C, 5% CO_2.
7. The next day, gently shake the plated flask to dislodge nonadherent debris.

3.10. Directed Differentiation of PSCs to OPCs: OPC Migration, Days 29–34

1. Change the medium every M-W-F.
2. Prepare 30 mL of GRM + 20 ng/mL EGF for each Matrigel™-coated flask.
3. Aspirate the old medium.
4. Add 30 mL of new GRM + EGF to the flask. Return to incubator.

3.11. Directed Differentiation of PSCs to OPCs: OPC Purification, Days 35–42

1. Prepare the required Matrigel™-coated 75 cm² flasks with 1:30 Matrigel™ in KO-DMEM 24 h in advance. Prepare enough flasks for splitting the cells at a 1:2 ratio.
2. Add 25 mL of GRM + 20 ng/ml EGF to each flask; prewarm and balance the pH of the medium in the CO_2 incubator for 1 h before use.
3. Prepare Matrigel™- (1:30) or laminin- (10 μg/ml/cm²) coated imaging slides or coverslips in wells one day in advance. Replace the coating solution with GRM (without EGF) and place the slides or dishes in the CO_2 incubator for temperature and pH balance 1 h before use.
4. Aspirate the medium from each cell-containing flask. Wash with 10 mL DPBS. Aspirate again.
5. Add 7 mL of warm trypsin/EDTA to the flask. Incubate 5–10 min at 37°C.
6. Add 7 mL of anti-trypsin solution to the flask. Collect the dispersed cells into a 15 mL conical tube.
7. Take a small sample of dissociated cells to count cells using a hemocytometer. Take a 50 μL sample of the cell suspension and mix with 50 μL of Trypan Blue. Count live (unsustained) and dead (blue-stained) cells.
8. Spin the cells at 250 × *g* for 5 min.
9. Panning for adherent cells: Aspirate medium and resuspend in 50 mL of GRM (without EGF) medium. Transfer cells to two 75 cm² or one 150 cm² uncoated tissue culture plastic flask(s) and incubate for 1 h at 37°C. This step allows astrocytes and other adherent cells to attach to the plastic bottom, while the less adherent OPCs will not attach.
10. Collect medium with a gentle shake of the flasks and transfer to 50-mL conical tube affixed with a 100 μm cell strainer.
11. Take a sample from the purified cell population for immunocytochemistry. Plate cells at 50,000 cells/cm² on imaging slides or coverslips. After 2 days, cells are ready to be fixed and stained.
12. Split the rest of the purified cells into two Matrigel™-coated 75 cm² flasks that were prepared earlier, containing GRM + 20 ng/mL EGF. After 7 days of growth, the cells are ready to be used for experiments.

3.12. General Protocol for Immunodetection

1. Fix cultures for 10 min at room temperature (RT) in 4% paraformaldehyde (pH 7.4).
2. Wash three times in phosphate-buffered saline (PBS) for 5 min each.
3. Incubate in 1% BSA + 0.1% Trition-X100 in PBS for 30 min at RT.
4. Wash three times in PBS for 5 min each.
5. Dilute primary antibodies in 1% BSA in PBS (see Note 6). Incubate overnight at 4°C.
6. Wash three times in PBS for 5 min each.
7. Block with 10% goat serum in PBS for 30 min at RT.
8. Wash three times PBS for 5 min each.
9. Add secondary antibody according to the primary host species (anti-mouse or anti-rabbit) 1:200 in PBS for 1 h at RT.
10. Wash three times in PBS for 5 min each.
11. Counter stain with Hoechst (1:5,000) for 5 min at RT.
12. Wash three times in PBS for 5 min each and one time with dH_2O.
13. Apply mounting medium and coverslip.

4. Notes

1. Matrigel™ is stored at −20°C and must be kept cold until ready to use. Matrigel™ forms a gel at RT. Diluted aliquots, as used here, can be prepared as long as the solution remains cold/on ice. This coating procedure can be applied to other culture flasks and plates.
2. Use Mitomycin-C-treated MEFs.
3. Human embryonic stem cells can be purchased from the National Stem Cell Bank or other vendors. Frozen aliquots referred to here contain approximately 1.5×10^6 cells.
4. Human embryonic stem cell colonies are grown and expanded on Matrigel™-coated flasks until the appropriate confluence for the differentiation protocol is obtained. We recommend using one six-well low-attachment plate for each 70–90% confluent 75 cm^2 flask to yield approximately $10–15 \times 10^6$ cells.
5. Use minimal light during feeding since RA is light sensitive. For RA, use 1 μL of each mL of medium from the 10 μM/mL stock solution. Discard the vial after use.

Table 2
Antibodies useful for characterizing the cells at different stages of the procedure

Undifferentiated hESCs	
SSEA-4	A glycolipid epitope that is used as a marker of many pluripotent cells
POU5F1/OCT4	A transcription factor characteristic of pluripotent cells
Yellow spheres/neural progenitors	
Pax6	A transcription factor indicative of neural commitment
Nestin	Intermediate filament often used as a marker for neural commitment
A2B5	Marker for early neural progenitors
Oligodendrocyte progenitors	
Olig1/2	Transcription factor expressed during oligodendroglial development (located both nuclear and cytoplasmic)
A2B5	Marker for early neural progenitors
NG2	A chondrotin sulfate proteoglycan
PDGFRα	Growth factor receptor
Markers of contaminating cells	
GFAP	Used as an indicator for astrocytes (typically less than 5%)
Neurofilaments (βTubulin, MAP2, Tuj1 etc.)	Used as indicators for neurons (typically less than 5%)
SMA (Smooth Muscle Actin)	Occasionally detected in single cells (less than 0.1%)

6. Table 2 lists suggestions for antibodies that are useful for characterizing the cells at different stages of the procedure. All of the antibodies require slightly different protocols and specific dilutions; refer to Subheading 3.12 of this chapter, and Chapter 15 of this book, and the manufacturer's suggestions for specific protocols.

References

1. Sharp J, Keirstead HS (2007). Therapeutic applications of oligodendrocyte precursors derived from human embryonic stem cells. *Curr Opin Biotechnol.* **18**: 434–440.
2. Nistor GI, Totoiu MO, Haque N, Carpenter MK, Keirstead HS (2005). Human embryonic stem cells differentiate into oligodendrocytes in high purity and remyelinate after spinal cord transplantation. *Glia* **49**: 385–396.
3. Izrael M, Zhang P, Kaufman R, Shinder V, Ella R, Amit M, Itskovitz-Eldor J, Chebath J, Revel M (2007). Human oligodendrocytes derived from embryonic stem cells: Effect of noggin on phenotypic differentiation in vitro and on myelination in vivo. *Mol Cell Neurosci.* **34**: 310–323.

Chapter 30

Directed Differentiation of Dopamine Neurons from Human Pluripotent Stem Cells

Lixiang Ma, Yan Liu, and Su-Chun Zhang

Abstract

Midbrain dopaminergic (mDA) neurons play a critical role in regulating postural reflexes and movement as well as modulating psychological processes. Dysfunction or degeneration of mDA neurons is involved in a number of neurological disorders including Parkinson's disease. Availability of large quantities of human mDA neurons would greatly enhance our ability to reveal pathological processes underlying mDA neuron degeneration and to identify treatments for these neurological conditions. Human pluripotent stem cells (PSCs), including embryonic stem cells (ESCs) and induced pluripotent stem cells (iPSCs), provide an unlimited source for mDA neurons. Here we describe a chemically defined protocol for mDA neuron differentiation. PSCs are first converted to neuroepithelia in a chemically defined medium without any growth factors, followed by patterning the neuroepithelia to midbrain progenitors with fibroblast growth factor 8 (FGF8) and sonic hedgehog (SHH) and subsequent differentiating to functional mDA neurons. This protocol typically yields about half of the neuronal population being mDA neurons, determined by expression of mDA markers, electrophysiological recordings, and the ability to reverse functional deficit in a rat model of Parkinson's disease.

Key words: human embryonic stem cells, induced pluripotent stem cells, neuroepithelia, dopaminergic neurons, neurodegeneration, parkinson's disease

1. Introduction

Midbrain dopaminergic (mDA) neurons are located in the substantia nigra, the ventral tegmental area, and the retrorubral field (1). During development, neuroepithelial cells (or neural stem cells) in the ventral midbrain differentiate to mDA neurons in response to sonic hedgehog (SHH) from the floor plate and fibroblast growth factor-8 (FGF8) from the mid–hind brain boundary (isthmus) (2). These mDA neurons project to the striatum and the limbic system, regulating locomotion and emotion. Degeneration or loss of mDA neurons is implicated in various

Philip H. Schwartz and Robin L. Wesselschmidt (eds.), *Human Pluripotent Stem Cells: Methods and Protocols*, Methods in Molecular Biology, vol. 767, DOI 10.1007/978-1-61779-201-4_30, © Springer Science+Business Media, LLC 2011

neurological conditions, most notably Parkinson's disease. It is presently unknown how these mDA neurons degenerate and how to rescue them. Availability of large quantities of human mDA neurons, including those from patients, would greatly enhance our ability to reveal pathological processes underlying mDA neuron degeneration and to identify treatments for these neurological conditions.

Human pluripotent stem cells (PSCs), including embryonic stem cells (ESCs) and induced pluripotent stem cells (iPSCs), have the potential for generating all cell types in the body, including mDA neurons. Two major approaches have been developed for differentiating human PSCs, mainly ESCs, to mDA neurons: the co-culture method and the embryoid body-based, serum-free culture system. Co-culture with mesenchymal stromal cells such as MS5 cells and PA6 cells (3–6) or with midbrain astrocytes (6), followed by treatment of the cultures with FGF8 and SHH generally yields a good population of DA neurons. The co-culture with stromal cells appears to mainly promote neuroepithelial differentiation and neural precursors are then treated with FGF8 and SHH for DA neuron differentiation. Signals from the stromal cells, particularly from midbrain astrocytes (6), may also contribute to the patterning of midbrain progenitors. The disadvantage of the co-culture system is the unknown nature of the factors produced by the stromal cells and astrocytes as well as the contamination of the DA neuron cultures with the immortalized stroma cells.

The embryoid body-based, serum-free differentiation culture, which we developed (7, 8) and will be described here, is relatively simple and yields DA neurons with an efficiency at least as high as that using the co-culture methods. The major advantage of this method is a chemically defined system, which allows dissection of molecules that influence the specification and differentiation of DA neurons. Another advantage is that it is free of immortalized cells which are often tumorigenic and thus influence subsequent experiments including transplantation. The method comprises three major steps, induction of neuroepithelial cells in the absence of growth factors, patterning the neuroepithelial cells to midbrain progenitors with FGF8 and SHH, and differentiation of the progenitors to DA neurons with neurotrophic factors. For induction of neuroepithelial cells, hESCs are aggregated and grown as free-floating hESC aggregates, often termed "embryoid bodies," for 6 days. This is followed by adherent colony culture at a low density until formation of neural tube-like rosettes by columnar epithelia by 2 weeks of differentiation. Neural rosettes are a readily identifiable sign of neuroepithelial differentiation, which determines the time for midbrain patterning and isolation of neuroepithelia. We have shown that when the neural tube-like rosettes form at around day 14–16, the cells become definitive neuroepithelial cells and express Pax6 and Sox1. Application of FGF8 (50–100 ng/mL) and SHH (200 ng/mL) before this time point results in more efficient patterning to midbrain progenitors. The patterned neural progenitors are then expanded as free-floating cultures for 1–2 weeks before they are differentiated to neurons in adherent cultures.

This method generally yields 32% DA neurons among total differentiated progenies, or 50% of all neurons at 5–6 weeks of differentiation (9). The DA neurons are defined by their expression of tyrosine hydroxylase (TH) but not dopamine β-hydroxylase (DβH). These DA neurons secrete DA in an activity-dependent manner. They also contribute to functional recovery of locomotion following transplantation into a rat model of Parkinson's disease (10). Therefore, the hESC-differentiated DA neurons are functional. The same method has been applied to the differentiation of DA neurons from human iPSCs. This technology may be useful in future development of stem cell-based therapy for a number of neurodegenerative disorders including Parkinson's disease.

2. Materials

2.1. Stock Solutions

1. DMEM/F12.
2. Neurobasal media.
3. N2 supplement (Invitrogen; cat. No. 17502–048).
4. B27 (Invitrogen; cat. No. 12587–010).
5. MEM nonessential amino acids 100× (NEAA, 10 mM solution).
6. Basic fibroblast growth factor (bFGF, R&D Systems; cat. No. 233-FB) 10 μg/mL; in sterilized PBS with 0.1% bovine serum albumin (BSA).
7. L-Glutamine solution (200 mM).
8. Dispase (1 unit/mL); dissolve 50 units of Dispase in 50 mL DMEM/F12 at 37°C for 15 min. Filter sterilize using a 50 mL filter. The solution can be stored at 4°C for up to 2 weeks (see Note 1) (11).
9. Knockout serum replacer (Invitrogen; cat. No. 10828–028) Make aliquots of 50 mL and store at −80°C.
10. Heparin (1 mg/mL); dissolve 10 mg heparin (Sigma; cat. No. H3149) in 10 mL DMEM.
11. Laminin from human placenta (Sigma; cat. No. L6274).
12. β-Mercaptoethanol (2-ME) at 14.3 M concentration.

13. Glial cell line-derived neurotrophic factor (GDNF), brain-derived neurotrophic factor (BDNF), insulin-like growth factor-1 (IGF1) (PeproTech Inc, 450–02, 450–10, 100–11); 100 μg/mL, in sterilize distilled water.
14. Ascorbic acid (AA, 200 μg/mL); 2 mg ascorbic acid in 10 mL PBS.
15. Cyclic AMP (cAMP, 1 mM, Sigma, cat. No. D-0260); in sterilized water. Aliquot and store the stock solutions are at −80°C.

2.2. Supplies

1. 6-well and 24-well tissue culture plates.
2. 60-mm tissue culture dishes.
3. 500 ml filter unit (PES, low-protein-binding).

2.3. Media

hESC growth medium: Add 3.5 μL β-mercaptoethanol to 2.5 ml l-glutamine solution, and then combine it with 392.5 mL DMEM/F12, 100 mL Knockout serum replacer and 5 mL Nonessential amino acids solution. Sterilize by filtering through a 500 mL filter unit and store at 4°C for up to 10 days (see Note 2).

Neural induction medium: Combine 489.5 mL DMEM/F12, 5 mL N2 supplement, 5 mL MEM nonessential amino acids solution, and 0.5 mL of 1 mg/mL heparin. Sterilize by filtering through a 500 mL filter unit and store at 4°C for up to 2 weeks.

Neural differentiation medium: Combine: 490 mL Neurobasal media, 5 mL N2 supplement, and 5 mL MEM nonessential amino acids solution. Filter through a 500 mL filter unit and store at 4°C for up to 2 weeks.

3. Methods

1. hESCs (see Fig. 1a) are grown to confluence (see Chapter 8). Aspirate off hESC medium and add 1 mL of fresh dispase (1 unit/mL) to each well of a six-well plate and incubate the cultures at 37°C for 2–5 min.
2. Check the cultures every 2 min and when the edges of hESC colonies begin to curl off of the plate, aspirate the dispase off gently, add 1 mL of hESC medium to each well, and gently swirl the plate and/or pipette the colonies off the well.
3. Collect the colonies into a 15-mL conical tube and centrifuge at 200 × *g* for 2 min. Aspirate off the supernatant gently.
4. Rinse the cells once by resuspending them in 5 mL of fresh ESC medium and then spin them down at 200 × *g* for 2 min.
5. Aspirate off supernatant, gently resuspend the colonies in hESC medium and transfer to 175 cm^2 flask. Add 40 mL of hESC medium without bFGF (see Note 3).

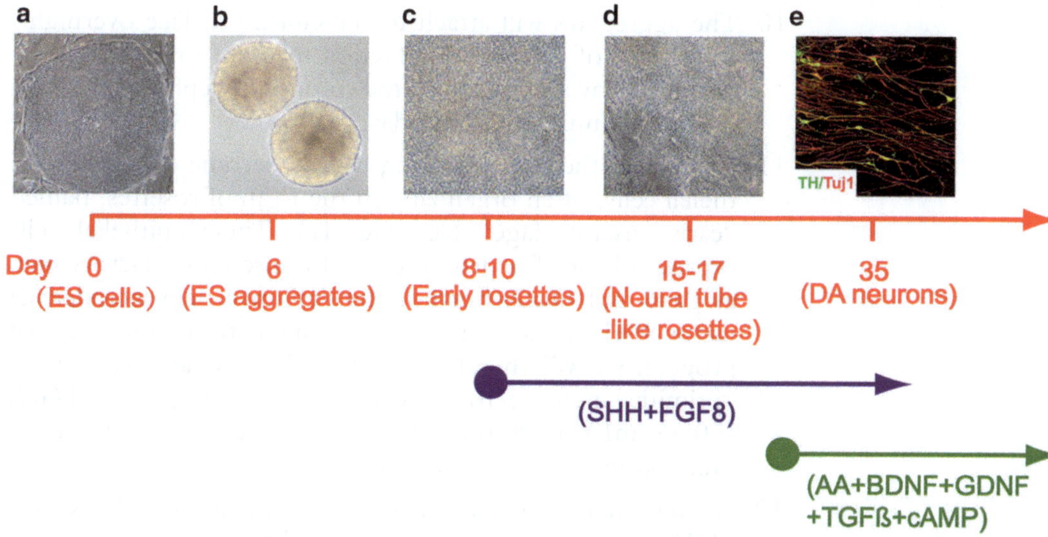

Fig. 1. Differentiation of dopaminergic neurons from human embryonic cells. hESCs (**a**) form ES aggregates (**b**) in suspension culture in days 4–6, they organize into early rosettes (**c**) in days 8–10, and neural tube-like rosettes (**d**) in days 15–17. In response to SHH and FGF8 from day 10 to day 35, neuroepithelial cells are patterned into dopaminergic neurons. Around 30% dopaminergic neurons which express TH (*green*) and Tuj1 (*red*) (**e**) are generated at the 5 weeks of differentiation.

6. Feed the cells with hESC medium every day for 4 days. The cells form aggregates (see Fig. 1b). To feed the aggregates: allow them to settle to the bottom of the flask by standing the flask on end, then aspirate off half of the medium, and add the same volume of fresh medium to the flask. Return to the incubator (see Note 4).
7. After 4 days, the aggregates are fed with the ***neural induction medium*** for another 2 days. Collect the aggregates, centrifuge for 2 min at 200 × *g* and wash once with 5 mL of neural induction medium. Resuspend aggregates in 35 mL of neural induction medium and transfer to a new 175 cm^2 flask.
8. After 2 days, the aggregates should become bright and are ready for attachment. In order to attach the aggregates, six-well plates need to be coated with laminin. Add 300 μL of neural induction medium containing 20 μg/mL laminin to each well of a six-well plate. Do not let the medium drain to the edge of the well (leave as a drop in the center of the well). Incubate the plate at 37°C overnight (see Note 5).
9. Collect the hESC aggregates to a 15-mL conical tube, centrifuge at 200 × *g* for 2 min. Aspirate off supernatant and resuspend hESC aggregates in the neural induction medium. Aspirate the laminin solution from the prepared plates in step 8 above and transfer 300 μL of aggregate solution containing 20–30 clusters. Be careful to let the aggregates distribute evenly.

10. The aggregates will attach to the culture surface overnight. Add 2 mL of the neural induction medium to each well the following day and then feed the cultures by replacing 50–60% of the medium every other day.
11. At day 10, the differentiating hESCs become columnar epithelial cells, often organizing in the form of rosettes, named "early rosette stage" (see Fig. 1c). These epithelial cells express a host of neuroectodermal transcription factors such as Pax6 and Otx2, but not the definitive neuroectodermal factor Sox1. These cells can be readily patterned to regional progenitors. We therefore refer to the cells at this stage as "primitive neuroepithelial cells." At this stage, add FGF8 (50 ng/mL) and SHH (100 ng/mL) to pattern ventral midbrain progenitors (see Notes 6 and 7).
12. Change half of the neural induction medium with FGF8/SHH every other day for the next 5 days.
13. At day 15–16, the neuroepithelial cells in the rosettes should have expanded substantially and formed multiple layers, giving a tube-like appearance which we term "neural tube-like rosette" stage (see Fig. 1d). Formation of neural tube-like rosettes suggests that neuroepithelial cells can be isolated. Rinse the rosettes once with DPBS. Detach the clusters by gently triturating the center part of the attached colonies. The center part of the clusters will easily detach from the culture surface while the peripheral flat cells remain largely attached (12).
14. Collect the rosette clumps in a 15-mL conical tube. Triturate the clumps with a 10-mL serological pipette twice, but do not break up the clumps. Centrifuge at $50 \times g$ for 2 min at room temperature.
15. Aspirate off the supernatant, resuspend the clusters in 5 mL of neural induction medium supplemented FGF8 (50 ng/mL), SHH (100 ng/mL), and B27, and transfer the culture to a 25-cm^2 flask and grow them as free-floating neuroepithelial clusters.
16. Feed the cultures by replacing 50–60% of the medium with the neural induction medium containing SHH/FGF8 every other day (do not need B27). The neuroepithelial clusters will form spheres after 1 or 2 days, typically 100–200 μm in diameter. If the spheres grow bigger than 200 μm in diameter, break them using a flame-polished Pasteur pipette.
17. At day 23–25 of hESC differentiation, plate the neuroepithelial clusters onto laminin-coated Petri dishes for neuronal differentiation. For immunostaining, the cells may be plated onto glass coverslips that are coated with polyornithine and laminin (2–4 clusters/coverslip). After attachment, feed the

cells with neural differentiation medium supplemented with cAMP (1 μM), ascorbic acid (0.2 mM), laminin (1 μg/mL), TGFβ3 (1 ng/mL), brain-derived neurotrophic factor (BDNF, 10 ng/mL), and glial cell line-derived neurotrophic factor (GDNF, 10 ng/mL).

18. Continue feeding the attached cells with neural differentiation medium as detailed in step 17 every other day (50% medium change).
19. Dopamine neurons, identified by immunostaining for tyrosine hydroxylase (TH) (see Fig. 1e), are readily observed by 5 weeks of differentiation. The number and proportion of TH-expressing dopamine neurons increase over time up to day-50. Additional dopaminergic markers and midbrain markers should be used to determine the midbrain dopaminergic identity. In particular, co-expression of TH and En-1, a midbrain transcription factor, is an important indicator of mDA.

4. Notes

1. The activity of dispase varies among lots. Therefore, attention should be paid to the activity of each lot. Also, the activity goes away quickly at 37°C. So, do not warm up in a water bath for a long period.
2. β-Mercaptoethanol is combustible, corrosive and toxic in case of ingestion and skin absorption; keep away from sources of ignition; and avoid direct contact, as it can penetrate some gloves.
3. Feed the hESC aggregates (embryoid bodies) for the first time within ~12 h and move the aggregates to a new flask in order to remove debris and any mouse embryonic fibroblasts that may have been carried along with the hESCs.
4. Gently triturate the hESC aggregates with a 10-mL pipette two to three times, let the aggregates settle down to the bottom of the tube for 2–3 min, and carefully aspirate off the medium without disturbing the aggregates. In this way, debris attached to the aggregate surface can be effectively removed.
5. For best results, plate the day-6 aggregates on laminin-coated surface overnight. Alternatively, one can avoid precoating the dish by adding 10% FBS to the culture medium when plating the aggregates. In this case, use serum as short as possible because serum inhibits neural induction. Remove the serum within 24 h of plating.
6. Heparin should always be added with FGFs to stabilize their activities.

7. If no typical rosettes are formed, the most probable causes are either the ESCs are partially differentiated before initiating the differentiation protocol or the ESC colonies were damaged when they were removed from the MEF feeders or during the aggregation process.

Acknowledgments

This study was supported by the National Institute of Neurological Diseases and Stroke (NS045926, NS046587 to SCZ), and in part by a core grant to the Waisman Center from the National Institute of Child Health and Human Development (P30 HD03352), the Ministry of Science and Technology, China (2006CB94700, 2006AA02A101), and the Shanghai Municipality (06dj14001).

References

1. Smidt MP, Burbach JP. (2007) How to make a mesodiencephalic dopaminergic neuron *Nat Rev Neurosci* **8**, 21–32.
2. Ye W, Shimamura K, Rubenstein JL, Hynes MA, Rosenthal A. (1998) FGF and Shh signals control dopaminergic and serotonergic cell fate in the anterior neural plate *Cell* **93**, 755–66.
3. Barberi T, Klivenyi P, Calingasan NY, et al. (2003) Neural subtype specification of fertilization and nuclear transfer embryonic stem cells and application in parkinsonian mice *Nat Biotechnol* **21**, 1200–7.
4. Perrier AL, Tabar V, Barberi T, et al.(2004) Derivation of midbrain dopamine neurons from human embryonic stem cells *Proc Natl Acad Sci USA* **101**, 12543–8.
5. Buytaert-Hoefen KA, Alvarez E, Freed CR.(2004) Generation of tyrosine hydroxylase positive neurons from human embryonic stem cells after coculture with cellular substrates and exposure to GDNF *Stem Cells* **22**, 669–74.
6. Roy NS, Cleren C, Singh SK, Yang L, Beal MF, Goldman SA.(2006) Functional engraftment of human ES cell-derived dopaminergic neurons enriched by coculture with telomerase-immortalized midbrain astrocytes *Nat Med* **12**, 1259–68.
7. Zhang SC, Wernig M, Duncan ID, Brustle O, Thomson JA.(2001) In vitro differentiation of transplantable neural precursors from human embryonic stem cells *Nat Biotechnol* **19**, 1129–33.
8. Li XJ, Du ZW, Zarnowska ED, et al.(2005) Specification of motoneurons from human embryonic stem cells *Nat Biotechnol* **23**, 215–21.
9. Yan Y, Yang D, Zarnowska ED, et al.(2005) Directed differentiation of dopaminergic neuronal subtypes from human embryonic stem cells *Stem Cells* **23**, 781–90.
10. Yang D, Zhang ZJ, Oldenburg M, Ayala M, Zhang SC. (2008) Human embryonic stem cell-derived dopaminergic neurons reverse functional deficit in parkinsonian rats *Stem Cells* **26**, 55–63.
11. Hu, B.Y., Zhang S.C.(2009) Differentiation of spinal motor neurons from pluripotent human stem cells. *Nature Protocols* **9**, 1295–304.
12. Xia, X.F., Zhang, S.C. (2009) Differentiation of neuroepithelia from human embryonic stem cells. *Methods in Molecular Biology.* **549**, 51–8 (Human Press, New York)

Chapter 31

Methods for the Derivation and Use of Cardiomyocytes from Human Pluripotent Stem Cells

Wei-Zhong Zhu, Benjamin Van Biber, and Michael A. Laflamme

Abstract

The availability of human cardiomyocytes derived from embryonic stem cells (ESCs) has generated considerable excitement, as these cells are an excellent model system for studying myocardial development and may have eventual application in cell-based cardiac repair. Cardiomyocytes derived from the related induced pluripotent stem cells (iPSCs) have similar properties, but also offer the prospects of patient-specific disease modeling and cell therapies. Unfortunately, the methods by which cardiomyocytes have been historically generated from pluripotent stem cells are unreliable and typically result in preparations of low cardiac purity (typically <1% cardiomyocytes). We detail here the methods for a recently reported directed cardiac differentiation protocol, which involves the serial application of two growth factors known to be involved in early embryonic heart development, activin A, and bone morphogenetic protein-4 (BMP-4). This protocol reliably yields preparations of 30–60% cardiomyocytes, which can then be further enriched to >90% cardiomyocytes using straightforward physical methods.

Key words: cardiomyocytes, directed differentiation, activin A, bone morphogenetic protein-4

1. Introduction

Cardiomyocytes from human embryonic stem cells (hESCs) and the related human-induced pluripotent stem cells (hiPSCs) have tremendous promise as a model system for heart development and disease, a platform for in vitro drug screening, and a potential source of cells for cardiac repair. Both of these pluripotent stem cell types have unquestioned cardiomyogenic potential, in contrast to many adult stem cell types for whom the capacity to differentiate into significant numbers of definitive cardiomyocytes is controversial (for a recent review, see (1)). Moreover, undifferentiated pluripotent stem cells (PSCs), ESCs, and iPSCs, as well as their differentiated cardiac progeny, show robust proliferative activity, which makes these cell types particularly attractive for

Philip H. Schwartz and Robin L. Wesselschmidt (eds.), *Human Pluripotent Stem Cells: Methods and Protocols*, Methods in Molecular Biology, vol. 767, DOI 10.1007/978-1-61779-201-4_31, © Springer Science+Business Media, LLC 2011

applications requiring large quantities of cells (e.g., replacing the ~1×10^9 host cardiomyocytes lost in a typical human myocardial infarct). Human PSC-derived cardiomyocytes have an unambiguous cardiac phenotype, exhibiting spontaneous contractile activity, cardiac-type mechanisms of excitation–contraction coupling, and expression of expected sarcomeric proteins, ion channels, and transcription factors (2–4). Moreover, we and others have shown that, following transplantation into rodent infarct models, hESC-derived cardiomyocytes form nascent human myocardium and help preserve cardiac function (5–7).

Despite this progress, the derivation of highly purified populations of cardiomyocytes from PSCs remains a significant challenge to the field, particularly for in vivo applications, in which the transplantation of undifferentiated cells can give rise to teratomas or other undesirable noncardiac derivatives (8, 9). The method by which cardiomyocytes have been historically generated from ESCs involves their spontaneous differentiation in high serum via embryoid bodies, a poorly controlled approach that typically results in preparations of <1% of cardiomyocytes. Our group and others have sought to develop more efficiently cardiogenic guided differentiation protocols, including the procedure described here, which reliably yields preparations of 30–60% cardiomyocytes (6). If a greater degree of cardiac purity is required, additional enrichment steps (e.g., Percoll gradient centrifugation (6, 10)) can be performed, which typically results in preparations of >90% human cardiomyocytes.

2. Materials

2.1. Cells

1. Primary mouse embryonic fibroblasts (MEFs), not mitotically inactivated.
2. WA07 (H7 hESC line) (Wicell Research Institute, Madison, WI) (see Note 1).

2.2. Stock Solutions

1. Dulbecco's phosphate-buffered saline (DPBS).
2. MEF medium: 89% (v/v) Dulbecco's modified Eagle medium (DMEM, Invitrogen; cat. no. 11965–092), 10% heat-inactivated fetal bovine serum (FBS), and 2 mM L-glutamine.
3. hESC medium: 79% (v/v) Knock-out DMEM (Invitrogen; cat. no. 10829–018), 20% Knock-out serum replacement (Invitrogen; cat. no. 10828–028), 1% nonessential amino acids solution, 1 mM L-glutamine, and 0.1 mM β-mercaptoethanol. Add 4 ng/mL bFGF stock solution immediately before use.
4. RPMI-B27 medium: 98% (v/v) RPMI 1640 (Invitrogen; cat. no. 21870–092), 2% B27 serum supplement (Invitrogen; cat. no. 17504–044), and 2 mM L-glutamine.

Table 1
Preparation of Percoll gradient solutions (for 100 mL final volumes)

	40.5% Percoll solution	58.5% Percoll solution
Percoll (GE Healthcare/Amersham; cat. no. 17-0891-02)	40.5 mL	58.5 mL
1.5 M NaCl (sterile-filtered)	10 mL	10 mL
1 M HEPES (Invitrogen; cat. no. 15630–080)	1 mL	1 mL
H_2O (WFI-quality)	48.5 mL	30.5 mL

5. Percoll (GE Healthcare/Amersham; cat. no. 17-0891-02) solutions: shortly before use, prepare 40.5 and 58.5% (v/v) solutions, using the reagents and quantities indicated in Table 1.

2.3. Growth Factors

1. Human basic fibroblast growth factor (bFGF) (PeproTech; cat. no. 100-18B): dissolve at 10 μg/mL in DPBS with 0.2% bovine serum albumin (BSA) carrier , aliquot and store at –20°C.
2. Activin A (R & D Systems; cat. no. 338-AC-025): dissolve at 10 μg/mL in DPBS with 0.2% BSA, aliquot and store at –20°C.
3. Bone morphogenetic protein-4 (BMP-4; R & D Systems; cat. no. 314-BP-010): dissolve at 1 μg/mL in DPBS with 0.2% BSA carrier, aliquot and store at –20°C.

2.4. Enzymes

1. Dispase (Invitrogen; cat. no. 17105–041): dilute to 0.1 U/mL in PBS, filter-sterilize, aliquot, and store at –20°C.
2. Liberase Blendzyme IV (Roche Applied Sciences; cat. no. 11-988-476-001): dilute to 0.56 U/mL in PBS, aliquot, and store at –20°C.
3. 0.05% Trypsin-EDTA solution.
4. Defined trypsin inhibitor, 1× (Cascade Biologics; cat. no. R-007-100).
5. DNase I (Invitrogen; cat. no. 18047–019).

2.5. Substrates

1. 0.5% gelatin solution: 2% bovine gelatin (Sigma-Aldrich; cat. no. G1393) warmed at 37°C for 10 min and then diluted to 0.5% (v/v) with DPBS.
2. Matrigel™ solution: dilute growth factor-reduced Matrigel™ (BD Biosciences; cat. no. 356231) 1:60 with cold (4°C) Knock-out DMEM. Diluted aliquots can be stored at –20°C, but should be thawed at 4°C overnight before use.

3. Polyethylenimine (PEI, Sigma-Aldrich; cat. no. P7239): dilute 0.1% (v/v) in sterile water.

2.6. Other Reagents

1. Versene (Invitrogen; cat. no. 15040–066).
2. Cryostor CS10 (BioLife Solutions; cat. no. 640221).

3. Methods

3.1. Compatible Methods for Maintaining Undifferentiated hESC Cultures

All cell cultures (i.e., MEFs, undifferentiated hESCs, and differentiated progeny) described below should be maintained at 37°C in a humidified incubator with 5% CO_2 and ambient O_2 (see Notes 2 and 3).

3.1.1. Preparation of MEF-Conditioned Medium After Xu et al. (11)

1. Coat tissue culture flasks with 0.5% gelatin and air dry.
2. Thaw and plate MEFs at 2.25×10^4 cells/cm^2 on gelatin-coated flasks in MEF medium. MEFs may be passaged with trypsin-EDTA treatment a maximum of four times prior to inactivation and use in generating MEF-CM. (While we generally purchase MEFs, MEFs generated "in-house" typically show more robust growth and less batch-to-batch variability.)
3. Harvest the expanded MEFs using trypsin-EDTA, and then inactivate them by irradiating the cell suspension. (The amount of irradiation needed to inactivate MEFs varies somewhat with the cell source and from lot-to-lot. We typically use ~4,000 rads. Mitomycin C can also be used to inactivate MEFs.)
4. Replate the inactivated MEFs on gelatin-coated flasks at a density of 5.6×10^4 cells/cm^2 (which corresponds to ~12.5×10^6 cells per T-225 flask). Gently agitate the flask to uniformly distribute the cells.
5. After allowing a minimum of 5 h for cell adherence, replace the MEF medium with hESC medium plus 4 ng/mL of bFGF. We generally use 90 mL of hESC medium per T-225 flask of confluent MEFs.
6. Collect the resultant MEF-CM daily, replacing with fresh hESC medium for up to a maximum of 7 days. MEF-CM medium can be filter-sterilized for immediate use or combined and sterile-filtered at the end of the 7-day conditioning run (preferred). MEF-CM can be stored at 4°C for short-term use or at −80°C for up to a year. Avoid repetitive heat–thaw cycles (make appropriate aliquots).

3.1.2. Preparation of Matrigel™-Coated Plates

1. Place diluted (1:60) Matrigel™ in an ice bucket.
2. Add cold, diluted Matrigel™ to tissue culture plates (at 1 mL per well of a 6-well plate, 0.5 mL per well of a 24-well plate,

and 50 μL per well of a 96-well plate), and then transfer the latter to storage at 4°C. (During this coating step, we keep plates in a clean, airtight plastic box in the refrigerator.) Plates should be coated by exposure to Matrigel™ for at least 24 h, but they can be stored for up to 2 weeks before use.

3. Remove Matrigel™ by aspiration immediately before use.

3.1.3. Routine Culture and Passage of Undifferentiated hESCs Under Feeder-Free Conditions

General information on the routine culture of undifferentiated hESCs can be found elsewhere in this volume, but specific instructions on how to transition undifferentiated hESC cultures on feeders to feeder-free conditions in MEF-CM can be found in Note 2. Once feeder-free hESC cultures are established, we strongly recommend maintaining these in a six-well plate format, as this provides maximal flexibility for both passaging and setting up cultures for cardiac induction.

1. Once feeder-free, undifferentiated hESC cultures are established, maintain these by feeding daily with MEF-CM supplemented with 4 ng/mL of fresh bFGF (at 4 mL/well of a six-well plate). Once the undifferentiated hESC colonies occupy ~75% of the well surface area, the cultures should be passaged as detailed in steps 2–8.
2. Aspirate MEF-CM medium and rinse with 2 mL/well DPBS.
3. Aspirate DPBS, replace with 1 mL/well dispase (0.1 U/mL) and incubate the cells in dispase at 37°C, until the edges of the hESC colonies begin to curl. This typically requires ~1.5–2 min incubation in the enzyme.
4. Gently aspirate the dispase without dislodging the cells, and replace the enzyme solution with 2 mL/well MEF-CM supplemented with 4 ng/mL bFGF.
5. Using a cell scraper, collect the hESCs, which should readily detach in small clumps. Gently triturate and break up the larger clumps until they no longer gravity-settle (Avoid over-triturating!).
6. Dilute the cell suspension in an appropriate volume of MEF-CM with 4 ng/mL bFGF, and then replate at 2 mL/well in Matrigel-coated six-well plates. We generally split our WA07 (H7) hESCs at a 1:3–1:6 ratio, passaging every 5–7 days.
7. To ensure an even distribution of the dispersed clumps of hESCs on the new plate surface, alternate sliding the plate left-to-right and front-to-back on the incubator shelf. (Do not swirl, which will place cells either at the periphery or center of the well.)
8. Allow at least 4–6 h for cell adhesion before moving the plate or refeeding. Maintain the newly passaged cultures in MEF-CM plus bFGF as described in step 1.

3.2. Directed Cardiac Differentiation of hESC Culture

3.2.1. Setting Up hESC Cultures for Cardiac Induction

1. Maintain undifferentiated hESC cultures in MEF-CM, as detailed in Subheading 3.1. When the undifferentiated colonies occupy ~75% of the well surface area, they are ready for either routine passaging to maintain undifferentiated cultures or set up for cardiac induction. (Our standard practice is to maintain our undifferentiated hESCs in a six-well plate. When the plate is ready for passaging, we split 1–2 wells at either 1:3 or 1:6, thereby generating a new plate. The remaining 4–5 wells are induced into cardiomyocytes following steps 2–7).
2. To set up cultures for cardiac induction, aspirate the MEF-CM, and gently rinse the cells with DPBS at 2 mL/well of a six-well plate.
3. Aspirate the DPBS, replace with 2 mL/well of Versene, and incubate the cells in Versene at 37°C until they become rounded up and loosely adherent, but not yet detached. This typically requires 3–7 min (see Note 4).
4. Gently aspirate the Versene and replace with 1 mL/well of prewarmed MEF-CM supplemented with 4 ng/mL fresh bFGF. Dislodge the cells by gently flowing MEF-CM over them with a 1,000 μL micropipette.
5. Collect the dispersed hESCs, gently triturate them into a single-cell suspension, and quantitate by hemacytometer. Add MEF-CM supplemented with 4 ng/mL bFGF to reach a final concentration of 4×10^5 cells/mL.
6. Replate the dispersed hESCs on Matrigel™-coated plates at a density of 4×10^5 cell/cm^2, which corresponds to roughly 100 μL of cell suspension per well of a 96-well plate well or 0.5 mL per well of a 24-well plate well (see Note 1).
7. Until the replated hESCs form a confluent monolayer with the compact appearance illustrated by Fig. 1 (Day 0), continue to feed the cultures daily with MEF-CM supplemented with 4 ng/mL bFGF. When refeeding, use MEF-CM plus bFGF at 100 μL of per well of a 96-well plate well or 1 mL per well of a 24-well plate.

3.2.2. Induction of Cardiac Differentiation with Serial Activin A and BMP-4

1. When the replated hESCs assume the compact appearance illustrated by Fig. 1 (Day 0), they are ready for induction with serial activin A and BMP-4. At this point (by convention, "Day 0"), aspirate the MEF-CM and replace with RPMI-B27 medium supplemented with 100 ng/mL activin A, using 100 μL per well of a 96-well plate well or 1 mL per well of a 24-well plate well.
2. On day 1 postinduction (i.e., 24 h later), aspirate the activin A-containing medium, and very gently replace it with an equivalent volume of RPMI-B27 medium supplemented with

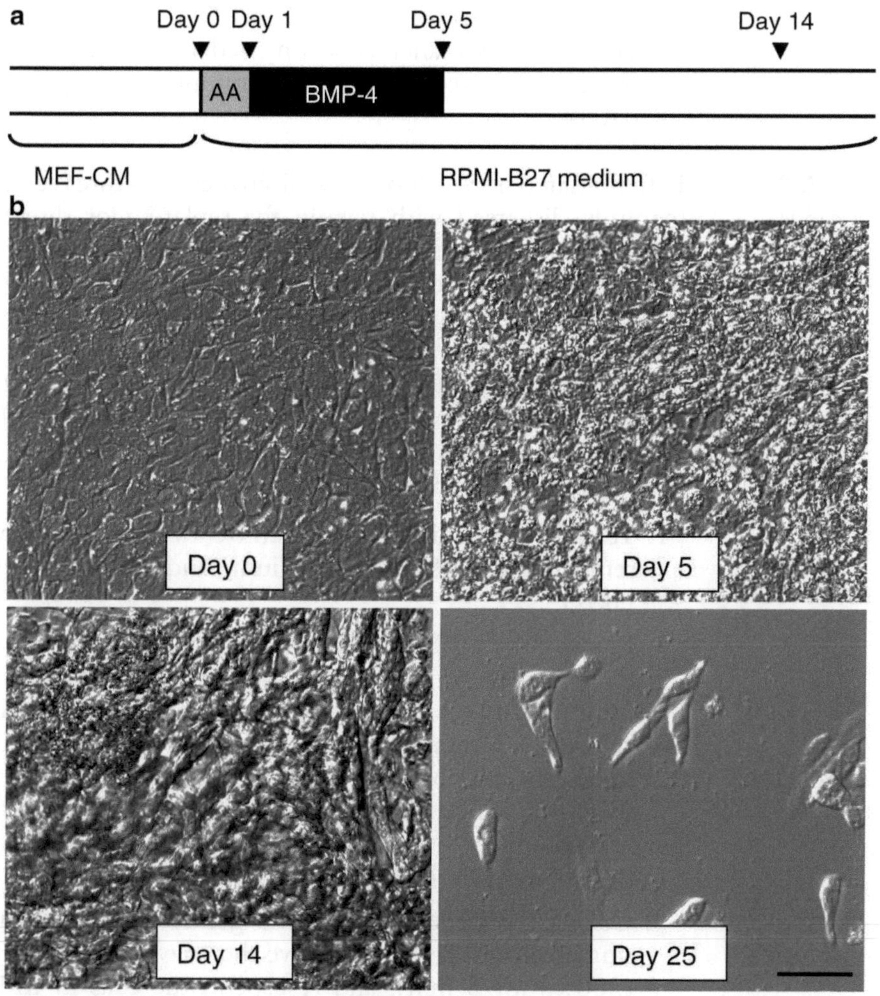

Fig. 1. Directed cardiac differentiation protocol. (**a**) Timeline for the protocol used to generate cardiomyocytes from human pluripotent stem cells. In brief, after growth to confluence in the undifferentiated state in MEF-CM plus bFGF, cultures are switched to differentiating conditions in RPMI-B27 medium and are serially pulsed with two growth factors, activin A (day 0) and BMP-4 (day 1). After day 5 postinduction with activin A, the differentiating cultures are grown in RPMI-B27 medium in the absence of exogenous factors. Spontaneous beating activity commences on ~day 9–11 postinduction, and cultures may be harvested after day 14. (**b**) Photomicrographs illustrating morphological changes in cultures during the protocol. Prior to treatment with activin A (day 0), the cultures should have formed a compact, 100% confluent monolayer that is composed of cells with a high nucleus-to-cytoplasm ratio but greater irregularity in cell shape than do hESCs in a usual undifferentiated colony. By day 5, cells in the monolayer are phase-bright, more loosely attached, and have a rounded morphology. Some cell death is to be expected at this stage, but the monolayer should still occupy >90% of the well surface area. By day 14, the cultures are far more heterogeneous and include widespread areas with spontaneous beating activity. After dispersion and replating at low density (day 25), the resultant differentiated cultures are comprised of individual and small clusters of contractile cardiomyocytes with either a spindled or triangular morphology. Scale bar = 50 μm.

10 ng/mL BMP-4. Incubate the differentiating cultures for an additional 4 days without a medium change (see Note 5).

3. On day 5 postinduction, gently replace the BMP-containing medium with an equivalent volume of RPMI-B27 medium without exogenous growth factors.

4. Thereafter, feed the cells with RPMI-B27 medium every other day. Spontaneous beating activity typically commences sometime between days 9–11 postinduction and peaks around day 14.

3.2.3. Replating hESC-Derived Cardiomyocyte Cultures for In Vitro Experiments

Differentiated hESC-derived cardiomyocyte cultures can be enzymatically dispersed with trypsin for replating or direct use in transplantation experiments anytime after day 14 postinduction with minimal loss of viability. Dispersed cardiomyocytes will replate particularly well on gelatin-coated tissue culture plastic (coated as described above in Subheading 3.1.1) or glass surfaces coated serially with PEI and gelatin. (To prepare the latter, coat the glass coverslips or glass-bottom dishes with 0.1% PEI at 4°C overnight, rinse thoroughly with sterile water, and then gelatin-coat as described in Subheading 3.1.1.)

1. To disperse the differentiated hESC-derived cardiomyocytes, remove the RPMI-B27 medium, and rinse the cells with DPBS.
2. Remove the DPBS, and replace it with 0.05% trypsin supplemented with 63 U/mL DNase I (using 100 μL/well of a 96-well plate or 1 mL/well of a 24-well plate). Incubate the cells for 3–5 min at 37°C, and monitoring them periodically under the microscope. Once the cells reach a point where they have rounded up but not yet detached, add an equivalent volume of 1× defined trypsin inhibitor, and gently pipette this solution over the cells to dislodge them.
3. Collect the dispersed cells, and gently triturate them into a single-cell suspension. Remove the enzyme by centrifugation (300 × g for 5 min), and resuspend the cells in RMPI-B27 medium supplemented with 20% FBS. Use a hemacytometer to determine the cell count, and then replate on gelatin-coated surfaces at density of $1–5 \times 10^4$ cells/cm^2.
4. Feed the replated cells the next day and every other day thereafter, using an appropriate volume of RPMI-B27 medium without FBS or exogenous growth factors. The resultant cultures are typically comprised of ~30–60% cardiomyocytes (see Note 6).

3.3. Further Enrichment for Cardiomyocytes by Percoll Gradient Centrifugation

If a greater degree of cardiac purity is required, the directly differentiated cultures can be further enriched by Percoll gradient centrifugation (6, 12). This procedure is best performed on differentiating cultures around day 14–18 postinduction with activin A. Percoll gradient centrifugation of cultures at other time points or after replating steps may result in reduced yield and cardiac purity.

1. Begin by removing the RPMI-B27 medium and rinsing the differentiated cell preparation with DPBS.

2. Remove the DPBS, add Liberase Blendzyme IV (0.56 U/mL) supplemented with DNase I (63 U/mL), and incubate at 37°C for 30 min. Gentle agitation or cell scraping is sometimes required to ensure cell detachment.
3. Collect the dispersed cells, and gently triturate them into a single-cell suspension. Remove the enzyme by centrifugation (300 × *g* for 5 min), and resuspend the pelleted cells in 10 mL of RMPI-B27 medium per starting plate.
4. Transfer each 10 mL cell suspension to a 50 mL tube and add a 12 mL layer of 40.5% Percoll solution to the bottom of the tube. Pipette very slowly (!) to avoid mixing the cells and Percoll solution.
5. Add a layer of 58.5% Percoll solution to the bottom of the tube, again pipetting slowly to avoid mixing of the layers. (A sharp interface should be visible between each of the three layers.)
6. Centrifuge at 1,500 × *g* for 30 min with the brake off. (The tubes should be carefully balanced!)
7. After centrifugation, the cell pellet at the bottom of the tube (i.e., at the bottom of the 58.5% Percoll layer) will be highly enriched for cardiomyocytes (typically 80–95% positive for cardiac markers). Collect these cells, add a vast excess of RPMI, and centrifuge (300 × *g* for 5 min) to remove the Percoll. These cells can then be replated (see Subheading 3.2.3) or cryopreserved as described below.

3.4. Cryostorage of hESC-Derived Cardiomyocyte Preparations

3.4.1. Cryopreservation of hESC-Derived Cardiomyocytes

1. Prepare a single-cell suspension of differentiated hESC-derived cardiomyocytes (as described in Subheadings 3.2.3 or 3.3). Wash thoroughly with RPMI-B27 medium to ensure removal of enzymes, Percoll, etc., and determine the total cell number by hemacytometer.
2. Centrifuge the cells at 300 × *g* for 5 min, aspirate the supernatant, and gently resuspend the pellet in 250 μL Cryostor CS10 per 10×10^6 cells, while slowly swirling in an ice-water bath.
3. Aliquot the resuspended cells into cryogenic vials, using volumes <500 μL to promote more uniform freezing and thawing. Transfer the cryogenic vials to a controlled-rate freezer, previously chilled to 0°C. Cool the samples from 0 to −7°C at a rate of 1°C/min, from −7 to −10°C at 0.75°C/min, and then finally from −10 to −80°C at 1°C/min. The frozen cardiomyocytes can then be transferred to liquid nitrogen for long-term storage (>1 year).

3.4.2. Thawing Cryopreserved hESC-Derived Cardiomyocytes

1. Transfer the cryogenic vial to a bucket of dry ice until ready for use. Preheat a 50 mL tube of RPMI-B27 medium to 37°C.

2. Thaw the vial in a 37°C water bath with gently agitation until the pellet is completely melted.
3. Add 1 mL of the preheated RPMI-B-27 medium to the cryogenic vial and mix by gently shaking. Slowly transfer the resultant cell suspension in a drop-wise fashion to the remaining medium in the 50 mL tube.
4. Remove the cryopreservative by centrifugation (300 × *g* for 5 min), and resuspend in an appropriate volume of RPMI-B27 medium. The thawed cardiomyocytes can then be replated on gelatin-coated surfaces as described in Subheading 3.2.3.

4. Notes

1. The directed cardiac differentiation protocol described here was developed with and optimized for the WA07 (H7) hESC line (13), so we recommend using this line when first trying this protocol. WA07 hESCs are also relatively straightforward to culture in the undifferentiated state and seem to be particularly cardiogenic. That said, this protocol has been tested with and proven successful in the reliable generation of cardiomyocytes from multiple other hESC and hiPSC lines. When adapting our standard protocol to stem cell lines other than WA07 hESCs, we have found three relatively minor modifications helpful:
   1. *Number of undifferentiated hESCs seeded per well*: When setting up cells for cardiac induction, the goal is to obtain an evenly distributed, 80–95% confluent monolayer by 24 h after dispersion with Versene and replating. Stem cell lines vary in terms of plating efficiency, but seeding in the range of $1–10 \times 10^4$ cells per well of a 96-well plate (plated out as 100 μL/well of suspended cells in MEF-CM plus bFGF) has sufficed for all of the lines we have tried. If one seeds too many cells per well, the cultures tend to pile up rather than form a uniform monolayer.
   2. *Interval between replating and induction with activin A*: Before the undifferentiated cultures are treated with activin A, they should be allowed to form a very compact, 100% confluent monolayer (see Fig. 1). Growth kinetics vary from line to line (and can even vary with increasing passage number within a single line), so it is not surprising that there is some variation in the amount of time the replated cells need to form an optimally compact monolayer. We recommend testing a range of intervals from 3 to 10 days when commencing work with a new line.

3. *Supplementation with bFGF or extra medium during BMP-4 treatment*: A significant amount of cell death normally occurs between days 1–5 postinduction with activin A, usually peaking at ~day 3 postinduction (during BMP-4 treatment). With WA07 hESCs, the rate of cell death is generally matched by cell proliferation, but we have found some stem cell lines may show an exaggerated death response. In such cases, we have found it sometimes helpful to either add 4 ng/mL bFGF on day 1 postinduction and/or an extra 25% volume of fresh RPMI-B27 medium on days 3–4 postinduction. If you do so, do not aspirate the medium already present, and add the supplement very gently to avoid dislodging the cells. Note that, by performing a pilot experiment in a 96-well plate, one can simultaneously vary these three parameters and quickly adapt the protocol to a new line. We have tested other variables (e.g., the timing or concentration of growth factors), but have never found these to significantly improve the yield of cardiomyocytes.

2. Many laboratories maintain their undifferentiated hESC cultures in direct contact with MEFs or other feeder cell types, but the directed cardiac differentiation protocol described here requires undifferentiated hESCs that have been maintained under feeder-free conditions for at least 2–3 passages. To wean undifferentiated hESCs on feeders to feeder-free growth in MEF-CM on Matrigel™-coated surface, serially passage the hESCs onto progressively sparser feeder cells in a mixture of hESC medium and MEF-CM. Throughout this transition, hESCs are fed daily with media supplemented with usual bFGF (as described in Subheading 3.1.2), and the cells are passaged with dispase (as described in Subheading 3.1.3). Start by passaging the hESCs onto a MEF layer at 75% normal density and feeding with MEF-CM: hESC medium mixed at a 1:3 ratio. With the next passage, replate the hESCs onto MEF feeders at 50% density and culture in 1:1 media, followed by a third passage onto feeders at 25% density and culture in 3:1 media. With the fourth passage, the hESCs are switched to feeder-free culture in undiluted MEF-CM plus bFGF on Matrigel™-coated plates.

3. Recently, several alternative feeder-free culture systems for maintaining undifferentiated hESCs have been reported, some of which employ defined media formulations with specific growth factors (14–19). We have tried some of the latter for compatibility with this directed cardiac differentiation protocol. Unfortunately, in our hands, the use of hESCs cultured in these media alternatives to MEF-CM either resulted

in differentiated preparations of somewhat lower cardiac purity or did not support cardiogenesis at all. We have had success when undifferentiated hESCs were expanded in mTeSR1 medium (Stemcell Technologies, Canada) and then switched back to MEF-CM plus bFGF shortly before induction with activin A. Our lab has ongoing efforts to adapt the protocol to hESCs maintained throughout in an animal-free, defined medium culture system, but, for now, we recommend continued use of MEF-CM.

4. When setting up hESCs for subsequent cardiac induction, their dispersion with Versene is a critical step that requires close monitoring under the microscope, and a bit of practice. The goal is to incubate the cells in Versene until they become rounded up and loosely adherent, but not yet fully detached. If this is done properly, the overall shape of the initial colony should still be apparent, and the hESCs are easily dislodged by gently flowing MEF-CM over them. A cell scraper should not be needed. On the other hand, excessive incubation in Versene adversely affects the viability and plating efficiency of the hESCs and may prevent successful cardiac induction. In general, lower-passage hESC cultures are more strongly adherent and require longer incubation times in Versene (~5–7 min), while higher-passage hESCs are easier to dislodge (~3–5 min). As noted above, replating the proper number of healthy hESCs should result in an 80–95% confluent monolayer by 24 h.
5. During this and other medium changes, take care not to dislodge the cell monolayer, which is particularly loosely adherent on days 1–5 postinduction. The best approach is often to aspirate only ~80% of the medium in any given well, leaving behind an undisturbed layer over the cells. Avoid creating turbulence when adding medium or handling the plates.
6. The cardiac purity of the resultant cell preparations can be assessed by immunocytochemistry or fluorescence-activated cell sorting (FACS), using antibodies against any of a number of cardiac or muscle markers (e.g., Nkx2.5, sarcomeric actins, cardiac troponin, etc.). We routinely use a commercially available monoclonal antibody (clone A4.951, which can be purchased as hybridoma from American Type Culture Collection, or as purified antibody from multiple vendors) against β-myosin heavy chain, a striated muscle marker which is strongly expressed by human cardiomyocytes.

References

1. Reinecke H, Minami E, Zhu WZ, Laflamme MA (2008) Cardiogenic differentiation and transdifferentiation of progenitor cells. *Circ Res.* **103**: 1058–1071.
2. Zhu WZ, Hauch KD, Xu C, Laflamme MA (2009) Human embryonic stem cells and cardiac repair. *Transplant Rev (Orlando)* **23**: 53–68.

3. Zhu WZ, Santana LF, Laflamme MA (2009) Local control of excitation-contraction coupling in human embryonic stem cell-derived cardiomyocytes. *PLoS ONE*. **4**: e5407.
4. Zhang J, Wilson GF, Soerens AG, Koonce CH, Yu J, et al. (2009) Functional cardiomyocytes derived from human induced pluripotent stem cells. *Circ Res*. **104**: e30-41.
5. Caspi O, Huber I, Kehat I, Habib M, Arbel G, et al. (2007) Transplantation of human embryonic stem cell-derived cardiomyocytes improves myocardial performance in infarcted rat hearts. *J Am Coll Cardiol*. **50**: 1884–1893.
6. Laflamme MA, Chen KY, Naumova AV, Muskheli V, Fugate JA, et al. (2007) Cardiomyocytes derived from human embryonic stem cells in pro-survival factors enhance function of infarcted rat hearts. *Nat Biotechnol*. **25**: 1015–1024.
7. van Laake LW, Passier R, Monshouwer-Kloots J, Verkleij AJ, Lips DJ, et al. (2007) Human embryonic stem cell-derived cardiomyocytes survive and mature in the mouse heart and transiently improve function after myocardial infarction. *Stem Cell Research*. **1**: 9–24.
8. Nussbaum J, Minami E, Laflamme MA, Virag JA, Ware CB, et al. (2007) Transplantation of undifferentiated murine embryonic stem cells in the heart: teratoma formation and immune response. *Faseb J*. **21**: 1345–1357.
9. Swijnenburg RJ, Tanaka M, Vogel H, Baker J, Kofidis T, et al. (2005) Embryonic stem cell immunogenicity increases upon differentiation after transplantation into ischemic myocardium. *Circulation*. **112**: I166-172.
10. Xu C, Police S, Rao N, Carpenter MK (2002) Characterization and enrichment of cardiomyocytes derived from human embryonic stem cells. *Circ Res*. **91**: 501–508.
11. Xu C, Inokuma MS, Denham J, Golds K, Kundu P, et al. (2001) Feeder-free growth of undifferentiated human embryonic stem cells. *Nat Biotechnol*. **19**: 971–974.
12. Xu C, Carpenter MK (2003) Human embryonic stem cell-derived cardiomyocytes: Derivation and characterization; Chiu A, Rao MS, editors. Totowa, NJ: Humana Press Inc.
13. Thomson JA, Itskovitz-Eldor J, Shapiro SS, Waknitz MA, Swiergiel JJ, et al. (1998) Embryonic stem cell lines derived from human blastocysts. *Science*. **282**: 1145–1147.
14. Ludwig TE, Levenstein ME, Jones JM, Berggren WT, Mitchen ER, et al. (2006) Derivation of human embryonic stem cells in defined conditions. *Nat Biotechnol*. **24**: 185–187.
15. Li Y, Powell S, Brunette E, Lebkowski J, Mandalam R (2005) Expansion of human embryonic stem cells in defined serum-free medium devoid of animal-derived products. *Biotechnol Bioeng*. **91**: 688–698.
16. Wang G, Zhang H, Zhao Y, Li J, Cai J, et al. (2005) Noggin and bFGF cooperate to maintain the pluripotency of human embryonic stem cells in the absence of feeder layers. *Biochem Biophys Res Commun*. **330**: 934–942.
17. Wang L, Li L, Menendez P, Cerdan C, Bhatia M (2005) Human embryonic stem cells maintained in the absence of mouse embryonic fibroblasts or conditioned media are capable of hematopoietic development. *Blood*. **105**: 4598–4603.
18. Wang L, Schulz TC, Sherrer ES, Dauphin DS, Shin S, et al. (2007) Self-renewal of human embryonic stem cells requires insulin-like growth factor-1 receptor and ERBB2 receptor signaling. *Blood*.
19. Sato N, Meijer L, Skaltsounis L, Greengard P, Brivanlou AH (2004) Maintenance of pluripotency in human and mouse embryonic stem cells through activation of Wnt signaling by a pharmacological GSK-3-specific inhibitor. *Nat Med*. **10**: 55–63.

Chapter 32

In Vivo Evaluation of Putative Hematopoietic Stem Cells Derived from Human Pluripotent Stem Cells

Melinda K. Hexum, Xinghui Tian, and Dan S. Kaufman

Abstract

Efficient derivation and isolation of hematopoietic stem cells (HSCs) from human pluripotent stem cell (hPSC) populations remains a major goal in the field of developmental hematopoiesis. These enticing pluripotent stem cells (comprising both human embryonic stem cells and induced pluripotent stem cells) have been successfully used to generate a wide array of hematopoietic cells in vitro, from primitive hematoendothelial precursors to mature myeloid, erythroid, and lymphoid lineage cells. However, to date, PSC-derived cells have demonstrated only limited potential for long-term multilineage hematopoietic engraftment in vivo – the test by which putative HSCs are defined. Successful generation and characterization of HSCs from hPSCs not only requires an efficient in vitro differentiation system that provides insight into the developmental fate of hPSC-derived cells, but also necessitates an in vivo engraftment model that allows identification of specific mechanisms that hinder or promote hematopoietic engraftment. In this chapter, we will describe a method that utilizes firefly luciferase-expressing hPSCs and bioluminescent imaging to noninvasively track the survival, proliferation, and migration of transplanted hPSC-derived cells. Combined with lineage and functional analyses of engrafted cells, this system is a useful tool to gain insight into the in vivo potential of hematopoietic cells generated from hPSCs.

Key words: hPSCs, hematopoiesis, HSCs, bioluminescent imaging, transplantation

1. Introduction

Human pluripotent stem cells (hPSCs) possess the ability to self-renew as undifferentiated cells in vitro, while maintaining the ability to differentiate into any of the cells found in the adult body (1, 2). The category of pluripotent stem cells includes embryonic stem cells (ESCs) derived from preimplantation embryos and induced pluripotent stem cells (iPSCs) derived from adult somatic cells (fibroblasts, cord blood cells, or other cell populations). In vitro differentiation of hPSCs has been shown to recapitulate events

Philip H. Schwartz and Robin L. Wesselschmidt (eds.), *Human Pluripotent Stem Cells: Methods and Protocols*, Methods in Molecular Biology, vol. 767, DOI 10.1007/978-1-61779-201-4_32,

that occur during human embryonic development. These findings make hESCs – and potentially iPSCs – a uniquely useful system in which to define the cellular and molecular mechanisms that mediate cellular differentiation toward specific lineages.

The potential utility of hPSCs also extends to diverse clinical applications. PSCs provide a promising resource for regenerative medicine therapies – particularly in the field of hematology where cellular therapies are relatively commonplace. For over three decades, hematopoietic stem cell (HSC) transplantation has been successfully performed in the clinic, using cells from adult bone marrow, umbilical cord blood, or mobilized peripheral blood (3–5). While these transplant therapies are well-established, the process of HSC transplantation remains complicated by a number of issues, including graft-vs.-host disease and lack of donors with suitable histocompatibility. PSCs could potentially serve as an alternative source of HSCs for clinical therapies.

Multiple studies have now focused on utilizing different in vitro systems that facilitate hematopoietic development from hPSCs. The earliest studies of hematopoietic differentiation showed that co-culture of hESCs with the murine bone marrow stromal line S17 in culture medium supplemented with fetal bovine serum (FBS), but no additional cytokines or growth factors, produced $CD34^{+}$ hematopoietic precursor cells (6). Selecting for $CD34^{+}$ cells from these cultures led to a significant enrichment of progenitor cells capable of forming characteristic myeloid, erythroid, and megakaryocytic lineages in vitro. Subsequent reports have shown generation of hematopoietic precursors from hESCs via co-culture with other stromal cell lines, such as OP9, M2-10B4, and AM20.1B4 or via embryoid body (EB) formation (7–10). While the bulk of the hematopoietic differentiation studies conducted over the past decade have utilized hESCs, unpublished work by our group and published reports from others has indicated that iPSCs can be differentiated in vitro to produce $CD34^{+}$ and $CD34^{+}CD45^{+}$ hematopoietic progenitors, and can also produce mature blood lineage cells, similar to hESCs (11, 12).

Co-culture methods of differentiation present the opportunity to characterize or modify stromal cells to define the niche-related factors that contribute to hematopoietic development. As we have previously shown, hESC co-culture with stromal lines engineered to express Wnt 1, a mediator of canonical Wnt signaling, increases the generation of $CD34^{bright}CD31^{+}Flk1^{+}$ hematoendothelial cells and $CD34^{dim}CD45^{+}$ hematopoietic progenitor cells (8). In contrast to co-culture methods, EB-mediated differentiation avoids issues related to complex interactions with mouse-derived stroma. While initial EB methods relied on fragmentation of undifferentiated hESC colonies – resulting in EBs of varying size and varying hematopoietic potential – Elefanty's group and others have developed forced aggregation EB-formation methods that seem to improve the efficiency and reproducibility of hematopoietic differentiation in these systems (13–16).

Although in vitro studies have indicated that hPSCs possess the ability to generate an array of hematopoietic cells, from primitive hematoendothelial precursors to mature myeloid and lymphoid lineage cells (17–20), demonstration of putative HSCs from hESC or iPSCs requires that these in vitro results translate to multilineage, long-term hematopoietic engraftment in in vivo models. Generally, cells with the ability to reconstitute the hematopoietic system of immunodeficient mice (typically NOD/SCID or related strains) – termed SCID-repopulating cells (SRC) – are considered to be a close surrogate of HSCs (21, 22). While such SRCs can be isolated from umbilical cord blood (UCB), adult bone marrow, and peripheral blood (23–28), hESC-derived cells have thus far demonstrated relatively limited potential for long-term hematopoietic engraftment in vivo (10, 29–34). Successful generation, isolation, and characterization of HSCs from hPSCs requires in vivo engraftment models that provide insight into the mechanisms that hinder or promote hematopoietic engraftment. Several murine transplantation models have been used to study hematopoietic engraftment of hESC-derived cells, including intra-bone marrow or intravenous injection (29, 34) into adult immunodeficient mice and intrahepatic (39) or facial vein injection (40) into neonates. Recently, we described an effective model that utilized hESCs that stably express firefly luciferase (luc) as a means to better understand the fate of hESC-derived cells posttransplantation (33). Using these luc⁺ cells allows noninvasive, serial bioluminescent imaging to track the survival, proliferation, and migration of hPSC-derived hematopoietic progenitors transplanted into immunodeficient, neonatal mice. Paired with the characterization of engrafted cells via flow cytometric analysis or further in vitro culture, this in vivo system is a useful tool with which the hematopoietic developmental potential of hPSCs can be evaluated. In this chapter, we will provide an overview of the in vitro differentiation systems we routinely use to generate hematopoietic progenitors in vitro, and will describe the in vivo, bioluminescent system we use to evaluate engraftment of hPSC-derived hematopoietic cells.

2. Materials

2.1. In Vitro Differentiation of luc⁺ PSCs

Refer to established protocols for maintenance of luc⁺hPSCs, stromal co-culture of hPSCs (e.g., Refs. 35, 37, EB differentiation of hPSCs, e.g., Refs. 13–16), or those protocols referenced in Subheading 3, for specifics of cell culture typically utilized.

2.2. Harvest and Dissociation of In Vitro-Differentiated PSCs

1. Collagenase passage medium: DMEM/F12 medium (Invitrogen; Cat. No. 11330–032) containing 1 mg/mL Collagenase type IV (Invitrogen; Cat. No. 17104–019). Collagenase medium is filter sterilized, 0.22 μm membrane.
2. Trypsin-ethylene diamine tetra-acetic acid (EDTA) ± 2% chicken serum: 0.05% trypsin-0.02% EDTA solution (Invitrogen; Cat. No. 25300–054) with 2% chicken serum (Sigma; Cat. No. C5405; see Note 1).
3. Dulbecco's phosphate-buffered saline (DPBS), Ca^{2+}- and Mg^{2+}-free.
4. R-10 medium (for washing): RPMI-1640 (Invitrogen; Cat. No. 11875–093) containing 10% FBS (Hyclone Laboratories; Logan, UT; Cat. No. SH30070.03) and 1% penicillin/streptomycin (P/S; Invitrogen Corporation/Gibco; Cat. No. 15140–122).
5. Disposable serological pipettes.
6. 70 μm cell strainer filter (Becton Dickinson/Falcon; Ref. No. 352350).
7. 0.4% Trypan blue stain.

2.3. Intrahepatic Transplantation of luc⁺ PSC-Derived Cells

1. Dulbecco's phosphate-buffered saline (DPBS), Ca^{2+}- and Mg^{2+}-free.
2. BD Ultra-Fine II insulin syringes, 31-G × 8 mm (Becton Dickinson; Cat. No. 328468).
3. X-RAD 320 irradiation system (Precision X-ray, Inc.) or other system suitable for mouse irradiation.

2.4. Bioluminescent Imaging

1. Approved anesthetic agent. We use Avertin (2,2,2-Tribromoethanol; TCI America; Wellesly Hills, MA; Cat. No. T1420). Stock solution prepared by dissolving 5 g of 2, 2, 2-tribromoethanol in 5 mL *tert*-amyl alcohol for a concentration of 1 g/mL. Store stock solution protected from light at 4°C. Prepare working solution by diluting 0.2 mL stock solution with 9.8 mL 10% ethanol; sterilize via filtration through a 0.22 μm syringe filter. Confirm that pH of working solution is >5 prior to each use. 0.25 mL working solution is administered to a 25 g mouse for a dose of 200 mg/kg.
2. D-luciferin sodium salt (Gold Biotechnology; St. Louis, MO; Cat. No. LUCNA-1G) working solution prepared by dissolving 1 g D-luciferin powder in 40 mL DPBS (pH adjusted to 6.3) for a concentration of 25 mg/mL. D-luciferin solution is filter sterilized using a 50 mL, 0.22 μm membrane. Store 0.3 mL and 0.6 mL aliquots of D-luciferin working solution at –80°C.
3. BD Ultra-Fine II insulin syringes (Becton Dickinson; Cat. No. 328468).

4. Black cardstock/construction paper.
5. I-5 medium for ex vivo organ imaging: Iscove's modified Dulbecco's medium (Cellgro/Mediatech; Cat. No. 10-016-CV) with 5% FBS and 1% P/S.
6. 35 mm cell culture dishes for ex vivo organ imaging.
7. Xenogen IVIS 100 imaging system (Caliper Life Sciences Corp.; Alameda, CA) or equivalent system.

2.5. Characterization of Engrafted Cells In Vivo

1. Liver dissociation medium: RPMI-1640 (Invitrogen; Cat. No. 11875–093) medium containing 10% fetal bovine serum, 1 mg/mL Collagenase type IV (Invitrogen; Cat. No. 17104–019) and 50 μg/mL DNase I (Roche Diagnostics; Ref. No.10104159001).
2. Ficoll-Paque plus (GE Healthcare; Cat. No. 17-1440-03).
3. 0.8% Ammonium chloride solution with 0.1 mM EDTA (StemCell Technologies, Cat. No. 07850).
4. 10 cm cell culture dishes.
5. 70 μm cell strainer filter (Becton Dickinson/Falcon; Ref. No. 352350).
6. 6 mL monoject syringes with luer lock tip (Kendall/Covidien; Ref. No. 8881516937).
7. BD PrecisionGlide™ Hypodermic needles: 16-G × 1½ in. (Cat. No. 305198); 20-G × 1½ in. (Cat. No. 305176).
8. 25-G × 5/8 in. hypodermic needles (Kendall/Covidien; Ref. No. 8881250313).
9. 0.4% Trypan blue stain.

3. Methods

3.1. Maintenance of Undifferentiated Luc-Expressing PSCs

The firefly luciferase (luc) transgene can be introduced into undifferentiated hPSCs as described previously (35). Undifferentiated luc⁺ hPSCs are maintained in hESC medium through co-culture with inactivated mouse embryonic fibroblasts (MEFs) (1, 36). Alternatively, undifferentiated PSCs can be cultured on Matrigel™-coated plates with MEF-conditioned hESC medium or mTeSR® medium (available from StemCell Technologies). See also (Chapters 9 and 10).

3.2. In Vitro Differentiation of luc⁺ PSCs

Co-culture of hPSCs with the murine bone marrow stromal cell line M2-10B4 (see Note 2) has been routinely used in our laboratory to generate hematopoietic progenitors for in vivo studies and further in vitro differentiation to mature hematopoietic lineage cells. hPSC/M2-10B4 co-culture can be set up as previously described (37).

While we have found the greatest success using M2-10B4 or S17 stromal cells for co-culture with hPSCs, a number of other stromal cell lines have been shown to support hematopoietic differentiation of hESCs (e.g., OP9, S17, AM20.1B4). The previously described co-culture method (37) could, in most cases, be easily adapted to use of alternative stromal cell lines. Our lab has also adapted the spin-EB system that was developed and described by Elefanty's group (13–15). This method involves forced aggregation of defined numbers of undifferentiated PSCs in 96-well plates by centrifugation to form EBs of uniform size (see Chapter 28) – however, this method, first requires, the adaptation of PSCs to single-cell passage technique (see Chapter 10). When cultured in a serum-free medium with a defined set of cytokines and growth factors, these "spin EBs" efficiently generate hematopoietic progenitor cells, based on flow cytometric analysis and CFU assay. In our hands, this system yields more robust production of $CD34^+$ and $CD34^+CD45^+$ hematopoietic precursor cells than the M2-10B4 stromal co-culture method: 20–50% $CD34^+$ and 10–15% $CD34^+CD45^+$ can be obtained with spin EBs, in contrast to the 10–20% $CD34^+$ and 0.5–5% $CD34^+CD45^+$ generated via stromal co-culture (Fig. 1).

3.3. Harvest and Dissociation of Differentiated PSCs

The optimal time for harvest and analysis of in vitro-differentiated hPSCs is dependent on the desired hematopoietic cell population. For example, in M2-10B4 co-culture, $CD34^+$ hESC-derived cells with hematoendothelial potential are best isolated at earlier differentiation time points (e.g., day 7–10 of differentiation), while $CD34^{dim}CD45^+$ that give rise only to hematopoietic cells appear on later days (e.g., day 17–21) of differentiation (8, 33). The course of differentiation will vary depending on the specific co-culture or EB system used, so we recommend conducting initial time-course experiments in which a sample of differentiating cells is analyzed every 2–3 days to determine the optimal time-point for derivation of the desired lineages from the PSC line under study.

Differentiated PSCs must be dissociated to single cells for in vivo injection, flow cytometric analysis, and colony-forming unit (CFU) assay. Although stromal cells will represent a small fraction (<10%) of the harvested cells in co-culture systems, mitotic inactivation of stromal cells prior to co-culture with hPSCs minimizes the chance that residual stromal cells will significantly interfere with further culture or analyses. If desired, contaminating mouse cells (MEFs or stromal cells) can be removed from the suspension by magnetic sorting using anti-mouse CD29 antibody (38).

To dissociate stromal co-culture differentiated PSCs to a single-cell suspension:

1. Prewarm trypsin-EDTA + 2% chicken serum in a 37°C water bath.
2. Harvest PSC-derived cells from the stromal feeder cell layer via incubation with collagenase passage medium. Aspirate R-15

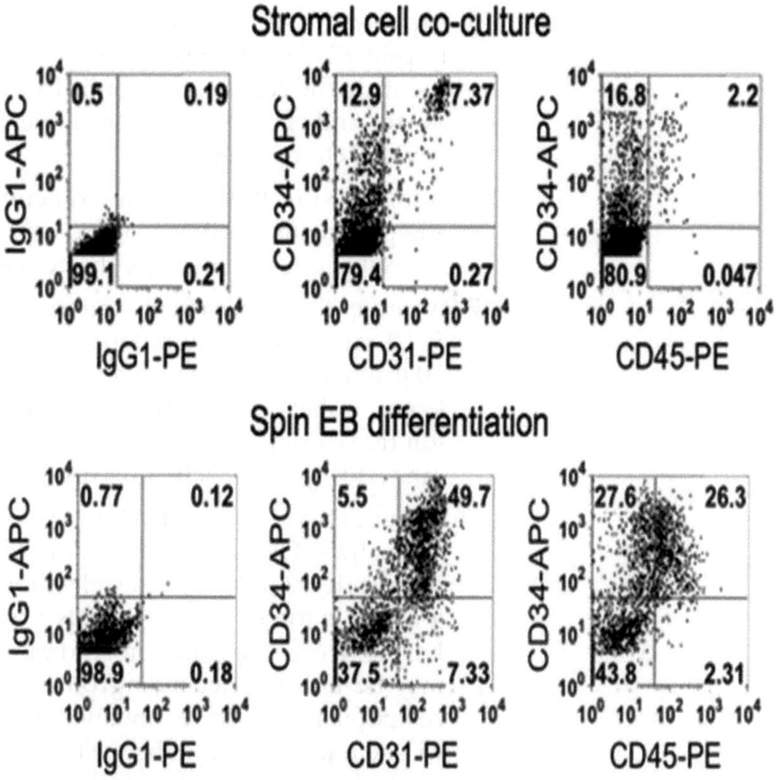

Fig. 1. Spin EB differentiation of hESCs yields higher percentages of $CD34^+$, $CD34^+CD31^+$ and $CD34^+CD45^+$ hematopoietic cells relative to M2-10B4 stromal co-culture differentiation. Flow cytometric analysis was conducted to characterize hematopoietic cells derived from M2-10B4 co-culture and spin EB differentiation of H9 hESCs. Shown are analyses of representative (*Top*) M2-10B4 co-culture and (*bottom*) spin EB differentiation experiments at day 21 and day 10 of differentiation, respectively. Stromal co-culture methods typically yield a maximum of 10–20% $CD34^+$, 5–10% $CD34^+CD31^+$, and 0.5–5% $CD34^+CD45^+$ cells. Spin EB differentiation routinely yields markedly higher percentages of all three subsets.

differentiation medium and add 1.5 mL/well collagenase passage medium. Incubate 5–10 min at 37°C, checking wells at 5 min intervals. Cells are ready to be harvested when the edges of the differentiated colonies start to become more defined and the stromal layer begins to break up.

3. Use a 5 mL glass pipette to scrape and wash cells off the plate. Transfer the cell suspension to a 15 or 50 mL conical tube, depending on the number of wells harvested. Pool 2–3 plates, maximum, per conical tube. Add an equal volume of DPBS to the cell suspension and pipette up and down to further break apart cell clumps. Centrifuge the cell suspension at 320–350 × *g* for 5 min.
4. Aspirate the supernatant and wash the cells with 5–10 mL DPBS. Repeat centrifugation at 320–350 × *g*, 5 min.

5. Aspirate the supernatant. Resuspend the cells in prewarmed trypsin-EDTA + 2% chicken serum, 1.5 mL per well collected. Incubate the cell solution for 5–15 min in a 37°C water bath. Vortex and check the cell solution every 3–5 min until few, if any, cell clumps remain.
6. Add 5–10 mL R-10 medium to neutralize the trypsin-EDTA and pipette up and down vigorously to dissociate any remaining cell clumps. Centrifuge the cells at 400 × *g*, 5 min.
7. Aspirate the supernatant and resuspend the cell pellet in 5–10 mL R-10 medium. Filter the cell suspension through a 70 μm cell strainer filter to remove any remaining cell clumps. Count viable cells via trypan dye exclusion (0.4% trypan blue solution) using a hemacytometer. One near-confluent well can yield $1–2 \times 10^6$ cells after dissociation.
8. Aliquot the cells as needed for in vivo injection, flow cytometric analysis, CFU assay, or RNA isolation. Depending on the cell density, 2–3 harvested wells should provide enough cells for flow cytometric analysis, RNA isolation, and CFU assay. For in vivo transplantation studies, various numbers of cells may be needed, depending on whether the heterogeneous cell population directly obtained from the stromal co-culture will be injected or whether cells will be sorted to enrich for a particular phenotype (e.g., $CD34^+$ cells).

Harvest of differentiating cells from EB systems also requires dissociation of cells with trypsin-EDTA or a recombinant trypsin replacement enzyme. Reference established EB formation protocols for detailed dissociation procedures (13, 15, 37).

3.4. Intrahepatic Transplantation of luc⁺ PSC-Derived Cells

The engraftment system we describe here and in previous work (33) involves intrahepatic transplantation of hPSC-derived cells in neonatal NOD.Cg-Prkdc^{scid}Il2rg^{tmIWjl}/SzJ (NOD/SCID/$\gamma c^{-/-}$, NSG) mice (see Note 3). Although other injection methods can be used for hematopoietic engraftment studies, including intravenous (tail vein) and intra-bone marrow injection in adult mice (29, 34), the neonatal liver is potentially a more efficient site of hematopoietic cell development (39) and is the method that we will focus on in this chapter. Alternatively, facial vein injection has been described as a method for transplantation of cells into neonatal mice (40). We specifically utilize the NSG mouse strain as several other reports indicate that use of this "more immunodeficient" model improves engraftment of hematopoietic cells relative to the NOD/SCID strain (10, 41, 42). Our studies have shown that NOD/SCID mice retain residual natural killer (NK) cell activity, which likely hinders efficient engraftment of hPSC-derived hematopoietic cells (29). The NSG mouse is more completely NK-deficient, potentially making it a better model for in vivo analysis of PSC-derived hematopoietic cells.

In vitro differentiation of luc⁺ PSCs for hematopoietic engraftment studies should be planned so that the cells are ready to harvest for injection within 1–2 days of the birth of a NSG litter (see Note 3). The single-cell suspension of differentiated luc⁺ cells prepared as described in Subheading 3.3 can be transplanted directly into immunodeficient mice as a heterogeneous population. In many instances, however, enrichment for a particular hematopoietic cell population by magnetic sorting or other selection method is preferable. If setting up in vivo engraftment studies that require isolation/enrichment of a specific hematopoietic cell population, sort the single-cell suspension prepared in Subheading 3.3 for the desired population using established protocols before injection.

1. Irradiate NSG neonates at 100 cGy within 24–48 h of birth (see Note 4). We have used both cesium-137 and X-ray sources for neonate irradiation (see Note 5). For irradiation, it is easiest to place the pups in a sterile, covered beaker lined with sterile gauze or nesting material. After irradiation, immediately return the pups to the mother and wait 3–6 h before transplantation.
2. Harvest stromal- or EB-differentiated luc⁺ PSCs and dissociate to a single-cell suspension, as described in Subheading 3.3. If enriching for a specific cell population (e.g., $CD34^+$ cells), complete magnetic or FACS procedures.
3. Aliquot the desired numbers of luc⁺ PSC-derived cells to sterile 1.5 mL microcentrifuge tubes for injection. For an enriched population, $0.3–1.0 \times 10^6$ cells can be injected into each pup. Centrifuge cells at $400 \times g$, 5 min. Resuspend cells in 25 μL DPBS per pup to be injected.
4. Inject 25 μL of cell solution directly into the liver of the irradiated pups using 31-G × 8 mm BD Ultra-Fine II insulin syringes. Mice injected with DPBS only are used as negative control for bioluminescent imaging.

3.5. Bioluminescent Imaging

To track transplanted luc⁺ PSC-derived cells, we perform bioluminescent imaging of animals immediately postinjection and at regular intervals using the Xenogen IVIS 100 imaging system. See Fig. 2 for a representative bioluminescent image of a neonate shortly posttransplantation. The optimal exposure time and settings for bioluminescent imaging may need to be determined in initial tests, though a 1–2 min exposure is generally sufficient to obtain significant signal.

1. Anesthetize mice by i.p. injection of 200 mg/kg Avertin, or other approved anesthesia method (see Note 6). To decrease the fatality rate of the neonatal mice, a quick image (30 s–1 min) can be acquired without anesthesia.

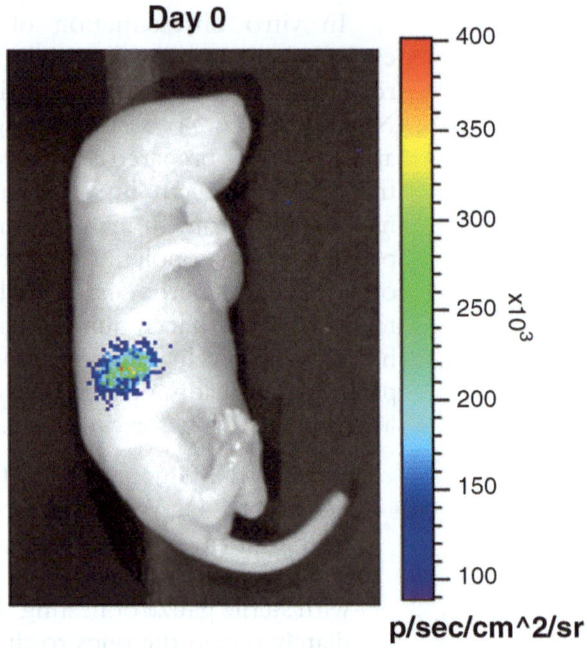

Fig. 2. Bioluminescent image of a neonatal mouse shortly posttransplantation. The 24 h NSG neonatal mouse received intrahepatic injection of 10^5 $CD34^+CD45^+$ cells derived from luc + hESCs after 22 days in M2-10B4 co-culture. The neonate was imaged within 30 min of transplantation, after i.p. injection of 150 mg/kg D-luciferin and a 2 min acquisition using the Xenogen IVIS 100 system.

2. Inject mice intraperitoneally (i.p.) with the D-luciferin working solution at a dose of 150 mg/kg (see Note 7). Wait 10 min after D-luciferin injection before imaging the mice.
3. Secure mice to be imaged onto their backs on clean black paper. Weanlings and adult mice can be secured by gently taping their paws to the black paper. For neonates and young pups, fold the black paper accordion style and nestle pups within the valleys of the folds. For all images, make sure that mice are spread sufficiently far apart to prevent bleeding of the luminescent signal from one mouse to its neighbor (see Note 8).
4. Acquire a bioluminescent image of mice using the Xenogen IVIS. A 5 min exposure/acquisition is usually suitable for in vivo tracking of engraftment in adult mice.
5. Analyze optical images using appropriate software (e.g., Xenogen Living Image software).

Ex vivo bioluminescent imaging of organs (e.g., liver, spleen, or femur) can also be performed using the Xenogen IVIS system. Place harvested organs individually in 35 mm culture dishes containing 2–3 mL I-5 medium. Add 300 μg/mL (12 μL of

25 mg/mL solution in 1 mL I-5 medium) D-luciferin 5 min before imaging. Acquire bioluminescent images as described above.

3.6. Characterization of Engrafted Cells

In order to better define the engrafted cell populations detected through bioluminescent imaging, mononuclear cells can be isolated from various organs post-sacrifice for further analysis – including flow cytometry, secondary transplant, in vitro culture (e.g., CFU assay), and nucleic acid analyses. The following is a brief explanation of methods for isolating mononuclear cells from liver, bone marrow, spleen, and peripheral blood to allow for assessment of hematopoietic lineage engraftment.

3.6.1. To Isolate Mononuclear Cells from the Liver

1. Before beginning cell isolation, prepare fresh liver dissociation medium, and prewarm in 37°C water bath. If Ficoll-Paque is stored at 4–8°C, allow it to warm to room temperature before using in step 5.
2. Transfer the harvested liver to a 10-cm culture dish containing 7 mL prewarmed liver dissociation medium. Using a sterile scalpel, chop the liver into small fragments. Further dissociate the tissue by passing it, five to ten times, through a 16-G × 1½ inch needle attached to a 6-mL syringe. Switch to a 20-G × 1½ inch needle to further break up the tissue fragments.
3. Transfer the liver solution from the culture dish to a 50-mL conical tube. Use 8 mL of liver dissociation medium to wash the culture dish and add this wash to the conical tube. Incubate the liver solution for 30 min in a 37°C water bath, vortexing every 10 min.
4. Centrifuge the liver cell solution at 400 × *g* for 5 min. Aspirate the supernatant. Resuspend the pellet in 15–20 mL I-5 medium.
5. Gently overlay the liver cell solution onto 15 mL Ficoll-Paque in a 50 mL conical tube. Pipette slowly to avoid disturbing the liver solution/Ficoll interface. Centrifuge at 400 × *g* at room temperature for 20–30 min, without brake.
6. After centrifugation, liver-derived mononuclear cells will be located in the interface between the Ficoll and I-5 medium, with residual liver tissue and red blood cells pelleted at the bottom of the tube. Collect the mononuclear cells at the Ficoll/I-5 interface and transfer to a 15-mL conical tube. Dilute with an equal volume I-5 medium and centrifuge at 400 × *g* for 10 min to wash away residual Ficoll.
7. Resuspend liver mononuclear cells in 3–5 mL I-5 medium. Count viable cells using a hemacytometer and trypan dye exclusion.

3.6.2. To Isolate Cells from Bone Marrow and Spleen

1. Transfer harvested tibia and femurs into a 10-cm culture dish with 12 mL I-5 medium. Using sterile forceps and scissors, remove excess tissue from the bone. Fill a 3-mL syringe attached to a 25-G × 5/8 inch needle with I-5 medium and use it to flush the bone marrow from tibias and femurs into the 10-cm culture dish. Pass the bone marrow media solution through the needle and syringe two to three times to break apart blood clumps, if any.
2. Transfer harvested spleen into a separate 10-cm culture dish with 12 mL I-5 medium. Homogenize spleen tissue by crushing between the frosted ends of two clean glass microscope slides. Pipette spleen solution up and down to break up any remaining tissue clumps.
3. Filter bone marrow and spleen solutions through 70 μm cell strainers into conical tubes to ensure single cell suspensions. Centrifuge cells at 400 × *g* for 5 min.
4. Resuspend cells in ice-cold ammonium chloride solution to lyse the red blood cells. Incubate on ice for 5 min. Centrifuge at 400 × *g* for 5 min.
5. Aspirate the supernatant. Resuspend cells in I-5 medium for counting.

3.6.3. To Isolate Cells from Peripheral Blood

1. At sacrifice, collect peripheral blood using standard methods.
2. Lyse red blood cells in ammonium chloride solution as described above for bone marrow and spleen (steps 4–5). It may take two rounds of lysis with ammonium chloride to completely get rid of the red cells in peripheral blood.
3. Centrifuge at 400 × *g* for 5 min.

3.6.4. Preparation of Cells for Analysis

Incubate mononuclear cells isolated from liver, spleen, bone marrow, and peripheral blood in blocking solution for 20–30 min on ice before proceeding to flow cytometric analysis of cells for human-specific hematopoietic lineage markers. Perform flow cytometric analysis, hematopoietic CFU assay, and nucleic acid analyses using standard methods (example protocols in Ref. 37).

4. Notes

1. Chicken serum added to trypsin-EDTA provides a protein source to improve viability of PSC-derived cells. Chicken serum is used because, in contrast to FBS, it does not contain trypsin inhibitors.

2. M2-10B4 cells can be purchased from ATCC (Manassas, VA). M2-10B4 cells are typically passed at a low ratio (1:3 or 1:4) to minimize loss of hematopoietic support in prolonged culture. Lower passages of M2-10B4 provide better support for hematopoietic differentiation of hPSCs than do higher passages. Generally, we do not use M2-10B4 cells beyond passage 25.
3. In our experience, NOD/SCID/$\gamma c^{-/-}$ (NSG) mice are good breeders, with breeding pairs regularly producing litters every ~21 days once established. NSG breeders can routinely produce eight healthy pups per litter and demonstrate strong mothering instincts. However, the stress caused by repeatedly removing pups for procedures (i.e., irradiation, injection, imaging) can cause behavior in the mother that is detrimental to the health of the current litter or can negatively impact the health of future litters. Minimize the time that pups are kept separate from the mother and establish a rotating schedule for use of breeder pairs to reduce stress and the chance for problems with animal health.
4. To maximize survival, neonates should have visible milk spots before being separated from the mother for the irradiation procedure.
5. We recommend conducting tests to determine the optimal irradiation parameters (e.g., dosage, distance from source, filters used) for a particular machine. If irradiation dose is too high or is delivered too fast, pups may display delayed growth and die 1–2 weeks postirradiation.
6. Some Xenogen IVIS and equivalent bioimaging systems have built-in ports for gas anesthesia. Where available and approved, gas anesthesia (e.g., isofluorane) may be an easier option for mouse sedation during luminescent imaging.
7. Prior to injection, keep D-luciferin working solution on ice and protected from light. Avoid repeated freeze–thaw of D-luciferin solution, and check for precipitate before injecting mice.
8. If luminescent carryover from neighboring mice becomes an issue, placing black paper dividers between mice during image acquisition can minimize the signal carryover.

References

1. Thomson, J. A., Itskovitz-Eldor, J., Shapiro, S. S., Waknitz, M. A., Swiergiel, J. J., Marshall, V. S., and Jones, J. M. (1998) Embryonic stem cell lines derived from human blastocysts, *Science* **282**, 1145–1147.
2. Odorico, J. S., Kaufman, D. S., and Thomson, J. A. (2001) Multilineage differentiation from human embryonic stem cell lines, *Stem Cells* **19**, 193–204.
3. Thomas, E. D. (1999) Bone marrow transplantation: a review, *Semin Hematol* **36**, 95–103.
4. Korbling, M., and Anderlini, P. (2001) Peripheral blood stem cell versus bone marrow

allotransplantation: does the source of hematopoietic stem cells matter?, *Blood* **98**, 2900–2908.

5. Brunstein, C. G., and Wagner, J. E. (2006) Umbilical cord blood transplantation and banking, *Annu Rev Med* **57**, 403–417.
6. Kaufman, D. S., Hanson, E. T., Lewis, R. L., Auerbach, R., and Thomson, J. A. (2001) Hematopoietic colony-forming cells derived from human embryonic stem cells, *Proc Natl Acad Sci USA* **98**, 10716–10721.
7. Tian, X., Morris, J. K., Linehan, J. L., and Kaufman, D. S. (2004) Cytokine requirements differ for stroma and embryoid body-mediated hematopoiesis from human embryonic stem cells, *Exp Hematol* **32**, 1000–1009.
8. Woll, P. S., Morris, J. K., Painschab, M. S., Marcus, R. K., Kohn, A. D., Biechele, T. L., Moon, R. T., and Kaufman, D. S. (2008) Wnt signaling promotes hematoendothelial cell development from human embryonic stem cells, *Blood* **111**, 122–131.
9. Vodyanik, M. A., Bork, J. A., Thomson, J. A., and Slukvin, II. (2005) Human embryonic stem cell-derived CD34+ cells: efficient production in the coculture with OP9 stromal cells and analysis of lymphohematopoietic potential, *Blood* **105**, 617–626.
10. Ledran, M. H., Krassowska, A., Armstrong, L., Dimmick, I., Renstrom, J., Lang, R., Yung, S., Santibanez-Coref, M., Dzierzak, E., Stojkovic, M., Oostendorp, R. A., Forrester, L., and Lako, M. (2008) Efficient hematopoietic differentiation of human embryonic stem cells on stromal cells derived from hematopoietic niches, *Cell Stem Cell* **3**, 85–98.
11. Lengerke, C., Grauer, M., Niebuhr, N. I., Riedt, T., Kanz, L., Park, I. H., and Daley, G. Q. (2009) Hematopoietic development from human induced pluripotent stem cells, *Ann N Y Acad Sci* **1176**, 219–227.
12. Choi, K. D., Vodyanik, M. A., and Slukvin, II. (2009) Generation of mature human myelomonocytic cells through expansion and differentiation of pluripotent stem cell-derived lin-CD34 + CD43 + CD45+ progenitors, *J Clin Invest* **119**, 2818–2829.
13. Ng, E. S., Davis, R., Stanley, E. G., and Elefanty, A. G. (2008) A protocol describing the use of a recombinant protein-based, animal product-free medium (APEL) for human embryonic stem cell differentiation as spin embryoid bodies, *Nat Protoc* **3**, 768–776.
14. Ng, E. S., Davis, R. P., Azzola, L., Stanley, E. G., and Elefanty, A. G. (2005) Forced aggregation of defined numbers of human embryonic stem cells into embryoid bodies fosters robust, reproducible hematopoietic differentiation, *Blood* **106**, 1601–1603.
15. Ng, E. S., Davis, R. P., Hatzistavrou, T., Stanley, E. G., and Elefanty, A. G. (2008) Directed differentiation of human embryonic stem cells as spin embryoid bodies and a description of the hematopoietic blast colony forming assay, *Curr Protoc Stem Cell Biol Chapter 1*, Unit 1D 3.
16. Bauwens, C. L., Peerani, R., Niebruegge, S., Woodhouse, K. A., Kumacheva, E., Husain, M., and Zandstra, P. W. (2008) Control of human embryonic stem cell colony and aggregate size heterogeneity influences differentiation trajectories, *Stem Cells* **26**, 2300–2310.
17. Olivier, E. N., Qiu, C., Velho, M., Hirsch, R. E., and Bouhassira, E. E. (2006) Large-scale production of embryonic red blood cells from human embryonic stem cells, *Exp Hematol* **34**, 1635–1642.
18. Anderson, J. S., Bandi, S., Kaufman, D.S. (2006) Derivation of normal macrophages from human embryonic stem (hES) cells for applications in HIV gene therapy, *Retrovirology* **3**.
19. Gaur, M., Kamata, T., Wang, S., Moran, B., Shattil, S. J., and Leavitt, A. D. (2006) Megakaryocytes derived from human embryonic stem cells: a genetically tractable system to study megakaryocytopoiesis and integrin function, *J Thromb Haemost* **4**, 436–442.
20. Woll, P. S., Martin, C. H., Miller, J. S., and Kaufman, D. S. (2005) Human embryonic stem cell-derived NK cells acquire functional receptors and cytolytic activity, *J Immunol* **175**, 5095–5103.
21. Vormoor, J., Lapidot, T., Pflumio, F., Risdon, G., Patterson, B., Broxmeyer, H. E., and Dick, J. E. (1994) SCID mice as an in vivo model of human cord blood hematopoiesis, *Blood Cells* **20**, 316–320; discussion 320–312.
22. Lapidot, T., Pflumio, F., Doedens, M., Murdoch, B., Williams, D. E., and Dick, J. E. (1992) Cytokine stimulation of multilineage hematopoiesis from immature human cells engrafted in SCID mice, *Science* **255**, 1137–1141.
23. Bhatia, M., Wang, J. C., Kapp, U., Bonnet, D., and Dick, J. E. (1997) Purification of primitive human hematopoietic cells capable of repopulating immune-deficient mice, *Proc Natl Acad Sci USA* **94**, 5320–5325.
24. Larochelle, A., Vormoor, J., Hanenberg, H., Wang, J. C., Bhatia, M., Lapidot, T., Moritz, T., Murdoch, B., Xiao, X. L., Kato, I., Williams, D. A., and Dick, J. E. (1996) Identification of primitive human hematopoietic cells capable of repopulating NOD/SCID mouse bone marrow: implications for gene therapy, *Nat Med* **2**, 1329–1337.

25. Bhatia, M., Bonnet, D., Murdoch, B., Gan, O. I., and Dick, J. E. (1998) A newly discovered class of human hematopoietic cells with SCID-repopulating activity, *Nat Med* **4**, 1038–1045.

26. Kimura, T., Asada, R., Wang, J., Morioka, M., Matsui, K., Kobayashi, K., Henmi, K., Imai, S., Kita, M., Tsuji, T., Sasaki, Y., Ikehara, S., and Sonoda, Y. (2007) Identification of long-term repopulating potential of human cord blood-derived CD34-flt3- severe combined immunodeficiency-repopulating cells by intra-bone marrow injection, *Stem Cells* **25**, 1348–1355.

27. Hogan, C. J., Shpall, E. J., and Keller, G. (2002) Differential long-term and multilineage engraftment potential from subfractions of human CD34+ cord blood cells transplanted into NOD/SCID mice, *Proc Natl Acad Sci USA* **99**, 413–418.

28. de Wynter, E. A., Buck, D., Hart, C., Heywood, R., Coutinho, L. H., Clayton, A., Rafferty, J. A., Burt, D., Guenechea, G., Bueren, J. A., Gagen, D., Fairbairn, L. J., Lord, B. I., and Testa, N. G. (1998) CD34+AC133+ cells isolated from cord blood are highly enriched in long-term culture-initiating cells, NOD/SCID-repopulating cells and dendritic cell progenitors, *Stem Cells* **16**, 387–396.

29. Tian, X., Woll, P. S., Morris, J. K., Linehan, J. L., and Kaufman, D. S. (2006) Hematopoietic engraftment of human embryonic stem cell-derived cells is regulated by recipient innate immunity, *Stem Cells* **24**, 1370–1380.

30. Narayan, A. D., Chase, J. L., Lewis, R. L., Tian, X., Kaufman, D. S., Thomson, J. A., and Zanjani, E. D. (2006) Human embryonic stem cell-derived hematopoietic cells are capable of engrafting primary as well as secondary fetal sheep recipients, *Blood* **107**, 2180–2183.

31. Lu, S. J., Feng, Q., Caballero, S., Chen, Y., Moore, M. A., Grant, M. B., and Lanza, R. (2007) Generation of functional hemangioblasts from human embryonic stem cells, *Nat Methods* **4**, 501–509.

32. Ji, J., Vijayaragavan, K., Bosse, M., Menendez, P., Weisel, K., and Bhatia, M. (2008) OP9 stroma augments survival of hematopoietic precursors and progenitors during hematopoietic differentiation from human embryonic stem cells, *Stem Cells* **26**, 2485–2495.

33. Tian, X., Hexum, M. K., Penchev, V. R., Taylor, R. J., Shultz, L. D., and Kaufman, D. S. (2009) Bioluminescent imaging demonstrates that transplanted human embryonic stem cell-derived CD34(+) cells preferentially develop into endothelial cells, *Stem Cells* **27**, 2675–2685.

34. Wang, L., Menendez, P., Shojaei, F., Li, L., Mazurier, F., Dick, J. E., Cerdan, C., Levac, K., and Bhatia, M. (2005) Generation of hematopoietic repopulating cells from human embryonic stem cells independent of ectopic HOXB4 expression, *J Exp Med* **201**, 1603–1614.

35. Wilber, A., Linehan, J. L., Tian, X., Woll, P. S., Morris, J. K., Belur, L. R., McIvor, R. S., and Kaufman, D. S. (2007) Efficient and stable transgene expression in human embryonic stem cells using transposon-mediated gene transfer, *Stem Cells* **25**, 2919–2927.

36. Xu, C., Inokuma, M. S., Denham, J., Golds, K., Kundu, P., Gold, J. D., and Carpenter, M. K. (2001) Feeder-free growth of undifferentiated human embryonic stem cells, *Nat Biotechnol* **19**, 971–974.

37. Tian, X., and Kaufman, D. S. (2008) Hematopoietic development of human embryonic stem cells in culture, *Methods Mol Biol* **430**, 119–133.

38. Vodyanik, M. A., and Slukvin, II. (2007) Hematoendothelial differentiation of human embryonic stem cells, *Curr Protoc Cell Biol Chapter 23*, Unit 23 26.

39. Pearson, T., Greiner, D. L., and Shultz, L. D. (2008) Creation of "humanized" mice to study human immunity, *Curr Protoc Immunol Chapter 15*, Unit 15 21.

40. Johnson, S. A., and Yoder, M. C. (2005) Reconstitution of hematopoiesis following transplantation into neonatal mice, *Methods Mol Med* **105**, 95–106.

41. Hiramatsu, H., Nishikomori, R., Heike, T., Ito, M., Kobayashi, K., Katamura, K., and Nakahata, T. (2003) Complete reconstitution of human lymphocytes from cord blood CD34+ cells using the NOD/SCID/gammacnull mice model, *Blood* **102**, 873–880.

42. Yahata, T., Ando, K., Nakamura, Y., Ueyama, Y., Shimamura, K., Tamaoki, N., Kato, S., and Hotta, T. (2002) Functional human T lymphocyte development from cord blood CD34+ cells in nonobese diabetic/Shi-scid, IL-2 receptor gamma null mice, *J Immunol* **169**, 204–209.

Chapter 33

Differentiation of Dendritic Cells from Human Embryonic Stem Cells

Kathryn M. Silk, Su-Yi Tseng, Kevin P. Nishimoto, Jane Lebkowski, Anita Reddy, and Paul J. Fairchild

Abstract

Improving our understanding of the interactions between human dendritic cells (DCs) and T cells may contribute to the development of therapeutic strategies for a variety of immune-mediated disorders. The possibility of using DCs themselves as tools to manipulate immune responses opens even greater therapeutic avenues. Current methods of generating human DCs are both inadequate and susceptible to high levels of variability between individuals. DCs differentiated from human embryonic stem cells (hESCs) could provide a more reliable, consistent solution. DCs have now successfully been differentiated from hESCs and more recently this has been repeated using protocols that avoid the inclusion of animal products, an important modification for clinical use. We have developed a novel method for the generation of DCs from hESCs in the absence of animal products that does not necessitate a separate embryoid body (EB) generation step. The technique involves the use of four growth factors and their successive removal from culture, resulting in accumulation of DCs with phenotypic, morphological, and immunostimulatory properties comparable to those of classical human monocyte-derived DCs. In addition to the application of hESC-derived DCs in basic research and novel approaches to cancer immunotherapy, they may also play a central role in the field of regenerative medicine. Tolerogenic DCs differentiated from hESCs may be used to persuade the immune system of the recipients of cell replacement therapy to tolerate allogeneic tissues differentiated from the same hESC line. Such an approach may help to address the immunological barriers that threaten to derail the clinical application of hESCs.

Key words: dendritic cells, human embryonic stem cells, ESC, differentiation of human ESC, regulatory T cells, treg, feeder-free culture, Serum-free culture

1. Introduction

Research into the function of dendritic cells (DCs) has identified them as cells of pivotal importance at the interface between innate and adaptive immunity. DCs integrate "danger signals" and down-modulatory cues in order to direct the adaptive immune response. They are considered unique in their ability to prime

Philip H. Schwartz and Robin L. Wesselschmidt (eds.), *Human Pluripotent Stem Cells: Methods and Protocols*, Methods in Molecular Biology, vol. 767, DOI 10.1007/978-1-61779-201-4_33, © Springer Science+Business Media, LLC 2011

T cells, emphasising their critical role in the immune system at the decision-making stage. Strategically, DCs have become key targets for the manipulation of immune responses, for example as vaccines delivering antigen in the desired context to T cells, to promote a robust anti-tumour response in the case of cancer immunotherapy, or to dampen immune responses in the context of transplantation and autoimmunity (1). A reliable source of human DCs would also enable further research into interactions between DCs and regulatory T cells (Tregs). Tregs regulate immune responses in vivo and an absence of functional Treg leads to the development of severe autoimmune conditions (2, 3). A better understanding of how DCs prime and expand Tregs could lead to the generation of therapeutic strategies for autoimmune diseases. Inhibition of Treg, on the other hand, can promote immune responses in cancer and tumour regression (4).

Conventional methods of obtaining human DCs involve culture of human monocytes separated from peripheral blood mononuclear cells (PBMCs) in medium containing granulocyte macrophage-colony stimulating factor (GM-CSF) and interleukin-4 (IL-4). A high level of inter-donor variation creates considerable problems in both in vitro experiments and clinical trials in which DCs are generated from monocytes following patient leukophoresis. Following the successful in vitro differentiation of DCs from mouse embryonic stem cells (mESCs) (5, 6), the potential of applying this protocol to human ESCs was investigated. DCs differentiated from hESCs offer consistency, expandability, and higher yields compared to monocyte-derived DCs (moDCs). Downstream clinical applications necessitate the generation of human embryonic stem cell-derived DCs (esDCs) using culture conditions that exclude the use of animal products. Recently, we have described the successful differentiation of DCs from hESCs under such stringent culture conditions (7). The protocol that we describe avoids a separate embryoid body (EB) generation step but allows EBs to form spontaneously in cultures on ultra-low attachment plates. The four growth factors bone morphogenetic protein-4 (BMP-4), vascular endothelial growth factor (VEGF), stem cell factor (SCF), and GM-CSF are vital in this method and their inclusion results in the emergence of copious numbers of cells with surface marker expression and morphology comparable to moDCs. hESC-derived DCs, differentiated under these culture conditions, efficiently process and present antigen and stimulate vigorous T-cell responses in vitro (7).

The availability of hESC-derived DCs may also have an important role in the field of regenerative medicine (8–11). Although studies performed in the mouse suggest that tissues derived from ESCs possess a degree of immune privilege, perhaps by virtue of their origin from blastocysts that are themselves immune privileged, tissues differentiated from ESCs with more than a single minor histocompatibility mismatch with the recipient are promptly rejected by the immune system (12). This presents a significant barrier to progress, given that tissues differentiated from hESCs will inevitably be allogeneic to the recipients of cell replacement therapy. While immune suppression is conventionally applied under such circumstances, its protracted use is associated with severe side effects, the risks of which may exceed those of diseases amenable to this form of intervention. A promising alternative is, therefore, the administration of tolerogenic DCs differentiated from the same hESC line as the therapeutic cells, thereby coaxing the immune system into accepting the allograft in an antigen-specific manner (8–11). Here we describe protocols for the generation of DCs in culture from well-characterised hESC lines under conditions conducive to their subsequent clinical use.

2. Materials

2.1. Coating Tissue Culture Plates with Matrigel™ Matrix

1. Matrigel™ (phenol red-free, growth factor-reduced, BD Biosciences) thawed on ice.
2. Ice-cold, Knock-Out Dulbecco's Modified Eagle's Medium (KO-DMEM, Invitrogen).
3. 50 mL centrifuge tubes on ice.
4. Culture vessels to be coated with Matrigel™.

2.2. Culture of Human ES Cells and Routine Passage/Harvesting

1. hESC culture medium: XVIVO-10 supplemented with 80 ng/mL recombinant human basic fibroblast growth factor (rhbFGF) (R&D Systems) and 0.5 ng/mL recombinant human transforming growth factor-β (rhTGF-β) (R&D Systems).
2. XVIVO-10 medium without gentamycin or phenol red (Lonza).
3. Collagenase IV (Invitrogen).
4. Dulbecco's Phosphate-Buffered Saline (DPBS).
5. Cell scrapers or 5-mL pipettes, depending on the culture vessels.
6. Culture vessels coated with Matrigel™.

2.3. Counting hESCs

1. Collagenase IV.
2. Dulbecco's Phosphate-Buffered Saline (DPBS).
3. TrypLE™ Express (Invitrogen) at room temperature.
4. Cell Culture Medium containing 10% FCS (any medium suitable for cell culture may be used).

2.4. Differentiation of ESCs into Dendritic Cells and Feeding of Differentiation Cultures

1. Culture medium for the differentiation of hESCs into DC consists of room temperature XVIVO-15 (with phenol red and gentamycin, Lonza), supplemented with the following:
   (a) 1 mM Sodium Pyruvate
   (b) 1× non-essential amino acids
   (c) 2 mM L-glutamine
   (d) 50 μM 2-mercaptoethanol
   (e) Recombinant human bone morphogenetic protein-4 (rhBMP-4) (R&D Systems) to give a final concentration of 50 ng/mL; 50 ng/mL recombinant human vascular endothelial growth factor (rhVEGF) (R&D Systems); 20 ng/mL recombinant human stem cell factor (rhSCF) (R&D Systems); and 50 ng/mL recombinant human granulocyte macrophage-colony stimulating factor (rhGM-CSF) (R&D Systems). On day 5, BMP-4 is removed from feeding medium, followed by VEGF on day 10 and SCF on day 15.
2. XVIVO-10 medium at room temperature.
3. Dulbecco's Phosphate-Buffered Saline (DPBS).
4. Collagenase IV.
5. Cell scrapers or 5-mL pipettes.

2.5. Differentiation of Monocytes into Immature DCs

1. Room temperature XVIVO-15 supplemented with 50 ng/mL rhGM-CSF and 100 ng/mL recombinant human interleukin-4 (rhIL-4) (R&D Systems).
2. Trypan blue.
3. Dulbecco's Phosphate-Buffered Saline (DPBS).
4. 70 μm cell strainers.

2.6. Maturation of DC Differentiated from hESCs

1. Recombinant human interferon-γ (rhIFN-γ) (R&D Systems).
2. Prostaglandin E_2 (PGE_2) (Sigma).
3. Recombinant human tumour necrosis factor-α (rhTNF-α) (R&D Systems).
4. Recombinant human interleukin-1β (rhIL-1β) (R&D Systems).
5. Recombinant human granulocyte macrophage-colony stimulating factor (rhGM-CSF) (R&D Systems).

3. Methods

It is important to note that, in our experience, it becomes increasingly difficult to direct cells along a haematopoietic route when hESCs have been cultured for more than 40 passages, although DCs have been generated from hESCs at higher passages.

3.1. Culture of hESCs

1. Human ESC lines can be cultured using different methods. We have adapted the culture of H1, H7, and H9 hESC lines to feeder- and serum-free conditions (13, 14) (see Fig. 1). Defined culture conditions are more reliable, have important implications in downstream clinical studies, facilitate scale-up of cultures, and avoid the time and expense required to maintain feeder cells.
2. hESCs are cultured in XVIVO-10 medium supplemented with 80 ng/mL bFGF and 0.5 ng/mL TGF-β on Matrigel™. XVIVO-10 medium is first warmed before adding bFGF and TGF-β (see Note 1).
3. A complete change of medium is performed daily except on the day immediately following either passage of the cells or thawing of the cells.

3.2. Coating Tissue Culture Plates with Matrigel™

1. Culture vessels coated with Matrigel™ need to be prepared in advance of passaging hESCs. To a 10 mL vial of phenol red-free, growth factor-reduced Matrigel™, add 10 mL ice-cold KO-DMEM. Keep Matrigel™ on ice and work quickly. Be careful not to introduce excess bubbles. The diluted Matrigel™ can be aliquoted and stored at −20°C.
2. It is important to avoid the generation of bubbles when handling Matrigel™ to prevent uneven coating of the tissue

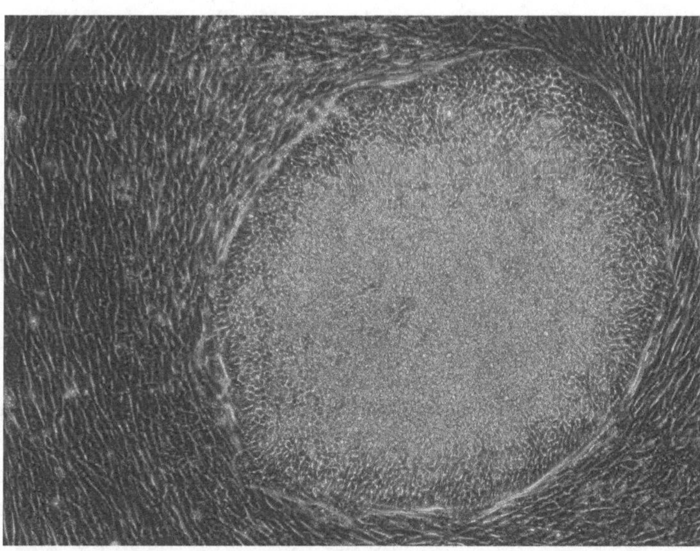

Fig. 1. A typical H1 hESC colony cultured under the feeder and serum-free conditions described in this chapter. Human ESCs have clear borders and at higher magnifications the cells can be seen to have a high nucleus to cytoplasm ratio. The fibroblast-like cells surrounding this colony are not feeder cells but have differentiated spontaneously from the hESCs. This has been reported before in the H1 line and the fibroblast-like cells in these cultures have been shown to support the undifferentiated growth of H1 hESCs (15). Objective magnification: 10×.

culture surface. If Matrigel™ starts to warm, it gels very quickly; for this reason it is best to work with Matrigel™ on ice and keep KO-DMEM ice-cold.

3. Thaw aliquots of Matrigel™ on ice. Transfer Matrigel™ from aliquots into sterile 50 mL centrifuge tubes. To each 1–3 mL of Matrigel™, add 5 mL ice-cold KO-DMEM using a 5-mL pipette and mix thoroughly. Top up with media to give a final volume of 15 mL per 1 mL of thawed Matrigel™ (includes initial volume of thawed Matrigel™ and volume of media used to mix). The final dilution of Matrigel™ is 1:30.
4. A six-well tissue culture plate can be coated with 1 mL per well of diluted Matrigel™, a 25 cm^2 flask with 3 mL and a 75 cm^2 flask with 10 mL. Calculate the volume required according to the surface area of other culture vessels using these volumes as a guide. Tap the sides of culture vessels to distribute the Matrigel™ evenly over the surface.
5. Culture vessels can be coated by leaving them at room temperature for at least 1 h or stored immediately at 4°C. Use parafilm to cover the caps of vented tissue culture flasks and seal tissue culture plates with micro-pore tape and wrap in cling-film to prevent evaporation. Discard culture vessels if Matrigel™ solution is no longer covering the entire surface.

3.3. Routine Passage of hESCs/Harvesting hESCs

1. In our hands, using xeno-free culture conditions, H1 hESCs can be routinely passaged every 4–6 days. hESCs are passaged as clusters of cells using collagenase IV to loosen the colonies from the tissue culture surface and scraping with a cell scraper or 5-mL pipette to dislodge and break up the colonies.
2. Using the same counting method as that used to estimate the number of hESCs (see Subheading 3.4), cells can be seeded at a density of approximately 1×10^5 cells per cm^2. In practice, once culture vessels reach 50% confluency, H1 hESCs can be passaged at a 1:5 dilution (see Note 2).
3. If hESCs are being expanded, the volume of culture medium required can be prepared in advance and the hESCs passaged using supplemented XVIVO-10 medium. If hESCs are being maintained then the volume of unsupplemented room temperature XVIVO-10 used to passage hESCs can be subtracted from the final volume required; TGF-β and bFGF can be added to this amount to give the correct final concentration for culture and used to top up the suspension of harvested hESC clusters.
4. Remove culture medium from hESC cultures and incubate the cells with pre-warmed collagenase at 37°C for the time period calculated while counting hESCs (see Subheading 3.4, step 2). Alternatively, if using a pre-determined dilution, observe cultures after 4–7 min: when the majority of stromal cells, which are differentiated hESCs, have lifted off the tissue culture

surface and hESC colonies are beginning to round at the edges, immediately remove collagenase. Wash gently with DPBS, being careful not to scrape off colonies with the pipette.

5. With XVIVO-10 medium, cover the tissue culture surface and gently scrape off hESC colonies. A 5-mL pipette can be used to scrape the surface of wells from a six-well plate or a cell scraper for flasks. It is critical to maintain clusters of hESC and prevent generation of a single-cell suspension that will result in loss of viability.
6. Using a 5-mL pipette, generate a suspension of hESC clusters.
7. Top up the suspension of hESC clusters to give the correct final volume and concentration of bFGF and TGF-β necessary for culture and pipette into Matrigel™-coated culture vessels (remove Matrigel™ solution immediately before adding hESCs; there is no need to rinse the culture vessel first).
8. Gently rock the culture vessel backwards and forwards and side-to-side to distribute the clusters of hESCs evenly over the tissue culture surface. Incubate at 37°C, 5% CO_2 in a humidified atmosphere (see Note 3).

3.4. Counting hESCs

1. In order to plate hESCs at the correct density for differentiation cultures, it is important to first count them.
2. Remove culture medium from one representative flask or well. Add warm collagenase IV so that the surface is covered and incubate at 37°C for 4–7 min. When the majority of fibroblast-like cells (see Fig. 2) have lifted off the surface and

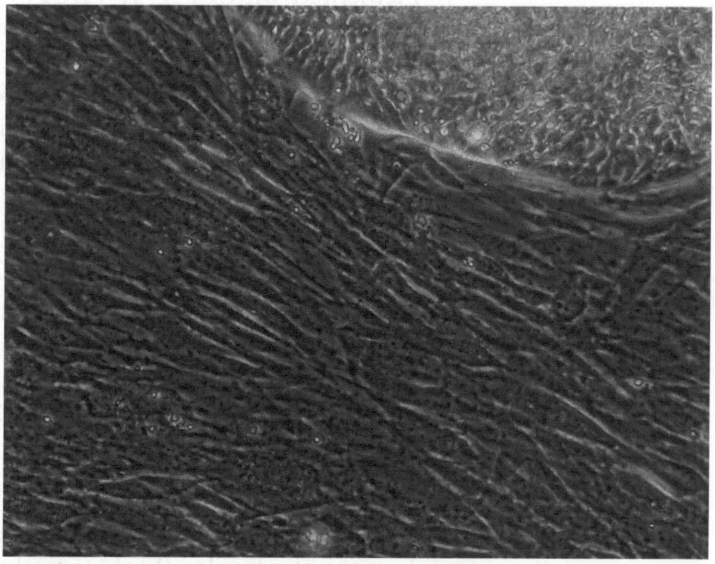

Fig. 2. Fibroblast-like cells surround colonies of H1 hESCs grown under the described feeder and serum-free conditions. Objective magnification: 20×.

colonies of hESC are beginning to round up at the edges, immediately remove collagenase, and gently wash with DPBS (see Subheading 3.3, step 4). Record the time of collagenase incubation as this will be required later.

3. Add room temperature TrypLE™ Express to cover the surface of the dish and incubate at 37°C (for example: 1 mL/well of six-well plate, 5 mL/25 cm^2 flask). After 5 min, shake the culture vessel in quick, sharp motions to dislodge all the hESC colonies. Pipette to create a single-cell suspension and wash any remaining cells from the tissue culture surface. Quickly add the cell suspension to the same volume of culture medium containing 10% FBS to give a 1:2 dilution and pipette further, if necessary, to achieve a single-cell suspension.
4. The cells are counted without trypan blue exclusion to monitor viability, as trypsin treatment and producing a single-cell suspension will generate significant cell death. It is not possible to distinguish definitively between ESC cells and stromal cells so each cell is counted and used as an estimate of hESC number. Removing the majority of stromal cells during collagenase treatment will alleviate this problem.
5. Trypsin-treated hESCs can be stained for Oct-4, SSEA-4, and Tra-1-60 and analysed by flow cytometry to monitor their pluripotency (see Chapter 16).

3.5. Directed Differentiation of hESCs into Dendritic Cells

1. After counting hESCs, it is possible to calculate how many wells to harvest for differentiation culture and the volume of medium required (see Note 4). Cells are plated at 3×10^6 cells per well of a six-well plate. Prepare XVIVO-15 medium supplemented with BMP-4, VEGF, SCF, and GM-CSF. These growth factors are successively removed from the differentiation culture leaving only GM-CSF in the final feed. Prepare sufficient medium to set up the differentiation culture and for the first feed (6 mL per well of a six-well plate). XVIVO-10, XVIVO-15 and DPBS should be at room temperature.
2. Harvest hESC cultures using collagenase IV treatment for the time period determined when counting hESCs as described in Subheading 3.4, step 2.
3. Wash the cells gently using DPBS as before. Add sufficient XVIVO-10 medium to cover the culture vessel surface and gently scrape off colonies, being careful not to create a single-cell suspension (see Subheading 3.3, step 5).
4. Wash the surface of culture vessels with XVIVO-10 medium to ensure all hESC colonies have been removed.
5. Pool colonies and allow them to settle for 10–20 min at the bottom of a sterile 50 mL conical tube.

6. Gently remove medium without disturbing the loose, settled colonies. Add some of the prepared differentiation medium; a similar volume or less than that used to harvest the colonies. Using a 5-mL pipette, create a suspension of cell clusters.
7. Calculate the volume of differentiation medium required to prepare the determined number of wells. Use 4 mL of medium per well of a six-well plate. Dilute cell clusters to give this final volume.
8. Due to the nature of differentiation, variation is often observed between cultures. For this reason, it is important to be as accurate as possible when pipetting the number of cells per well. Use a 10-mL pipette to prevent further breakdown of cell clusters and keep mixing the suspension as you pipette. It is best to aspirate and dispense 4 mL per well of the suspension, in order to distribute cell clusters as accurately as possible. Pipette into ultra-low attachment six-well plates.
9. Optional: Seal six-well plates with micro-pore tape to reduce potential contamination during long-term culture, and incubate in a humidified 37°C incubator at 5% CO_2.

3.6. Feeding Differentiation Cultures

1. Cultures of hESCs differentiating into DCs need to be fed every 2–3 days. This is particularly important during the early stages of differentiation. In practice, cells can be fed on Mondays, Wednesdays, and Fridays. Every 5 days, a growth factor is removed from the differentiation medium until only GM-CSF remains. Concentrations of the added growth factors are therefore effectively diluted throughout the course of the experiment. BMP-4 is removed from the differentiation culture first, followed by VEGF and then SCF.
2. For the first feed, warm XVIVO 15 containing BMP-4, VEGF, SCF, and GM-CSF and top-up wells with an extra 2 mL, giving a final volume of 6 mL per well. For successive feeds, gently replace 2 mL or 3 mL of culture medium using a 10-mL pipette, being careful not to remove cells or, at later stages of culture, the embryoid bodies (EBs) that spontaneously form (see Note 5). Replace with warm medium containing the appropriate growth factors.
3. Differentiation cultures contain significant debris during the early stages of differentiation due to high levels of cell death, which is normal during differentiation, and due to the inability of intermediates to adhere to the ultra-low attachment (ULA) surface. Around days 14–19 of culture, small, round, non-adherent haematopoietic cells should start to appear and later to accumulate (see Fig. 3). From day 19 onwards, "monocyte-like" cells should become apparent. These cells look morphologically like human blood monocytes and

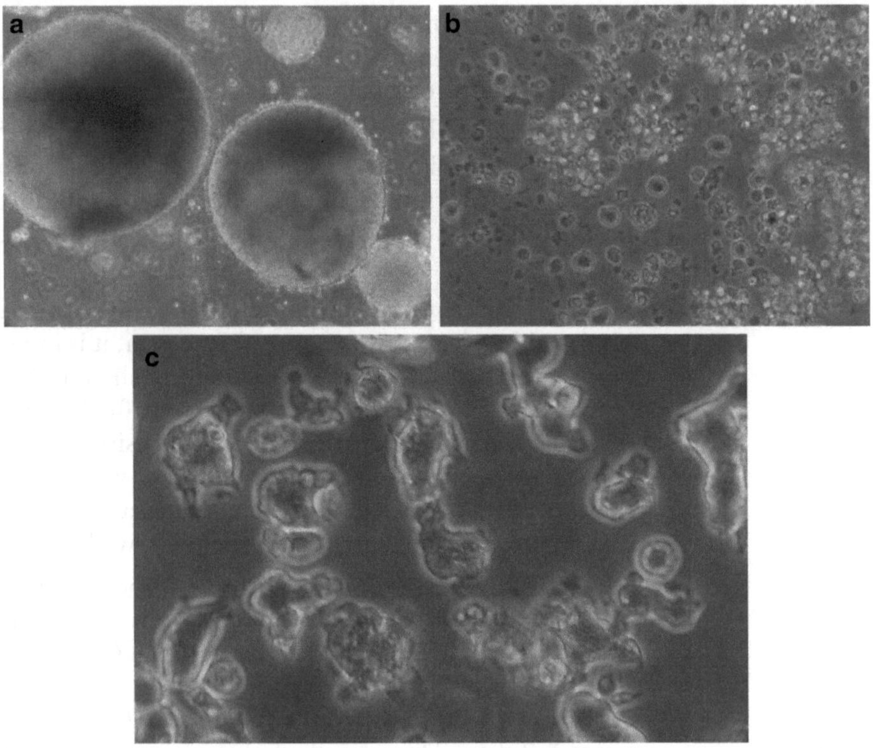

Fig. 3. Different stages of differentiation of H1 hESCs into DCs. (**a**) Embryoid bodies form early on in differentiation cultures. Objective magnification: 10×. (**b**) A significant amount of debris is seen in the initial stages of differentiation. In this image, haematopoietic cells are starting to appear. Objective magnification: 20× (**c**) Monocyte-like cells are evident from day 19 onwards. Objective magnification: 40×.

express high levels of CD14 as well as other myeloid markers, such as CD11b. Cells with monocytic morphology accumulate in number towards the later stages of culture.

4. Monocytes are usually ready to be harvested for DC differentiation between days 30–35. The cultures can be monitored for the appearance of monocytes and their percentages assessed using CD14 expression as determined by flow cytometry.

3.7. Differentiation of Monocytes into iDCs

1. Monocytes are harvested by gently pipetting cultures using a pipettor set on slow and a 10-mL pipette. The aim is to remove the monocytes, which are non-adherent, while leaving adherent macrophages in the culture plate (see Note 6). Transfer cells to 50-mL centrifuge tubes. EBs can be left to settle at the bottom of the tube (approximately 2–5 min) and removed using a pipette before cells also start to pellet.
2. Once the EBs have been removed, the cell suspension can be passed through a 70 μm cell strainer. This excludes any large clumps of cell debris (mostly created by EBs breaking up). Rinse the cell strainer with DPBS.

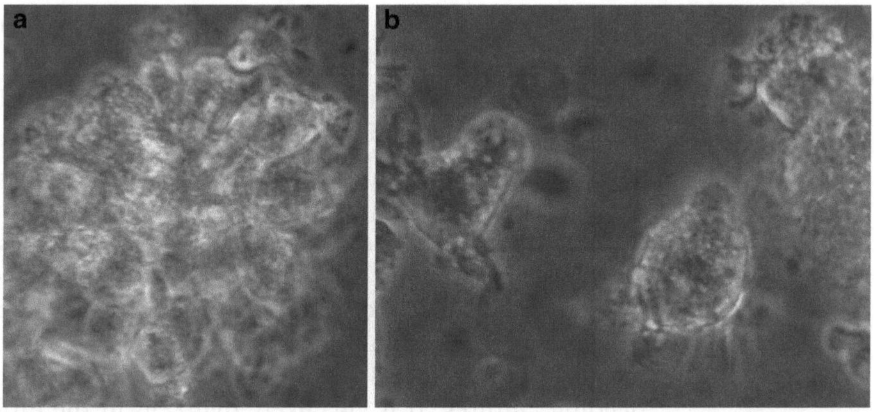

Fig. 4. Immature DCs differentiated from H1 hESCs. (**a**) Immature DCs frequently form tight clusters of cells. Typical veils of cytoplasm can be seen on DCs at the edge of the cluster. (**b**) Immature DC morphology showing veils of cytoplasm characteristic of DCs. Objective magnification: 40×.

3. Monocytes can then be washed by centrifuging at 200 × *g* for 5 min at 4°C. Discard the cell supernatant and resuspend cells in XVIVO-15 medium supplemented with 50 ng/mL GM-CSF and 100 ng/mL IL-4.
4. Pipette 1–1.5 × 10^6 monocytes per well of a six-well ULA plate and incubate for 6–8 days to differentiate monocytes into iDCs (see Note 7) (see Fig. 4).

3.8. Maturation of DCs from iDCs

iDCs differentiated from hESCs can be matured using a cocktail of cytokines including TNFα, IFNγ, PGE_2, and IL-1β. This can be made up in medium already supplemented with GM-CSF and IL-4 and added to cultures of iDCs for the last 48 h. Alternatively, IL-4 can be removed by washing the cells and replacing medium with the maturation cocktail (GM-CSF must be included throughout) (see Notes 8 and 9).

4. Notes

1. Human bFGF is extremely heat labile and therefore addition to pre-warmed medium instead of warming medium that has already been supplemented with bFGF prolongs its half-life in culture.
2. When cells are initially thawed and are therefore more fragile, dilutions of 1:3 or 1:4 can be used, depending on the appearance and recovery of the cells.
3. Under these culture conditions, fibroblast-like stromal cells, differentiated from hESCs are seen. Evidence has shown that these fibroblast-like cells support the pluripotent growth of

hESCs; however, it is necessary to prevent them from overwhelming the cultures.

4. It is best not to move the cultures for 48 h to allow hESCs to adhere to the Matrigel™-coated surface.
5. EBs become cystic and have a tendency to float, making it particularly difficult to avoid aspirating them with the pipette. Haematopoietic cells can often be found inside these cystic EBs when examining the cultures under a microscope, so they are highly likely to be a source of haematopoietic cells in the differentiation cultures. It is, therefore, important to avoid losing these EBs as much as possible when routinely feeding cultures. It is also likely that cells in the differentiation cultures are producing growth factors that may promote the differentiation process and therefore we try not to completely remove the conditioned medium. Also try not to excessively disturb the differentiating cells by only removing medium from the top of cultures.
6. DCs normally weakly adhere to tissue culture plastic whereas it is known that macrophages can stick to the surface of plates that have not been tissue culture treated (for example bacteriological petri dishes). This is because macrophages attempt to phagocytose the surface and therefore can usually only be removed at temperatures below 37°C (to remove macrophages, it is usual to add cold DPBS and incubate at 4°C for a short period of time). By pipetting the surface gently and using reagents at room temperature, it is possible to select for non-adherent or weakly adherent DCs while leaving behind any macrophages, which are stuck to the surface.
7. By counting large cells using a haemocytometer it is possible to distinguish between monocytes and precursor cells when establishing cultures.
8. Monocyte-derived DCs (moDCs) are commonly matured by treating them with the toll-like receptor 4 (TLR4) ligand lipopolysaccharide (LPS). (TLRs bind to pathogen-associated molecular patterns (PAMPs) that are conserved between microbes. TLRs are important in alerting the immune system to the presence of pathogens in the internal milieu.) However, DCs differentiated from H1 hESC do not express TLR4 and are, therefore, unable to respond to stimulation with LPS. For this reason, we have used a maturation cocktail of cytokines in order to mature DCs differentiated from hESC.
9. In our hands, hESC-derived DCs already produce IL-6; for this reason IL-6 is not included in the maturation cocktail.

References

1. Silk, K. M., and Fairchild, P. J. (2009) Harnessing dendritic cells for the induction of transplantation tolerance. *Curr Opin Organ Transplant* *14*, 344–50.
2. Gavin, M. A., Torgerson, T. R., Houston, E., DeRoos, P., Ho, W. Y., Stray-Pedersen, A., Ocheltree, E. L., Greenberg, P. D., Ochs, H. D., and Rudensky, A. Y. (2006) Single-cell analysis of normal and FOXP3-mutant human T cells: FOXP3 expression without regulatory T cell development. *Proc Natl Acad Sci USA* ***103***, 6659–64.
3. Wing, K., Onishi, Y., Prieto-Martin, P., Yamaguchi, T., Miyara, M., Fehervari, Z., Nomura, T., and Sakaguchi, S. (2008) CTLA-4 control over Foxp3+ regulatory T cell function. *Science* *322*, 271–5.
4. Wang, L., Pino-Lagos, K., de Vries, V. C., Guleria, I., Sayegh, M. H., and Noelle, R. J. (2008) Programmed death 1 ligand signaling regulates the generation of adaptive Foxp3+CD4+ regulatory T cells. *Proc Natl Acad Sci USA* ***105***, 9331–6.
5. Fairchild, P. J., Brook, F. A., Gardner, R. L., Graca, L., Strong, V., Tone, Y., Tone, M., Nolan, K. F., and Waldmann, H. (2000) Directed differentiation of dendritic cells from mouse embryonic stem cells. *Curr Biol* ***10***, 1515–8.
6. Fairchild, P. J., Nolan, K. F., Cartland, S., Graca, L., and Waldmann, H. (2003) Stable lines of genetically modified dendritic cells from mouse embryonic stem cells. *Transplantation* *76*, 606–8.
7. Tseng, S. Y., Nishimoto, K. P., Silk, K. M., Majumdar, A. S., Dawes, G. N., Waldmann, H., Fairchild, P. J., Lebkowski, J. S., and Reddy, A. (2009) Generation of immunogenic dendritic cells from human embryonic stem cells without serum and feeder cells. *Regen Med* *4*, 513–26.
8. Fairchild, P. J., Nolan, K. F., Cartland, S., and Waldmann, H. (2005) Embryonic stem cells: a novel source of dendritic cells for clinical applications. *Int Immunopharmacol* *5*, 13–21.
9. Fairchild, P. J., Robertson, N. J., Cartland, S., Nolan, K. F., and Waldmann, H. (2005) Cell replacement therapy and the evasion of destructive immunity. *Stem Cell Rev* *1*, 159–67.
10. Fairchild, P. J., Robertson, N. J., Minger, S. L., and Waldmann, H. (2007) Embryonic stem cells: protecting pluripotency from alloreactivity. *Curr Opin Immunol* *19*, 596–602.
11. Fairchild, P. J., Cartland, S., Nolan, K. F., and Waldmann, H. (2004) Embryonic stem cells and the challenge of transplantation tolerance. *Trends Immunol* *25*, 465–70.
12. Robertson, N. J., Brook, F. A., Gardner, R. L., Cobbold, S. P., Waldmann, H., and Fairchild, P. J. (2007) Embryonic stem cell-derived tissues are immunogenic but their inherent immune privilege promotes the induction of tolerance. *Proc Natl Acad Sci USA* ***104***, 20920–5.
13. Li, Y., Powell, S., Brunette, E., Lebkowski, J., and Mandalam, R. (2005) Expansion of human embryonic stem cells in defined serum-free medium devoid of animal-derived products. *Biotechnol Bioeng* *91*, 688–98.
14. Xu, C., Inokuma, M. S., Denham, J., Golds, K., Kundu, P., Gold, J. D., and Carpenter, M. K. (2001) Feeder-free growth of undifferentiated human embryonic stem cells. *Nat Biotechnol* *19*, 971–4.
15. Bendall, S. C., Stewart, M. H., Menendez, P., George, D., Vijayaragavan, K., Werbowetski-Ogilvie, T., Ramos-Mejia, V., Rouleau, A., Yang, J., Bossé, M., Lajoie, G. and Bhatia, M. (2007) IGF and FGF cooperatively establish the regualtory stem cell niche of pluripotent human cells in vitro. *Nature* *448*, 1015–21.

References

[illegible]

Index

Philip H. Schwartz and Robin L. Wesselschmidt (eds.), *Human Pluripotent Stem Cells: Methods and Protocols*, Methods in Molecular Biology, vol. 767, DOI 10.1007/978-1-61779-201-4, © Springer Science+Business Media, LLC 2011

MIX
Papier aus verantwortungsvollen Quellen
Paper from responsible sources
FSC® C105338

If you have any concerns about our products,
you can contact us on
ProductSafety@springernature.com

In case Publisher is established outside the EU,
the EU authorized representative is:
Springer Nature Customer Service Center GmbH
Europaplatz 3, 69115 Heidelberg, Germany

Printed by Libri Plureos GmbH
in Hamburg, Germany